Mariusz Urbański, Mario Roy, Sara Munday
Non-Invertible Dynamical Systems

De Gruyter Expositions in Mathematics

Volume 69/3

Mariusz Urbański, Mario Roy, Sara Munday

Non-Invertible Dynamical Systems

Volume 3: Analytic Endomorphisms of the Riemann Sphere

DE GRUYTER

Mathematics Subject Classification 2010
37A05, 37A25, 37A30, 37A35, 37A40, 37B10, 37B25, 37B40, 37B65, 37C05, 37C20, 37C40, 37D20, 37D35, 37E05, 37E10

Authors
Prof. Dr. Mariusz Urbański
University of North Texas
Department of Mathematics
1155 Union Circle #311430
Denton, TX 76203-5017
USA
mariusz@urbanskimath.com

Dr. Sara Munday
John Cabot University
Via della Lungara 233
00165 Rome
Italy
smunday@johncabot.edu

Prof. Dr. Mario Roy
York University
Glendon College
2275 Bayview Avenue
Toronto, M4N 3M6
Canada
mroy@glendon.yorku.ca

ISBN 978-3-11-076984-5
e-ISBN (PDF) 978-3-11-076987-6
e-ISBN (EPUB) 978-3-11-076989-0
ISSN 0938-6572

Library of Congress Control Number: 2022950737

Bibliographic information published by the Deutsche Nationalbibliothek
The Deutsche Nationalbibliothek lists this publication in the Deutsche Nationalbibliografie; detailed bibliographic data are available on the Internet at http://dnb.dnb.de.

Typesetting: VTeX UAB, Lithuania
Printing and binding: CPI books GmbH, Leck

www.degruyter.com

Mariusz Urbański dedicates this book to his wife, Irena.
À mes parents Thérèse et Jean-Guy, à ma famille et à mes amis, sans qui ce livre n'aurait pu voir la vie... du fond du coeur, merci! Mario

Preface

Dynamical systems and ergodic theory is a rapidly evolving field of mathematics with a large variety of subfields, which use advanced methods from virtually all areas of mathematics. These subfields comprise but are by no means limited to: abstract ergodic theory, topological dynamical systems, symbolic dynamical systems, smooth dynamical systems, holomorphic/complex dynamical systems, conformal dynamical systems, one-dimensional dynamical systems, hyperbolic dynamical systems, expanding dynamical systems, thermodynamic formalism, geodesic flows, Hamiltonian systems, KAM theory, billiards, algebraic dynamical systems, iterated function systems, group actions, and random dynamical systems.

All of these branches of dynamical systems are mutually intertwined in many involved ways. Each of these branches nonetheless also has its own unique methods and techniques, in particular embracing methods which arise from the fields of mathematics the branch is closely related to. For example, complex dynamics borrows advanced methods from complex analysis, both of one and several variables; geodesic flows utilize methods from differential geometry; and abstract ergodic theory and thermodynamic formalism rely heavily on measure theory and functional analysis.

Indeed, it is truly fascinating how large the field of dynamical systems is and how many branches of mathematics it overlaps with. In this book, we focus on some selected subfields of dynamical systems, primarily noninvertible ones.

In the first volume, we give introductory accounts of topological dynamical systems acting on compact metrizable spaces, of finite-state symbolic dynamical systems, and of abstract ergodic theory of measure-theoretic dynamical systems acting on probability measure spaces, the latter including the metric entropy theory of Kolmogorov and Sinai. More advanced topics include infinite ergodic theory, general thermodynamic formalism, and topological entropy and pressure. This volume also includes a treatment of several classes of dynamical systems, which are interesting on their own and will be studied at greater length in the second volume: we provide a fairly detailed account of distance expanding maps and discuss Shub expanding endomorphisms, expansive maps, and homeomorphisms and diffeomorphisms of the circle.

The second volume is somewhat more advanced and specialized. It opens with a systematic account of thermodynamic formalism of Hölder continuous potentials for open transitive distance expanding systems. One chapter comprises no dynamics but rather is a concise account of fractal geometry, treated from the point of view of dynamical systems. Both of these accounts are later used to study conformal expanding repellers. Another topic exposed at length is that of thermodynamic formalism of countable state subshifts of finite type. Relying on this latter, the theory of conformal graph directed Markov systems, with their special subclass of conformal iterated function systems, is described. Here, in a similar way to the treatment of conformal expanding repellers, the main focus is on Bowen's formula for the Hausdorff dimension of the limit set and

https://doi.org/10.1515/9783110769876-201

multifractal analysis. A rather short examination of Lasota–Yorke maps of an interval is also included in this second volume.

The third volume is entirely devoted to the study of the dynamics, ergodic theory, thermodynamic formalism, and fractal geometry of rational functions of the Riemann sphere. We present a fairly complete account of classical as well as more advanced topological theory of Fatou and Julia sets. Nevertheless, primary emphasis is placed on measurable dynamics generated by rational functions and fractal geometry of their Julia sets. These include the thermodynamic formalism of Hölder continuous potentials with pressure gaps, the theory of Sullivan's conformal measures, invariant measures and their dimensions, entropy, and Lyapunov exponents. We further examine in detail the classes of expanding, subexpanding, and parabolic rational functions. We also provide, with proofs, several of the fundamental tools from complex analysis that are used in complex dynamics. These comprise Montel's Theorem, Koebe's Distortion Theorems and Riemann–Hurwitz formulas, with their ramifications.

In virtually each chapter of this book, we describe a large number of concrete selected examples illustrating the theory and serving as examples in other chapters. Also, each chapter of the book is supplied with a number of exercises. These vary in difficulty, from very easy ones asking to verify fairly straightforward logical steps to more advanced ones enhancing largely the theory developed in the chapter.

This book originated from the graduate lectures Mariusz Urbański delivered at the University of North Texas in the years 2005–2010 and that Sara Munday took notes of. With the involvement of Mario Roy, the book evolved and grew over many years. The last 2 years (2020 and 2021) of its writing were most dramatic and challenging because of the COVID-19 pandemic. Our book borrows widely from many sources including the books [73, 88, 110]. We nevertheless tried to keep it as self-contained as possible, avoiding to refer the reader too often to specific results from special papers or books. Toward this end, an appendix comprising classical results, mostly from measure theory, functional analysis and complex analysis, is included. The book covers quite a many topics treated with various degrees of completeness, none of which are fully exhausted because of their sheer largeness and their continuous dynamical growth.

List of Figures

https://doi.org/10.1515/9783110769876-202

Introduction to Volume 3

In the first volume, we developed the general theory of thermodynamic formalism, that is, for general topological dynamical systems (i. e., continuous self-maps of a compact metrizable space) subject to general potentials (i. e., continuous potentials).

In the second volume, we carried on with the study of thermodynamic formalism and its dynamical and geometric applications, but this time for specific classes of systems subject to specific classes of potentials, including topologically transitive open and distance expanding maps under Hölder continuous potentials.

The third volume concentrates on the dynamical and fractal properties of endomorphisms of the Riemann sphere. We now describe in more detail the content of each chapter of this third volume, including their mutual dependence and interrelations.

Introduction

The theory of iteration of rational functions of the Riemann sphere $\widehat{\mathbb{C}}$ emerged in the second half of the 19th century with the works of Böttcher [14–18], Cayley [22–27], Farkas [47], Koenigs [64–69], Korkine [70], Leau [75–77] and Schröder [116–118] on the local behavior of rational functions near their attracting and indifferent fixed points and on Newton's method. The mature and modern treatment of the iteration of rational functions appeared at the beginning of the 20th century with the long series of works by Pierre Fatou (see [48] and the references in [3, 4]) and Gaston Julia (see [61] and the references in [3, 4]), which were prompted by the French Academy of Science Award competition announcement. The seminal work of Paul Montel (see the references in [3, 4]) on normal families of analytic functions rendered this modern, comprehensive and systematic approach to iteration of rational functions, both locally and globally, possible. Fatou and Julia introduced many concepts and proved several theorems significantly examining the sets that nowadays bear their names. Since then and until the late 1970s, only scattered isolated papers were published. Among others, let us mention the works of Cremer [29] and Siegel [121], who respectively proved the existence of what is now called Cremer points and Siegel disks. These latter are of profound importance in the KAM theory. This early period of development of the theory of iteration of rational functions is thoroughly presented and well documented in the beautiful books [3, 4].

The contemporary era in the theory of iteration of rational functions truly began when Dennis Sullivan proved in his breakthrough paper [131] Fatou's conjecture from [48] and [49] that rules out the existence of wandering domains. The importance of this paper also lies in the introduction to complex dynamics of the powerful tool of quasiconformal mappings, which have been widely used in rational and transcendental complex dynamics ever since. We mention here only a few other milestones which highly influenced and accelerated the progress and development of complex dynamics. Namely, Mandelbrot's original definition and numerical studies of his namesake set [81]; Douady

https://doi.org/10.1515/9783110769876-203

and Hubbard's pioneering works [43, 44] on the theoretical understanding of the topological and conformal structure of the Mandelbrot set; Sullivan's works [127, 129] introducing conformal measures to the realm of rational functions; and Rufus Bowen's work [19] which launched the use of thermodynamic formalism in conformal dynamical systems (more specifically, quasi-Fuchsian groups).

Since then, research on complex dynamics has exploded. It has covered a large variety of topics, such as genericity of expanding maps, the topological and fractal structure of the Mandelbrot set and quadratic polynomials, the topological and fractal structure of Julia sets, the structure of Fatou sets and the boundaries of their connected components, Newton's method and the measurable dynamics and thermodynamic formalism induced by rational functions. Part of this development has been documented in book monographs such as [8, 21, 94, 150] as well as in Lyubich's expository article [80] and Urbański's paper [141].

In this volume, we present the core of the theory of iteration of rational functions in its topological, conformal, and fractal aspects. We base it on fundamental and powerful tools and methods from complex analysis, such as the Riemann–Hurwitz formula, normal families of analytic functions, Montel's theorem and Koebe's distortion theorems. Because of their continual use, and in fact their indispensability, we prove them all in Chapters 22–23 for the sake of completeness. Normal families, Montel's theorem and Koebe's distortion theorems are classic; the Riemann–Hurwitz formula, too, but following Beardon's approach in [8], we present it in a form tailored for the needs of complex dynamics, that is, to handle the topological structure of images and preimages of open connected sets in $\widehat{\mathbb{C}}$.

In Chapter 24, we prove all the standard properties and characterizations of the Julia and Fatou sets, including the density of repelling periodic points in the Julia set, the perfectness of the Julia set, the topological exactness of the map restricted to its Julia set, as well as a detailed analysis of the components of the Fatou set and their boundaries, with a complete proof of a classification of their periodic components. We also provide a thorough account of the local behavior of the map around its periodic points. We pay particular attention to rationally indifferent fixed and periodic points. In addition to qualitative properties such as the Fatou–Leau flower theorem, we give a rigorous quantitative description of the local and asymptotic behavior of the iterates of points in attracting and repelling petals.

Chapter 25 comprises a number of technical properties of rational functions, in particular, their behavior around their critical points. It also includes Lyubich's geometric lemma and special types of partitions such as Mañé's partition.

Chapter 26 introduces the three classes of rational functions whose topological, fractal, conformal and thermodynamic properties are studied in this chapter and from Chapter 30 on, namely the expanding, subexpanding and parabolic rational functions. The expanding functions form a subclass of the subexpanding ones, whereas the classes of expanding and parabolic functions are disjoint. The union of these latter two classes constitutes the family of expansive rational functions. We first treat these three classes from a topological viewpoint, establishing among others their dynamical and topological characterizations and proving the continuity of the Julia sets of expanding and subexpanding maps in the Hausdorff metric of compact subsets of $\widehat{\mathbb{C}}$. Using the technique of quasisymmetric and quasiconformal mappings developed in Chapter 16, we demonstrate the structural Julia stability of expanding maps, and as a consequence show that the Julia sets of the quadratic maps $z \mapsto z^2 + c$, where c belongs to the main cardioid of the Mandelbrot set, are quasicircles and in particular Jordan curves.

We then move on to topics that originate from the paper [111] but that have not had any book presentation yet (except for one chapter in [110], which partially overlaps with some of them).

Chapter 27 pertains to thermodynamic formalism for all rational functions and Hölder continuous potentials with a pressure gap as described in [36, 106], including further advances due to [60, 133, 134]. Our exposition of this topic is improved and simplified via the combination and further development of the methods from all those papers. The main result is the existence and uniqueness of equilibrium states, their particular form and some stochastic properties such as K-mixing.

Chapter 28 treats of invariant measures and fractal and dynamical properties of rational functions, including the nonnegativity of Lyapunov exponents due to Przytycki, Ruelle's inequality, Pesin's theory, volume lemmas, and conformal Katok's theory. We also introduce three significant subsets of the Julia set, namely the radial (conical), expanding radial and uniformly expanding radial subsets. We study them at length in all subsequent chapters, especially in Chapter 29.

In Chapter 29, we give a general and detailed treatment of Sullivan's conformal measures. Our approach arises from [34, 35] and comprises the progress made in [110]. By means of a fairly effective construction, we prove as the main result the existence of a Sullivan conformal measure with a minimal exponent and equate this exponent with not only the Hausdorff dimension but also many other fractal characteristics/dimensions of the Julia set, such as its expanding (hyperbolic) dimension, its dynamical dimension, and the Hausdorff dimension of radial, expanding radial and uniformly expanding radial subsets of the Julia set. For the first time ever, the use of the otherwise very powerful technique of $K(V)$ sets published in [35] is avoided.

In the remaining chapters, we focus on the fractal properties and measurable dynamics generated by the three classes of functions examined in Chapter 26. The most transparent ones are found in the class of expanding maps scrutinized in Section 26.1 and especially in Chapter 30: the classical version of Bowen's formula, a unique Sullivan's conformal measure (whose exponent is necessarily equal to the Hausdorff dimension h of the Julia set), the finiteness and positivity of the h-dimensional Hausdorff and packing measures, their equality (up to a multiplicative constant) with the conformal measure and their Ahlfors (i. e., geometric) property. As a repercussion, all fractal dimensions (commonly used in complex dynamics) of the Julia set of an expanding map T are equal. The Gibbs/equilibrium state of the Hölder continuous potential $-h\log|T'|$ is the unique invariant Borel probability measure absolutely continuous with respect to the unique conformal measure. Furthermore, those two measures are equivalent with log-bounded Radon–Nikodym derivatives and, in particular, the invariant Gibbs measure is geometric.

As mentioned earlier, the expanding and parabolic classes are mutually disjoint but their union coincides with the family of expansive maps. Although there are some similarities, the portrait for parabolic rational functions, given in Section 26.2 and especially in Chapter 31, is quite different from that for expanding and subexpanding maps. The potential $-h\log|T'|$ is well-defined and even Hölder continuous but because of the presence of rationally indifferent periodic points, it does not exhibit a pressure gap. In particular, the results proved in Chapter 27 do not apply. Nevertheless, there is a neat version of Bowen's formula, which identifies the Hausdorff dimension h of the Julia set as the least zero of the pressure function $\mathrm{P}(t)$ but, unlike in the expanding case, $\mathrm{P}(t) = 0$ for all $t > h$. We also show that h is equal to the minimal exponent for which a conformal measure exists. As in the subexpanding case in Chapter 32, there exists a unique Sullivan's h-conformal measure and that measure is atomless; it is even ergodic and conservative. Similarly to the subexpanding case, there exist t-conformal measures for all $t > h$. All of those are purely atomic and are convex combinations of atomic measures supported on the backward orbits of rationally indifferent periodic points. In Section 31.4, we establish the existence and uniqueness of an invariant measure which is absolutely continuous with respect to the h-conformal measure, and we provide a simple criterion for this measure to be either finite or infinite. In particular, this measure is finite for the quadratic polynomial $z \mapsto z^2 + \frac{1}{4}$, while it is infinite for parabolic Blaschke products. In the last section of Chapter 31, we provide a complete description of the h-dimensional Hausdorff and packing measures on the Julia sets of parabolic rational functions. These are sometimes finite, sometimes positive, and sometimes infinite, their nature depending exclusively on the value of the Hausdorff dimension h. Chapter 31 combines and develops the results proved in several of Denker and Urbański's papers [1, 37, 38, 40].

For subexpanding maps, the picture is in some respects quite similar to that of expanding maps, but is considerably different in some other respects. Despite the possible presence of critical points in the Julia set and the classical topological pressure of the potential $-h\log|T'|$ not necessarily being well-defined, the equality of all commonly used fractal dimensions holds. There also exists a unique Sullivan's h-conformal measure and that measure is atomless. As in the case of expanding maps, the h-dimensional Hausdorff and packing measures are finite, positive and equal to the h-conformal measure, up to a multiplicative constant. They also share the Ahlfors property. We further show, and this is the most involved proof in Chapter 32, the existence and uniqueness of an invariant Borel probability measure absolutely continuous with respect to the unique h-conformal measure. Moreover, these two measures are equivalent though this time not necessarily with log-bounded Radon–Nikodym derivatives. As in the expanding case in Chapter 30 and the parabolic case in Chapter 31, there are no t-conformal measures for

$t < h$. The drastic difference between subexpanding and expanding maps is that if critical points are found in the Julia set of a subexpanding map, then there exist t-conformal measures for all $t > h$. All of those are purely atomic and are convex combinations of atomic measures supported on the backward orbits of critical points lying in the Julia set. Our presentation of subexpanding maps, given in Section 26.3 and especially in Chapter 32, stems partly from the paper [39] but both improves it substantially and goes beyond it.

The field of complex dynamics is beautiful, deep and wide. It has many subfields that, to some extent, live on their own, have their respective goals, methods, problems and results. We cover quite a bit of those but still much less than we left aside. We do not deeply deal with genericity of expandingness, the Mandelbrot set, external rays, Yoccoz puzzles, quadratic polynomials, quasiconformal surgeries, issues of local connectedness, the existence of Siegel disks (and rotation numbers for which these exist), the existence of Herman rings, parabolic implosion, Newton's method or polynomials with a positive plane Lebesgue measure. Nor do we deal with topics quite close to Chapters 28–32 such as harmonic measure, ergodic theory and fractal geometry of nonrecurrent rational functions and topological Collet–Eckmann maps. We do not dive very deep into the topic of geometric pressure, i. e., for potentials of the form $-t \log |T'|$ and all rational functions, although we get quite close to it in Chapter 29. Many of these topics would deserve an entire book or even several books.

Euclidean metric/derivative versus spherical metric/derivative

Whenever possible, we will work with the standard Euclidean metric and the ordinary Euclidean derivative. This is especially so when local topics are considered since one can then always, by means of a Möbius change of coordinates (i. e., via a Möbius conjugacy), move the problem from $\widehat{\mathbb{C}}$ to a map from an open subset of the complex plane $\mathbb{C}$ to $\mathbb{C}$. But when dealing with global issues, the use of the spherical metric and of the spherical derivative is actually necessary. Although this use is inconvenient and somewhat awkward, the essentially only alternative is to work with local holomorphic charts, which may be even more cumbersome.

In addition to local issues, the other clear cut context in which one can rely upon the Euclidean metric/derivative is when the Julia set of a rational map is not the whole Riemann sphere $\widehat{\mathbb{C}}$ and one is only interested in the dynamics on a sufficiently small open neighborhood of the Julia set. One can then similarly, i. e., by means of a Möbius conjugacy, send the Julia set outside of an open neighborhood of infinity. For instance, this is the case for any polynomial and any expansive rational function.

Since the spherical metric and the spherical derivative behave frequently quite well, and in particular since there are dependable distortion theorems about them (see Section 23.1), we will sometimes not explicitly state which metric or derivative is being considered; we will just write $|T'(z)|$ for the modulus (scaling factor) of either the

Euclidean or the spherical derivative of a meromorphic map T. But if distinguishing between these two derivatives is truly substantive and choosing the spherical one is actually necessary, we will then clearly indicate that we work with the spherical metric ρ and the spherical derivative $|T'(z)|_\rho$.

Contents

Volume 1

https://doi.org/10.1515/9783110769876-204

Volume 2

Volume 3

22 The Riemann–Hurwitz formula

In this chapter, we present a relation between holomorphic maps, their degree, their critical points and the topological structure of their images and preimages. This relation is known as the Riemann–Hurwitz formula. This formula has a long history and is treated in many textbooks on Riemann surfaces and algebraic geometry. It is usually formulated for compact Riemann surfaces. However, as we need greater generality, we provide here a complete, self-contained exposition which fulfills our needs.

The above-mentioned topological structure enters our considerations in terms of the Euler characteristic. We briefly introduce this latter, in a restricted scope and in the most elementary way; in particular, we only speak of triangulations rather than dealing with the more general and flexible concept of CW-complexes.

The results of this chapter, which are various versions of the Riemann–Hurwitz formula and their repercussions, will be instrumental throughout this volume. These formulas most likely constitute the best tool to handle the topological structure of connected components of preimages of open connected sets under holomorphic maps, especially to make sure that such connected components are simply connected.

Our exposition stems from that of Beardon [8]. We begin with a preparatory section devoted to proper maps and their topological degree.

22.1 Proper analytic maps and their degree

In this short section, we examine the concept of proper holomorphic maps and their degree. We closely follow Section 4.2 of Forster's book [51].

Let X and Y be topological spaces. A continuous map $f : X \to Y$ is said to be *proper* if $f^{-1}(K)$ is compact for every compact set $K \subseteq Y$.

Proposition 22.1.1. *Every proper map between locally compact Hausdorff topological spaces is closed, i. e., the image of every closed set is closed.*

Proof. Let X and Y be locally compact Hausdorff spaces and let $f : X \to Y$ be a proper map. Let F be a closed subset of X and $w \in \overline{f(F)}$. As Y is a locally compact T_2 space, there exists an open neighborhood W of w in Y such that $\overline{W}$ is compact. Since f is proper, the preimage $f^{-1}(\overline{W})$ is a compact subset of X. As F is closed in X, the set $F \cap f^{-1}(\overline{W})$ is thus compact. The continuity of f then ensures that the set $f(F \cap f^{-1}(\overline{W}))$ is compact, and hence closed in the Hausdorff space Y.

If V is any open neighborhood of w, then so is $W \cap V$. As $w \in \overline{f(F)}$, we have

$$(f(F) \cap W) \cap V = f(F) \cap (W \cap V) \neq \emptyset.$$

Therefore,

$$w \in \overline{f(F) \cap W} \subseteq \overline{f(F) \cap \overline{W}}.$$

https://doi.org/10.1515/9783110769876-022

Since $f(F) \cap \overline{W} = f(F \cap f^{-1}(\overline{W}))$ is a closed set, we deduce that

$$w \in \overline{f(F \cap f^{-1}(\overline{W}))} = f(F \cap f^{-1}(\overline{W})) \subseteq f(F).$$

So, $\overline{f(F)} \subseteq f(F)$, i. e., the set $f(F)$ is closed. □

A continuous map $f : X \to Y$ is called *discrete* if for every $y \in Y$ the preimage $f^{-1}(y)$ is discrete in X. More precisely, this means that $f^{-1}(y)$, when viewed as a topological subspace of X endowed with its relative topology, is a discrete topological space. Equivalently, for every point $x \in f^{-1}(y)$ there exists an open set $U \subseteq X$ such that $U \cap f^{-1}(y) = \{x\}$.

Lemma 22.1.2. *Suppose that X, Y are locally compact Hausdorff spaces and $f : X \to Y$ is a proper, discrete map. The following statements hold:*
(a) *For every point $y \in Y$, the preimage $f^{-1}(y)$ is a finite set.*
(b) *If $y \in Y$ and U is a neighborhood of $f^{-1}(y)$, then there exists a neighborhood V of y such that $f^{-1}(V) \subseteq U$.*
(c) *If $D \subseteq X$ is a closed discrete set, then so is $f(D)$.*

Proof. Item (a) follows from the fact that $f^{-1}(y)$ is a compact discrete subset of X.

For item (b), we may assume that U is open. As the set $X \setminus U$ is closed, its image $f(X \setminus U)$ is closed according to Proposition 22.1.1. Observe also that $y \notin f(X \setminus U)$. Thus, $V := Y \setminus f(X \setminus U)$ is an open neighborhood of y such that $f^{-1}(V) \subseteq U$.

To prove (c), fix $x \in D$. By (a), the set $f^{-1}(f(x))$ is finite. Since D is closed and discrete and $\# f^{-1}(f(x)) < \infty$, there exists an open neighborhood U of $f^{-1}(f(x))$ such that

$$U \cap D = f^{-1}(f(x)) \cap D.$$

Then $f(x) \notin f(X \setminus U)$, and $f(X \setminus U)$ is a closed set per Proposition 22.1.1. So, $Y \setminus f(X \setminus U)$ is an open neighborhood of $f(x)$ and

$$f(x) \in (Y \setminus f(X \setminus U)) \cap f(D) \subseteq f(U \cap D) \subseteq \{f(x)\} \cap f(D) = \{f(x)\}.$$

Thus, $(Y \setminus f(X \setminus U)) \cap f(D) = \{f(x)\}$, hence proving that $f(D)$ is discrete. The closedness of $f(D)$ follows from Proposition 22.1.1. □

We now turn our attention to holomorphic maps between Riemann surfaces. Recall that given a nonconstant holomorphic map $f : X \to Y$ between two Riemann surfaces X and Y, the local degree of f at z, denoted by $\deg_z(f)$, is the unique $k \in \mathbb{N}$ such that (using any pair of charts around z and $f(z)$)

$$f(\zeta) = f(z) + a(\zeta - z)^k + O((\zeta - z)^{k+1}) \tag{22.1}$$

for some $a \in \mathbb{C} \setminus \{0\}$. Observe that $\deg_z(f) = 1$ if and only if z is not a critical point of f.

Theorem 22.1.3. *Suppose that X and Y are Riemann surfaces and $f : X \to Y$ is a nonconstant proper holomorphic map. Then there exists $n \in \mathbb{N}$ such that f takes every value $w \in Y$ exactly n times (counting multiplicities), i. e.,*

$$\sum_{z \in f^{-1}(w)} \deg_z(f) = n.$$

In particular, the map f is surjective. The number n is called the (topological) degree *of f and is denoted by* $\deg(f)$.

Proof. It is a simple consequence of Taylor's theorem that the zeros of a nonzero holomorphic map are isolated. Consequently, the map f is discrete.

It also ensues from this and the continuity of f' that the set $\mathrm{Crit}(f)$ of critical (branch) points of f is closed and discrete. It is called the *critical set* of f. Since f is proper and discrete, it follows from Lemma 22.1.2 that the *set of critical values* $f(\mathrm{Crit}(f))$ of f is also closed and discrete. Let

$$X' := X \setminus f^{-1}(f(\mathrm{Crit}(f))) \subseteq X \setminus \mathrm{Crit}(f) \quad \text{and} \quad Y' := Y \setminus f(\mathrm{Crit}(f)).$$

Since $f|_{X'} : X' \to Y'$ is a proper discrete covering map (as a proper local homeomorphism between locally compact Hausdorff spaces) and since the space X' is pathwise connected, it follows from Lemma 22.1.2(a) that there exists $n \in \mathbb{N}$ such that

$$\#(f|_{X'})^{-1}(y) = n, \quad \forall y \in Y'. \tag{22.2}$$

It only remains to deal with the critical values $b \in f(\mathrm{Crit}(f))$. There then are mutually disjoint neighborhoods U_x of the preimages $x \in f^{-1}(b)$ as well as neighborhoods V_x of b such that $V_x \setminus \{b\} \subseteq Y'$ and for which there exist $q_x(f) \in \mathbb{N}$ such that

$$\#(f^{-1}(w) \cap U_x) = q_x(f), \quad \forall w \in V_x \setminus \{b\}.$$

It is clear from Taylor's theorem that $q_x(f) = \deg_x(f)$. By Lemma 22.1.2(b), there is a neighborhood

$$V \subseteq \bigcap_{x \in f^{-1}(b)} V_x$$

of b such that

$$f^{-1}(V) \subseteq \bigcup_{x \in f^{-1}(b)} U_x.$$

For every point $w \in V \setminus \{b\} \subseteq Y'$, we then have

$$\#(f^{-1}(w)) = \sum_{x \in f^{-1}(b)} q_x(f).$$

Combining this with (22.2), we conclude that

$$\sum_{x \in f^{-1}(b)} q_x(f) = n. \qquad \square$$

22.2 The Euler characteristic of plane bordered surfaces

We devote this section to a brief, restricted in scope, introduction to the Euler characteristic. More about this concept can be found in virtually any book on algebraic topology.

A domain (i. e., an open connected set) in the Riemann sphere $\widehat{\mathbb{C}}$ is said to be an *open bordered surface* if its boundary consists of a finite number (possibly 0) of mutually disjoint simple closed (i. e. Jordan) curves. Note that the complex plane $\mathbb{C}$ is not an open bordered surface but the Riemann sphere $\widehat{\mathbb{C}}$ is. An open bordered surface is called *proper* if it is different from $\widehat{\mathbb{C}}$.

A set $S \subseteq \widehat{\mathbb{C}}$ is said to be a *closed bordered surface* if it is the closure of an open bordered surface. Observe again that the boundary (border) of S can be empty and then $S = \widehat{\mathbb{C}}$. In fact, $\widehat{\mathbb{C}}$ is the only set which is simultaneously an open and closed bordered surface.

A *triangulation* T of S is a finite partition of S into subsets called *vertices*, *edges* and *faces*, denoted by V, E and F respectively, with the following properties:

(1) each vertex $v \in V$ is a point of S;

(2) for each edge $e \in E$, there are a compact interval $[a, b] \subset \mathbb{R}$ and a homeomorphism $\varphi : [a, b] \to \overline{e}$ such that

$$\varphi((a, b)) = e \quad \text{and} \quad \varphi(\{a, b\}) \subseteq V;$$

(3) for each face $f \in F$, there are a closed triangle $\Delta \subseteq \mathbb{C}$ and a homeomorphism $\varphi : \Delta \to \overline{f}$ which maps the edges and vertices of Δ (in the usual sense) into E and V, respectively, and such that

$$\varphi(\mathrm{Int}(\Delta)) = f.$$

Of course, if (2) holds for some compact interval (resp., if (3) holds for some triangle) then (2) holds for any compact interval (resp., (3) holds for any triangle). We stress once again that T partitions S into mutually disjoint subsets of S. Each such subset is either a vertex, an edge or a face, and we call each of these a *simplex* of T of *dimension* 0, 1 and 2, respectively. For any simplex s of dimension m, the *Euler characteristic* $\chi(s)$ is defined to be

$$(-1)^m.$$

More generally, if S' is any subset of S comprising a union of simplices, say $s_1, \ldots, s_k$, where s_j has dimension m_j, then S' is called a *subcomplex* of S relative to T, and we define

$$\chi(S',T) := \sum_{j=1}^{k} \chi(s_j) = \sum_{j=1}^{k} (-1)^{m_j}. \tag{22.3}$$

In particular, if a triangulation T of S consists of faces F, edges E and vertices V, then the Euler characteristic $\chi(S)$ of S is, by definition,

$$\chi(S) = \chi(S,T) := \#F - \#E + \#V.$$

The crucial fact that is well known in algebraic topology (which we accept here without proof) is that $\chi(S,T)$ is independent of the triangulation T. In particular, it is a topological invariant, meaning that two homeomorphic closed bordered surfaces have the same Euler characteristic. So, we can compute $\chi(S)$ using the triangulation of our choice. For this, we need to know, and we indeed do by obvious geometry, that every such surface admits a triangulation.

For each edge $e \in E$, the closure $\overline{e}$ is a subcomplex of S relative to T with

$$\chi(\overline{e},T) = 1. \tag{22.4}$$

The boundary ∂S of S is a subcomplex of S relative to any triangulation T and it is again evident that

$$\chi(\partial S,T) = 0,$$

as every connected component of ∂S is a Jordan curve, which is a subcomplex of any triangulation of S and whose Euler characteristic is clearly equal to zero. The independence from the triangulation T permits us to write

$$\chi(\partial S) = 0. \tag{22.5}$$

We say that the Euler characteristic of ∂S is zero.

Calculations of χ can often be simplified by making use of the above and the following simple idea extending it. If T is a triangulation of S, and S_1 and S_2 are two subcomplexes of S relative to T, then $S_1 \cup S_2$ and $S_1 \cap S_2$ are subcomplexes of S relative to T as well, and we obtain from (22.3) that

$$\chi(S_1 \cup S_2,T) + \chi(S_1 \cap S_2,T) = \chi(S_1,T) + \chi(S_2,T). \tag{22.6}$$

Let us illustrate this idea with examples. The interior $\mathrm{Int}(S) = S \setminus \partial S$ of S is a subcomplex of S relative to any triangulation T of S. By (22.6)–(22.5),

$$\chi(S) = \chi(\mathrm{Int}(S),T) + \chi(\partial S) = \chi(\mathrm{Int}(S),T).$$

Hence, $\chi(\mathrm{Int}(S),T)$ is independent of T and we can speak of $\chi(\mathrm{Int}(S))$, calling it the Euler characteristic of $\mathrm{Int}(S)$. We thus have

$$\chi(\mathrm{Int}(S)) = \chi(S), \tag{22.7}$$

and for open bordered surfaces $V \subseteq \widehat{\mathbb{C}}$ we define

$$\chi(V) := \chi(\overline{V}). \tag{22.8}$$

However, we must issue a warning. We have not yet established that $\chi(\mathrm{Int}(S))$ is a topological invariant. Indeed, we do not know yet whether $\chi(\mathrm{Int}(\widehat{S})) = \chi(\mathrm{Int}(S))$ for every bordered surface $\widehat{S}$ such that $\mathrm{Int}(\widehat{S})$ is homeomorphic to $\mathrm{Int}(S)$. We know this, though, when $\widehat{S}$ and S are homeomorphic. The more general case will follow from Proposition 22.4.4.

By constructing explicit triangulations, it is immediate that $\chi(\widehat{\mathbb{C}}) = 2$ and $\chi(\overline{D}) = 1$ for any open topological disk D (i. e., a set homeomorphic to $\mathbb{D}$).

By its very definition, each closed bordered surface S is the complement in $\widehat{\mathbb{C}}$ of a finite number, say $k \geq 0$, of mutually disjoint open topological disks $D_1, \ldots, D_k$, whose boundaries are Jordan curves. So, S is of connectivity k, meaning that its complement has k connected components. We can triangulate the sphere so that each of the sets S, $D_1, \ldots, D_k$ is a subcomplex of $\widehat{\mathbb{C}}$. Then (22.6) yields

$$2 = \chi(\widehat{\mathbb{C}}) = \chi(S) + \sum_{j=1}^{k} \chi(D_j) = \chi(S) + k.$$

Hence,

$$\chi(S) = 2 - k. \tag{22.9}$$

So, for any such closed bordered surface S of the sphere $\widehat{\mathbb{C}}$, we have that:
(a) $\chi(S) = 2$ if and only if S is the sphere $\widehat{\mathbb{C}}$, i. e., $k = 0$;
(b) $\chi(S) = 1$ if and only if S is simply connected but is not $\widehat{\mathbb{C}}$, i. e., S is a closed topological disk (i. e., S is homeomorphic to $\overline{\mathbb{D}}$), or equivalently $k = 1$;
(c) $\chi(S) = 0$ if and only if S is doubly connected, i. e., $k = 2$;
(d) $\chi(S) < 0$ in all other cases, i. e., if and only if $k \geq 3$.

22.3 The Riemann–Hurwitz formula for bordered surfaces in $\widehat{\mathbb{C}}$

Our first main goal is to prove a formula commonly called Riemann–Hurwitz formula, in the following setting. The ambient space will always be the Riemann sphere $\widehat{\mathbb{C}}$ and the boundary ∂A of a subset A of $\widehat{\mathbb{C}}$ will always be with respect to the topological space $\widehat{\mathbb{C}}$, i. e., $\partial A = \overline{A} \cap \overline{\widehat{\mathbb{C}} \setminus A}$. If $A \subseteq B \subseteq \widehat{\mathbb{C}}$, the boundary of A with respect to the topological space B will be denoted by $\partial_B A$. Note that

$$\partial_B A \subseteq \partial A.$$

Recall that a domain in $\widehat{\mathbb{C}}$ is (by definition) an open connected subset of $\widehat{\mathbb{C}}$.

Observation 22.3.1. If S is a domain in $\widehat{\mathbb{C}}$ and K is a compact subset of S, then

$$\operatorname{dist}(K, \partial S) > 0.$$

Let S_1 and S_2 be domains in $\widehat{\mathbb{C}}$ and let

$$R : S_1 \to S_2$$

be an analytic map which has a continuous extension from $\overline{S_1}$ to $\overline{S_2}$. That extension is unique and will generally be denoted by the same symbol R.

Theorem 22.3.2. *Let S_1 and S_2 be domains in $\widehat{\mathbb{C}}$. If an analytic map $R : S_1 \to S_2$ admits a (unique) continuous extension to $\overline{S_1}$, then R is proper if and only if*

$$R(\partial S_1) \subseteq \partial S_2. \tag{22.10}$$

Proof. For this proof, we reserve the symbol R for the map from S_1 to S_2 and denote by $\overline{R}$ the continuous extension of R to $\overline{S_1}$. By way of contradiction, suppose that the map $R : S_1 \to S_2$ is proper but that (22.10) does not hold. This means that there exists a point $z \in \partial S_1$ such that $\overline{R}(z) \in S_2$. As S_2 is open, there is $r > 0$ such that

$$\overline{B(\overline{R}(z), r)} \subseteq S_2.$$

Since R is proper, the preimage $R^{-1}(\overline{B(\overline{R}(z), r)})$ is a compact subset of S_1. As $z \in \partial S_1$, Observation 22.3.1 guarantees that there exists $\varepsilon > 0$ for which

$$B(z, \varepsilon) \cap R^{-1}(\overline{B(\overline{R}(z), r)}) = \emptyset. \tag{22.11}$$

Given that $\overline{R} : \overline{S_1} \to \overline{S_2}$ is continuous, there is $\delta \in (0, \varepsilon)$ such that

$$\overline{R}(B(z, \delta) \cap \overline{S_1}) \subseteq B(\overline{R}(z), r). \tag{22.12}$$

But since $z \in \overline{S_1}$, there exists a point $w \in B(z, \delta) \cap S_1$. It then follows from (22.11) that $R(w) \notin B(\overline{R}(z), r)$ while it ensues from (22.12) that $R(w) \in B(\overline{R}(z), r)$. This contradiction finishes the proof of the direct implication.

Now suppose that (22.10) holds but that the map R is not proper. Then there exists a compact set $K \subseteq S_2$ such that $R^{-1}(K)$ is not compact. But since $\overline{R}^{-1}(K)$ is compact (by the continuity of $\overline{R}$ and the compactness of $\overline{S_1}$), we get that $\overline{R}^{-1}(K) \backslash R^{-1}(K) \neq \emptyset$. However, $\overline{R}^{-1}(K) \backslash R^{-1}(K) \subseteq \partial S_1$, whence $\partial S_2 \supseteq \overline{R}(\partial S_1) \supseteq \overline{R}(\overline{R}^{-1}(K) \backslash R^{-1}(K)) \subseteq K \subseteq S_2$. This means that $\overline{R}(\overline{R}^{-1}(K) \backslash R^{-1}(K)) = \emptyset$, and hence $\overline{R}^{-1}(K) \backslash R^{-1}(K) = \emptyset$. This contradiction completes the proof of the converse implication. □

Theorem 22.3.3. *Let S_1 and S_2 be domains in $\widehat{\mathbb{C}}$ and $R : S_1 \to S_2$ be a proper analytic map which has a (unique) continuous extension to $\overline{S_1}$. If V is a subdomain of S_2 and U is a connected component of $R^{-1}(V)$, then the restriction $R|_U : U \to V$ is a proper analytic map satisfying*

$$R(U) = V, \quad R(\partial U) = \partial V \quad \text{and} \quad \overline{U} \cap R^{-1}(\partial V) = \partial U. \tag{22.13}$$

In particular, $R(\partial S_1) = \partial S_2$.

Proof. We first prove that the map $R|_U : U \to V$ is proper. Let $K \subseteq V$ be a compact set. Let $\mathcal{C}$ be the collection of all connected components of $R^{-1}(V)$. Then

$$R|_U^{-1}(K) = R^{-1}(K) \cap \Big[\widehat{\mathbb{C}} \setminus \bigcup_{C \in \mathcal{C}\setminus\{U\}} C\Big].$$

As the set $R^{-1}(K)$ is compact (given that $R : S_1 \to S_2$ is proper) and each $C \in \mathcal{C}$ is an open subset of $\widehat{\mathbb{C}}$, we conclude that the set $R|_U^{-1}(K)$ is compact. Thus, the map $R|_U : U \to V$ is proper.

Moreover, it follows from Theorem 22.3.2 that

$$R(\partial U) \subseteq \partial V$$

and from Theorem 22.1.3 that

$$R(U) = V.$$

This immediately implies that $U \cap R^{-1}(\partial V) = \emptyset$. Hence,

$$\overline{U} \cap R^{-1}(\partial V) = \partial U. \tag{22.14}$$

In order to complete the proof, it remains to show that

$$\partial V \subseteq R(\partial U). \tag{22.15}$$

Let $w \in \partial V$. There exists a sequence $(w_n)_{n=1}^{\infty}$ of points in V converging to w. For every $n \in \mathbb{N}$, choose a point $z_n \in U \cap R^{-1}(w_n)$. Passing to a subsequence if necessary, we may assume that the sequence $(z_n)_{n=1}^{\infty}$ converges in $\widehat{\mathbb{C}}$. Denote its limit by z. Then $z \in \overline{U}$ and

$$R(z) = \lim_{n\to\infty} R(z_n) = \lim_{n\to\infty} w_n = w.$$

So, $z \in \overline{U} \cap R^{-1}(\partial V)$, and thus $z \in \partial U$ by (22.14). Consequently, $w = R(z) \in R(\partial U)$ and (22.15) holds. □

From now on, we assume throughout the rest of this section that S_1 and S_2 are domains in $\widehat{\mathbb{C}}$ and $R : S_1 \to S_2$ is a proper analytic map having a (unique) continuous extension to S_1. We also assume that V is a subdomain of S_2 such that $\overline{V} \subseteq S_2$ and

$$\partial V \cap R(\mathrm{Crit}(R)) = \emptyset, \tag{22.16}$$

where Crit(R) is the set of critical points of R. So, $R(\mathrm{Crit}(R))$ is the set of critical values of R. Furthermore, we let U be a connected component of $R^{-1}(V)$. We will establish some properties of the map R in relation to V and U that will turn out to be useful in the proof of the Riemann–Hurwitz formula.

In light of Theorems 22.1.3 and 22.3.3, we may speak of the degree of the map $R|_U : U \to V$ and state that $\deg(R|_U) \geq 1$.

Lemma 22.3.4. *Let S_1 and S_2 be domains in $\widehat{\mathbb{C}}$ and let $R : S_1 \to S_2$ be a proper analytic map which admits a continuous extension to $\overline{S_1}$. If $V \subseteq S_2$ is an open bordered surface, then any connected component U of $R^{-1}(V)$ is also an open bordered surface. (In particular, $\overline{U}$ and $\overline{V}$ are closed bordered surfaces in $\widehat{\mathbb{C}}$ and $\mathrm{Int}(\overline{U}) = U$ and $\mathrm{Int}(\overline{V}) = V$.) In addition, the restriction $R|_{\partial U} : \partial U \to \partial V$ is a proper covering map of finite degree.*

Proof. Because of Theorem 22.3.3 and relation (22.16), the map

$$R|_{\partial U} : \partial U \to \partial V$$

is a locally homeomorphic surjection. As ∂U is compact, the map $R|_{\partial U} : \partial U \to \partial V$ is thereby a covering map of finite degree, and we write $q := \deg(R|_{\partial U})$. If Γ is a connected component of ∂V, then Γ is a Jordan curve, and $R : R^{-1}(\Gamma) \to \Gamma$ is also a covering map of degree at most q. Let L be a connected component of $R^{-1}(\Gamma)$. We claim that $R(L) = \Gamma$. Otherwise, there would exist a point $x \in L$ such that $R(x) \in \partial_\Gamma R(L)$. But, because of (22.16), there exists an open topological arc α containing x such that $R(\alpha) \subseteq \Gamma$ and $R(\alpha)$ is an open topological arc containing $R(x)$. But then $L \cup \alpha$ is connected and $R(L \cup \alpha) \subseteq \Gamma$. Hence, $L \cup \alpha = L$ and, therefore, $R(x) \in R(\alpha) \subseteq \mathrm{Int}_\Gamma R(L)$. This contradiction gives that

$$R(L) = \Gamma.$$

Thus, the map $R : L \to \Gamma$ is a covering surjection. In consequence, it is at most q–to–1 and $R^{-1}(\Gamma)$ has at most q connected components. We are thus left to show that L is a Jordan curve. But L is compact and $R : L \to \Gamma$ is a local homeomorphism, so L is a compact connected 1-dimensional topological manifold. It is well known that L is then a Jordan curve. Another argument, less topological and more analytic, goes as follows. Fix a point $\xi \in \Gamma$. Then $\#(L \cap R^{-1}(\xi)) \leq q$ and each point $z \in L \cap R^{-1}(\xi)$ gives rise to a local inverse map R_z^{-1} from a sufficiently small neighborhood of ξ onto a neighborhood of z; in particular, $R_z^{-1}(\xi) = z$. Extending R_z^{-1} analytically from ξ to ξ along Γ, we traverse a path Γ_z in L whose other endpoint $\hat{z}$ belongs to $R^{-1}(\xi)$. The path Γ_z is either a closed topological arc or a Jordan curve depending on whether $\hat{z} \neq z$ or $\hat{z} = z$, respectively. Moreover, the map

$$R^{-1}(\xi) \ni z \longmapsto \widehat{z} \in R^{-1}(\xi)$$

is a bijection. We repeatedly extend analytically, successively getting points $z, \widehat{z}, \widehat{\widehat{z}}, \ldots$, until we reach z again, through at most q iterations. The consecutive closed topological arcs $\Gamma_z, \Gamma_{\widehat{z}}, \Gamma_{\widehat{\widehat{z}}}, \ldots$, have exactly one common endpoint, $\widehat{z}, \widehat{\widehat{z}}, \ldots$, respectively. Their union, up to the second-last one of them, is a closed topological arc as well. The union of this latter with the last closed topological arc is a Jordan curve since these two arcs have two common endpoints and no other common points. □

Lemma 22.3.5. *Under the hypotheses of Lemma* 22.3.4, *we have*

$$\deg(R|_{\partial U}) = \deg(R|_U).$$

Proof. Fix $\xi \in \partial V$. As ∂V contains no critical value of R, there exists $\varepsilon > 0$ such that all analytic branches of R^{-1} are defined on $B(\xi, \varepsilon)$. Since ∂U consists of finitely many disjoint Jordan curves (by Lemma 22.3.4) and since for every $z \in \partial U \cap R^{-1}(\xi)$, the map $R : R_z^{-1}(B(\xi, \varepsilon)) \to B(\xi, \varepsilon)$ is a homeomorphism, there exists an open (relative to $\overline{U}$) neighborhood of z in $\overline{U}$, denoted by $\Gamma_z \subseteq R_z^{-1}(B(\xi, \varepsilon)) \cap \overline{U}$, such that $R(\Gamma_z)$ is an open (relative to $\overline{V}$) neighborhood of ξ in $\overline{V}$. Hence,

$$V_\xi := \bigcap_{z \in \partial U \cap R^{-1}(\xi)} R(\Gamma_z)$$

is also an open (relative to $\overline{V}$) neighborhood of ξ in $\overline{V}$. Fix $y \in V_\xi \cap V$. As each family

$$\mathcal{F}_\xi := \{R_z^{-1}(B(\xi, \varepsilon)) \mid z \in \partial U \cap R^{-1}(\xi)\} \quad \text{and} \quad \mathcal{F}_y := \{R_x^{-1}(B(\xi, \varepsilon)) \mid x \in U \cap R^{-1}(y)\}$$

consists of mutually disjoint sets, we know that

$$\deg(R|_{\partial U}) = \#\mathcal{F}_\xi \quad \text{and} \quad \deg(R|_U) = \#\mathcal{F}_y.$$

Thus, in order to complete the proof, it suffices to show that

$$\mathcal{F}_\xi = \mathcal{F}_y. \tag{22.17}$$

If $z \in \partial U \cap R^{-1}(\xi)$, then there exists a (unique) point $x \in \Gamma_z$ such that $R(x) = y$. So, $x \in \overline{U} \cap R^{-1}(V) = U$. Hence, $x \in U \cap R^{-1}(y)$ and $R_x^{-1}(y) = x = R_z^{-1}(y)$. Consequently, $R_z^{-1}(B(\xi, \varepsilon)) = R_x^{-1}(B(\xi, \varepsilon)) \in \mathcal{F}_y$, and the inclusion

$$\mathcal{F}_\xi \subseteq \mathcal{F}_y \tag{22.18}$$

is proved.

For the opposite inclusion, suppose that $x \in U \cap R^{-1}(y)$. Then $R_x^{-1}(V_\xi \cap V) \subseteq U$, whence

$$z := R_x^{-1}(\xi) \in R_x^{-1}(\overline{V_\xi \cap V}) = \overline{R_x^{-1}(V_\xi \cap V)} \subseteq \overline{U}.$$

Therefore, $z \in \partial U \cap R^{-1}(\xi)$ and $R_z^{-1}(\xi) = z = R_x^{-1}(\xi)$. Thus, $R_z^{-1}(B(\xi, \varepsilon)) = R_x^{-1}(B(\xi, \varepsilon))$. So, $R_x^{-1}(B(\xi, \varepsilon)) \in \mathcal{F}_\xi$ and the inclusion $\mathcal{F}_y \subseteq \mathcal{F}_\xi$ is proved. Along with (22.18), this completes the proof of formula (22.17), and, simultaneously, the proof of Lemma 22.3.5. □

We now introduce the *deficiency* of R at a point $z \in S_1$ as

$$\delta_R(z) := \deg_z(R) - 1 \geq 0, \tag{22.19}$$

where $\deg_z(R)$ is the local degree of R at z (see (22.1)). Note that $\delta_R(z) > 0$ if and only if z is a critical point of R. For a set $A \subseteq S_1$, the *total deficiency* of R *over* A is naturally defined as

$$\delta_R(A) := \sum_{z \in A} \delta_R(z) \geq 0.$$

The function $A \mapsto \delta_R(A)$ is nondecreasing and additive, i. e., $\delta_R(A) \leq \delta_R(B)$ whenever $A \subseteq B \subseteq S_1$ while for disjoint sets A and B in S_1,

$$\delta_R(A \cup B) = \delta_R(A) + \delta_R(B).$$

We are now ready to relate the quantities $\chi(U), \chi(V), \deg(R|_U)$ and $\delta_R(U)$.

Theorem 22.3.6 (Riemann–Hurwitz formula for bordered surfaces in $\widehat{\mathbb{C}}$). *Let S_1 and S_2 be domains in $\widehat{\mathbb{C}}$ and $R : S_1 \to S_2$ be a proper analytic map which admits a continuous extension to $\overline{S_1}$. Let V be a subdomain of S_2 such that $\overline{V} \subseteq S_2$ and such that $\overline{V}$ is a bordered surface in $\widehat{\mathbb{C}}$ for which*

$$\partial V \cap R(\mathrm{Crit}(R)) = \emptyset. \tag{22.20}$$

If U is a connected component of $R^{-1}(V)$, then

$$\chi(U) + \delta_R(U) = \deg(R|_U) \cdot \chi(V). \tag{22.21}$$

Proof. We can triangulate $\overline{V}$ in such a way that all critical values of R in V are vertices of the triangulation. Indeed, since each connected component of ∂V is a Jordan curve, given any triangulation, we can connect via a closed topological arc each critical value of R with at least two distinct vertices of the triangulation, bounding a 1-dimensional simplex on one connected component of ∂V. This way we form a larger triangulation with the required property. Denote such a triangulation by T. Its sets of vertices, edges and faces are respectively denoted by V_T, E_T and F_T.

We construct a triangulation T_U of ∂U in the following way. Let $m = \deg(R|_U)$. Let $\Delta \in F_T$. Since Δ is an open topological disk, there exists an open topological disk $\Delta' \subseteq S_2$ containing $\overline{\Delta} \setminus V$ and disjoint from the union of the critical values of R and the vertices of $\overline{\Delta}$. Then, according to the monodromy theorem (Theorem A.3.4), there are exactly m distinct analytic branches $R_j^{-1} : \Delta' \to S_1$ of R^{-1} such that $R_j^{-1}(\Delta) \subseteq U$. Define

$$F_{T_U} := \{R_j^{-1}(\Delta) \mid 1 \le j \le m,\ \Delta \in F_T\}.$$

The elements of F_{T_U} will be the faces of the triangulation T_U. The edges of every face $R_j^{-1}(\Delta)$ correspond with the sets $R_j^{-1}(e_i)$, $i = 1, 2, 3$, where the e_i's are the three edges of Δ. So, define

$$E_{T_U} := \{R_j^{-1}(e) \mid 1 \le j \le m,\ \Delta \in F_T,\ e \in E_T \text{ and } e \subseteq \partial\Delta\}.$$

Finally, the set V_{T_U} consists of the endpoints of the closures of all elements of E_{T_U}. Note that the endpoints of the closure of each $e \in E_{T_U}$ are distinct since they are mapped by R onto two different points, namely the endpoints of $R(e) \in E_T$. It readily follows from this construction that the sets F_{T_U} labeled as faces, E_{T_U} labeled as edges and V_{T_U} labeled as vertices, form a triangulation of $\overline{U}$. We denote that triangulation by T_U.

It is immediate from this construction that

$$\#F_{T_U} = m \cdot \#F_T, \quad \#E_{T_U} = m \cdot \#E_T$$

and

$$\#V_{T_U} = m \cdot \#V_T - \sum_{c \in V_{T_U} \cap \mathrm{Crit}(R)} (\deg_c(R) - 1) = m \cdot \#V_T - \delta_R(U).$$

Therefore,

$$\begin{aligned}\chi(U) = \chi(\overline{U}, T_U) &= \#F_{T_U} - \#E_{T_U} + \#V_{T_U}\\ &= m(\#F_T - \#E_T + \#V_T) - \delta_R(U)\\ &= m\chi(\overline{V}) - \delta_R(U) = m\chi(V) - \delta_R(U).\end{aligned}$$ □

We now derive a series of consequences of this theorem.

Corollary 22.3.7. *Under the hypotheses of Theorem 22.3.6 and the additional assumption that V is simply connected but different from $\widehat{\mathbb{C}}$ (i. e., V is conformally equivalent to the unit disk $\mathbb{D}$ due to Koebe's uniformization theorem (Theorem 23.2.2)), it turns out that $U \ne \widehat{\mathbb{C}}$, and hence $\chi(U) \le 1$ and*

$$\delta_R(U) \ge \deg(R|_U) - 1,$$

with equality holding if and only if U is simply connected (i. e., is conformally equivalent to $\mathbb{D}$). In this latter case,

$$\#\mathrm{Crit}(R|_U) \le \deg(R|_U) - 1. \tag{22.22}$$

Corollary 22.3.8. *Under the hypotheses of Theorem 22.3.6 and the additional assumptions*
(1) *that V is simply connected (i. e., is either equal to $\widehat{\mathbb{C}}$ or conformally equivalent to the unit disk $\mathbb{D}$); and*
(2) *that U contains no critical point of R,*
the map $R|_U : U \to V$ is a conformal homeomorphism. In consequence, U has the same nature as V (i. e., either $U = V = \widehat{\mathbb{C}}$, or U and V are both conformally equivalent to $\mathbb{D}$).

Proof. If $V = \widehat{\mathbb{C}}$, then $\chi(V) = 2$. By (2), $\delta_R(U) = 0$. As $\chi(U) \le 2$, we deduce from Theorem 22.3.6 that $\deg(R|_U) \le 1$. Thus, $\deg(R|_U) = 1$, whence $R|_U$ is a conformal homeomorphism. Therefore, $U = \widehat{\mathbb{C}}$.

If $V \ne \widehat{\mathbb{C}}$, then $U \ne \widehat{\mathbb{C}}$, for otherwise V would be compact, and thus equal to $\widehat{\mathbb{C}}$. Hence, $\chi(U) \le 1$. We therefore infer from Theorem 22.3.6 that $\deg(R|_U) \le 1$. Consequently, $\deg(R|_U) = 1$, whence $R|_U$ is a conformal homeomorphism. This immediately yields that U, like V, is conformally equivalent to $\mathbb{D}$. □

Corollary 22.3.9. *Under the hypotheses of Theorem 22.3.6 and the additional assumptions*
(1) *that V is conformally equivalent to the unit disk $\mathbb{D}$; and*
(2) *that the map $R|_U : U \to V$ has a unique critical point c,*
it holds that:
(a) *U is conformally equivalent to $\mathbb{D}$; and*
(b) *$\deg(R|_U) = \deg_c(R)$.*

Proof. Obviously, $\deg(R|_U) \ge \deg_c(R)$. Since $U \ne \widehat{\mathbb{C}}$ (as otherwise V would be compact and thus equal to $\widehat{\mathbb{C}}$), we know that $\chi(U) \le 1$. It then follows from Theorem 22.3.6 that $\deg(R|_U) \le \deg_c(R)$. Hence, $\deg(R|_U) = \deg_c(R)$, proving (b). Applying Theorem 22.3.6 once more, we conclude that $\chi(U) = 1$ and (a) holds. □

22.4 Euler characteristic: the general $\widehat{\mathbb{C}}$ case

The Euler characteristic of a domain $D \subseteq \widehat{\mathbb{C}}$ has been defined in Section 22.2 whenever its boundary ∂D consists of finitely many Jordan curves. In general, however, the boundary of a domain D is much more complicated. We may not be able to triangulate the closure of D and in such circumstances, $\chi(D)$ has yet to be defined. As this is likely to be so in the case of main interest to us (when D is a component of the Fatou set), this presents us with a problem which we will now tackle. We propose to show that given any domain D, we can define $\chi(D)$ as the limiting value of the Euler characteristic of smooth subdomains which exhaust D. Once this is done, we can use the Euler characteristic as a tool to study the way in which a rational function maps one component of the Fatou set onto another. The following exposition is closely related to the construction of the ideal boundary components of a Riemann surface.

Let $D \subseteq \widehat{\mathbb{C}}$ be a domain, i. e., an open connected set. A subdomain Δ of D is said to be a *regular subdomain* of D if:

(1) Δ is an open bordered surface whose closure $\overline{\Delta}$ is contained in D, i. e. $\overline{\Delta} \subseteq D$, and $\partial\Delta$ is a finite (possibly empty) union of mutually disjoint Jordan curves $\gamma_1, \dots, \gamma_n$,

and

(2) if $\Gamma_1, \dots, \Gamma_n$ denote all the connected components of $\widehat{\mathbb{C}} \setminus \Delta$ enumerated so that their boundaries are respectively $\gamma_1, \dots, \gamma_n$, then $\Gamma_j \cap (\widehat{\mathbb{C}} \setminus D) \neq \emptyset$ for every $1 \leq j \leq n$.

For example, the open unit disk $\mathbb{D} = \{z \in \mathbb{C} : |z| < 1\}$ is a regular subdomain of $\mathbb{C}$, whereas the annulus $\{z \in \mathbb{C} : 1 < |z| < 2\}$ is not.

Observation 22.4.1. The only regular subdomain of the Riemann sphere $\widehat{\mathbb{C}}$ is $\widehat{\mathbb{C}}$ itself.

Per the previous section, the Euler characteristic $\chi(\Delta)$ is defined for each regular subdomain Δ of D.

If a nonregular subdomain Δ of D satisfies condition (1) but not (2) (for example, an annulus), we can form a regular subdomain Δ_* of D by adjoining to Δ those sets Γ_j which do not intersect $\widehat{\mathbb{C}} \setminus D$. Obviously, $\chi(\Delta_*) \geq \chi(\Delta)$; in fact, $\chi(\Delta_*) = \chi(\Delta) + k$, where $k \geq 0$ is the number of adjoined sets Γ_j.

We want to represent D as a limit of regular subdomains. Given that no canonical sequence of regular subdomains of D presents itself, it is best to reject the idea of a sequential limit and to consider instead convergence with respect to the directed set (or net) of regular subdomains. There is no need for great generality here and the details are quite simple and explicit. First, we prove the following.

Lemma 22.4.2. *If D is a domain in $\widehat{\mathbb{C}}$, then*

(a) *any compact subset of D lies in a regular subdomain of D*

and

(b) *if Δ_1 and Δ_2 are regular subdomains of D, then there is a regular subdomain Δ of D which contains $\Delta_1 \cup \Delta_2$.*

Proof. If $D = \widehat{\mathbb{C}}$, the lemma is immediate by virtue of Observation 22.4.1. So suppose that D is a proper subset of $\widehat{\mathbb{C}}$. If ξ is an arbitrary point in D, then the Möbius map $\widehat{\mathbb{C}} \ni z \mapsto (z - \xi)^{-1} \in \widehat{\mathbb{C}}$ sends ξ to ∞ and we may assume without loss of generality that $\infty \in D$. Let $n \in \mathbb{N}$. Cover the plane $\mathbb{C}$ with squares whose sides of length $1/2^n$ are parallel to the axes and whose corners have coordinates that are integer multiples of $1/2^n$. Let K_n be the union of those squares in the cover that intersect $\widehat{\mathbb{C}} \setminus D$. If $n \in \mathbb{N}$ is large enough, say $n \geq N$, then $\infty \notin K_n$ and let D_n be the connected component of $\widehat{\mathbb{C}} \setminus K_n$ that contains ∞. It is easy to see that $(D_n)_{n=N}^{\infty}$ is an ascending sequence of regular subdomains of D such that $\cup_{n=N}^{\infty} D_n = D$.

Having this, the rest of the proof is straightforward. Given any compact set $K \subseteq D$, the family $\{D_n\}_{n=N}^{\infty}$ is an open cover of K and so admits a finite subcover. As the sequence $(D_n)_{n=N}^{\infty}$ is ascending, this subcover encompasses a largest domain D_m which contains K. This proves (a).

Regarding (b), if Δ_1 and Δ_2 are regular subdomains of D, then $\overline{\Delta}_1 \cup \overline{\Delta}_2$ is a compact subset of D and so lies, by (a), in a regular subdomain of D. □

Lemma 22.4.2 states that the class $\mathcal{R}(D)$ of all regular subdomains of D is a net under the order determined by the (direct) inclusion relation.

Our next task is to show that the χ function is monotone decreasing on $\mathcal{R}(D)$ and so tends to a limit, which may be $-\infty$.

Lemma 22.4.3. *For every domain $D \subseteq \widehat{\mathbb{C}}$, the Euler characteristic function $\chi : \mathcal{R}(D) \to \{2, 1, 0, -1, -2, \dots, -\infty\}$ is decreasing. That is,*

$$\Delta_1, \Delta_2 \in \mathcal{R}(D),\ \ \Delta_1 \subseteq \Delta_2 \quad \Longrightarrow \quad \chi(\Delta_2) \le \chi(\Delta_1).$$

Proof. Let $V_1, \dots, V_m$ be the connected components of $\widehat{\mathbb{C}} \setminus \Delta_1$, and let $W_1, \dots, W_n$ be the connected components of $\widehat{\mathbb{C}} \setminus \Delta_2$. Since $\Delta_1 \subseteq \Delta_2$, we have

$$W_1 \cup \dots \cup W_n \subseteq V_1 \cup \dots \cup V_m.$$

For each $j \in \{1, \dots, m\}$, pick a point $z_j \in V_j \setminus D$. As z_j is not in D, it lies in some set W_k and so W_k (being connected) is contained in V_j. It follows that each set $V_j, j \in \{1, \dots, m\}$, contains some set W_k with $k \in \{1, \dots, n\}$. Hence, $m \le n$. Using (22.9), we deduce that

$$\chi(\Delta_2) = 2 - n \le 2 - m = \chi(\Delta_1).$$ □

An immediate repercussion of Lemma 22.4.3 is the following.

Proposition 22.4.4. *For any subdomain D of $\widehat{\mathbb{C}}$, the limit*

$$\chi(D) := \lim_{\Delta \in \mathcal{R}(D)} \chi(\Delta)$$

exists and

$$\chi(D) = \inf_{\Delta \in \mathcal{R}(D)} \chi(\Delta).$$

More precisely:

(1) Either $\chi(D) = -\infty$, and there are regular subdomains Δ_n of D with $\chi(\Delta_n) \to -\infty$ (equivalently, the connectivity of the Δ_n's diverges to ∞)

(2) or else there is a regular subdomain D_* of D such that

$$\chi(D_*) = \chi(D) > -\infty,$$

and $\chi(\Delta) = \chi(D)$ for every regular subdomain Δ of D which contains D_*.

If D is a simply connected domain, then ∂D is connected and each regular subdomain Δ of D has at most one complementary connected component, regardless of the nature of ∂D: it has no complementary component if $D = \widehat{\mathbb{C}}$ and exactly one if $D \neq \widehat{\mathbb{C}}$. Using the Riemann mapping theorem (Theorem A.3.14), we obtain the following.

Theorem 22.4.5. *A domain $D \subseteq \widehat{\mathbb{C}}$ is simply connected if and only if either $\chi(D) = 2$ or $\chi(D) = 1$. In the former case, $D = \widehat{\mathbb{C}}$. In the latter case, D is either $\mathbb{C}$ or is conformally equivalent to the unit disk $\mathbb{D}$.*

More generally, if D has connectivity k, then $\chi(D) = 2 - k$ for all sufficiently large regular subdomains Δ of D, and so

$$\chi(D) = 2 - k, \tag{22.23}$$

again irrespective of the complexity of ∂D.

Furthermore, the current definition of the Euler characteristic of a domain in $\widehat{\mathbb{C}}$ coincides with the one in (22.9) in the case of a bordered surface D. Also, we now know that the Euler characteristic of domains is a topological invariant.

22.5 Riemann–Hurwitz formula: the general $\widehat{\mathbb{C}}$ case

The main theorem of this section and of the entire Chapter 22 is a generalization of Theorem 22.3.6.

Theorem 22.5.1 (General Riemann–Hurwitz formula). *Let S_1 and S_2 be domains in $\widehat{\mathbb{C}}$ (i. e., open connected subsets of $\widehat{\mathbb{C}}$). Let $R : S_1 \to S_2$ be a proper analytic map which admits a (unique) continuous extension to $\overline{S_1}$. Let V be a subdomain of S_2. If U is a connected component of $R^{-1}(V)$, then*

$$\chi(U) + \delta_R(U) = \deg(R|_U) \cdot \chi(V). \tag{22.24}$$

Proof. Fix a point $w \in V \setminus R(\mathrm{Crit}(R))$. Because of Lemma 22.4.2(a) there exists a regular subdomain Δ_0 of U such that:

(1) $\mathrm{Crit}(R) \cap U \subseteq \Delta_0$

and

(2) $\emptyset \neq U \cap R^{-1}(w) \subseteq \overline{U} \cap R^{-1}(w) \subseteq \Delta_0$.

Again due to Lemma 22.4.2(a) and the compactness of $R(\overline{\Delta}_0) \subseteq V$, there exists a regular subdomain Δ_1 of V such that

$$R(\overline{\Delta}_0) \subseteq \Delta_1. \tag{22.25}$$

We shall prove the following.

Claim. *The set*

$$\Delta_2 := U \cap R^{-1}(\Delta_1)$$

is a regular subdomain of U *containing* Δ_0.

Proof. Since Δ_0 is connected and since $R(\Delta_0) \subseteq \Delta_1$, there exists a unique connected component Δ_2' of $R^{-1}(\Delta_1)$ such that

$$\Delta_0 \subseteq \Delta_2'. \tag{22.26}$$

As every connected component of $R^{-1}(\Delta_1)$ is either disjoint from U or is contained in U, and as any component Δ_2'' contained in U intersects $R^{-1}(w)$ (because $R(\Delta_2'') = \Delta_1$ by Theorem 22.3.3), it follows from property (2) of Δ_0 that $\Delta_2'' \cap \Delta_0 \neq \emptyset$. But then $\Delta_2'' \cap \Delta_2' \neq \emptyset$ by (22.26). Consequently, $\Delta_2'' = \Delta_2'$. We have thus proved that

$$\Delta_2' = U \cap R^{-1}(\Delta_1) =: \Delta_2.$$

So, Δ_2 is a subdomain of U, and $\Delta_0 \subseteq \Delta_2$ by (22.26). Moreover, $\Delta_2 = \Delta_2'$ is a bordered surface by Lemma 22.3.4. We are thus left to show that the subdomain Δ_2 of U is regular. For this, let W be a connected component of $\widehat{\mathbb{C}} \setminus \Delta_2$. Seeking a contradiction, suppose that $W \cap (\widehat{\mathbb{C}} \setminus U) = \emptyset$. This means that

$$W \subseteq U \cap (\widehat{\mathbb{C}} \setminus \Delta_2) = U \setminus \Delta_2 = U \setminus (U \cap R^{-1}(\Delta_1)) = U \cap R^{-1}(\widehat{\mathbb{C}} \setminus \Delta_1). \tag{22.27}$$

We know that $R(W)$ is connected and

$$R(W) \subseteq R(U \cap R^{-1}(\widehat{\mathbb{C}} \setminus \Delta_1)) \subseteq R(U) \cap R(R^{-1}(\widehat{\mathbb{C}} \setminus \Delta_1)) \subseteq V \cap (\widehat{\mathbb{C}} \setminus \Delta_1) = V \setminus \Delta_1.$$

Let W_1 be the connected component of $V \setminus \Delta_1$ containing $R(W)$ and let W_2 be the connected component of $\widehat{\mathbb{C}} \setminus \Delta_1$ containing W_1. In particular, W_2 is a connected component of $\widehat{\mathbb{C}} \setminus \Delta_1$ and, therefore, since Δ_1 is a regular subdomain of V, we deduce that

$$W_2 \cap (\widehat{\mathbb{C}} \setminus V) \neq \emptyset. \tag{22.28}$$

As $W_1 \subseteq V \cap W_2$, we know that

$$\partial W_1 \subseteq \partial V \cup \partial W_2.$$

We will show that $\partial W_1 \cap \partial V \neq \emptyset$. By way of contradiction, suppose that $\partial W_1 \cap \partial V = \emptyset$. Then $\partial W_1 \subseteq \partial W_2$. Hence, $W_2 = W_1 \cup (W_2 \setminus \overline{W_1})$. Since W_2 is connected and $W_1 \neq \emptyset$, this implies that $W_2 \setminus \overline{W_1} = \emptyset$. Equivalently, $W_2 \subseteq \overline{W_1}$. As both W_1 and W_2 are open, this yields $W_2 \subseteq W_1$. Thus, $W_2 = W_1$, whence $W_2 \subseteq V$, contrary to (22.28). We have thereby proved that

$$\partial W_1 \cap \partial V \neq \emptyset. \tag{22.29}$$

We now show that W is a connected component of $R^{-1}(W_1)$ contained in U. Indeed, by (22.27),

$$W \subseteq U \cap R^{-1}(W_1) \subseteq U \cap R^{-1}(V \setminus \Delta_1) = U \setminus \Delta_2 \subseteq \widehat{\mathbb{C}} \setminus \Delta_2.$$

Therefore, since W is a connected component of $\widehat{\mathbb{C}} \setminus \Delta_2$, it is also a connected component of $U \cap R^{-1}(W_1)$. As each connected component of $R^{-1}(W_1)$ is either contained in U or is disjoint from U, it follows that W is a connected component of $R^{-1}(W_1)$ contained in U.

It ensues from this and Theorem 22.3.3 that $R(\partial W) = \partial W_1$. Invoking (22.29), we infer that $R(\partial W) \cap \partial V \neq \emptyset$. Hence,

$$R(\partial W) \cap (\widehat{\mathbb{C}} \setminus V) \neq \emptyset. \tag{22.30}$$

But $\partial W \subseteq \partial \Delta_2 \subseteq \overline{\Delta_2} \subseteq \overline{U} \cap R^{-1}(\overline{\Delta_1}) \subseteq \overline{U} \cap R^{-1}(V) \subseteq U$. So, $R(\partial W) \subseteq R(U) = V$, contrary to (22.30). The proof of the claim is complete. ■

Now, since $\Delta_0 \subseteq \Delta_2 \subseteq U$ by the claim, it follows from Lemma 22.4.3 that

$$\chi(\Delta_0) \geq \chi(\Delta_2) \geq \chi(U). \tag{22.31}$$

From property (1) of the subdomain Δ_0, we have

$$\delta_R(\Delta_0) = \delta_R(\Delta_2) = \delta_R(U). \tag{22.32}$$

According to the claim, Theorem 22.3.6 applies and yields, with the use of properties (1)–(2) of Δ_0 along with (22.31)–(22.32), that

$$\deg(R|_U) \cdot \chi(\Delta_1) = \deg(R|_{\Delta_2}) \cdot \chi(\Delta_1) = \chi(\Delta_2) + \delta_R(\Delta_2) = \chi(\Delta_2) + \delta_R(U) \geq \chi(U) + \delta_R(U). \tag{22.33}$$

Recall that the only requirement on Δ_1 is (22.25). Thanks to Lemmas 22.4.2–22.4.3, the infimum of $\chi(\Delta_1)$ over all regular subdomains Δ_1 of V is $\chi(V)$. Therefore, (22.33) yields

$$\deg(R|_U) \cdot \chi(V) \geq \chi(U) + \delta_R(U). \tag{22.34}$$

On the other hand, (22.33) and (22.31), combined with Lemma 22.4.3, give

$$\deg(R|_U) \cdot \chi(V) \leq \deg(R|_U) \cdot \chi(\Delta_1) = \chi(\Delta_2) + \delta_R(U) \leq \chi(\Delta_0) + \delta_R(U). \tag{22.35}$$

Note that the sole requirements on Δ_0 are (1)–(2). In light of Lemmas 22.4.2–22.4.3, the supremum of $\chi(\Delta_0)$ over all such domains Δ_0 is $\chi(U)$. Consequently, (22.35) results in

$$\deg(R|_U) \cdot \chi(V) \leq \chi(U) + \delta_R(U).$$

Along with (22.34), this completes the proof of Theorem 22.5.1. □

Remark 22.5.2. It is important to observe that Corollaries 22.3.7–22.3.9 hold under the weaker hypotheses of Theorem 22.5.1, rather than those of Theorem 22.3.6. In particular, it is not necessary that the boundary consists of Jordan curves or that the critical values lie off the boundary. The proofs are essentially the same. We list these new corollaries below for the convenience of the reader and ease of reference.

Corollary 22.5.3. *Under the hypotheses of Theorem 22.5.1 and the additional assumption that $V \neq \widehat{\mathbb{C}}$ is simply connected (i. e., V is conformally equivalent to $\mathbb{D}$), we have that*

$$\delta_R(U) \geq \deg(R|_U) - 1,$$

with equality holding if and only if U is simply connected (i. e., is conformally equivalent to $\mathbb{D}$). In this latter case,

$$\#\mathrm{Crit}(R|_U) \leq \deg(R|_U) - 1. \tag{22.36}$$

Corollary 22.5.4. *Under the hypotheses of Theorem 22.5.1 and the additional assumptions (1) that V is simply connected and (2) that U contains no critical point of R, the map $R|_U : U \to V$ is a conformal homeomorphism. In particular, U is simply connected.*

Corollary 22.5.5. *Under the hypotheses of Theorem 22.5.1 and the additional assumptions (1) that $V \neq \widehat{\mathbb{C}}$ is simply connected and (2) that the map $R|_U : U \to V$ has a unique critical point c, we know that:*
(a) *U is simply connected.*
(a′) *If V is conformally equivalent to the unit disk $\mathbb{D}$, then so is U.*
(b) *$\deg(R|_U) = \deg_c(R)$.*

Item (a′) readily follows from item (a) since the possibility that U be $\mathbb{C}$ is ruled out by Liouville's theorem while the possibility that U be $\widehat{\mathbb{C}}$ is excluded by the non-compactness of V.

We will frequently use, often without explicitly invoking it, the following reformulation of Corollary 22.5.4.

Corollary 22.5.6. *Under the hypotheses of Theorem 22.5.1, if V is simply connected, $\xi \in S_1$, $R(\xi) \in V$ and the (unique) connected component of $R^{-1}(V)$ containing ξ contains no critical point of R, then there exists a unique holomorphic map $\psi : V \to S_1$ such that*

$$R \circ \psi = \mathrm{Id}_V \quad \textit{and} \quad \psi(R(\xi)) = \xi.$$

Furthermore, $\psi(V)$ is the connected component of $R^{-1}(V)$ containing ξ, and both maps $\psi : V \to \psi(V)$ and $R|_{\psi(V)} : \psi(V) \to V$ are conformal homeomorphisms.

In the sequel, the map $\psi : V \to \psi(V)$ will be referred to as the unique holomorphic branch of R^{-1} defined on V that sends $R(\xi)$ to ξ. It will be denoted by R_ξ^{-1}.

23 Selected tools from complex analysis

23.1 Koebe distortion theorems

This section is devoted to the formulation and the proof of various versions of the celebrated Koebe distortion theorem. They are truly amazing features of univalent holomorphic functions and constitute one of the most indispensable tools when dealing with the dynamical and especially geometric aspects of meromorphic maps in the complex plane. Koebe distortion theorems will be among the most frequently invoked theorems in this volume, particularly from Chapter 27 onward.

Koebe's $\frac{1}{4}$–theorem (Theorem 23.1.3) was conjectured by Paul Koebe in 1907 and was proved by Ludwig Bieberbach in 1916. The proof is a fairly easy consequence of Bieberbach's coefficient inequality (Theorem 23.1.2) obtained in [10]. Likewise, analytic Koebe distortion theorems, such as Theorems 23.1.5–23.1.6 follow from Bieberbach's coefficient inequality. Our proofs are standard and closely follow the exposition in [59].

We start by demonstrating Theorems 23.1.1–23.1.2. The latter will form a crucial ingredient in the proof of a first version of Koebe's distortion theorem (Theorem 23.1.3).

Theorem 23.1.1 (Area theorem). *Let $g : B(0,1) \to \widehat{\mathbb{C}}$ be a univalent meromorphic function with a simple pole at 0. Assume that the residue of g at 0 is equal to 1, so that the function g can be represented in the form of a Laurent series*

$$g(z) = \frac{1}{z} + \sum_{n=0}^{\infty} b_n z^n.$$

Then

$$\sum_{n=1}^{\infty} n|b_n|^2 \le 1.$$

Proof. For every $0 < r < 1$, let $D_r = \mathbb{C} \setminus g(B(0,r))$. By Green's theorem, the area of D_r satisfies

$$S(D_r) = \iint_{D_r} dx\,dy = \frac{1}{2i} \int_{\partial D_r} \bar{z}\,dz = -\frac{1}{2i} \int_{\partial B(0,r)} \bar{g}\,dg. \tag{23.1}$$

Recall that

$$\frac{1}{2\pi i} \int_{\partial B(0,r)} z^k \bar{z}^l\,dz = \delta_{kl}.$$

Substituting the power series expansions of g and g' into (23.1), and performing the integration, we obtain

https://doi.org/10.1515/9783110769876-023

$$S(D_r) = \pi\left(\frac{1}{r^2} - \sum_{n=1}^{\infty} n|b_n|^2 r^{2n}\right).$$

Since $S(D_r) \geq 0$, taking the limit as $r \nearrow 1$ yields the desired result. □

Let $\mathcal{S}$ denote the class of all univalent holomorphic functions $f : B(0,1) \to \mathbb{C}$ such that

$$f(0) = 0 \quad \text{and} \quad f'(0) = 1.$$

Theorem 23.1.2 (Bieberbach's coefficient inequality). *If $f(z) = z + \sum_{n=2}^{\infty} a_n z^n \in \mathcal{S}$, then $|a_2| \leq 2$.*

Proof. Observe that the formula

$$h(z) := \frac{f(z^2)}{z^2} = 1 + \sum_{n=2}^{\infty} a_n z^{2n-2} = 1 + \sum_{n=2}^{\infty} a_n z^{2(n-1)}$$

defines a holomorphic function from $B(0,1)$ to $\mathbb{C} \setminus \{0\}$ such that $h(0) = 1$. Let

$$\sqrt{h} : B(0,1) \to \mathbb{C} \setminus \{0\}$$

be the square root of h uniquely determined by the requirement that $\sqrt{h}(0) = 1$. Let

$$g(z) := \frac{1}{z\sqrt{h(z)}} = \frac{1}{z} - \frac{1}{2}a_2 z + \cdots.$$

This series contains only odd powers of z as the function g is odd. If $g(z_1) = g(z_2)$, then $(g(z_1))^2 = (g(z_2))^2$, whence $f(z_1^2) = f(z_2^2)$. We infer from the univalence of f that $z_1^2 = z_2^2$, i. e., $z_1 = \pm z_2$. But since g is odd, we deduce that $z_1 = z_2$. Thus, g is univalent. Theorem 23.1.1 thereby gives that $|a_2| \leq 2$. □

Theorem 23.1.3 (Koebe's $\frac{1}{4}$-theorem). *If $w \in \mathbb{C}$, $r > 0$ and $H : B(w,r) \to \mathbb{C}$ is a univalent analytic function, then*

$$H(B(w,r)) \supseteq B(H(w), 4^{-1}|H'(w)|r).$$

Proof. Precomposing H with the scaled (by a factor $1/r$) translation moving 0 to w and $B(0,r)$ onto $B(w,r)$, and postcomposing H with a scaled (by a factor r) translation moving $H(w)$ to 0, we may assume without loss of generality that $H \in \mathcal{S}$. Fix $c \in \mathbb{C} \setminus H(B(0,1))$. An inspection shows that the function

$$B(0,1) \ni z \longmapsto \frac{cH(z)}{c - H(z)} \in \mathbb{C}$$

belongs to $\mathcal{S}$. Writing $H(z) = z + \sum_{n=2}^{\infty} a_n z^n$, this function takes on the form

$$B(0,1) \ni z \longmapsto z + \left(a_2 + \frac{1}{c}\right)z^2 + \cdots, \tag{23.2}$$

where a_2 is the coefficient of z^2 in the Taylor series expansion of H about 0. Applying Theorem 23.1.2 twice (once to the function in (23.2) and once to H), we obtain

$$\frac{1}{|c|} \le |a_2| + \left|a_2 + \frac{1}{c}\right| \le 2 + 2 = 4. \qquad \square$$

In the above theorem, the number 1/4 is optimal as the Koebe function

$$B(0,1) \ni z \longmapsto \frac{z}{(1-z)^2} = \sum_{n=1}^{\infty} nz^n \in \mathbb{C}$$

maps univalently the unit ball $B(0,1)$ to the slit plane $\mathbb{C} \setminus (-\infty, -1/4]$.

We now prove a series of Koebe distortion theorems. First, we establish Lemma 23.1.4, which will form a crucial ingredient in the proof of Theorem 23.1.5.

Lemma 23.1.4. *If $f \in \mathcal{S}$, then*

$$\left|\frac{1}{2}(1-|z|^2)\frac{f''(z)}{f'(z)} - \overline{z}\right| \le 2, \quad \forall z \in B(0,1). \tag{23.3}$$

Proof. Fix $z \in B(0,1)$. The Möbius transformation

$$w \longmapsto \frac{w+z}{1+\overline{z}w}$$

maps $B(0,1)$ onto itself, sending 0 to z. It follows that for any choice of constants $C_1 \in \mathbb{C}$ and $C_2 \in \mathbb{C} \setminus \{0\}$ the function

$$g(w) = C_1 + C_2 f\left(\frac{w+z}{1+\overline{z}w}\right)$$

is univalent on $B(0,1)$. Choose C_1 and C_2 so that $g(0) = 0$ and $g'(0) = 1$, i. e., so that $g \in \mathcal{S}$. These constraints yield

$$C_1 = -\frac{f(z)}{f'(z)(1-|z|^2)} \quad \text{and} \quad C_2 = \frac{1}{f'(z)(1-|z|^2)}.$$

Direct calculations give

$$g'(w) = \frac{1}{f'(z)}\frac{1}{(1+\overline{z}w)^2}f'\left(\frac{w+z}{1+\overline{z}w}\right)$$

and

$$g''(w) = -\frac{1}{f'(z)}\frac{2\overline{z}}{(1+\overline{z}w)^3}f'\left(\frac{w+z}{1+\overline{z}w}\right) + \frac{1}{f'(z)}\frac{1-|z|^2}{(1+\overline{z}w)^4}f''\left(\frac{w+z}{1+\overline{z}w}\right).$$

It ensues that

$$g(w) = w + \left[\frac{1}{2}(1-|z|^2)\frac{f''(z)}{f'(z)} - \bar{z}\right]w^2 + \cdots.$$

As $g \in \mathcal{S}$, the coefficient of w^2 has modulus at most 2 by Theorem 23.1.2. □

We now formulate and prove the central theorem among Koebe distortion theorems.

Theorem 23.1.5 (Koebe's distortion theorem, analytic version I). *Let $0 < r < 1$. If $f \in \mathcal{S}$, then*

$$\frac{1-r}{(1+r)^3} \le |f'(z)| \le \frac{1+r}{(1-r)^3}, \quad \forall z \in \overline{B(0,r)} \tag{23.4}$$

and there is a choice of argument for $f'(z)$ such that

$$|\arg f'(z)| \le 2\log\frac{1+r}{1-r}, \quad \forall z \in \overline{B(0,r)}. \tag{23.5}$$

Proof. By Lemma 23.1.4, there exists a continuous function $\eta : B(0,1) \to \overline{B(0,1)}$ such that

$$\frac{f''(t)}{f'(t)} - \frac{2\bar{t}}{1-|t|^2} = 4\frac{\eta(t)}{1-|t|^2}, \quad \forall t \in B(0,1).$$

Integrating this expression along the line segment from 0 to z in $B(0,1)$ yields

$$\log f'(z) + \log(1-|z|^2) = 4\int_0^z \frac{\eta(t)}{1-|t|^2}dt, \quad \forall z \in B(0,1), \tag{23.6}$$

where

$$\left|4\int_0^z \frac{\eta(t)}{1-|t|^2}dt\right| \le 4\int_0^r \frac{ds}{1-s^2} = 2\log\frac{1+r}{1-r}, \quad \forall z \in \overline{B(0,r)}. \tag{23.7}$$

Equating the real parts of both sides of (23.6) and using (23.7), we get the inequality

$$2\log\frac{1-r}{1+r} \le \log[(1-|z|^2)|f'(z)|] \le 2\log\frac{1+r}{1-r}, \quad \forall z \in \overline{B(0,r)}.$$

Since the function $\overline{B(0,r)} \ni z \mapsto \log[(1-|z|^2)|f'(z)|]$ is harmonic, it assumes its maximum and minimum values on $\partial B(0,r)$. This yields (23.4). Likewise, formula (23.5) results from equating the imaginary parts in (23.6) and using (23.7). □

Relation (23.5) is called the *rotation theorem*. It was discovered by Bieberbach in 1919. The estimates in (23.4) are the best possible; equality is reached for the Koebe functions

$$f_\beta(z) = \frac{z}{(1 + e^{i\beta} z)^2}, \tag{23.8}$$

where β is any real number.

Let $\mathcal{U}$ denote the class of all univalent holomorphic functions from the open unit disk $B(0,1)$ into $\mathbb{C}$. With obvious translations and rescalings, we deduce the following consequence of Theorem 23.1.5.

Theorem 23.1.6 (Koebe's distortion theorem, analytic version II). *Let $0 < r < 1$. If $f \in \mathcal{U}$, then*

$$\frac{1-r}{(1+r)^3} \le \frac{|f'(z)|}{|f'(0)|} \le \frac{1+r}{(1-r)^3}, \quad \forall z \in \overline{B(0,r)} \tag{23.9}$$

and there is a holomorphic branch of argument for $f'(z)$ such that

$$\left|\arg \frac{f'(z)}{f'(0)}\right| \le 2\log\frac{1+r}{1-r}, \quad \forall z \in \overline{B(0,r)}. \tag{23.10}$$

As an immediate consequence of Theorem 23.1.6, we obtain the following two facts.

Theorem 23.1.7 (Koebe's distortion theorem, analytic version III). *There exists an increasing continuous function*

$$K : [0,1) \to [1,\infty)$$

such that $K(0) = 1$ and with the following property: if $w \in \mathbb{C}$, $R > 0$, and $H : B(w,R) \to \mathbb{C}$ is any univalent analytic function, then

$$\left| \frac{|H'(z)|}{|H'(w)|} - 1 \right| \le K(r/R)|z - w| \tag{23.11}$$

for every $r \in [0,R)$ and for all $z \in \overline{B(w,r)}$.

Theorem 23.1.8 (Koebe's distortion theorem, Euclidean version). *There exists an increasing continuous function*

$$k : [0,1) \to [1,\infty)$$

such that $k(0) = 1$ and with the following property: if $w \in \mathbb{C}$, $r > 0$, and $H : B(w,r) \to \mathbb{C}$ is any univalent analytic function, then for every $t \in [0,1)$ we have

$$\sup\{|H'(z)| : z \in \overline{B(w,tr)}\} \le k(t) \inf\{|H'(z)| : z \in \overline{B(w,tr)}\}.$$

Denote by ρ the spherical metric on the Riemann sphere $\widehat{\mathbb{C}}$.

Theorem 23.1.9 (Koebe's distortion theorem, spherical version). *Given $R, s \in (0, \pi)$, there exists an increasing continuous function $k_{R,s} : [0,1) \to [1, \infty)$ with the following property: if $w \in \widehat{\mathbb{C}}$, $r \in (0, R]$, and $H : B_\rho(w,r) \to \widehat{\mathbb{C}}$ is any univalent analytic function whose range's complement $\widehat{\mathbb{C}} \setminus H(B_\rho(w,r))$ contains a spherical ball of radius s, then for every $t \in [0,1)$ we have*

$$\sup\{|H'(z)|_\rho : z \in \overline{B_\rho(w,tr)}\} \le k_{R,s}(t) \inf\{|H'(z)|_\rho : z \in \overline{B_\rho(w,tr)}\},$$

where $|H'(\cdot)|_\rho$ refers to the scaling factor of the spherical derivative.

Proof. Let $M_w : \widehat{\mathbb{C}} \to \widehat{\mathbb{C}}$ be the Möbius transformation corresponding via the stereographic projection to the rotation of the sphere $\mathbb{S}^2 := \{x \in \mathbb{R}^3 : \|x\| = 1\}$ that moves the point corresponding to 0 to the point corresponding to w. Then

$$M_w^{-1}(B_\rho(w,R)) = B_\rho(0,R) = B(0,R_1) \quad \text{and} \quad M_w^{-1}(B_\rho(w,r)) = B_\rho(0,r) = B(0,r_1)$$

for some appropriate radius $R_1 \in (0,\infty)$ depending only on R, and some radius $r_1 \in (0, R_1]$ depending only on r. Note that for every $t \in [0,1)$ there exists $t' \in (0,1)$ such that

$$M_w^{-1}(B_\rho(w,tr)) \subseteq B(0,t'r_1). \tag{23.12}$$

Let $B_\rho(\xi, s)$ be a spherical ball disjoint from $H(B_\rho(w,r))$. Let $M_\xi : \widehat{\mathbb{C}} \to \widehat{\mathbb{C}}$ be the Möbius transformation corresponding via the stereographic projection to the rotation of the sphere $\mathbb{S}^2$ that moves the point corresponding to ξ to the point corresponding to ∞. Then

$$M_\xi(\widehat{\mathbb{C}} \setminus B_\rho(\xi,s)) = \widehat{\mathbb{C}} \setminus B_\rho(\infty,s) = B(0,s_2)$$

for some appropriate radius $s_2 \in (0,\infty)$ depending only on s. Then

$$M_\xi \circ H \circ M_w : B(0,r_1) \longrightarrow B(0,s_2)$$

is a univalent analytic function. Note that the moduli of the Euclidean and spherical derivatives $|(M_\xi \circ H \circ M_w)|'$ and $|(M_\xi \circ H \circ M_w)'|_\rho$ are uniformly comparable on $B(0,r_1)$ to (multiplicative) comparability constants depending only on R and s. Observe also that the moduli of the spherical derivatives of M_w and M_ξ are everywhere equal to 1 as these maps are isometries with respect to spherical metric on $\widehat{\mathbb{C}}$. Then the result ensues from Theorem 23.1.8 by taking into account (23.12). □

Employing the mean value inequality, the next two lemmas are respective consequences of the latter two distortion theorems and Koebe's $\frac{1}{4}$–theorem (Theorem 23.1.3).

Lemma 23.1.10. *Suppose that $D \subseteq \mathbb{C}$ is an open set, $z \in D$ and $H : D \to \mathbb{C}$ is an analytic function which has an analytic inverse H_z^{-1} defined on $B(H(z), 2R)$ for some $R > 0$. Then for every $0 \le r \le R$ we have*

$$B\left(z, \frac{1}{4}r|H'(z)|^{-1}\right) \subseteq H_z^{-1}(B(H(z), r)) \subseteq B(z, Kr|H'(z)|^{-1}).$$

Lemma 23.1.11. *Suppose that $D \subseteq \widehat{\mathbb{C}}$ is an open set, $z \in D$ and $H : D \to \widehat{\mathbb{C}}$ is an analytic map which has an analytic inverse H_z^{-1} defined on $B_\rho(H(z), 2R)$ for some $R > 0$ such that $H_z^{-1}(B_\rho(H(z), 2R))$ avoids a spherical ball of some radius s. Then for every $0 \le r \le R$ we have*

$$B_\rho(z, [k_{2R,s}(1/2)]^{-1} r|H'(z)|_\rho^{-1}) \subseteq H_z^{-1}(B_\rho(H(z), r)) \subseteq B_\rho(z, k_{2R,s}(1/2) r|H'(z)|_\rho^{-1}).$$

23.2 Normal families and Montel's theorem

Introduced at the beginning of the 20th century, the beautiful concept of normal families is powerful. Its treatment can be found in many textbooks on complex analysis. The book by Schiff [115] is devoted to its systematic exposition.

Definition 23.2.1. Let $\mathcal{F}$ be a family of meromorphic functions, all defined on some open subset U of $\widehat{\mathbb{C}}$. The family $\mathcal{F}$ is called *normal* on U if it is relatively compact on U.

An immediate reformulation of normality is that from any sequence of functions in $\mathcal{F}$ one can extract a subsequence which converges uniformly on compact subsets of U. The limit of such a subsequence is a meromorphic function, possibly a constant function identically equal to an element of $\widehat{\mathbb{C}}$ (potentially ∞).

Shortly, we will state and prove Montel's theorem, which is an extremely helpful tool for establishing normality, and which makes normality so important and so useful in complex analysis. It is also absolutely indispensable and frequently used in the theory of iteration of rational functions. We first formulate and use several times, including in the proof of Montel's theorem, another great theorem of complex analysis.

Theorem 23.2.2 (Koebe's uniformization theorem). *Every simply connected Riemann surface (i. e., two-dimensional holomorphic manifold) is conformally equivalent to either the open unit disk $\mathbb{D} = \{z \in \mathbb{C} : |z| < 1\}$, the complex plane $\mathbb{C}$ or the Riemann sphere $\widehat{\mathbb{C}}$.*

This theorem was proved by Henri Poincaré and Paul Koebe in 1907. Its proof can be found in several books on complex analysis and Riemann surfaces; for instance, see Theorem 7.4 in Kodaira [63] or Theorem 27.9 (the Riemann mapping theorem) in Forster [51]. As a consequence of Koebe's uniformization theorem and of the great Picard's theorem, we prove the following intermediate result.

Theorem 23.2.3. *If a, b, c are three distinct points in $\widehat{\mathbb{C}}$, then the open unit disk $\mathbb{D}$ is a universal covering space of $\widehat{\mathbb{C}} \setminus \{a, b, c\}$ and there exists a meromorphic covering map $\pi : \mathbb{D} \to \widehat{\mathbb{C}} \setminus \{a, b, c\}$.*

Proof. Let D be a universal, i. e., simply connected, covering space of $\widehat{\mathbb{C}} \setminus \{a, b, c\}$ and let $P : D \to \widehat{\mathbb{C}} \setminus \{a, b, c\}$ be a covering map. This map uniquely induces a two-dimensional holomorphic manifold structure on D (a Riemann surface). In particular, the map P becomes meromorphic. The surface D cannot be conformally equivalent to $\widehat{\mathbb{C}}$ since then $\widehat{\mathbb{C}} \setminus \{a, b, c\}$ would be compact, and it cannot be conformally equivalent to $\mathbb{C}$ because of the great Picard's theorem. According to Koebe's uniformization theorem, D is conformally equivalent to $\mathbb{D}$. This means that there exists a conformal homeomorphism $h : \mathbb{D} \to D$. Then $\pi := P \circ h : \mathbb{D} \to \widehat{\mathbb{C}} \setminus \{a, b, c\}$ is the sought holomorphic map. □

We can now prove the ultimate theorem of this section.

Theorem 23.2.4 (Montel's theorem). *If $\mathcal{F}$ is a family of meromorphic functions, all defined on some open subset U of $\widehat{\mathbb{C}}$, that omits three points in $\widehat{\mathbb{C}}$, i. e., if $\widehat{\mathbb{C}} \setminus \bigcup_{f \in \mathcal{F}} f(U)$ contains at least three points, then the family $\mathcal{F}$ is normal on U.*

Proof. We first look at the following weaker version of Montel's theorem, whose demonstration is a fairly straightforward consequence of Cauchy's integral formula (Theorem A.3.2) and Arzelà–Ascoli's theorem (Theorem A.2.3).

Claim. *If $U \subseteq \mathbb{C}$ and the family $\mathcal{F}$ is bounded on U, i. e., if*

$$M := \sup\{|f(z)| : f \in \mathcal{F},\ z \in U\} < \infty,$$

then $\mathcal{F}$ is normal on U.

Proof. By Arzelà–Ascoli, it suffices to show that the family $\mathcal{F}_w := \{f|_{\overline{B(w,R)}} : f \in \mathcal{F}\}$ is equicontinuous (with respect to the Euclidean metric on $\mathbb{C}$) for any $w \in U$ and $R > 0$ such that $B(w, 4R) \subseteq U$. Let $\gamma : [0, 1] \to B(w, 4R)$ be the circle (a simple, closed, smooth and rectifiable curve)

$$\gamma(t) := w + 3Re^{2\pi i t}.$$

For any function $f \in \mathcal{F}$ and for every $z \in \overline{B(w, 2R)}$, Cauchy's integral formula affirms that

$$|f'(z)| = \left|\frac{1}{2\pi i}\int_\gamma \frac{f(z)}{(z-w)^2}\,dz\right| \le \frac{1}{2\pi}\int_\gamma \frac{|f(z)|}{|z-w|^2}\,|dz| \le \frac{M}{2\pi R^2}\int_\gamma |dz| = \frac{3M}{R} < \infty.$$

Thus, all functions $f|_{\overline{B(w,2R)}}$, $f \in \mathcal{F}$, are Lipschitz continuous with Lipschitz constant $3M/R$. So, the family $\{f|_{\overline{B(w,2R)}} : f \in \mathcal{F}\}$ is equicontinuous. Since it is uniformly bounded (by M) and since the set $\overline{B(w, R)}$ is compact, this family is relatively compact by the Arzelà–Ascoli theorem and the proof of the claim is complete. ■

Passing to a more general case, keep $U \subseteq \mathbb{C}$ but assume only that $\mathcal{F}$ avoids three points in $\widehat{\mathbb{C}}$, say a, b, c. By Theorem 23.2.3, there exists a meromorphic covering map $\pi : \mathbb{D} \to \widehat{\mathbb{C}} \backslash \{a, b, c\}$. As $\mathbb{D}$ is simply connected, for every $f \in \mathcal{F}$ there exists a continuous lift $\tilde{f} : U \to \mathbb{D}$ such that $f = \pi \circ \tilde{f}$, i. e. the following diagram commutes:

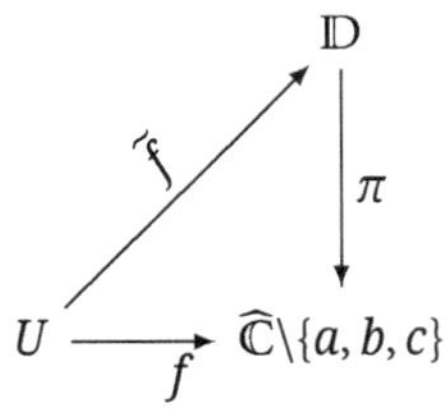

Since the meromorphic covering map π is locally one-to-one, we deduce that the map $\tilde{f}$ is holomorphic. As the set $\mathbb{D}$ is bounded, it follows from the claim that the family $\{\tilde{f} : U \to \mathbb{D} \mid f \in \mathcal{F}\}$ is normal. Hence, the family $\mathcal{F} = \{\pi \circ \tilde{f} : U \to \widehat{\mathbb{C}} \backslash \{a, b, c\} \mid f \in \mathcal{F}\}$ is normal, too.

The general case ensues from the fact that normality is a local property. If $\infty \notin U$, then the family $\mathcal{F}$ is normal on $U \backslash \{\infty\}$ according to the previous case. If $\infty \in U$, then there exists $R > 0$ such that $\widehat{\mathbb{C}} \backslash \overline{B(0, R)} \subseteq U$. If $\mathcal{I} : \widehat{\mathbb{C}} \to \widehat{\mathbb{C}}$ is the reciprocal map $\mathcal{I}(z) = 1/z$, then the family $\{f \circ \mathcal{I}|_{B(0,1/R)} : f \in \mathcal{F}\}$ avoids three points and is hence normal per the previous case. Then $\mathcal{F}$ is normal, too. $\square$

Invoking again the fact that normality is a local property, we obtain a direct yet effective consequence of Theorem 23.2.4.

Corollary 23.2.5. *If $U \subseteq \widehat{\mathbb{C}}$ is an open set and $\mathcal{F}$ is a family of meromorphic functions from U to $\widehat{\mathbb{C}}$ such that for every $w \in U$, there exists a ball $B(w, r) \subseteq U$ for which the family $\{f|_{B(w,r)} : f \in \mathcal{F}\}$ omits three points in $\widehat{\mathbb{C}}$, then the family $\mathcal{F}$ is normal.*

24 Dynamics and topology of rational functions: their Fatou and Julia sets

24.1 Fatou set, Julia set, and periodic points

Let $T : \widehat{\mathbb{C}} \to \widehat{\mathbb{C}}$ be an analytic endomorphism of the Riemann sphere $\widehat{\mathbb{C}}$, i. e. an holomorphic map from the Riemann sphere, considered as a Riemann surface, into itself. After looking at the zeros and the poles of T, it turns out that T is a rational function, i. e., the ratio of two polynomials

$$T = \frac{P}{Q},$$

where we may assume that P and Q are relatively prime, i. e., have no common factors.

We would like to distinguish the region of $\widehat{\mathbb{C}}$ where the dynamics of T are stable (or tame) from that where the dynamics of T are chaotic (or random). These regions are respectively called the Fatou and Julia sets of T. Recall from Chapter 23 the concept of normality of a family of analytic functions and Montel's theorem. The Fatou set of a rational function T is the set of all points that admit a neighborhood on which the iterates of T form a normal family while the Julia set is its complement.

Definition 24.1.1. Let $T : \widehat{\mathbb{C}} \to \widehat{\mathbb{C}}$ be a rational function. The *Fatou set* $\mathcal{F}(T)$ of T is the set of all points $z \in \widehat{\mathbb{C}}$ for each of which there exists an open neighborhood $U_z \subseteq \widehat{\mathbb{C}}$ of z such that the family of functions $\{T^n|_{U_z} : n \in \mathbb{N}\}$ is normal on U_z. Its complement $\mathcal{J}(T) = \widehat{\mathbb{C}} \setminus \mathcal{F}(T)$ is called the *Julia set* of T.

From its very definition, the Fatou set is open, and hence the Julia set is a closed, and thereby compact, subset of $\widehat{\mathbb{C}}$. Moreover, both sets are fully invariant.

Theorem 24.1.2. *Both the Julia set $\mathcal{J}(T)$ and the Fatou set $\mathcal{F}(T)$ are completely T-invariant, i. e.,*

$$T^{-1}(\mathcal{J}(T)) = \mathcal{J}(T) = T(\mathcal{J}(T)) \quad \text{and} \quad T^{-1}(\mathcal{F}(T)) = \mathcal{F}(T) = T(\mathcal{F}(T)).$$

Proof. It is enough to show that the Fatou set $\mathcal{F}(T)$ is completely invariant.

Let $z \in T^{-1}(\mathcal{F}(T))$. Then $T(z) \in \mathcal{F}(T)$, and thus the sequence of iterates $(T^n)_{n=1}^{\infty}$ is normal on an open neighborhood $U_{T(z)} \subseteq \widehat{\mathbb{C}}$ of $T(z)$. Let $(T^{n_k}|_{T^{-1}(U_{T(z)})})_{k=1}^{\infty}$ be an arbitrary sequence of iterates of T restricted to the open neighborhood $T^{-1}(U_{T(z)})$ of z. Then there exists a subsequence $(n_{k_j})_{j=1}^{\infty}$ of $(n_k)_{k=1}^{\infty}$ such that the sequence $(T^{n_{k_j}-1})_{j=1}^{\infty}$ converges uniformly on compact subsets of $U_{T(z)}$. But then the sequence $(T^{n_{k_j}} = T^{n_{k_j}-1} \circ T)_{j=1}^{\infty}$ converges uniformly on compact subsets of $T^{-1}(U_{T(z)})$. Hence, $z \in \mathcal{F}(T)$. So $T^{-1}(\mathcal{F}(T)) \subseteq \mathcal{F}(T)$.

To prove the opposite inclusion, fix $z \in \mathcal{F}(T)$. Let U_z be an open neighborhood of z upon which the iterates of T form a normal family. Let $(T^{n_k})_{k=1}^{\infty}$ be an arbitrary sequence

https://doi.org/10.1515/9783110769876-024

of iterates of T. There then exists a subsequence $(n_{k_j})_{j=1}^{\infty}$ of $(n_k)_{k=1}^{\infty}$ such that the sequence of iterates $(T^{n_{k_j}+1})_{j=1}^{\infty}$ converges uniformly on compact subsets of U_z. This implies that the sequence $(T^{n_{k_j}})_{j=1}^{\infty}$ converges uniformly on compact subsets of $T(U_z)$, which is an open neighborhood of $T(z)$. In consequence, $T(z) \in \mathcal{F}(T)$. Thus $z \in T^{-1}(\mathcal{F}(T))$, and the inclusion $\mathcal{F}(T) \subseteq T^{-1}(\mathcal{F}(T))$ holds.

In summary, we showed that

$$T^{-1}(\mathcal{F}(T)) = \mathcal{F}(T).$$

As T is surjective, this immediately yields that $T(\mathcal{F}(T)) = \mathcal{F}(T)$. □

Example 24.1.3. For any polynomial $P : \widehat{\mathbb{C}} \to \widehat{\mathbb{C}}$, the singleton ∞ is completely P-invariant, i. e., $P^{-1}(\{\infty\}) = \{\infty\}$. This property is a characterization of polynomials within the class of rational functions of the Riemann sphere.

If P is of degree at least two, then $T(\widehat{\mathbb{C}} \setminus B(0, r(T))) \subseteq \widehat{\mathbb{C}} \setminus B(0, r(T))$ for any $r(T) > 0$ large enough. So, $T^n(\widehat{\mathbb{C}} \setminus B(0, r(T))) \subseteq \widehat{\mathbb{C}} \setminus B(0, r(T))$ for all $n \in \mathbb{N}$ and by Montel's theorem (Theorem 23.2.4) the family $\{T^n : n \in \mathbb{N}\}$ is normal on $\widehat{\mathbb{C}} \setminus B(0, r(T))$. This means that $\widehat{\mathbb{C}} \setminus B(0, r(T)) \subseteq \mathcal{F}(T)$. In particular, $\infty \in \mathcal{F}(T)$.

According to Theorem 22.1.3, we can speak of the topological degree $\deg(T)$ of a rational function $T : \widehat{\mathbb{C}} \to \widehat{\mathbb{C}}$ as being the number of preimages of any regular point of T, i. e., of any point not belonging to the set of critical values $T(\mathrm{Crit}(T))$. It is then obvious that $\deg(T)$ is the maximum of the algebraic degrees of the polynomials P and Q. The concept of degree permits us to prove the following fundamental theorem (cf. Exercise 24.8.2).

Theorem 24.1.4. *If $T : \widehat{\mathbb{C}} \to \widehat{\mathbb{C}}$ is a rational function with* $\deg(T) \geq 2$, *then* $\mathcal{J}(T) \neq \emptyset$.

Proof. Assume, by way of contradiction, that $\mathcal{J}(T) = \emptyset$. Then the iterates $\{T^n : n \in \mathbb{N}\}$ form a normal family on $\widehat{\mathbb{C}}$, and hence there is a subsequence $(n_k)_{k=1}^{\infty}$ of $(n)_{n=1}^{\infty}$ such that the sequence $(T^{n_k})_{k=1}^{\infty}$ converges uniformly on $\widehat{\mathbb{C}}$ to some analytic function $S : \widehat{\mathbb{C}} \to \widehat{\mathbb{C}}$. Note that S cannot be constant since T is surjective. Moreover, as the map $R \mapsto \deg(R)$ is continuous on the space of analytic maps of the Riemann sphere endowed with the topology of uniform convergence, it follows that

$$\deg(S) = \lim_{n\to\infty} \deg(T^{n_k}) = \lim_{n\to\infty} (\deg(T))^{n_k} = \infty.$$

This is impossible since $\deg(S) < \infty$ per Theorem 22.1.3. □

From this point on, we will always assume that $\deg(T) \geq 2$, so that the Julia set of T is always nonempty. For a while, we will repeat this assumption but eventually we may omit it.

In light of the above two theorems, it makes sense to study the topological dynamical systems

$$T : \mathcal{F}(T) \to \mathcal{F}(T) \quad \text{and} \quad T : \mathcal{J}(T) \to \mathcal{J}(T).$$

We begin our investigation with a slew of fundamental properties.

Theorem 24.1.5. *If $T : \widehat{\mathbb{C}} \to \widehat{\mathbb{C}}$ is a rational function with $\deg(T) \geq 2$, then*

$$\mathcal{F}(T^k) = \mathcal{F}(T) \quad \textit{and} \quad \mathcal{J}(T^k) = \mathcal{J}(T), \quad \forall k \in \mathbb{N}.$$

Proof. Fix $k \in \mathbb{N}$. If the sequence of iterates of T is normal on some open set, then so is the sequence of iterates of T^k. Therefore, $\mathcal{F}(T) \subseteq \mathcal{F}(T^k)$.

On the other hand, let $z \in \mathcal{F}(T^k)$ and U_z a neighborhood of z on which the sequence $(T^{kn})_{n=1}^{\infty}$ is normal. Fix an arbitrary strictly increasing sequence of positive integers $(n_j)_{j=1}^{\infty}$. For each $j \in \mathbb{N}$, write $n_j = kn'_j + r_j$, where n'_j and r_j are nonnegative integers and $r_j < k$. There then exist $0 \leq r < k$ and a strictly increasing sequence of positive integers $(j_i)_{i=1}^{\infty}$ such that $r_{j_i} = r$ for all $i \in \mathbb{N}$. Due to the normality of $(T^{kn})_{n=1}^{\infty}$ on U_z, there is a strictly increasing sequence $(i_h)_{h=1}^{\infty}$ of positive integers such that the sequence $(T^{kn'_{j_{i_h}}})_{h=1}^{\infty}$ converges uniformly on compact subsets of U_z. Hence, $(T^{n_{j_{i_h}}} = T^r \circ T^{kn'_{j_{i_h}}})_{h=1}^{\infty}$ also converges uniformly on compact subsets of U_z. Thus, $z \in \mathcal{F}(T)$, and the inclusion $\mathcal{F}(T^k) \subseteq \mathcal{F}(T)$ holds.

In summary, $\mathcal{F}(T^k) = \mathcal{F}(T)$. So $\mathcal{J}(T^k) = \widehat{\mathbb{C}} \setminus \mathcal{F}(T^k) = \widehat{\mathbb{C}} \setminus \mathcal{F}(T) = \mathcal{J}(T)$. □

The proof of the following result is left to the reader as an exercise.

Lemma 24.1.6. *If a rational function $T : \widehat{\mathbb{C}} \to \widehat{\mathbb{C}}$ is conjugate, via a Möbius transformation $M : \widehat{\mathbb{C}} \to \widehat{\mathbb{C}}$, to a rational function $\tilde{T} : \widehat{\mathbb{C}} \to \widehat{\mathbb{C}}$, i. e., if $T = M^{-1} \circ \tilde{T} \circ M$, then $M(\mathcal{F}(T)) = \mathcal{F}(\tilde{T})$ and $M(\mathcal{J}(T)) = \mathcal{J}(\tilde{T})$.*

Since each rational function $T : \widehat{\mathbb{C}} \to \widehat{\mathbb{C}}$ is proper, the following fact is an immediate consequence of Theorem 22.3.3.

Lemma 24.1.7. *Let $T : \widehat{\mathbb{C}} \to \widehat{\mathbb{C}}$ be a rational function with $\deg(T) \geq 2$. If V is a domain in $\widehat{\mathbb{C}}$ and U is any connected component of $T^{-1}(V)$, then the map $T|_U : U \to V$ is proper and*

$$T(U) = V, \quad T(\partial U) = \partial V, \quad \textit{and} \quad \overline{U} \cap T^{-1}(\partial V) = \partial U. \tag{24.1}$$

Having this, we can easily prove the following.

Theorem 24.1.8. *If $T : \widehat{\mathbb{C}} \to \widehat{\mathbb{C}}$ is a rational function with $\deg(T) \geq 2$ and V is a connected component of the Fatou set $\mathcal{F}(T)$, then:*

(a) *$T(V)$ is a connected component of $\mathcal{F}(T)$.*
(b) *Each connected component U of $T^{-1}(V)$ is a connected component of $\mathcal{F}(T)$.*
(c) *For any connected component U of $T^{-1}(V)$, it holds that*

$$T(U) = V, \quad T(\partial U) = \partial V, \quad \textit{and} \quad \overline{U} \cap T^{-1}(\partial V) = \partial U. \tag{24.2}$$

Proof. Because $\mathcal{F}(T)$ is open in $\widehat{\mathbb{C}}$, every connected component of $\mathcal{F}(T)$ is open and so is a domain in $\widehat{\mathbb{C}}$. Therefore, part (c) is a special case of (24.1) in Lemma 24.1.7.

Regarding (a), note that $T(V)$ is open since T is an open map and V is open; it is also connected as T is continuous and V is connected. So, $T(V)$ is a domain in $\widehat{\mathbb{C}}$. By Theorem 24.1.2, there is a unique connected component $\widetilde{V}$ of $\mathcal{F}(T)$ that contains the domain $T(V)$. By that same theorem, we know that $T^{-1}(\widetilde{V}) \subseteq \mathcal{F}(T)$. Then V is a connected component of $T^{-1}(\widetilde{V})$ since it is a connected component of $\mathcal{F}(T)$. By (c), it follows that $T(V) = \widetilde{V}$ is a connected component of $\mathcal{F}(T)$.

Concerning (b), we know thanks to (c) that $T(U) = V$. By Theorem 24.1.2, the set U is a connected subset of $\mathcal{F}(T)$. So, there exists a unique connected component $\widetilde{U}$ of $\mathcal{F}(T)$ containing U. As $U \subseteq T^{-1}(V)$, the set $T(\widetilde{U})$ is a connected subset of $\mathcal{F}(T)$ intersecting V and, therefore, is contained in the connected component V. Hence, $V = T(U) \subseteq T(\widetilde{U}) \subseteq V$. Thus, $T(\widetilde{U}) = V$. Consequently, $\widetilde{U}$ is contained in a connected component of $T^{-1}(V)$. But since $\widetilde{U}$ intersects U (it even contains it) and since U is a connected component of $T^{-1}(V)$, we conclude that $\widetilde{U} = U$. Hence, U is a connected component of $\mathcal{F}(T)$. □

Due to Montel's theorem (Theorem 23.2.4), for every point $z \in \mathcal{J}(T)$ the set

$$E_z(T) = \widehat{\mathbb{C}} \setminus \bigcup_{n=0}^{\infty} T^n(U_z).$$

can consist of no more than two points for every neighborhood U_z of z, and is actually the same set for every such z and every small enough U_z. The following theorem gives a complete description of that exceptional set.

Theorem 24.1.9. *If $T : \widehat{\mathbb{C}} \to \widehat{\mathbb{C}}$ is a rational function with* $\deg(T) \geq 2$, *then:*
(a) *The set $E_z(T)$ is independent of $z \in \mathcal{J}(T)$ and shall henceforth be denoted by $E(T)$. The set $E(T)$ is called the* exceptional set *of T.*
(b) *If $E(T)$ is a singleton, then T is conjugate, via a Möbius transformation $M : \widehat{\mathbb{C}} \to \widehat{\mathbb{C}}$, to a polynomial P different from $z \mapsto az^d$, $d \geq 2$, i. e. $T = M^{-1} \circ P \circ M$, and then $E(P) = \{\infty\}$ and $E(T) = M^{-1}(\{\infty\})$.*
(c) *If $E(T)$ consists of exactly two points, then T is conjugate, via a Möbius transformation M, to a rational map R of the form $z \mapsto az^d$, $|d| \geq 2$, i. e. $T = M^{-1} \circ R \circ M$, and then $E(R) = \{0, \infty\}$ and $E(T) = M^{-1}(\{0, \infty\})$.*
(d) $E(T) \subseteq \mathcal{F}(T)$.

Proof. Fix $z \in \mathcal{J}(T)$. By definition, the set $E_z(T)$ is backward T-invariant, i. e.

$$T^{-1}(E_z(T)) \subseteq E_z(T). \tag{24.3}$$

Assume that $E_z(T)$ is a singleton, say $E_z(T) = \{b\}$. The above inclusion and the surjectivity of T imply that $T^{-1}(\{b\}) = \{b\}$. Let $M : \widehat{\mathbb{C}} \to \widehat{\mathbb{C}}$ be a Möbius transformation such that $M(b) = \infty$. Setting $\widetilde{T} := M \circ T \circ M^{-1}$, we deduce that $\widetilde{T}^{-1}(\{\infty\}) = \{\infty\}$ and $E_{M(z)}(\widetilde{T}) =$

$M(E_z(T)) = \{\infty\}$. Recalling Example 24.1.3, we infer that $\tilde{T} : \hat{\mathbb{C}} \to \hat{\mathbb{C}}$ is a polynomial which differs from $z \mapsto az^d$, $d \geq 2$ per Exercise 24.8.4. Since ∞ is completely $\tilde{T}$-invariant and $\infty \in \mathcal{F}(\tilde{T})$, it immediately follows that $E_w(\tilde{T}) = \{\infty\}$ for every $w \in \mathcal{J}(\tilde{T})$. Therefore, $E_\zeta(T) = M^{-1}(E_{M(\zeta)}(\tilde{T})) = \{b\}$ for all $\zeta \in \mathcal{J}(T)$ per Lemma 24.1.6. Thus, (b) is proved and (a) is true when (b) holds.

Moreover,

$$E(T) = \{b\} = M^{-1}(\{\infty\}) \subseteq M^{-1}(\mathcal{F}(\tilde{T})) = \mathcal{F}(T).$$

Thus, (d) is true when (b) holds.

Now assume that $E_z(T)$ consists of exactly two points. Let $M : \hat{\mathbb{C}} \to \hat{\mathbb{C}}$ be a Möbius transformation such that $M(E_z(T)) = \{0, \infty\}$. Setting $\tilde{T} := M \circ T \circ M^{-1}$ and invoking (24.3), we deduce that $E_z(\tilde{T}) = \{0, \infty\}$ and $\tilde{T}^{-1}(\{0, \infty\}) = \{0, \infty\}$. Either

$$\tilde{T}(0) = 0 \quad \text{and} \quad \tilde{T}(\infty) = \infty, \tag{24.4}$$

or

$$\tilde{T}(0) = \infty \quad \text{and} \quad \tilde{T}(\infty) = 0. \tag{24.5}$$

In the case of (24.4), the map $\tilde{T}$ must be a polynomial since $\tilde{T}^{-1}(\{\infty\}) = \{\infty\}$, and must be of the form $z \mapsto az^d$, $d \geq 2$, since $\tilde{T}^{-1}(\{0\}) = \{0\}$. In the case of (24.5), the only zero of the rational function $\tilde{T}$ is ∞, so $\tilde{T} = 1/Q$, where Q is a polynomial with $\deg(Q) \geq 2$. But since $\tilde{T}^{-1}(\{\infty\}) = \{0\}$, the polynomial Q must be of the form az^d, with $d \geq 2$. In both cases, $E_w(\tilde{T}) = \{0, \infty\}$ for every $w \in \mathcal{J}(\tilde{T})$, and hence $E_\zeta(T)$ consists of the same two points $M^{-1}(\{0, \infty\})$ for all $\zeta \in \mathcal{J}(T)$. Thus, (c) is proved and (a) is true when (c) holds. As in (b), it is easy to show that (d) is true when (c) holds. □

According to the preceding theorem, the exceptional set is empty unless, up to a conjugation by a Möbius transformation, $T(z) = az^d$ for some $|d| \geq 2$.

Since the exceptional set is contained in the Fatou set and since the Julia set is compact and completely invariant, we deduce the following result.

Corollary 24.1.10. *If $T : \hat{\mathbb{C}} \to \hat{\mathbb{C}}$ is a rational function with $\deg(T) \geq 2$, then for every open set $U \subseteq \hat{\mathbb{C}}$ intersecting the Julia set $\mathcal{J}(T)$, there exists a nonnegative integer $n(U)$ such that*

$$\bigcup_{k=0}^{n(U)} T^k(U \cap \mathcal{J}(T)) = \mathcal{J}(T).$$

In other words, $T : \mathcal{J}(T) \to \mathcal{J}(T)$ is very strongly transitive.

We now classify the periodic points of T.

Definition 24.1.11. Let $T : \widehat{\mathbb{C}} \to \widehat{\mathbb{C}}$ be a rational function with $\deg(T) \geq 2$. Let $w \in \mathbb{C}$ be a periodic point of T with prime period $\ell \in \mathbb{N}$. The periodic point w is said to be
(a) *attracting* if $|(T^\ell)'(w)| < 1$; if $(T^\ell)'(w) = 0$, then w is also commonly referred to as a *superattracting* periodic point.
(b) *repelling* if $|(T^\ell)'(w)| > 1$.
(c) *rationally indifferent* if $|(T^\ell)'(w)| = 1$ and $((T^\ell)'(w))^n = 1$ for some $n \in \mathbb{N}$.
(d) *irrationally indifferent* if $|(T^\ell)'(w)| = 1$ but $((T^\ell)'(w))^n \neq 1$ for all $n \in \mathbb{N}$.
Indifferent periodic points (i. e., (c) and (d) together) are sometimes called *neutral.*

If ∞ is a periodic point of T, we conjugate T by a Möbius transformation $M : \widehat{\mathbb{C}} \to \widehat{\mathbb{C}}$ which sends ∞ to a complex number $\xi \in \mathbb{C}$ and classify the periodic point ∞ of T as ξ is classified for the conjugate map $M \circ T \circ M^{-1}$.

The number $(T^\ell)'(w)$ is called the *multiplier* of w with respect to T.

Finally, a point $z \in \widehat{\mathbb{C}}$ is said to be *preperiodic* if there is $k \in \mathbb{N}$ such that $T^k(z)$ is a periodic point of T.

All the points in the orbit $(T^j(w))_{j=0}^{\ell-1}$ of w are periodic points of T with the same prime period ℓ and the same multiplier as w. Hence, they all have the same nature and it makes sense to talk about attracting, repelling and neutral (rationally or irrationally indifferent) orbits.

In Theorem 24.3.4, we will see that the set of periodic points of T is dense in $\mathcal{J}(T)$, and thus is infinite. At this stage, it suffices to establish that this set is infinite.

Lemma 24.1.12. *Every rational function* $T : \widehat{\mathbb{C}} \to \widehat{\mathbb{C}}$ *with* $d := \deg(T) \geq 2$ *admits infinitely many periodic points whose prime periods are prime numbers.*

Proof. Suppose that T has no point of period p, where p is a prime number. This means that every solution z' to the equation $T^p(z) = z$ is of period k' for some factor k' of p. That is, $k' = 1$ and every z' is actually a fixed point of T. As T and T^p have $d+1$ and $d^p + 1$ fixed points respectively (counting multiplicities), we deduce that there is a fixed point z^+ of T such that T^p has more fixed points at z^+ than T does. Comparing the Taylor expansions of T and T^p around z^+ reveals that this can only happen if $T'(z^+) \neq 1$ but $[T'(z^+)]^p = 1$ (see Exercise 24.8.5). Given any fixed point z of T, the equation $[T'(z)]^p = 1$ can be satisfied by at most one prime number $p(z)$. As T has $d+1$ fixed points (counting multiplicities), we conclude that the assumption that T has no point of period p can hold for at most $d+1$ prime numbers p. That is, T must have periodic points with prime period p for all but at most $d+1$ prime numbers p. □

As the number of periodic points of any fixed period is finite, let z_1 and z_2 be periodic points of T with prime periods at least 3 and disjoint forward orbits.

We end this section with a very useful technical fact. In the following lemma, all distances, derivatives and diameters are considered with respect to the spherical metric ρ on $\widehat{\mathbb{C}}$. Set

$$\delta_T := \frac{1}{8}\mathrm{dist}_\rho(\mathcal{O}_+(z_1), \mathcal{O}_+(z_2)) \in \left(0, \frac{1}{8}\mathrm{diam}_\rho(\widehat{\mathbb{C}})\right).$$

Lemma 24.1.13. *Let $T : \widehat{\mathbb{C}} \to \widehat{\mathbb{C}}$ be a rational function with $\deg(T) \geq 2$. Fix $\delta \in (0, \delta_T)$. For every $k \in \mathbb{N}$ and $z \in \mathcal{J}(T)$, let $T_*^{-k}(z)$ denote the set of all $w \in T^{-k}(z)$ for which there exists a (unique) holomorphic branch $T_w^{-k} : B_\rho(z, 4\delta) \to \widehat{\mathbb{C}}$ of T^{-k} that maps z to w. Thereafter, let $F_k(\delta)$ be the subset of all points $z \in \mathcal{J}(T)$ such that the set $T_*^{-k}(z)$ is nonempty. Then*

$$\lim_{k\to\infty} \inf\{|(T^k)'(w)|_\rho : z \in F_k(\delta),\ w \in T_*^{-k}(z)\} = \infty$$

and

$$\lim_{k\to\infty} \sup\{\mathrm{diam}_\rho(T_w^{-k}(B_\rho(z, \delta))) : z \in F_k(\delta),\ w \in T_*^{-k}(z)\} = 0,$$

with the convention that $\inf \emptyset = \infty$ *while* $\sup \emptyset = 0$.

Proof. We first show the second formula. Suppose for a contradiction that this formula fails. Then there exist $\varepsilon > 0$ and a strictly increasing sequence of positive integers $(n_k)_{k=1}^\infty$ such that for every $k \in \mathbb{N}$ there are $y_k \in F_{n_k}(\delta)$ and $x_k \in T_*^{-n_k}(y_k)$ for which

$$\mathrm{diam}_\rho(T_{x_k}^{-n_k}(B_\rho(y_k, \delta))) > \varepsilon. \tag{24.6}$$

Passing to a subsequence if necessary, we may assume that both sequences $(y_k)_{k=1}^\infty$ and $(x_k)_{k=1}^\infty$ converge. Denote their respective limits by y and x. Both of them belong to $\mathcal{J}(T)$. Disregarding finitely many terms of the sequence $(n_k)_{k=1}^\infty$, we then have that

$$B_\rho(y_k, \delta) \subseteq B_\rho(y, 3\delta/2) \quad \text{and} \quad B_\rho(y, 3\delta) \subseteq B_\rho(y_k, 4\delta), \quad \forall k \in \mathbb{N}.$$

Consider the family of inverse branches

$$\mathcal{F}_* := \{T_{x_k}^{-n_k}|_{B_\rho(y,3\delta)} : B_\rho(y, 3\delta) \to \widehat{\mathbb{C}}\}_{k\in\mathbb{N}}.$$

By the choice of δ, for every $k \in \mathbb{N}$ the set $T_{x_k}^{-n_k}(B_\rho(y, 3\delta))$ is disjoint from the forward orbit of at least one periodic point used to define the number δ_T. Passing to yet another subsequence, we may assume that this is the same periodic orbit for all $k \in \mathbb{N}$. But then Montel's theorem (Theorem 23.2.4) affirms that the family $\mathcal{F}_*$ is normal. So, passing yet again to another subsequence, we may assume that the sequence

$$\{T_{x_k}^{-n_k}|_{B_\rho(y,2\delta)} : B_\rho(y, 2\delta) \to \widehat{\mathbb{C}}\}_{k\in\mathbb{N}}$$

converges uniformly to analytic function $f : B_\rho(y, 2\delta) \to \widehat{\mathbb{C}}$. Therefore,

$$f(y) = \lim_{k\to\infty} T_{x_k}^{-n_k}(y_k) = \lim_{k\to\infty} x_k = x.$$

By Hurwitz's theorem (Theorem A.3.9), the function f is either constant or univalent. In the former case,

$$\begin{aligned}\limsup_{k\to\infty} \operatorname{diam}_\rho(T_{x_k}^{-n_k}(B_\rho(y_k,\delta))) &\le \limsup_{k\to\infty} \operatorname{diam}_\rho(T_{x_k}^{-n_k}(B_\rho(y,3\delta/2)))\\ &\le \operatorname{diam}_\rho(f(B_\rho(y,3\delta/2)))\\ &= 0,\end{aligned}$$

contrary to (24.6). In the latter case, $f(B_\rho(y,\delta))$ is an open set containing x and by Theorem A.3.11 there exists $N_\delta \in \mathbb{N}$ such that

$$T_{x_k}^{-n_k}(B_\rho(y,2\delta)) \supseteq f(\overline{B_\rho(y,\delta)}), \quad \forall k \ge N_\delta.$$

Consequently,

$$T^{n_k}(f(B_\rho(y,\delta))) \subseteq B_\rho(y,2\delta), \quad \forall k \ge N_\delta.$$

Thus, the sequence $(T^{n_k}|_{f(B_\rho(y,\delta))})_{k=1}^{\infty}$ is normal, contrary to the fact that $x \in \mathcal{J}(T)$. The second formula in this lemma is hence proved. The first formula is now an immediate consequence of Koebe's distortion theorem (Theorem 23.1.9). □

24.1.1 Attracting periodic points

If $w \in \mathbb{C}$ is an attracting periodic point of T with prime period $\ell \in \mathbb{N}$, then there exists $R > 0$ such that

$$\sup_{\xi\in B(w,R)} |(T^\ell)'(\xi)| \le \frac{1+|(T^\ell)'(w)|}{2} := \chi < 1.$$

By the mean value inequality,

$$|T^\ell(z) - w| \le |(T^\ell)'(\zeta(z))| \cdot |z-w| \le \chi|z-w|$$

for all $z \in B(w,R)$ and some $\zeta(z) \in [w,z] \subseteq B(w,R)$. For all $k \in \mathbb{N}$, we infer by recurrence that

$$|T^{k\ell}(z) - w| \le \chi^\ell |z-w|, \quad \forall z \in B(w,R).$$

So, $\lim_{k\to\infty} T^{k\ell}(z) = w$ for all $z \in B(w,R)$. Note further that

$$T^\ell(B(w,R)) \subseteq B\left(w, \frac{1+|(T^\ell)'(w)|}{2}R\right) \subseteq B(w,R).$$

By Montel's theorem (Theorem 23.2.4), it follows that $B(w,R) \subseteq \mathcal{F}(T)$. Letting

$$A(w) := \left\{ z \in \widehat{\mathbb{C}} \,\middle|\, \lim_{k\to\infty} T^{k\ell}(z) = w \right\},$$

it ensues from the complete invariance of the Fatou set that $B(w,R) \subseteq A(w) \subseteq \mathcal{F}(T)$. The set $A(w)$ is called the *basin of attraction of* w.

The connected component of $A(w)$ that contains w is called the *immediate basin of attraction of* w and will be denoted by $A^*(w)$. Obviously, $B(w,R) \subseteq A^*(w) \subseteq A(w) \subseteq \mathcal{F}(T)$.

We collect these observations in the following theorem.

Theorem 24.1.14. *Let $T : \widehat{\mathbb{C}} \to \widehat{\mathbb{C}}$ be a rational function. If $w \in \widehat{\mathbb{C}}$ is an attracting periodic point of T with prime period ℓ, then $A^*(w)$ and $A(w)$ are open sets such that*

$$w \in A^*(w) \subseteq A(w) \subseteq \mathcal{F}(T).$$

Moreover, $A^*(w)$ is forward T^ℓ-invariant whereas $A(w)$ is completely T^ℓ-invariant.

The open sets

$$A_p^*(w) := \bigcup_{j=0}^{\ell-1} A^*(T^j(w)) \quad \text{and} \quad A_p(w) := \bigcup_{j=0}^{\ell-1} A(T^j(w)) \tag{24.7}$$

are respectively coined *immediate basin of attraction* and *basin of attraction of the periodic orbit* $\{w, T(w), \ldots, T^{\ell-1}(w)\}$. The set $A_p^*(w)$ is forward T-invariant while $A_p(w)$ is completely T-invariant. Moreover,

$$\{w, T(w), \ldots, T^{\ell-1}(w)\} \subseteq A_p^*(w) \subseteq A_p(w) \subseteq \mathcal{F}(T).$$

In particular, this asserts that every attracting periodic orbit is in the Fatou set.

This observation is the last ingredient needed to establish the following remarkable topological property of the Julia set.

Theorem 24.1.15. *Let $T : \widehat{\mathbb{C}} \to \widehat{\mathbb{C}}$ be a rational function with* $\deg(T) \geq 2$. *The Julia set $\mathcal{J}(T)$ is perfect.*

Proof. Suppose by way of contradiction that w is an isolated point of $\mathcal{J}(T)$. Then there exists some open neighborhood U of w such that $U \cap \mathcal{J}(T) = \{w\}$. It follows from Corollary 24.1.10 that $\mathcal{J}(T) = \{T^n(w) : 0 \leq n \leq n(U)\}$ for some $n(U) < \infty$. So $\mathcal{J}(T)$ is finite. Hence, w is a preperiodic or periodic point of T, i. e., there exist a least $k \geq 0$ and a least $\ell \in \mathbb{N}$ such that $T^\ell(T^k(w)) = T^k(w)$. But since the map $T : \mathcal{J}(T) \to \mathcal{J}(T)$ is surjective, it ensues that $k = 0$, i. e., w is a periodic point, and thus $\mathcal{J}(T)$ consists of a single periodic orbit. It follows from the complete T-invariance of $\mathcal{J}(T)$ (cf. Theorem 24.1.2) that $T^{-1}(\{w\}) = \{T^{\ell-1}(w)\}$ is a singleton. On the other hand, no point in the periodic orbit

$(T^j(w))_{j=0}^{\ell-1}$ can be a critical point of T since that periodic orbit would then be superattracting, and hence, by Theorem 24.1.14, would be in $\mathcal{F}(T)$. This contradicts the fact that $\deg(T) \geq 2$. □

Returning to the Fatou set, the local dynamics around attracting but not superattracting fixed points is entirely understood thanks to G. Koenigs' result that dates back to 1884 (see [64–69]).

Theorem 24.1.16. *Let $T : \widehat{\mathbb{C}} \to \widehat{\mathbb{C}}$ be a rational function with $\deg(T) \geq 2$. If w is an attracting but not superattracting fixed point of T, then there exist a neighborhood U of w, a neighborhood V of 0, and a unique (up to multiplication by a nonzero constant) conformal homeomorphism $H : U \to V$ such that*

$$H \circ T \circ H^{-1}(z) = T'(w)z, \quad \forall z \in V.$$

Proof. We may assume without loss of generality that $w = 0$. For every $z \in \widehat{\mathbb{C}}$ and $n \in \mathbb{N}$, define $H_n(z) := (T'(0))^{-n} \cdot T^n(z)$. Then H_n satisfies

$$H_n \circ T = T'(0)H_{n+1}, \quad \forall n \in \mathbb{N}.$$

Assume momentarily that $(H_n)_{n=1}^{\infty}$ converges pointwise to a function H on some neighborhood of 0. Then $H \circ T = T'(0)H$ on that neighborhood and $H(0) = 0$. It remains to show that H is a univalent holomorphic function on some neighborhood of 0.

To obtain holomorphicity of H, we will show that the convergence is uniform on a compact neighborhood of 0. We know from the Maclaurin expansion of T that for some $\delta > 0$ there is a constant $C \geq 0$ such that

$$|T(z) - T'(0)z| \leq C|z|^2, \quad \forall z \in \overline{B(0,\delta)}.$$

Thus, $|T(z)| \leq |T'(0)|\,|z| + C|z|^2 \leq (|T'(0)| + C\delta)|z|$. By reducing δ if necessary, we may assume that $|T'(0)| + C\delta < 1$ and we then obtain by induction that

$$|T^n(z)| \leq (|T'(0)| + C\delta)^n |z|, \quad \forall z \in \overline{B(0,\delta)}.$$

In particular, this guarantees that T maps $\overline{B(0,\delta)}$ into itself, and so does all of its iterates. More precisely, $T^n(z) \to 0$ uniformly and exponentially fast on $\overline{B(0,\delta)}$. By reducing δ even further if necessary, we may assume that $\rho := (|T'(0)| + C\delta)^2/|T'(0)| < 1$. Then

$$|H_{n+1}(z) - H_n(z)| = \left|\frac{T(T^n(z)) - T'(0)T^n(z)}{(T'(0))^{n+1}}\right| \leq \frac{C|T^n(z)|^2}{|T'(0)|^{n+1}} \leq \frac{C\rho^n|z|^2}{|T'(0)|}, \quad \forall z \in \overline{B(0,\delta)}.$$

Hence, $(H_n)_{n=1}^{\infty}$ is uniformly Cauchy, and thus uniformly converges to some function H on $\overline{B(0,\delta)}$. We can set $U = B(0,\delta)$ and $V = H(U)$.

Let us now show that H is univalent. Shrinking δ if needed ensures that T is univalent on $\overline{B(0,\delta)}$. As T maps $\overline{B(0,\delta)}$ into itself, every iterate T^n is univalent on $\overline{B(0,\delta)}$. It follows from Hurwitz's theorem (Theorem A.3.9) that H is univalent on $U = B(0, \delta)$. (H cannot be constant on U since $H \circ T = T'(0)H$ on that neighborhood of 0.)

It remains to demonstrate the uniqueness of H up to a nonzero multiplicative constant. Suppose that H and $\widetilde{H}$ are two such conjugacies. Then $H \circ T \circ H^{-1} = T'(0)\mathrm{Id} = \widetilde{H} \circ T \circ \widetilde{H}^{-1}$. Write $h := H \circ \widetilde{H}^{-1}$ and observe that

$$(T'(0)\mathrm{Id}) \circ h = H \circ T \circ \widetilde{H}^{-1} = h \circ (T'(0)\mathrm{Id}).$$

Let $h(z) = \sum_{n=0}^{\infty} a_n z^n$ be the Maclaurin series of h. Substituting this series into the above conjugation equation and equating coefficients of same power, we deduce that $T'(0)a_n = a_n(T'(0))^n$ for all $n \geq 0$. As $T'(0)$ is neither zero nor a root of unity, we must have $a_n = 0$ for all $n \neq 1$. Therefore, $h(z) = a_1 z$ with $a_1 \neq 0$. Hence, $H(z) = h(\widetilde{H}(z)) = a_1 \cdot \widetilde{H}(z)$. □

The local dynamics around superattracting fixed points is also entirely understood, this time due to L. Böttcher's result published in 1904 (see [14–18]).

Theorem 24.1.17. *Let $T : \widehat{\mathbb{C}} \to \widehat{\mathbb{C}}$ be a rational function with $\deg(T) \geq 2$. If w is a superattracting fixed point of T, then there exist a neighborhood U of w, a neighborhood V of 0, and a unique (up to multiplication by a $(\deg_w(T) - 1)$ root of unity) conformal homeomorphism $H : U \to V$ such that $H(w) = 0$ and*

$$H \circ T \circ H^{-1}(z) = z^{\deg_w(T)}, \quad \forall z \in V.$$

Proof. Without losing generality, we may assume that $w = 0$ and let $d = \deg_0(T)$. Let

$$\mathbb{C}_+ := \{z \in \mathbb{C} \mid \mathrm{Re}(z) > 0\}.$$

Let $\log_0 : \mathbb{C}_+ \to \mathbb{C}$ be the principal branch of the logarithm function, i. e., the holomorphic branch of logarithm uniquely determined by the requirement that $\log_0(1) = 0$. We know that

$$T(z) = z^d(1 + zf(z)),$$

where f is a holomorphic function defined on $B(0, 2R)$ for some $R \in (0, 1/4)$. Set

$$B := \|f|_{B(0,R)}\|_\infty < \infty.$$

Decreasing $R > 0$ if needed, we may assume that

$$|e^z - 1| \leq \frac{4}{3}|z|, \quad |\log_0(1+z)| \leq \frac{4}{3}|z|, \quad \forall z \in B(0, R), \tag{24.8}$$

$$\log_0 \frac{1+z_2}{1+z_1} = \log_0(1+z_2) - \log_0(1+z_1) \;\&\; \left|\log_0 \frac{1+z_2}{1+z_1}\right| \leq 1, \; \forall z_1, z_2 \in B\Big(0, R\max\Big\{1, \tfrac{4B}{3d-4}\Big\}\Big) \tag{24.9}$$

and

$$|T(z)| \leq \frac{1}{2}|z|, \quad \forall z \in B(0,R). \tag{24.10}$$

Fix $r \in (0,R)$ so small that

$$rB < 1/2 \quad \text{and} \quad \frac{4rB}{3d}\left(\frac{4}{3d-4}+1\right) < R. \tag{24.11}$$

Define inductively a sequence of holomorphic functions $G_n : B(0,r) \to \mathbb{C}$, $n \geq 0$, satisfying

$$|G_n(z)| \leq \frac{4B}{3d-4}, \quad \forall z \in B(0,r) \tag{24.12}$$

as follows. Let

$$G_0 \equiv 0.$$

For the inductive step, suppose that G_n has been defined satisfying (24.12). Define $G_{n+1}(z)$ for $z \in B(0,r) \setminus \{0\}$ by the formula

$$G_{n+1}(z) := z^{-1}\left[\exp\left(\frac{1}{d}\log_0(1+T(z)G_n(T(z))) + \frac{1}{d}\log_0(1+zf(z))\right) - 1\right], \tag{24.13}$$

which is well-defined thanks to the first inequality in (24.11), (24.10) and (24.12). By the second inequality in (24.8), (24.10), (24.12) and the second inequality in (24.11), we get

$$\frac{1}{d}\left|\log_0(1+T(z)G_n(T(z)))+\log_0(1+zf(z))\right| \leq \frac{4|z|}{3d}(|G_n(T(z))|+|f(z)|) \leq \frac{4r}{3d}\left(\frac{4B}{3d-4}+B\right) < R.$$

Using this and the first inequality in (24.8), we deduce that

$$\begin{aligned}|G_{n+1}(z)| &\leq \frac{4}{3}|z|^{-1}\left|\frac{1}{d}\log_0(1+T(z)G_n(T(z))) + \frac{1}{d}\log_0(1+zf(z))\right| \\ &\leq \frac{4}{3d}(|G_n(T(z))| + |f(z)|) \leq \frac{4B}{3d}\left(\frac{4}{3d-4}+1\right) = \frac{4B}{3d-4}.\end{aligned}$$

Hence, (24.12) holds for all $z \in B(0,r)$ with n replaced by $n+1$. Thus, the limit

$$\lim_{z\to 0} G_{n+1}(z)$$

exists, and denoting its value by $G_{n+1}(0)$ completes the inductive step of the proof establishing a sequence $(G_n)_{n=1}^{\infty}$ of holomorphic functions satisfying (24.12).

For every $n \geq 0$, define the holomorphic function $H_n : B(0,r) \to \mathbb{C}$ by

$$H_n(z) := z(1+zG_n(z)). \tag{24.14}$$

Using (24.13), we obtain that

$$\begin{aligned}
(H_{n+1}(z))^d &= z^d(1+zG_{n+1}(z))^d \\
&= z^d\left[\exp\left(\frac{1}{d}\log_0(1+T(z)G_n(T(z)))+\frac{1}{d}\log_0(1+zf(z))\right)\right]^d \\
&= z^d\exp\big(\log_0(1+T(z)G_n(T(z)))+\log_0(1+zf(z))\big) \\
&= z^d\exp(\log_0(1+T(z)G_n(T(z))))\exp(\log_0(1+zf(z))) \\
&= z^d(1+T(z)G_n(T(z)))(1+zf(z)) \\
&= T(z)(1+T(z)G_n(T(z))) \\
&= H_n\circ T(z).
\end{aligned}$$

Consequently,

$$H_n\circ T(z)=(H_{n+1}(z))^d,\quad \forall z\in B(0,r). \tag{24.15}$$

By induction,

$$H_n\circ T^k(z)=(H_{n+k}(z))^{d^k},\quad \forall z\in B(0,r),\ \forall n,k\geq 0. \tag{24.16}$$

For every $n\geq 0$, define the function $Q_n : B(0,r)\to\mathbb{C}$ by

$$Q_n(z):=1+zG_n(z).$$

Claim. *The sequence of functions* $(\log_0 Q_n)_{n=1}^{\infty}$ *converges uniformly on* $B(0,r)$.

Proof. Given $n,k\geq 0$, it follows from (24.16) that

$$\left(\frac{H_{n+k+1}(z)}{H_{n+k}(z)}\right)^{d^k}=\frac{H_{n+1}(T^k(z))}{H_n(T^k(z))}.$$

According to (24.14), this means that

$$\left(\frac{1+zG_{n+k+1}(z)}{1+zG_{n+k}(z)}\right)^{d^k}=\frac{1+T^k(z)G_{n+1}(T^k(z))}{1+T^k(z)G_n(T^k(z))}.$$

By taking the logarithm of both sides, we get

$$d^k\log_0\frac{1+zG_{n+k+1}(z)}{1+zG_{n+k}(z)}=\log_0\frac{1+T^k(z)G_{n+1}(T^k(z))}{1+T^k(z)G_n(T^k(z))}.$$

It follows from (24.12), the second inequality in (24.11) and the first inequality in (24.9) that

$$\log_0(1 + zG_{n+k+1}(z)) - \log_0(1 + zG_{n+k}(z)) = d^{-k} \log_0 \frac{1 + T^k(z)G_{n+1}(T^k(z))}{1 + T^k(z)G_n(T^k(z))}.$$

Hence, using the second inequality in (24.9) we obtain that

$$\left|\log_0(1 + zG_{n+k+1}(z)) - \log_0(1 + zG_{n+k}(z))\right| \le d^{-k}.$$

So, for all $m, n \ge 0$ we deduce that

$$\begin{aligned} \left|\log_0(1 + zG_{n+m}(z)) - \log_0(1 + zG_n(z))\right| &\le \sum_{k=n}^{n+m-1} \left|\log_0(1 + zG_{k+1}(z)) - \log_0(1 + zG_k(z))\right| \\ &\le \sum_{k=n}^{n+m-1} d^{-k} \le \sum_{k=n}^{\infty} d^{-k} = \frac{d^{-n}}{1 - d^{-1}}. \end{aligned}$$

Consequently, the sequence $(\log_0 Q_n)_{n=1}^{\infty}$ is uniformly Cauchy, and thus converges uniformly. The proof of the claim is finished. ■

Denote by L the uniform limit of the sequence $(\log_0 Q_n)_{n=1}^{\infty}$. Then $L : B(0, r) \to \mathbb{C}$ is holomorphic and

$$L(0) = \lim_{n\to\infty} \log_0 Q_n(0) = \lim_{n\to\infty} \log_0 1 = 0.$$

Let

$$Q := e^L : B(0, r) \to \mathbb{C}.$$

Then Q is holomorphic and

$$Q(0) = e^{L(0)} = e^0 = 1. \tag{24.17}$$

Let the function $H : B(0, r) \to \mathbb{C}$ be given by

$$H(z) = zQ(z).$$

Then H is a holomorphic function and, by (24.17),

$$H(z) = z + O(z^2), \quad \forall z \in B(0, r). \tag{24.18}$$

By continuity of the exponential function $\exp : \mathbb{C} \to \mathbb{C}$ and the dth power, along with the above claim, we obtain

$$
\begin{aligned}
H(T(z)) &= T(z)Q(T(z)) = T(z)\exp(L(T(z))) = T(z)\exp\Big(\lim_{n\to\infty}\log_0 Q_n(T(z))\Big)\\
&= \lim_{n\to\infty}[T(z)\exp(\log_0 Q_n(T(z)))] = \lim_{n\to\infty}[T(z)Q_n(T(z))] = \lim_{n\to\infty}H_n(T(z))\\
&= \lim_{n\to\infty}(H_{n+1}(z))^d = \Big[\lim_{n\to\infty}H_{n+1}(z)\Big]^d = \Big[\lim_{n\to\infty}(zQ_{n+1}(z))\Big]^d\\
&= \Big[z\lim_{n\to\infty}\exp(\log_0 Q_{n+1}(z))\Big]^d = \Big[z\exp\Big(\lim_{n\to\infty}\log_0 Q_{n+1}(z)\Big)\Big]^d\\
&= (ze^{L(z)})^d = (zQ(z))^d = (H(z))^d.
\end{aligned}
$$

Thus, $H\circ T = H^d$ on $B(0,r)$ and along with (24.18), the proof of the existence of the sought conjugation is complete.

It remains to demonstrate the uniqueness of H up to multiplication by a $(\deg_w(T)-1)$ root of unity. In a similar manner to the uniqueness part of the proof of Theorem 24.1.16, it suffices to show that any conjugation of z^d to itself is a multiple of z^d and that multiple is a $(d-1)$th root of unity. Suppose that h is such a conjugation, i. e., $(h(z))^d = h(z^d)$. Let $h(z) = \sum_{n=0}^{\infty} a_n z^n$ be the Maclaurin series of h. Substituting this series into the above conjugation equation and equating coefficients of same power, we deduce that $a_d^{d-1} = 1$ while $a_n = 0$ for all $n \neq d$. □

As an immediate consequence of Theorems 24.1.16 and 24.1.17, we get the following two respective results.

Theorem 24.1.18. *Let $T : \widehat{\mathbb{C}} \to \widehat{\mathbb{C}}$ be a rational function with* $\deg(T) \geq 2$. *If w is an attracting but not superattracting periodic point of T of period $p \in \mathbb{N}$, then there exist a neighborhood U of w, a neighborhood V of 0, and a unique (up to multiplication by a nonzero constant) conformal homeomorphism $H : U \to V$ such that*

$$H \circ T^p \circ H^{-1}(z) = (T^p)'(w)z, \quad \forall z \in V.$$

Theorem 24.1.19. *Let $T : \widehat{\mathbb{C}} \to \widehat{\mathbb{C}}$ be a rational function with* $\deg(T) \geq 2$. *If w is a superattracting periodic point of T of period $p \in \mathbb{N}$, then there exist a neighborhood U of w, a neighborhood V of 0 and a unique (up to multiplication by a $(\deg_w(T^p) - 1)$ root of unity) conformal homeomorphism $H : U \to V$ such that*

$$H \circ T^p \circ H^{-1}(z) = z^{\deg_w(T^p)}, \quad \forall z \in V.$$

We end our examination of attracting periodic orbits with a demonstration that their immediate basin of attraction must comprise at least one critical point.

Theorem 24.1.20. *Let $T : \widehat{\mathbb{C}} \to \widehat{\mathbb{C}}$ be a rational function with* $\deg(T) \geq 2$. *If $w \in \widehat{\mathbb{C}}$ is an attracting periodic point of T, then the immediate basin of attraction $A_p^*(w)$ of the orbit of w contains at least one critical point of T.*

Proof. Assume first that w is an attracting fixed point of T. Suppose for a contradiction that $A^*(w)$ contains no critical point of T. We shall prove by induction that there exists an ascending sequence $(U_n)_{n=0}^{\infty}$ of open simply connected subsets of $A^*(w)$ such that

(a) $w \in U_0$.
(b) For every $n \in \mathbb{N}$, there exists $T_n^{-1} : U_{n-1} \to U_n$, a unique surjective holomorphic inverse branch of T such that $T_n^{-1}(w) = w$.
(c) $T_{n+1}^{-1}|_{U_{n-1}} = T_n^{-1}$, $\forall n \in \mathbb{N}$.

Indeed, let U_0 be an arbitrary open simply connected subset of $A^*(w)$ containing w. For the inductive step, suppose that open simply connected subsets $U_0 \subseteq U_1 \subseteq \cdots \subseteq U_n \subseteq A^*(w)$ satisfying conditions (a) and (b) have been defined for some $n \geq 0$. Since $A^*(w) \cap T^{-1}(U_n)$ contains no critical point of T, the inverse function theorem (Theorem A.3.13) and the monodromy theorem (Theorem A.3.4) yield the existence of a holomorphic inverse branch $T_{n+1}^{-1} : U_n \to A^*(w)$ uniquely determined by the requirement that $T_{n+1}^{-1}(w) = w$. Set $U_{n+1} := T_{n+1}^{-1}(U_n) \subseteq A^*(w)$. Readily from this definition, we see that U_{n+1} is a open simply connected subset of $\widehat{\mathbb{C}}$ and that $T_{n+1}^{-1} : U_n \to U_{n+1}$ is the unique surjective holomorphic inverse branch of T such that $T_{n+1}^{-1}(w) = w$. As $T_n^{-1} : U_{n-1} \to U_n$ is a holomorphic inverse branch of the same map T such that $T_n^{-1}(w) = w$ and as $U_{n-1} \subseteq U_n$ by the inductive hypothesis, we infer that $T_{n+1}^{-1}|_{U_{n-1}} = T_n^{-1}$. Hence, $U_{n+1} = T_{n+1}^{-1}(U_n) \supseteq T_n^{-1}(U_{n-1}) = U_n$. The inductive construction of the sequence $(U_n)_{n=0}^{\infty}$ is thus complete.

It follows from the defining properties of the sequence $(U_n, T_{n+1})_{n=0}^{\infty}$ that for all $n \in \mathbb{N}$, the composite function

$$T_w^{-n} := T_n^{-1} \circ T_{n-1}^{-1} \circ \cdots \circ T_2^{-1} \circ T_1^{-1} : U_0 \to U_n \subseteq A^*(w)$$

is a well-defined, surjective and holomorphic inverse branch of T^n such that $T_w^{-n}(w) = w$. Since $A^*(w) \subseteq \mathcal{F}(T)$ and since the Julia set $\mathcal{J}(T)$ contains at least three points (because it is perfect by Theorem 24.1.15), it follows from Montel's criterion (Theorem 23.2.4) that the family of maps $\{T_w^{-n} : U_0 \to A^*(w)\}_{n=1}^{\infty}$ is normal. This, however, produces a contradiction since

$$\lim_{n\to\infty} (T_w^{-n})'(w) = \lim_{n\to\infty} \left((T^n)'(w)\right)^{-1} = \lim_{n\to\infty} \left(T'(w)\right)^{-n} = \infty.$$

We are thus done when w is a fixed point of T.

In general, if w is a periodic point of T with prime period ℓ, then w is a fixed point of T^ℓ. By the case just treated, we know that $A_{T^\ell}^*(w)$ contains a critical point c_ℓ of T^ℓ. This critical point of T^ℓ has a forward iterate $T^k(c_\ell)$ which is a critical point of T (in fact, $0 \leq k < \ell$). But observe that $A_p^*(w) \supseteq A_{T^\ell}^*(w) \ni c_\ell$. Given that $A_p^*(w)$ is forward T-invariant, we conclude that $A_p^*(w) \ni T^k(c_\ell)$. □

As the number of critical points of T (counted by multiplicities) is $2\deg(T) - 2$, the previous theorem yields an upper estimate on the number of attracting cycles.

Corollary 24.1.21. *Any rational function $T : \widehat{\mathbb{C}} \to \widehat{\mathbb{C}}$ with $\deg(T) \geq 2$ has at most $2\deg(T) - 2$ attracting periodic cycles.*

24.1.2 Nonattracting periodic points

Recall that there are three different types of nonattracting periodic points: repelling, rationally indifferent, and irrationally indifferent points.

24.1.2.1 Repelling and rationally indifferent periodic points

For repelling periodic points, the situation is opposite to that of attracting periodic points (cf. Theorem 24.1.14).

Theorem 24.1.22. *If $T : \widehat{\mathbb{C}} \to \widehat{\mathbb{C}}$ is a rational function with $\deg(T) \geq 2$, then every repelling periodic point of T belongs to the Julia set $\mathcal{J}(T)$.*

Proof. Suppose for a contradiction that a repelling periodic point w of T with prime period ℓ belongs to the Fatou set $\mathcal{F}(T)$. Conjugating by a Möbius map if necessary, we may assume without loss of generality that $w \in \mathbb{C}$. Then there exists an open connected neighborhood $U \subseteq \mathbb{C}$ of w such that the family $\{T^n\}_{n=1}^{\infty}$ of iterates of T is normal on U. Hence, the sequence $(T^{\ell n}|_U)_{n=1}^{\infty}$ contains a subsequence $(T^{\ell n_j}|_U)_{j=1}^{\infty}$ converging uniformly on compact subsets of U to some meromorphic map $S : U \to \widehat{\mathbb{C}}$. Then $S(w) = w$. Shrinking U if necessary, we may assume that $S(U) \subseteq \mathbb{C}$, i. e., S is holomorphic. On one hand, $S'(w) \in \mathbb{C}$, i. e., $|S'(w)| \neq \infty$. On the other hand,

$$|S'(w)| = \lim_{j\to\infty} |(T^{\ell n_j})'(w)| = \lim_{j\to\infty} |(T^{\ell})'(w)|^{n_j} = \infty.$$

This is a contradiction. Hence, $w \in \mathcal{J}(T)$. □

The proof of the corresponding result for rationally indifferent periodic points is somewhat more involved.

Theorem 24.1.23. *If $T : \widehat{\mathbb{C}} \to \widehat{\mathbb{C}}$ is a rational function with $\deg(T) \geq 2$, then every rationally indifferent periodic point of T belongs to the Julia set $\mathcal{J}(T)$.*

Proof. As in the proof of the previous theorem, through an appropriate Möbius conjugation we may assume without loss of generality that the rationally indifferent periodic point is 0. In view of Theorem 24.1.5, passing to a sufficiently high iterate of T, we may further assume that $T(0) = 0$ and $T'(0) = 1$. Then the Maclaurin series of T is of the form

$$T(z) = z + az^{p+1} + O(z^{p+2}), \tag{24.19}$$

where $p \in \mathbb{N}$ and $a \in \mathbb{C} \setminus \{0\}$. We prove by induction that

$$T^n(z) = z + naz^{p+1} + O(z^{p+2}). \tag{24.20}$$

Indeed, (24.20) reduces to (24.19) when $n = 1$ and so holds in this case. For the inductive step, assume that (24.20) holds for some $n \in \mathbb{N}$. Then

$$\begin{aligned} T^{n+1}(z) = T^n(T(z)) &= T(z) + na(T(z))^{p+1} + O((T(z))^{p+2}) \\ &= z + az^{p+1} + O(z^{p+2}) + na(z + az^{p+1} + O(z^{p+2}))^{p+1} + O((z + az^{p+1} + O(z^{p+2}))^{p+2}) \\ &= z + az^{p+1} + naz^{p+1} + O(z^{p+2}) = z + (n+1)az^{p+1} + O(z^{p+2}). \end{aligned}$$

Thus, (24.20) is valid. Then the $(p+1)$th derivative of T^{n+1} at 0 is

$$(T^n)^{(p+1)}(0) = (p+1)!na,$$

and hence

$$\lim_{n\to\infty} (T^n)^{(p+1)}(0) = \infty.$$

Suppose for a contradiction that $0 \in \mathcal{F}(T)$. Then there exist an open connected neighborhood $U \subseteq \mathbb{C}$ of 0, a strictly increasing sequence $(n_k)_{k=1}^{\infty}$ of positive integers, and a meromorphic function $S : U \to \widehat{\mathbb{C}}$, such that

$$\lim_{k\to\infty} T^{n_k} = S$$

uniformly on U. Since $S(0) = 0$, shrinking U if necessary, we may assume that $S(U) \subseteq \mathbb{C}$, i. e., that S is holomorphic. But then

$$\infty \neq S^{(p+1)}(0) = \lim_{k\to\infty} (T^{n_k})^{(p+1)}(0) = \infty.$$

This contradiction shows that $0 \in \mathcal{J}(T)$. □

24.1.2.2 Irrationally indifferent periodic points

Peculiarly, irrationally indifferent periodic points are sometimes found in the Fatou set (they are then called Siegel points), sometimes in the Julia set (they are then called Cremer points).

Definition 24.1.24. Let $w \in \mathbb{C}$ be an irrationally indifferent periodic point of a rational function $T : \widehat{\mathbb{C}} \to \widehat{\mathbb{C}}$ with prime period ℓ. Then w is called a *Siegel point* if there exists a holomorphic function H mapping a neighborhood of 0 to $\mathbb{C}$ with the following properties:

(a) $H(0) = w$

(b) $H'(0) = 1$

(c) $H^{-1} \circ T^{\ell} \circ H(z) = (T^{\ell})'(w)z$.

Property (c) is commonly referred to as the *Schröder equation.*

Compare this definition with Theorem 24.1.18. As $H'(0) \neq 0$, we know that H is injective on a neighborhood of 0. In fact, H has the following striking property.

Proposition 24.1.25. *If $w \in \mathbb{C}$ is a Siegel point of T, then H is injective on any ball that is centered at 0 and contained in the domain of H.*

Proof. Suppose that $B(0,r)$ is contained in the domain of H. Set $\lambda := (T^\ell)'(w)$. Assume that $H(z_1) = H(z_2)$ for some points $z_1, z_2 \in B(0,r)$. For every $n \in \mathbb{N}$, property (c) of a Siegel point yields that

$$H(\lambda^n z_2) = T^{n\ell}(H(z_2)) = T^{n\ell}(H(z_1)) = H(\lambda^n z_1).$$

Since $\{\lambda^n\}_{n=0}^{\infty}$ is a dense subset of the unit circle $\mathbb{S}^1 = \{z \in \mathbb{C} : |z| = 1\}$, it follows that $H(zz_1) = H(zz_2)$ for all $z \in \mathbb{S}^1$. Therefore, the holomorphic functions $B(0, r/|z_1|) \ni z \mapsto H(zz_1) \in \mathbb{C}$ and $B(0, r/|z_2|) \ni z \mapsto H(zz_2) \in \mathbb{C}$ agree on $\mathbb{S}^1$. (Note: If $z_i = 0$, define $r/|z_i| = \infty$ and $B(0, r/|z_i|) = \mathbb{C}$.) Consequently, these functions must agree on $B(0,1) \subseteq B(0, r/\max\{|z_1|, |z_2|\})$, i. e.,

$$H(zz_1) = H(zz_2), \quad \forall z \in B(0,1).$$

Taking the derivative of both sides at 0 and using property (b) of a Siegel point results in

$$z_1 = z_1 H'(0) = z_2 H'(0) = z_2.$$

We have thus demonstrated that H is injective on $B(0,r)$. □

This proposition means that in the vicinity of a Siegel point w of T with prime period $\ell \in \mathbb{N}$, the map T^ℓ is topologically conjugate to the rotation by the multiplier of w, i. e., T^ℓ is conjugate to $z \mapsto (T^\ell)'(w)z$.

There is a simple but fundamental characterization of Siegel points.

Theorem 24.1.26. *An irrationally indifferent periodic point of T is a Siegel point if and only if it belongs to the Fatou set $\mathcal{F}(T)$.*

Proof. Set $\lambda = (T^\ell)'(w)$. If w is a Siegel point, it follows from properties (a) and (c) and Proposition 24.1.25 that $T^{\ell n}(z) = H(\lambda^n H^{-1}(z))$ on some neighborhood of w and for all $n \in \mathbb{N}$. Therefore, the iterates of T^ℓ form a bounded sequence on some sufficiently small neighborhood of w. Montel's criterion (Theorem 23.2.4) and Theorem 24.1.5 assert that $w \in \mathcal{F}(T^\ell) = \mathcal{F}(T)$.

Conversely, if $w \in \mathcal{F}(T)$ then as $T^\ell(w) = w$ there exists a neighborhood U of w on which the iterates of T^ℓ are uniformly bounded, i. e., there is $M > 0$ such that

$$|T^{\ell n}(z)| \leq M, \quad \forall z \in U, \ \forall n \geq 0.$$

As usual, we may assume that $w = 0$. Consider the functions $\widehat{H}_n : U \to \mathbb{C}$ defined by

$$\widehat{H}_n(z) := \frac{1}{n}\sum_{j=0}^{n-1} \lambda^{-j} T^{\ell j}(z).$$

Then $|\widehat{H}_n(z)| \leq M$ for all $z \in U$ and all $n \in \mathbb{N}$. Therefore, the sequence $(\widehat{H}_n)_{n=1}^{\infty}$ is normal on U. Let

$$\widehat{H} := \lim_{k\to\infty} \frac{1}{n_k}\sum_{j=0}^{n_k-1} \lambda^{-j} T^{\ell j}$$

be one of its limit functions. Then $\widehat{H} : U \to \mathbb{C}$ is analytic,

$$\widehat{H}(0) = \lim_{k\to\infty} \frac{1}{n_k}\sum_{j=0}^{n_k-1} \lambda^{-j} T^{\ell j}(0) = \lim_{k\to\infty} \frac{1}{n_k}\sum_{j=0}^{n_k-1} \lambda^{-j}\cdot 0 = 0, \tag{24.21}$$

$$\widehat{H}'(0) = \lim_{k\to\infty} \frac{1}{n_k}\sum_{j=0}^{n_k-1} \lambda^{-j} (T^{\ell j})'(0) = \lim_{k\to\infty} \frac{1}{n_k}\sum_{j=0}^{n_k-1} \lambda^{-j}\lambda^{j} = 1, \tag{24.22}$$

and

$$\widehat{H}\circ T^{\ell}(z) = \lim_{k\to\infty} \frac{1}{n_k}\sum_{j=0}^{n_k-1} \lambda^{-j} T^{\ell j}(T^{\ell}(z)) = \lambda \lim_{k\to\infty} \frac{1}{n_k}\sum_{j=0}^{n_k-1} \lambda^{-(j+1)} T^{\ell(j+1)}(z) = \lambda\widehat{H}(z). \tag{24.23}$$

Relations (24.21)–(24.23) directly imply that the holomorphic function $H := \widehat{H}^{-1}$ satisfies conditions (a)–(c) defining 0 as a Siegel point. □

We now turn our attention to the other class of irrationally indifferent periodic points. Irrationally indifferent periodic points that are not Siegel points are called Cremer points. Obviously, they are characterized by their belonging to the Julia set.

Definition 24.1.27. An irrationally indifferent periodic point of a rational function $T : \widehat{\mathbb{C}} \to \widehat{\mathbb{C}}$ is called a *Cremer point* of T if it lies in the Julia set $\mathcal{J}(T)$.

Per Theorem 24.1.26, if ξ is a Cremer point of T with prime period $\ell \in \mathbb{N}$, then in the vicinity of ξ the map T^{ℓ} is not topologically conjugate to the rotation by the multiplier of ξ. From all of the above, we have the following.

Theorem 24.1.28. *All attracting and Siegel periodic points of a rational function $T : \widehat{\mathbb{C}} \to \widehat{\mathbb{C}}$ are in the Fatou set of T, while all repelling, rationally indifferent, and Cremer periodic points are in the Julia set of T.*

Carl Ludwig Siegel [121] proved the existence of Siegel points, which is also of profound importance for the KAM theory; the proof can be found in [21]. The study of Siegel points, particularly the nature of the rotational numbers for which they exist, has been the topic of extensive research.

About the existence of Cremer points, we shall prove the following fact.

Theorem 24.1.29. *There exists $\theta \in (0,1)$ such that 0 is a Cremer fixed point for every polynomial of the form $\widehat{\mathbb{C}} \ni z \longmapsto e^{2\pi i\theta} z(1+Q(z))$, where Q is a nonconstant polynomial.*

Proof. The statement is equivalent to asserting that there exists $\lambda = e^{2\pi i\theta}$ so that the Schröder equation has no solution for any polynomial P of the form $P(z) = a_d z^d + \cdots + \lambda z$. Suppose for a contradiction that there is a conjugation H defined on $B(0,\delta)$ for some $\delta > 0$. Consider the d^n fixed points of P^n, i. e., the roots of

$$P^n(z) - z = z^{d^n} + \cdots + (\lambda^n - 1)z = 0.$$

One root is 0. Label the others $z_1, \ldots, z_{d^n-1}$. As $P^n(z) = H(\lambda^n H^{-1}(z))$ has only one zero in $B(0,\delta)$, we know that $z_j \notin B(0,\delta)$ for all $1 \le j \le d^n - 1$. Thus,

$$\delta^{d^n} \le \prod_{j=1}^{d^n-1} |z_j| = |1 - \lambda^n|.$$

We now construct a λ for which this is impossible, hence contradicting the existence of H. Take any strictly increasing sequence $(q_k)_{k=1}^{\infty}$ of positive integers. Set $\theta = \sum_{k=1}^{\infty} 2^{-q_k}$ and $\lambda = e^{2\pi i\theta}$. Then there is a constant $\Delta \ge 1$ such that

$$|1 - \lambda^{2^{q_k}}| \le \Delta \cdot 2^{q_k - q_{k+1}}, \quad \forall k \in \mathbb{N}.$$

After using the two relations above and taking logarithms, it can be proved by induction that

$$q_{k+1} \le C(\delta) d^{2^{q_k}}, \quad \forall k \in \mathbb{N},$$

where $C(\delta) = \max\{q_2/d^{2^{q_1}}, \frac{\log(\Delta/\delta)}{\log 2} \frac{d}{d-1}\}$. If the q_k's are defined inductively to grow very rapidly, e. g. if $\log q_{k+1} \ge k 2^{q_k}$, this inequality is violated for any d and $\delta > 0$. □

An example as above was first given by G. A. Pfeiffer [104] in 1917. His work was continued by Hubert Cremer [30], who proved in 1938 that if λ is not a root of unity but $|\lambda| = 1$ and

$$\liminf_{n\to\infty} |\lambda^n - 1|^{1/n} = 0,$$

then there exists an entire function of the form $z \longmapsto \lambda z(1 + Q(z))$, where Q is a nonconstant entire function. In a seminal paper in 1942, Siegel [121] gave the first example of a θ for which Schröder's equation is solvable.

24.2 Rationally indifferent periodic points

In the previous section, we observed that all rationally indifferent periodic points of a rational function lie in the Julia set of that function. We now look at the dynamics around such points not only for rational functions but more generally for holomorphic functions.

24.2.1 Local and asymptotic behavior of holomorphic functions around rationally indifferent periodic points: part I

In this section, we present basic results about the qualitative and especially the quantitative asymptotic behaviors of holomorphic functions around their parabolic (i. e., rationally indifferent) periodic points.

Definition 24.2.1. A holomorphic map φ defined on a neighborhood of a point $\omega \in \mathbb{C}$ is said to be a *locally holomorphic parabolic map* at ω (or *rationally indifferent* at ω) if:
(i) ω is a periodic point of φ, i. e., $\varphi^{\ell}(\omega) = \omega$ for some $\ell \in \mathbb{N}$,
(ii) its multiplier $(\varphi^{\ell})'(\omega)$ is a root of unity, and
(iii) no iterate of φ is the identity map.

The periodic point ω itself is then called parabolic (or rationally indifferent).

Definition 24.2.2. A locally holomorphic parabolic map φ (or the point ω) is called simple parabolic (resp., simple parabolic fixed point) if
(a) $\varphi(\omega) = \omega$,
(b) $\varphi'(\omega) = 1$, and
(c) φ is not the identity map.

Note that some sufficiently high iterate of any locally holomorphic parabolic map is simple. Therefore, the behavior of locally holomorphic parabolic maps can essentially be analyzed by studying simple parabolic ones. Accordingly, throughout this section the map φ (and its point ω) is always assumed to be locally holomorphic simple parabolic.

Within the realm of ideas related to Fatou's flower theorem (see [1, 8, 21, 94, 123] for historical information), we aim to analyze the behavior of φ in a sufficiently small neighborhood of the simple parabolic fixed point ω.

On a sufficiently small neighborhood of ω, the map φ has a Taylor series expansion around ω of the form

$$\varphi(z) = z + a(z - \omega)^{p+1} + O\big((z - \omega)^{p+2}\big) \tag{24.24}$$

for some $p = p(\omega) \in \mathbb{N}$ and $a = a(\omega) \in \mathbb{C} \setminus \{0\}$.

The rays coming out of ω and forming the set

$$\{z \in \mathbb{C} : a(z-\omega)^p < 0\}$$

are called *attracting directions* while the rays forming the set

$$\{z \in \mathbb{C} : a(z-\omega)^p > 0\}$$

are called *repelling directions*. See Figures 24.1–24.2.

Indeed, note that

$$\varphi'(z) = 1 + (p+1)a(z-\omega)^p + O((z-\omega)^{p+1}),$$

and hence $|\varphi'(z)| < 1$ for all z close enough to ω that belong to an attracting direction while $|\varphi'(z)| > 1$ for all z close enough to ω that lie along a repelling direction.

Fix an attracting direction, say

$$A := \omega + \sqrt[p]{-a^{-1}}(0,\infty), \tag{24.25}$$

where $\sqrt[p]{\cdot}$ is a holomorphic branch of the pth radical function defined on $\mathbb{C} \setminus a^{-1}(0,\infty)$. In order to simplify our analysis, let us change the coordinate system via a conjugation through the affine map

$$\rho_A(z) = \omega + \sqrt[p]{-(ap)^{-1}} \cdot z.$$

That is, we define the function

$$\varphi_{A,0} := \rho_A^{-1} \circ \varphi \circ \rho_A$$

whose Taylor series expansion around 0 is of the form

$$\varphi_{A,0}(z) = z - \frac{z^{p+1}}{p} + \sum_{n=2}^{\infty} a_n z^{p+n} \tag{24.26}$$

for some $a_n \in \mathbb{C}$, $n \geq 2$. So

$$\varphi_{A,0}(0) = 0 \quad \text{and} \quad (\varphi_{A,0})'(0) = 1,$$

i. e., 0 is a simple parabolic fixed point of $\varphi_{A,0}$. In addition, $\rho_A^{-1}(A) = (0,\infty)$ is an attracting direction for $\varphi_{A,0}$, as in Figures 24.1–24.2.

We want to analyze the behavior of $\varphi_{A,0}$ on sufficiently small neighborhoods of 0. In order to do this, similar to the previous section, we conjugate $\varphi_{A,0}$ on $\mathbb{C} \setminus (-\infty, 0]$ to a map defined "near" infinity. More precisely, we consider $\sqrt[p]{\cdot}$, the holomorphic branch of

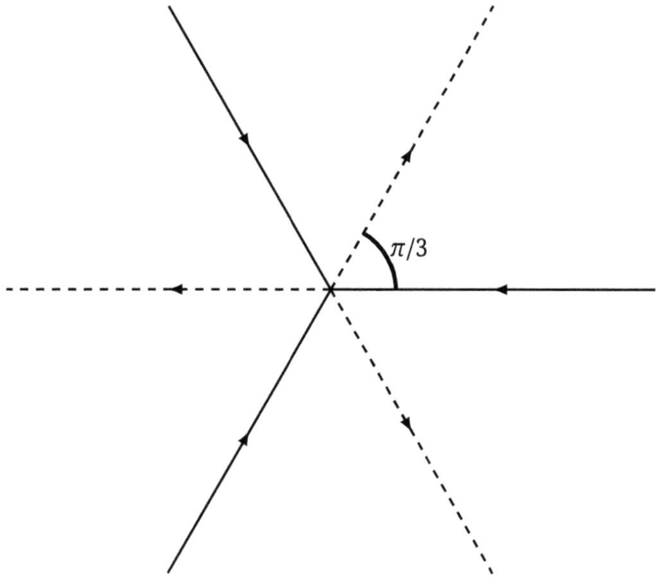

Figure 24.1: Case $p = 3$ (with $a = -1$). There are 3 attracting directions (indicated by rays and vectors pointing toward the parabolic point) and 3 repelling directions (indicated by dashed rays and vectors pointing away from the parabolic point).

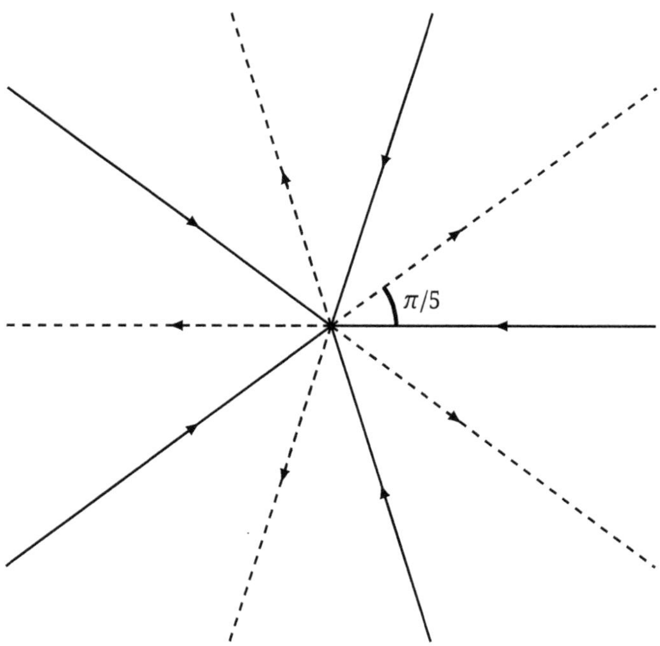

Figure 24.2: Case $p = 5$ (with $a = -1$). There are 5 attracting directions (indicated by rays and vectors pointing toward the parabolic point) and 5 repelling directions (indicated by dashed rays and vectors pointing away from the parabolic point).

the pth-root function defined on $\mathbb{C} \setminus (-\infty, 0]$ which leaves the point 1 fixed. We will also frequently denote this branch by $z^{1/p}$. Then we define the map

$$H(z) = \frac{1}{\sqrt[p]{z}} = z^{-1/p}.$$

This map has a meromorphic inverse $H^{-1} : \widehat{\mathbb{C}} \to \widehat{\mathbb{C}}$ defined by the formula

$$H^{-1}(z) = z^{-p}.$$

Consider the conjugate map

$$\widetilde{\varphi}_A = H^{-1} \circ \varphi_{A,0} \circ H = (\rho_A \circ H)^{-1} \circ \varphi \circ (\rho_A \circ H), \tag{24.27}$$

defined on $D(\widetilde{\varphi}_A) := (\mathbb{C} \setminus (-\infty, 0]) \cap H^{-1}(V_\varphi)$, where $V_\varphi \subseteq \mathbb{C}$ is a sufficiently small neighborhood of 0 on which $\varphi_{A,0}$ is defined.

24.2.1.1 Properties of $\widetilde{\varphi}_A$

We now study the function $\widetilde{\varphi}_A$, its iterates, their derivatives and the limit of those. We shall first prove the following technical but useful result about $\widetilde{\varphi}_A$ and its derivative.

Lemma 24.2.3. *If φ is a locally holomorphic simple parabolic map and A is an attracting direction of φ, then (shrinking the neighborhood V_φ if necessary) there exists a holomorphic function $B : V_\varphi \to \mathbb{C}$ such that*

$$\widetilde{\varphi}_A(z) = z + 1 + B(H(z)) \quad \text{with} \quad |B(H(z))| \le M|H(z)| = M|z|^{-\frac{1}{p}}, \quad \forall z \in D(\widetilde{\varphi}_A),$$

for some constant $M > 0$, where $D(\widetilde{\varphi}_A) := (\mathbb{C} \setminus (-\infty, 0]) \cap H^{-1}(V_\varphi)$.

Reducing V_φ if needed, this yields that

$$\widetilde{\varphi}_A'(z) = 1 + (B \circ H)'(z) \quad \text{with} \quad |(B \circ H)'(z)| \le Mp^{-1}|z|^{-\frac{p+1}{p}} < 1, \quad \forall z \in D(\widetilde{\varphi}_A).$$

In particular, this implies that $\widetilde{\varphi}_A$ is univalent and

$$\widetilde{\varphi}_A(z) \approx z + 1 \quad \text{and} \quad \widetilde{\varphi}_A'(z) \approx 1 \quad \text{near } \infty,$$

where $f \approx g$ near z_0 means that $\lim_{z \to z_0} |f(z) - g(z)| = 0$.

Proof. For all $z \in D(\widetilde{\varphi}_A)$,

$$\begin{aligned}
\widetilde{\varphi}_A(z) &= H^{-1}(\varphi_{A,0}(H(z))) = H^{-1}\left(H(z) - \frac{(H(z))^{p+1}}{p} + \sum_{n=2}^{\infty} a_n (H(z))^{p+n}\right) \\
&= H^{-1}\left(z^{-\frac{1}{p}} - \frac{z^{-\frac{p+1}{p}}}{p} + \sum_{n=2}^{\infty} a_n z^{-\frac{p+n}{p}}\right)
\end{aligned}$$

$$= H^{-1}\left(z^{-\frac{1}{p}}\left[1 - \frac{z^{-1}}{p} + \sum_{n=2}^{\infty} a_n z^{-\frac{p+n-1}{p}}\right]\right)$$

$$= z\left[1 - \frac{z^{-1}}{p} + \sum_{n=2}^{\infty} a_n z^{-\frac{p+n-1}{p}}\right]^{-p}. \tag{24.28}$$

Set $\zeta = H(z) = z^{-\frac{1}{p}}$ and

$$G(\zeta) := 1 - \frac{\zeta^p}{p} + \sum_{n=2}^{\infty} a_n \zeta^{p+n-1}, \quad \forall \zeta \in V_\varphi. \tag{24.29}$$

Clearly,

$$G(0) = 1, \quad \left.\frac{d^k G}{d\zeta^k}\right|_0 = 0, \quad \forall 0 \le k < p, \quad \text{and} \quad \left.\frac{d^p G}{d\zeta^p}\right|_0 = -(p-1)!$$

Therefore, the power function G^{-p} satisfies

$$G^{-p}(0) = 1, \quad \left.\frac{d^k (G^{-p})}{d\zeta^k}\right|_0 = 0, \quad \forall 0 \le k < p, \quad \text{and} \quad \left.\frac{d^p (G^{-p})}{d\zeta^p}\right|_0 = p!$$

Thus, the Taylor series expansion of G^{-p} about 0 is

$$G^{-p}(\zeta) = 1 + \zeta^p + \sum_{n=1}^{\infty} b_n \zeta^{p+n} \tag{24.30}$$

for some appropriate coefficients b_n, $n \in \mathbb{N}$, where the tail of the series converges absolutely uniformly on some sufficiently small neighborhood of 0; shrinking V_φ if needed, we may identify this neighborhood with V_φ. Then the associated series

$$B(\zeta) := \sum_{n=1}^{\infty} b_n \zeta^n$$

converges absolutely uniformly on V_φ, and hence constitutes an holomorphic function. Shrinking V_φ even more if necessary, there is a constant $M \in (0, \infty)$ such that

$$|B(\zeta)| \le M|\zeta| \quad \text{and} \quad |B'(\zeta)| \le M, \quad \forall \zeta \in V_\varphi. \tag{24.31}$$

Reverting back to the variable $z = \zeta^{-p} \in D(\tilde{\varphi}_A) := (\mathbb{C} \setminus (-\infty, 0]) \cap H^{-1}(V_\varphi)$, we get from (24.28)–(24.30) that

$$\tilde{\varphi}_A(z) = zG^{-p}(H(z)) = z\left[1 + (H(z))^p + \sum_{n=1}^{\infty} b_n (H(z))^{p+n}\right]$$

$$= z\left[1 + z^{-1} + z^{-1}\sum_{n=1}^{\infty} b_n (H(z))^n\right] = z + 1 + B(H(z)).$$

It also follows from (24.31) that

$$|B(H(z))| \le M|H(z)| = M|z|^{-\frac{1}{p}} \quad \text{and} \quad |(B \circ H)'(z)| = |B'(H(z))| \cdot |H'(z)| \le Mp^{-1}|z|^{-\frac{p+1}{p}}.$$

The expression of $\widetilde{\varphi}_A'$ simply ensues from differentiation of $\widetilde{\varphi}_A$. □

Given a point $x \in (0,\infty)$ and an angle $\alpha \in (0,\pi)$, let

$$S(x,\alpha) := \{z \in \mathbb{C} \setminus \{x\} \mid -\alpha < \arg(z-x) < \alpha\}$$

be the open sector with vertex x, central angle α and symmetric about the real axis. By Lemma 24.2.3, for all $\alpha \in (0,\pi)$ and all $\kappa \in (0,1)$ there is $x(\alpha,\kappa) \in (0,\infty)$ such that

$$\widetilde{\varphi}_A(S(x,\alpha)) \subseteq S(x+1-\kappa,\alpha) \subseteq S(x,\alpha), \quad \forall x \ge x(\alpha,\kappa), \tag{24.32}$$

and

$$\mathrm{Re}(\widetilde{\varphi}_A(z)) \ge \mathrm{Re}(z) + 1 - \kappa, \quad \forall z \in S(x(\alpha,\kappa),\alpha). \tag{24.33}$$

By induction, we obtain the following two statements. They affirm that every sector $S(x,\alpha)$ is forward $\widetilde{\varphi}_A$-invariant and describe how the iterates of $\widetilde{\varphi}_A$ map each such sector into itself.

Lemma 24.2.4. *Let φ be a locally holomorphic simple parabolic map and let A be an attracting direction for φ. Fix $\alpha \in (0,\pi)$ and $\kappa \in (0,1)$. Then there is $x(\alpha,\kappa) \in (0,\infty)$ such that all the iterates $\widetilde{\varphi}_A^n : S(x(\alpha,\kappa),\alpha) \to \mathbb{C}$ are well-defined and satisfy*

$$\widetilde{\varphi}_A^n(S(x,\alpha)) \subseteq S(x+n(1-\kappa),\alpha) \subseteq S(x,\alpha), \quad \forall x \ge x(\alpha,\kappa), \quad \forall n \in \mathbb{N}.$$

In particular, $\widetilde{\varphi}_A^n \to \infty$ uniformly on $S(x(\alpha,\kappa),\alpha)$.

Lemma 24.2.5. *If φ is a locally holomorphic simple parabolic map and A is an attracting direction for φ, then for all $\alpha \in (0,\pi)$ and $\kappa \in (0,1)$ we have*

$$|\widetilde{\varphi}_A^n(z)| \ge \mathrm{Re}(\widetilde{\varphi}_A^n(z)) \ge \mathrm{Re}(z) + n(1-\kappa), \quad \forall z \in S(x(\alpha,\kappa),\alpha), \quad \forall n \in \mathbb{N},$$

where $x(\alpha,\kappa)$ arises from (24.32)–(24.33).

A consequence of the previous results and yet another description of the behavior of the iterates of $\widetilde{\varphi}_A$ on the sector $S(x(\alpha,\kappa),\alpha)$, is the following.

Lemma 24.2.6. *If φ is a locally holomorphic simple parabolic map and A is an attracting direction for φ, then for all $\alpha \in (0,\pi)$, $\kappa \in (0,1)$, $z \in S(x(\alpha,\kappa),\alpha)$ and $n \in \mathbb{N}$, we have*

$$\widetilde{\varphi}_A^n(z) = z + n + O_{\mathrm{Re}(z)}(\max\{n^{1-\frac{1}{p}}, \log n\}),$$

where O_t *is a big* O *symbol that depends decreasingly on* $t \in \mathbb{R}$ *and that converges to* 0 *when* $t \to \infty$.

Proof. Fix $z \in S(x(\alpha,\kappa),\alpha)$. By Lemmas 24.2.3–24.2.5, for every $n \in \mathbb{N}$ we have

$$\tilde{\varphi}_A^{n+1}(z) = \tilde{\varphi}_A^n(z) + 1 + O_{\mathrm{Re}(z)}(n^{-\frac{1}{p}}).$$

By induction,

$$\tilde{\varphi}_A^n(z) = z + n + O_{\mathrm{Re}(z)}\left(\sum_{k=1}^{n-1} k^{-\frac{1}{p}}\right) = z + n + O_{\mathrm{Re}(z)}(\max\{n^{1-\frac{1}{p}}, \log n\}). \qquad \square$$

As an immediate ramification of this lemma, we get the following bounds on the distance between the iterates of a point in the sector $S(x(\alpha,\kappa),\alpha)$ and the point itself.

Lemma 24.2.7. *If* φ *is a locally holomorphic simple parabolic map and* A *is an attracting direction for* φ*, then for all* $\alpha \in (0,\pi)$, $\kappa \in (0,1)$ *and* $t \in \mathbb{R}$*, there exists a constant* $C = C(\alpha,\kappa,t)$ *such that for all* $n \in \mathbb{N}$ *and all* $z \in S(x(\alpha,\kappa),\alpha)$ *with* $\mathrm{Re}(z) \ge t$*, we have*

$$C^{-1} n \le |\tilde{\varphi}_A^n(z) - z| \le Cn.$$

For every $x \in (0,\infty)$, $\alpha \in (0,\pi)$ and $R > 0$, let

$$S(x,\alpha,R) := S(x,\alpha) \cap B(0,R).$$

We now establish properties of the derivatives of the iterates of $\tilde{\varphi}_A$.

Lemma 24.2.8. *Let* φ *be a locally holomorphic simple parabolic map and let* A *be an attracting direction for* φ*. Fix* $\alpha \in (0,\pi)$, $\kappa \in (0,1)$ *and* $R > 0$*. Then*

$$\begin{aligned} 0 &< \inf\{|(\tilde{\varphi}_A^n)'(z)| : z \in S(x(\alpha,\kappa),\alpha,R),\ n \in \mathbb{N}\} \\ &\le \sup\{|(\tilde{\varphi}_A^n)'(z)| : z \in S(x(\alpha,\kappa),\alpha,R),\ n \in \mathbb{N}\} < \infty. \end{aligned}$$

Furthermore, for every $\gamma > 1$,

$$\lim_{x\to\infty} \sup\{|(\tilde{\varphi}_A^n)'(z) - 1| : z \in S(x,\alpha,\gamma x),\ n \in \mathbb{N}\} = 0$$

and

$$\lim_{x\to\infty} \sup\{|(\tilde{\varphi}_A^n)'(z) - 1| : z \in S(x,\pi/2),\ n \in \mathbb{N}\} = 0.$$

Proof. For every $z \in S(x(\alpha,\kappa),\alpha)$, let

$$g(z) = \tilde{\varphi}_A'(z) - 1.$$

By the chain rule, for every $n \in \mathbb{N}$ we have

$$(\tilde{\varphi}_A^n)'(z) = \prod_{j=0}^{n-1} \tilde{\varphi}_A'(\tilde{\varphi}_A^j(z)) = \prod_{j=0}^{n-1} [1 + g(\tilde{\varphi}_A^j(z))]. \tag{24.34}$$

According to Exercise 24.8.6 (with $a_n = 1 - |g(\tilde{\varphi}_A^n(z))|$ and $b_n = 1 + |g(\tilde{\varphi}_A^n(z))|$) and Lemma 24.2.3, it suffices to show that the series $\sum_{j=0}^{\infty} |g(\tilde{\varphi}_A^j(z))|$ has a finite sum which is independent of $z \in S(x(\alpha,\kappa),\alpha,R)$.

By Lemmas 24.2.3–24.2.4, we know that

$$|g(\tilde{\varphi}_A^j(z))| \le Mp^{-1} |\tilde{\varphi}_A^j(z)|^{-\frac{p+1}{p}}.$$

For every $x > 0$, let $r = r(x,\alpha)$ be the radius of the largest ball centered at 0 which is disjoint from $S(x,\alpha)$. One can show that

$$r(x,\alpha) = ux$$

for some constant $u \in (0,1]$ which depends solely on α (in fact, $u = 1$ when $0 < \alpha \le \pi/2$ and $u = \sin\alpha$ when $\pi/2 \le \alpha < \pi$). Let $k_R \ge 0$ be the least integer such that

$$-R + (k_R + 1)(1 - \kappa) \ge R. \tag{24.35}$$

Then $-R + k_R(1-\kappa) < R$ and, therefore,

$$k_R < \frac{2R}{1-\kappa}.$$

Fix an arbitrary $x \ge x(\alpha,\kappa)$. By Lemma 24.2.4, for all $j \ge 0$ and all $z \in S(x,\alpha)$, we have

$$|\tilde{\varphi}_A^j(z)| \ge r(x,\alpha) = ux.$$

In particular, this holds for all $0 \le j \le k_R$. By virtue of Lemma 24.2.5, for all $j > k_R$ and all $z \in S(x(\alpha,\kappa),\alpha,R)$ we get

$$|\tilde{\varphi}_A^j(z)| \ge R + (j - (k_R + 1))(1-\kappa).$$

Therefore,

$$\begin{aligned}
\sum_{j=0}^{\infty} |g(\tilde{\varphi}_A^j(z))| &= \sum_{j=0}^{k_R} |g(\tilde{\varphi}_A^j(z))| + \sum_{j=k_R+1}^{\infty} |g(\tilde{\varphi}_A^j(z))| \\
&\le Mp^{-1} \left[\sum_{j=0}^{k_R} |\tilde{\varphi}_A^j(z)|^{-\frac{p+1}{p}} + \sum_{j=k_R+1}^{\infty} |\tilde{\varphi}_A^j(z)|^{-\frac{p+1}{p}} \right]
\end{aligned}$$

$$
\begin{aligned}
&\le Mp^{-1}\left[(ux)^{-\frac{p+1}{p}}\left(\frac{2R}{1-\kappa}+1\right)+\sum_{j=k_R+1}^{\infty}[R+(j-(k_R+1))(1-\kappa)]^{-\frac{p+1}{p}}\right]\\
&= Mp^{-1}\left[(ux)^{-\frac{p+1}{p}}\left(\frac{2R}{1-\kappa}+1\right)+\sum_{j=0}^{\infty}[R+j(1-\kappa)]^{-\frac{p+1}{p}}\right]\\
&\le Mp^{-1}\left[\left(\frac{2R}{1-\kappa}+1\right)(ux)^{-\frac{p+1}{p}}+\int_0^{\infty}[R+t(1-\kappa)]^{-\frac{p+1}{p}}\,dt\right]\\
&= Mp^{-1}\left[\left(\frac{2R}{1-\kappa}+1\right)(ux)^{-\frac{p+1}{p}}+\frac{1}{1-\kappa}\int_R^{\infty}s^{-\frac{p+1}{p}}\,ds\right]\\
&= Mp^{-1}\left[\left(\frac{2R}{1-\kappa}+1\right)(ux)^{-\frac{p+1}{p}}+\frac{p}{1-\kappa}R^{-\frac{1}{p}}\right].
\end{aligned}
$$

This upper estimate proves the first assertion in the lemma.

For the second assertion, setting $R := \gamma x$ in the above estimate, we get

$$
\sum_{j=0}^{\infty}|g(\tilde{\varphi}_A^j(z))| \le Mp^{-1}\left[\left(\frac{2\gamma}{1-\kappa}+x^{-1}\right)u^{-\frac{p+1}{p}}+\frac{p\gamma^{-\frac{1}{p}}}{1-\kappa}\right]x^{-\frac{1}{p}} \to 0 \quad \text{as} \quad x\to\infty.
$$

Taking well-defined branches around 1 of the logarithms of both sides of (24.34) and using the fact that $|\log(1+z)| = O(|z|)$, the second assertion ensues from the above convergence to 0.

For the third assertion, if $z \in S(x,\pi/2)$ then $\mathrm{Re}(z) \ge x$. By Lemma 24.2.5, we obtain (in a similar way as in the first part)

$$
\begin{aligned}
\sum_{j=0}^{\infty}|g(\tilde{\varphi}_A^j(z))| &\le Mp^{-1}\sum_{j=0}^{\infty}|\tilde{\varphi}_A^j(z)|^{-\frac{p+1}{p}}\\
&\le Mp^{-1}\sum_{j=0}^{\infty}(x+j(1-\kappa))^{-\frac{p+1}{p}}\\
&\le Mp^{-1}\left(x^{-1}+\frac{p}{1-\kappa}\right)x^{-1/p} \to 0 \quad \text{as} \quad x\to\infty.
\end{aligned}
$$

The third assertion is thus proved. □

Finally, we examine the limit of the derivatives of the iterates of $\tilde{\varphi}_A$.

Lemma 24.2.9. *Let φ be a locally holomorphic simple parabolic map and let A be an attracting direction for φ. Fix $\alpha \in (0,\pi)$ and $\kappa \in (0,1)$. The following statements hold:*

(a) *For any $z \in S(x(\alpha,\kappa),\alpha)$, the limit*

$$
\tilde{\varphi}'_{A,\infty}(z) := \lim_{n\to\infty}(\tilde{\varphi}_A^n)'(z)
$$

exists and belongs to $\mathbb{C}\setminus\{0\}$.

(b) *The convergence is uniform on every set* $S(x(\alpha,\kappa),\alpha)\cap\{\zeta\in\mathbb{C}:\mathrm{Re}(\zeta)\geq t\}$, $t\in\mathbb{R}$, *and in particular on every set* $S(x(\alpha,\kappa),\alpha,R)$, $R>0$.

(c) *The function* $S(x(\alpha,\kappa),\alpha)\ni z\longmapsto\tilde{\varphi}'_{A,\infty}(z)\in\mathbb{C}\setminus\{0\}$ *is holomorphic.*

(d)

$$\lim_{\substack{z\in S(x(\alpha,\kappa),\alpha)\\ \mathrm{Re}(z)\to\infty}}\tilde{\varphi}'_{A,\infty}(z)=1.$$

(e) *In particular, if* $\alpha\in(0,\pi/2)$ *then*

$$\lim_{S(x(\alpha,\kappa),\alpha)\ni z\to\infty}\tilde{\varphi}'_{A,\infty}(z)=1.$$

(f)

$$\tilde{\varphi}'_{A,\infty}(z)=\tilde{\varphi}'_{A,\infty}(\tilde{\varphi}_A(z))\tilde{\varphi}'_A(z),\quad\forall z\in S(x(\alpha,\kappa),\alpha).\tag{24.36}$$

(g) *The holomorphic function* $\tilde{\varphi}'_{A,\infty}$ *and its modulus are not constant.*

Proof. Let $\varepsilon>0$ and $t\in\mathbb{R}$. By Lemma 24.2.8, there exists $x>0$ such that

$$\left|(\tilde{\varphi}^n_A)'(\xi)-1\right|<\varepsilon,\quad\forall\xi\in S(x,\pi/2),\ \forall n\in\mathbb{N}.\tag{24.37}$$

By Lemma 24.2.5, there exists $N\in\mathbb{N}$ so large that $\mathrm{Re}(\tilde{\varphi}^n_A(z))\geq x$, i. e.,

$$\tilde{\varphi}^n_A(z)\in S(x,\pi/2),\quad\forall z\in S(x(\alpha,\kappa),\alpha)\cap\{\zeta\in\mathbb{C}:\mathrm{Re}(\zeta)\geq t\},\ \forall n\geq N.$$

Then for every $i\geq N$ and every $j\geq0$ we have

$$\left|\frac{(\tilde{\varphi}^{i+j}_A)'(z)}{(\tilde{\varphi}^i_A)'(z)}-1\right|=\left|(\tilde{\varphi}^j_A)'(\tilde{\varphi}^i_A(z))-1\right|<\varepsilon,\quad\forall z\in S(x(\alpha,\kappa),\alpha)\cap\{\zeta\in\mathbb{C}:\mathrm{Re}(\zeta)\geq t\}.$$

This means that the sequence $((\tilde{\varphi}^n_A)'(z))_{n=0}^\infty$ is uniformly quotient Cauchy on $S(x(\alpha,\kappa),\alpha)\cap\{\zeta\in\mathbb{C}:\mathrm{Re}(\zeta)\geq t\}$ and statements (a) and (b) hold. Assertion (c) follows immediately from (b), while (d) is a direct consequence of (24.37), and (e) is an immediate implication of (d). Statement (f) follows from a straightforward calculation using the definitions of $\tilde{\varphi}'_{A,\infty}(z)$ and $\tilde{\varphi}'_{A,\infty}(\tilde{\varphi}_A(z))$. Finally, (g) clearly ensues from (f) and the openness of holomorphic maps. □

24.2.1.2 Properties of φ

We now return to our original and main goal, namely, the examination of the function φ and its iterates, their derivatives and the limit of those.

For every $x\in(0,\infty)$, $\alpha\in(0,\pi)$ and $R>0$, let

$$S_0(x,\alpha):=H(S(x,\alpha)),\tag{24.38}$$

$$S_{\varphi}^{A}(x,\alpha) := \rho_A(S_0(x,\alpha)) = \rho_A \circ H(S(x,\alpha)) \tag{24.39}$$

$$S_0(x,\alpha,R) := S_0(x,\alpha) \setminus \overline{B(0,R)} = H(S(x,\alpha,R^{-p})) \tag{24.40}$$

$$S_{\varphi}^{A}(x,\alpha,R) := S_{\varphi}^{A}(x,\alpha) \setminus \overline{B(\omega,R)} = \rho_A \circ H(S(x,\alpha,(|a|p)^{-1}R^{-p})). \tag{24.41}$$

The regions $S_0(x,\alpha)$ and $S_{\varphi}^{A}(x,\alpha)$ look like flower petals (see Figures 24.3–24.4) that are

(1) symmetric about the rays $(0,\infty)$ and $A = \omega + \sqrt[p]{-a^{-1}}(0,\infty)$, respectively,

(2) contain initial segments of these rays, and

(3) whose tangents at the respective points 0 and ω form a central angle of measure α/p.

Indeed, observe that

$$\lim_{\mathbb{R}\ni t\to\infty} \arg H(x+te^{\pm i\alpha}) = \lim_{\mathbb{R}\ni t\to\infty} \arg\frac{1}{\sqrt[p]{x+te^{\pm i\alpha}}} = \lim_{t\to\infty} \arg\frac{1}{e^{\pm i\frac{\alpha}{p}}\sqrt[p]{1+\frac{x}{te^{\pm i\alpha}}}} = \mp\frac{\alpha}{p}.$$

The following result is the counterpart of Lemma 24.2.4 for the function φ. It asserts that every flower petal $S_{\varphi}^{A}(x,\alpha)$ is forward φ-invariant and gives a description of the behavior of the iterates of φ on such a petal.

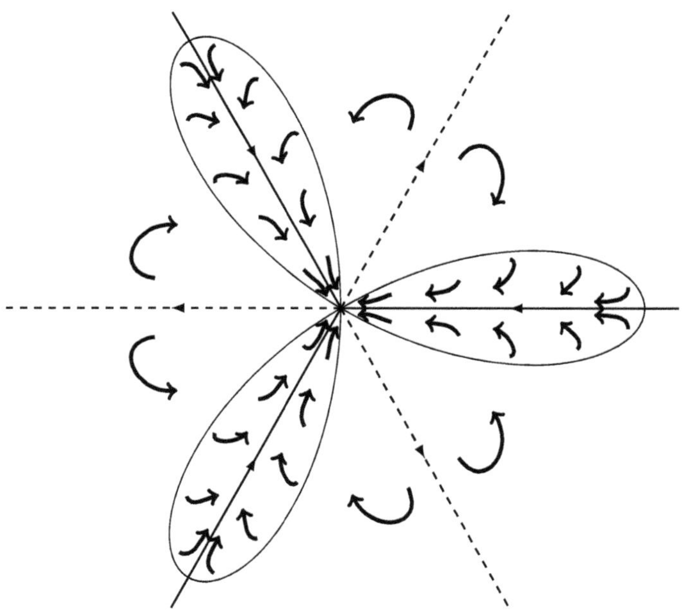

Figure 24.3: Three petals, case $p = 3$ (with $a = -1$). Some dynamics are depicted by arrows. There are 3 attracting directions (indicated by rays and vectors pointing toward the parabolic point) with respect to which the petals are symmetric and 3 repelling directions (indicated by dashed rays and vectors pointing away from the parabolic point).

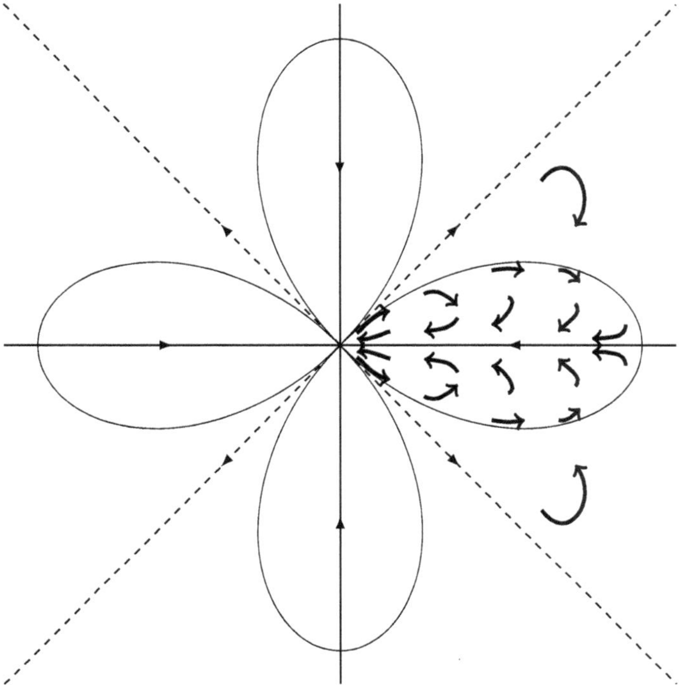

Figure 24.4: Four petals, case $p = 4$ (with $a = -1$). There are 4 attracting directions (indicated by rays and vectors pointing toward the parabolic point) with respect to which the petals are symmetric and 4 repelling directions (indicated by dashed rays and vectors pointing away from the parabolic point). Some dynamics in and near one of the petals are depicted by the arrows. Reproduce them in and near the other three petals.

Lemma 24.2.10. *Let φ be a locally holomorphic simple parabolic map at $\omega \in \mathbb{C}$ and let A be an attracting direction for φ. Fix $\alpha \in (0,\pi)$ and $\kappa \in (0,1)$. Then all the iterates $\varphi^n : S_\varphi^A(x(\alpha,\kappa),\alpha) \to \mathbb{C}$ are well-defined and satisfy*

$$\varphi^n(S_\varphi^A(x,\alpha)) \subseteq S_\varphi^A(x+n(1-\kappa),\alpha) \subseteq S_\varphi^A(x,\alpha), \quad \forall x \geq x(\alpha,\kappa), \quad \forall n \in \mathbb{N}.$$

In particular, the sequence $(\varphi^n)_{n=1}^\infty$ converges to ω uniformly on $S_\varphi^A(x(\alpha,\kappa),\alpha)$.

An elementary calculation based on (24.24) yields the next result.

Lemma 24.2.11. *If φ is a locally holomorphic simple parabolic map at $\omega \in \mathbb{C}$ and A is an attracting direction for φ, then for every $0 \leq \alpha < \pi/p$ and $\kappa \in (0,1)$ there exists $R_{\alpha,\kappa} > 0$ such that*

$$|\varphi(z) - \omega| < |z-\omega| \quad \text{and} \quad |\varphi'(z)| < 1, \quad \forall z \in S_\varphi^A(x(\alpha,\kappa),\alpha) \cap B(\omega, R_{\alpha,\kappa}) \setminus \{\omega\}.$$

The next statement specifies the asymptotic decay of the distance that separates successive iterates of points in the petal $S_\varphi^A(x(\alpha,\kappa),\alpha)$, as well as the asymptotic decay of the distance between these iterates and the point ω.

Proposition 24.2.12. *Let φ be a locally holomorphic simple parabolic map at $\omega \in \mathbb{C}$ and let A be an attracting direction for φ. Fix $\alpha \in (0,\pi)$ and $\kappa \in (0,1)$. For all $z \in S_\varphi^A(x(\alpha,\kappa),\alpha)$, we have*

$$\lim_{n\to\infty} n^{\frac{p+1}{p}} |\varphi^{n+1}(z) - \varphi^n(z)| = (|a|p)^{-\frac{1}{p}} p^{-1} \tag{24.42}$$

$$\lim_{n\to\infty} n^{\frac{1}{p}} |\varphi^n(z) - \omega| = (|a|p)^{-\frac{1}{p}}. \tag{24.43}$$

In addition, in both limits, the convergence is uniform on $S_\varphi^A(x(\alpha,\kappa),\alpha,R)$ for every $R > 0$.

Proof. Let $z \in S_\varphi^A(x(\alpha,\kappa),\alpha,R)$. Then $\xi = \rho_A^{-1}(z) \in S_0(x(\alpha,\kappa),\alpha,(|a|p)^{\frac{1}{p}}R)$ and hence $H^{-1}(\xi) \in S(x(\alpha,\kappa),\alpha,(|a|p)^{-1}R^{-p})$. By Lemma 24.2.6, we get that

$$\begin{aligned} |\varphi_{A,0}^n(\xi)| &= |H \circ \tilde{\varphi}_A^n \circ H^{-1}(\xi)| = |\tilde{\varphi}_A^n(H^{-1}(\xi))|^{-\frac{1}{p}} \\ &= |H^{-1}(\xi) + n + \mathrm{O}_{\mathrm{Re}(H^{-1}(\xi))}(\max\{n^{1-\frac{1}{p}}, \log n\})|^{-\frac{1}{p}}. \end{aligned}$$

It ensues that

$$\lim_{n\to\infty} n^{\frac{1}{p}} |\varphi_{A,0}^n(\xi)| = \lim_{n\to\infty} |n^{-1}H^{-1}(\xi) + 1 + \mathrm{O}_{-(|a|p)^{-1}R^{-p}}(\max\{n^{-\frac{1}{p}}, n^{-1}\log n\})|^{-\frac{1}{p}} = 1,$$

and the convergence is uniform on $\xi \in S_0(x(\alpha,\kappa),\alpha,(|a|p)^{\frac{1}{p}}R)$. As

$$\varphi^n(z) - \omega = \rho_A(\varphi_{A,0}^n(\rho_A^{-1}(z))) - \omega = \sqrt[p]{-(ap)^{-1}} \cdot \varphi_{A,0}^n(\xi),$$

relation (24.43) follows along with the corresponding uniform convergence on the set $S_\varphi^A(x(\alpha,\kappa),\alpha,R)$.

Turning our attention to relation (24.42), we have by (24.24) and (24.43) that

$$\begin{aligned} &\lim_{n\to\infty} n^{\frac{p+1}{p}} |\varphi^{n+1}(z) - \varphi^n(z)| \\ &= \lim_{n\to\infty} n^{\frac{p+1}{p}} |a(\varphi^n(z) - \omega)^{p+1} + \mathrm{O}(|\varphi^n(z) - \omega|^{p+2})| \\ &= \lim_{n\to\infty} n^{\frac{p+1}{p}} |a(\varphi^n(z) - \omega)^{p+1}| \cdot |1 + \mathrm{O}(|\varphi^n(z) - \omega|)| \\ &= |a| \lim_{n\to\infty} (n^{\frac{1}{p}} |\varphi^n(z) - \omega|)^{p+1} \cdot \left|1 + \lim_{n\to\infty} n^{-\frac{1}{p}} \cdot \lim_{n\to\infty} \mathrm{O}(n^{\frac{1}{p}} |\varphi^n(z) - \omega|)\right| \\ &= |a|((|a|p)^{-\frac{1}{p}})^{p+1} \cdot |1 + 0 \cdot \mathrm{O}((|a|p)^{-\frac{1}{p}})| \\ &= (|a|p)^{-\frac{1}{p}} p^{-1}. \end{aligned}$$

As for relation (24.43), the convergence is uniform on $z \in S^A_\varphi(x(\alpha,\kappa),\alpha,R)$. □

For every $z \in S^A_\varphi(x(\alpha,\kappa),\alpha)$, let

$$\varphi'_{A,\infty}(z) := \tilde{\varphi}'_{A,\infty}((\rho_A \circ H)^{-1}(z)). \tag{24.44}$$

The last result about the iterates of φ concerns their derivatives.

Proposition 24.2.13. *Let φ be a locally holomorphic simple parabolic map at $\omega \in \mathbb{C}$ and let A be an attracting direction for φ at ω. Fix $\alpha \in (0,\pi)$ and $\kappa \in (0,1)$. Then for every $z \in S^A_\varphi(x(\alpha,\kappa),\alpha)$ we have*

$$\lim_{n\to\infty} n^{\frac{p+1}{p}} |(\varphi^n)'(z)| = (|a|p)^{-\frac{p+1}{p}} |z-\omega|^{-(p+1)} |\varphi'_{A,\infty}(z)|, \tag{24.45}$$

and the convergence is uniform on $S^A_\varphi(x(\alpha,\kappa),\alpha,R)$ for every $R>0$. If $\alpha \in (0,\pi/2)$, then

$$\lim_{S^A_\varphi(x(\alpha,\kappa),\alpha)\ni z\to\omega} \varphi'_{A,\infty}(z) = 1. \tag{24.46}$$

Proof. Let $z \in S^A_\varphi(x(\alpha,\kappa),\alpha,R)$. Then $\xi = \rho_A^{-1}(z) \in S_0(x(\alpha,\kappa),\alpha,(|a|p)^{\frac{1}{p}}R)$ and hence $H^{-1}(\xi) \in S(x(\alpha,\kappa),\alpha,(|a|p)^{-1}R^{-p})$. By Lemma 24.2.6, we get that

$$\begin{aligned}
|\xi|^{p+1} n^{\frac{p+1}{p}} |(\varphi^n_{A,0})'(\xi)| &= |\xi|^{p+1} n^{\frac{p+1}{p}} |H'(\tilde{\varphi}^n_A(H^{-1}(\xi))) \cdot (\tilde{\varphi}^n_A)'(H^{-1}(\xi)) \cdot (H^{-1})'(\xi)| \\
&= |\xi|^{p+1} n^{\frac{p+1}{p}} \frac{1}{p} |\tilde{\varphi}^n_A(H^{-1}(\xi))|^{-\frac{p+1}{p}} \cdot |(\tilde{\varphi}^n_A)'(H^{-1}(\xi))| \cdot p|\xi|^{-(p+1)} \\
&= n^{\frac{p+1}{p}} |\tilde{\varphi}^n_A(H^{-1}(\xi))|^{-\frac{p+1}{p}} |(\tilde{\varphi}^n_A)'(H^{-1}(\xi))| \\
&= n^{\frac{p+1}{p}} |(\tilde{\varphi}^n_A)'(H^{-1}(\xi))| \left|H^{-1}(\xi) + n + O_{\mathrm{Re}(H^{-1}(\xi))}(\max\{n^{1-\frac{1}{p}}, \log n\})\right|^{-\frac{p+1}{p}} \\
&= |(\tilde{\varphi}^n_A)'(H^{-1}(\xi))| \left|\frac{H^{-1}(\xi)}{n} + 1 + O_{-(|a|p)^{-1}R^{-p}}(\max\{n^{-\frac{1}{p}}, n^{-1}\log n\})\right|^{-\frac{p+1}{p}}.
\end{aligned}$$

Therefore,

$$\begin{aligned}
n^{\frac{p+1}{p}} |(\varphi^n)'(z)| &= n^{\frac{p+1}{p}} |(\varphi^n_{A,0})'(\rho_A^{-1}(z))| = n^{\frac{p+1}{p}} |(\varphi^n_{A,0})'(\xi)| \\
&= |\xi|^{-(p+1)} |(\tilde{\varphi}^n_A)'(H^{-1}(\xi))| \cdot \\
&\quad \cdot \left|\frac{H^{-1}(\xi)}{n} + 1 + O_{-(|a|p)^{-1}R^{-p}}(\max\{n^{-\frac{1}{p}}, n^{-1}\log n\})\right|^{-\frac{p+1}{p}} \\
&= |\rho_A^{-1}(z)|^{-(p+1)} |(\tilde{\varphi}^n_A)'(H^{-1}(\rho_A^{-1}(z)))| \cdot \\
&\quad \cdot \left|\frac{H^{-1}(\rho_A^{-1}(z))}{n} + 1 + O_{-(|a|p)^{-1}R^{-p}}(\max\{n^{-\frac{1}{p}}, n^{-1}\log n\})\right|^{-\frac{p+1}{p}}
\end{aligned}$$

$$= (|a|p)^{-\frac{p+1}{p}} |z-\omega|^{-(p+1)} \big|(\tilde{\varphi}_A^n)'((\rho_A \circ H)^{-1}(z))\big| \cdot$$
$$\cdot \left| \frac{(\rho_A \circ H)^{-1}(z)}{n} + 1 + O_{-(|a|p)^{-1}R^{-p}}(\max\{n^{-\frac{1}{p}}, n^{-1}\log n\}) \right|^{-\frac{p+1}{p}}.$$

Letting $n \to \infty$ and applying Lemma 24.2.9 leads to the result. □

24.2.1.3 Fatou coordinates

In the next theorem, we establish the existence of functions, called Fatou coordinates, around simple parabolic fixed points. Though this theorem is of local character, it will eventually and primarily be used to describe the basin of attraction of parabolic points.

For every $\alpha \in (0, \pi)$ and $t \in \mathbb{R}$, set

$$Q_\alpha := S(x(\alpha, 1/2), \alpha) \quad \text{and} \quad Q_{\alpha,t} := Q_\alpha \cap \{z \in \mathbb{C} \mid \operatorname{Re}(z) > t\}.$$

By Lemmas 24.2.4–24.2.5, notice that

$$\tilde{\varphi}_A(Q_\alpha) \subseteq Q_\alpha \quad \text{and} \quad \tilde{\varphi}_A(Q_{\alpha,t}) \subseteq Q_{\alpha,t}, \quad \forall t \in \mathbb{R}. \tag{24.47}$$

In particular, all iterates $(\tilde{\varphi}_A^n)_{n=1}^\infty$ are well-defined on Q_α and $Q_{\alpha,t}$.

Theorem 24.2.14. *Let φ be a locally holomorphic simple parabolic map at $\omega \in \mathbb{C}$ and let A be an attracting direction for φ at ω. Then for every $\alpha \in (0, \pi)$ there exists a unique (up to an additive constant) holomorphic function $\tilde{F}_\alpha : Q_\alpha \to \mathbb{C}$ such that*

$$\tilde{F}_\alpha \circ \tilde{\varphi}_A = \tilde{F}_\alpha + 1 \tag{24.48}$$

and

$$\lim_{Q_{\alpha,t} \ni z \to \infty} \frac{\tilde{F}_\alpha(z)}{z} = 1, \quad \forall t \in \mathbb{R}. \tag{24.49}$$

Furthermore, replacing $x(\alpha, 1/2)$ by a sufficiently large number and keeping the same symbol $x(\alpha, 1/2)$ for that number, renders the function $\tilde{F}_\alpha$ univalent.

The functions $\{\tilde{F}_\alpha\}_{\alpha \in (0,\pi)}$ are called the Fatou coordinates *of the parabolic map $\tilde{\varphi}_A$.*

Proof. To establish the uniqueness of Fatou coordinates, suppose that a holomorphic function $\tilde{F}_\alpha : Q_\alpha \to \mathbb{C}$ satisfies (24.48)–(24.49). Fix $\varepsilon > 0$. By definition of Q_α and (24.49), there exist $\beta \in (0, \alpha)$ small enough and $s > \max\{2, x(\alpha, 1/2)\}$ large enough that

$$\overline{B(z, |z|/2)} \subseteq Q_\alpha \quad \text{and} \quad \left| \frac{\tilde{F}_\alpha(z)}{z} - 1 \right| < \frac{\varepsilon}{3}, \quad \forall z \in Q_{\beta,s}.$$

Cauchy's integral formula then yields

$$
\begin{aligned}
|\widetilde{F}'_\alpha(z) - 1| &= \left| \frac{1}{2\pi i} \int\limits_{\partial B(z,|z|/2)} \frac{\widetilde{F}_\alpha(\xi)}{(\xi - z)^2}\, d\xi - 1 \right| = \left| \frac{1}{2\pi i} \int\limits_{\partial B(z,|z|/2)} \frac{\widetilde{F}_\alpha(\xi) - \xi}{(\xi - z)^2}\, d\xi \right| \\
&\leq \frac{1}{2\pi} \int\limits_{\partial B(z,|z|/2)} \frac{|\widetilde{F}_\alpha(\xi) - \xi|}{|\xi - z|^2}\, |d\xi| \\
&= \frac{1}{\pi(|z|/2)^2} \int\limits_{\partial B(z,|z|/2)} |\widetilde{F}_\alpha(\xi) - \xi|\, |d\xi| \\
&\leq \frac{2}{\pi|z|^2} \int\limits_{\partial B(z,|z|/2)} \frac{\varepsilon}{3}|\xi|\, |d\xi| \\
&\leq \frac{2}{\pi|z|^2} \cdot 2\pi\frac{|z|}{2} \cdot \frac{\varepsilon}{3} \cdot \frac{3|z|}{2} = \varepsilon, \quad \forall z \in Q_{\beta,s}.
\end{aligned} \tag{24.50}
$$

Now fix $z, \zeta \in Q_\alpha$. By Lemma 24.2.6 and (24.47), there is $N \in \mathbb{N}$ such that $\widetilde{\varphi}_A^n(z), \widetilde{\varphi}_A^n(\zeta) \in Q_{\beta,s}$ for all $n \geq N$. Using (24.48), we obtain that

$$
\begin{aligned}
\widetilde{F}_\alpha(z) - \widetilde{F}_\alpha(\zeta) &= \widetilde{F}_\alpha(\widetilde{\varphi}_A^n(z)) - \widetilde{F}_\alpha(\widetilde{\varphi}_A^n(\zeta)) \\
&= \int\limits_{I_n} \widetilde{F}'_\alpha(\xi)\, d\xi = \int\limits_{I_n} ((\widetilde{F}'_\alpha(\xi) - 1) + 1)\, d\xi \\
&= \widetilde{\varphi}_A^n(z) - \widetilde{\varphi}_A^n(\zeta) + \int\limits_{I_n} (\widetilde{F}'_\alpha(\xi) - 1)\, d\xi,
\end{aligned}
$$

where I_n is the oriented segment from $\widetilde{\varphi}_A^n(\zeta)$ to $\widetilde{\varphi}_A^n(z)$. So,

$$
|(\widetilde{F}_\alpha(z) - \widetilde{F}_\alpha(\zeta)) - (\widetilde{\varphi}_A^n(z) - \widetilde{\varphi}_A^n(\zeta))| \leq \int\limits_{I_n} |\widetilde{F}'_\alpha(\xi) - 1|\, |d\xi|.
$$

For all $n \geq N$, we know that $\widetilde{\varphi}_A^n(z), \widetilde{\varphi}_A^n(\zeta) \in Q_{\beta,s}$. The convexity of $Q_{\beta,s}$ guarantees that $I_n \subseteq Q_{\beta,s}$ for all $n \geq N$. Applying (24.50), it follows that

$$
|(\widetilde{F}_\alpha(z) - \widetilde{F}_\alpha(\zeta)) - (\widetilde{\varphi}_A^n(z) - \widetilde{\varphi}_A^n(\zeta))| \leq \varepsilon|I_n| = \varepsilon|\widetilde{\varphi}_A^n(z) - \widetilde{\varphi}_A^n(\zeta)|, \quad \forall n \geq N.
$$

Thus,

$$
\left| \frac{\widetilde{F}_\alpha(z) - \widetilde{F}_\alpha(\zeta)}{\widetilde{\varphi}_A^n(z) - \widetilde{\varphi}_A^n(\zeta)} - 1 \right| \leq \varepsilon, \quad \forall n \geq N.
$$

Hence, the limit of the sequence $(\widetilde{\varphi}_A^n(z) - \widetilde{\varphi}_A^n(\zeta))_{n=1}^\infty$ exists and

$$
\widetilde{F}_\alpha(z) - \widetilde{F}_\alpha(\zeta) = \lim_{n\to\infty} (\widetilde{\varphi}_A^n(z) - \widetilde{\varphi}_A^n(\zeta)).
$$

The uniqueness part is thus established. (Note that $\varepsilon \searrow 0 \Rightarrow s \nearrow \infty \Rightarrow N \nearrow \infty$.)

This argument also gives us a hint as to how to prove the existence of Fatou coordinates. Let $\beta \in (0,\alpha)$, $t \in \mathbb{R}$ and $\zeta \in Q_{\beta,t}$. Consider the functions

$$Q_\alpha \ni z \longmapsto \psi_n(z) := \tilde{\varphi}_A^n(z) - \tilde{\varphi}_A^n(\zeta), \quad n \in \mathbb{N}. \tag{24.51}$$

Using Lemma 24.2.3, the mean value inequality, (24.47) and Lemma 24.2.5, we obtain that

$$\begin{aligned}
|\psi_{n+1}(z) - \psi_n(z)| &= \left|[\tilde{\varphi}_A(\tilde{\varphi}_A^n(z)) - (\tilde{\varphi}_A^n(z)+1)] - [\tilde{\varphi}_A(\tilde{\varphi}_A^n(\zeta)) - (\tilde{\varphi}_A^n(\zeta)+1)]\right| \\
&= \left|B \circ H(\tilde{\varphi}_A^n(z)) - B \circ H(\tilde{\varphi}_A^n(\zeta))\right| \\
&\le Mp^{-1} \sup\{|\xi|^{-\frac{p+1}{p}} : \xi \in I_n\} |\tilde{\varphi}_A^n(z) - \tilde{\varphi}_A^n(\zeta)| \\
&\le Mp^{-1}(t + n/2)^{-\frac{p+1}{p}} |\tilde{\varphi}_A^n(z) - \tilde{\varphi}_A^n(\zeta)| \\
&\le C_t n^{-\frac{p+1}{p}} |\psi_n(z)|
\end{aligned}$$

for some constant $C_t > 0$ independent of z and n but depending on t in general. Equivalently,

$$\left|\frac{\psi_{n+1}(z)}{\psi_n(z)} - 1\right| \le C_t n^{-\frac{p+1}{p}}. \tag{24.52}$$

As the series $\sum_{n=1}^\infty n^{-\frac{p+1}{p}}$ converges, estimate (24.52) ensures that the sequence $(\psi_n)_{n=1}^\infty$ converges uniformly on $Q_{\alpha,t}$ as it is uniformly quotient Cauchy. So, its limit

$$\tilde{F}_\alpha := \lim_{n\to\infty} \psi_n \tag{24.53}$$

is a holomorphic function on $Q_{\alpha,t}$. Letting $t \to -\infty$, we conclude that (24.53) defines a holomorphic function $\tilde{F}_\alpha : Q_\alpha \to \mathbb{C}$. Moreover, due to (24.53) and Lemmas 24.2.3 and 24.2.5, for all $z \in Q_\alpha$ we obtain the following relation:

$$\begin{aligned}
\tilde{F}_\alpha(\tilde{\varphi}_A(z)) &= \lim_{n\to\infty} \psi_n(\tilde{\varphi}_A(z)) = \lim_{n\to\infty} [\tilde{\varphi}_A^{n+1}(z) - \tilde{\varphi}_A^n(\zeta)] \\
&= \lim_{n\to\infty} \left[(\tilde{\varphi}_A^{n+1}(z) - \tilde{\varphi}_A^{n+1}(\zeta)) + (\tilde{\varphi}_A^{n+1}(\zeta) - \tilde{\varphi}_A^n(\zeta))\right] \\
&= \lim_{n\to\infty} \psi_{n+1}(z) + 1 \\
&= \tilde{F}_\alpha(z) + 1.
\end{aligned}$$

This proves (24.48).

In order to establish (24.49), we must work a little harder. Keep again $\beta \in (0,\alpha)$, $t \in \mathbb{R}$ fixed and take $z \in Q_{\alpha,t}$. From Lemma 24.2.3, we deduce by induction that

$$|\tilde{\varphi}_A^n(z) - (z+n)| \le L_{t,n}|z|^{-1/p}, \tag{24.54}$$

where $L_{t,n} > 0$ is some constant independent of $z \in Q_{\alpha,t}$ that may depend on t and n. Let $\varepsilon > 0$. By (24.52)–(24.53), there exists $k \in \mathbb{N}$ so large that

$$\left| \frac{\widetilde{F}_\alpha(z)/z}{\psi_k(z)/z} - 1 \right| < \frac{\varepsilon}{2}, \quad \forall z \in Q_{\alpha,t}. \tag{24.55}$$

By (24.51) and (24.54), we infer that

$$\begin{aligned}
\lim_{Q_{\alpha,t} \ni z \to \infty} \left| \frac{\psi_k(z)}{z} - 1 \right| &= \lim_{Q_{\alpha,t} \ni z \to \infty} \left| \frac{\widetilde{\varphi}_A^k(z) - \widetilde{\varphi}_A^k(\zeta)}{z} - 1 \right| \\
&= \lim_{Q_{\alpha,t} \ni z \to \infty} \frac{|\widetilde{\varphi}_A^k(z) - z|}{|z|} \\
&\le \lim_{Q_{\alpha,t} \ni z \to \infty} \frac{k + L_{t,k}|z|^{-1/p}}{|z|} \\
&= 0.
\end{aligned}$$

So, $\lim_{Q_{\alpha,t} \ni z \to \infty} \frac{\psi_k(z)}{z} = 1$, and hence (24.55) yields

$$\left| \frac{\widetilde{F}_\alpha(z)}{z} - 1 \right| < \varepsilon$$

for all $z \in Q_{\alpha,t}$ with sufficiently large moduli. Since $\varepsilon > 0$ is arbitrary, the validity of (24.49) is confirmed.

It only remains to show that $\widetilde{F}_\alpha : Q_\alpha \to \mathbb{C}$ is univalent. As the main ingredient in the proof of this statement, we establish the following.

Claim. *The map $\widetilde{\varphi}_A|_{Q_\alpha} : Q_\alpha \to \mathbb{C}$ is univalent.*

Proof. Assume that $x(\alpha, 1/2) \ge 1$ is so large that $|z| > 1$ and $M|z|^{-\frac{1}{p}} < 1/2$ for every $z \in Q_\alpha$, where $M > 0$ is the constant in Lemma 24.2.3. In particular, recall that Q_α is convex. If $\zeta \in Q_\alpha$, then it follows from Lemma 24.2.3 that

$$|\widetilde{\varphi}_A'(\zeta) - 1| \le M|\zeta|^{-\frac{p+1}{p}} \le M|\zeta|^{-\frac{1}{p}} < \frac{1}{2}.$$

In particular, $\operatorname{Re}(\widetilde{\varphi}_A'(\zeta)) > 1/2$.

Now, if $\widetilde{\varphi}_A(\zeta) = \widetilde{\varphi}_A(\xi)$ for some $\zeta, \xi \in Q_\alpha$, then it ensues from Lemma 24.2.3 that

$$|\zeta - \xi| \le 2M(\min\{|\zeta|, |\xi|\})^{-\frac{1}{p}} < 1. \tag{24.56}$$

So, it suffices to prove the injectivity of $\widetilde{\varphi}_A$ on the convex set $B(\xi, 1) \cap Q_\alpha$ for an arbitrary $\xi \in Q_\alpha$. Let $\zeta_1, \zeta_2 \in B(\xi, 1) \cap Q_\alpha$ and denote by I the line segment from ζ_1 to ζ_2. Then

$$|\widetilde{\varphi}_A(\zeta_2) - \widetilde{\varphi}_A(\zeta_1)| = \left| \int_I \widetilde{\varphi}_A'(\zeta)\, d\zeta \right|$$

$$
\begin{aligned}
&= |\zeta_2 - \zeta_1| \cdot \left| \int_0^1 \widetilde{\varphi}_A'(\zeta_1 + t(\zeta_2 - \zeta_1))dt \right| \\
&\geq |\zeta_2 - \zeta_1| \cdot \operatorname{Re} \int_0^1 \widetilde{\varphi}_A'(\zeta_1 + t(\zeta_2 - \zeta_1))dt \\
&= |\zeta_2 - \zeta_1| \cdot \int_0^1 \operatorname{Re}[\widetilde{\varphi}_A'(\zeta_1 + t(\zeta_2 - \zeta_1))]dt \\
&\geq |\zeta_2 - \zeta_1| \cdot \frac{1}{2}.
\end{aligned}
$$

Therefore, the function $\widetilde{\varphi}_A$ is injective on $B(\xi, 1) \cap Q_\alpha$. This completes the proof of the claim. ■

In view of this claim and of the uniform convergence of the sequence $(\psi_n)_{n=1}^\infty$ on $Q_{\alpha,t}$, an application of Hurwitz's theorem (Theorem A.3.8) yields the univalence of the limit function $\widetilde{F}_\alpha|_{Q_{\alpha,t}}$ for any $t \in \mathbb{R}$. Hence, $\widetilde{F}_\alpha : Q_\alpha \to \mathbb{C}$ is univalent. □

We now establish an important property of the Fatou coordinates, namely that whenever $\alpha \geq \pi/2$ the $\widetilde{F}_\alpha$-image of every $Q_{\alpha,t}$ contains a vertical half-plane.

Lemma 24.2.15. *Let φ be a locally holomorphic simple parabolic map at $\omega \in \mathbb{C}$ and let A be an attracting direction for φ at ω. For every $\alpha \in [\pi/2, \pi)$ and every $t \in \mathbb{R}$, there exists $s = s(t) > 0$ such that*

$$\widetilde{F}_\alpha(Q_{\alpha,t}) \supseteq \{\zeta \in \mathbb{C} : \operatorname{Re}(\zeta) > s\}.$$

Proof. By (24.49), there exists $s > \max\{0, 4[x(\alpha, 1/2) + t]\}$ so large that

$$\left| \frac{\widetilde{F}_\alpha(z)}{z} - 1 \right| \leq \frac{1}{8}$$

whenever $\operatorname{Re}(z) > s/4$. Take any $\zeta \in \mathbb{C}$ with $\operatorname{Re}(\zeta) > s$. Observe that

$$B(\zeta, 2|\zeta|/3) \subseteq Q_{\alpha,t}. \tag{24.57}$$

If $z \in \partial B(\zeta, |\zeta|/2)$, then

$$
\begin{aligned}
|(\widetilde{F}_\alpha(z) - \zeta) - (z - \zeta)| &= |\widetilde{F}_\alpha(z) - z| = |z| \left| \frac{\widetilde{F}_\alpha(z)}{z} - 1 \right| \\
&\leq \frac{|z|}{8} \leq \frac{|\zeta| + |\zeta|/2}{8} = \frac{3}{16}|\zeta| < \frac{|\zeta|}{2} = |z - \zeta|.
\end{aligned}
$$

Thus, Rouché's theorem (Theorem A.3.7) asserts that the function $z \mapsto \widetilde{F}_\alpha(z) - \zeta$ has the same number of zeros in $B(\zeta, |\zeta|/2) \subseteq Q_{\alpha,t}$ as the function $z \mapsto z - \zeta$. But this latter function has exactly one zero, namely $z = \zeta$. Hence, there exists $z \in B(\zeta, |\zeta|/2) \subseteq Q_{\alpha,t}$ such that $\widetilde{F}_\alpha(z) = \zeta$. □

Fix $\alpha \in (0, \pi)$, $\kappa = 1/2$, and denote by

$$S_a^1(\omega, \alpha), \dots, S_a^p(\omega, \alpha) \tag{24.58}$$

the *attracting sectors/flower petals* $S_\varphi^A(x(\alpha, 1/2), \alpha)$ defined in (24.39) and corresponding to the attracting directions A generated by the p holomorphic branches of the pth root function $\sqrt[p]{\cdot}$ defined in (24.25). See Figures 24.3–24.4 as well as 24.5.

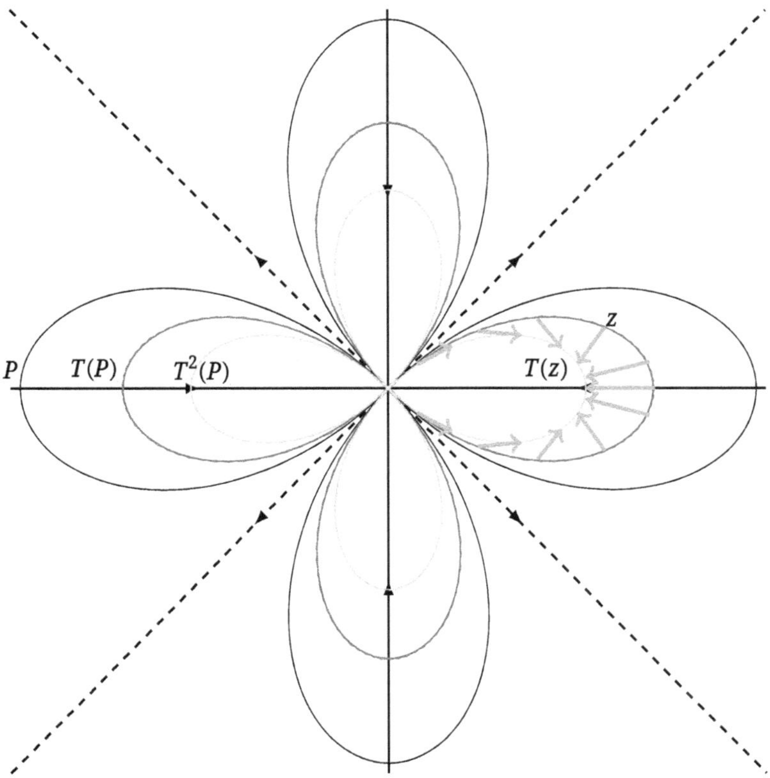

Figure 24.5: Four petals, case $p = 4$ (with $a = -1$). There are 4 attracting directions (indicated by rays and vectors pointing towards the parabolic fixed point) with respect to which the petals are symmetric and 4 repelling directions (indicated by dashed rays and vectors pointing away from the point). Some dynamics within one of the petals are depicted by the arrows. Reproduce them in the other three petals.

Denote also by

$$S_r^1(\omega, \alpha), \dots, S_r^p(\omega, \alpha) \tag{24.59}$$

the analogous attracting sectors for φ^{-1}. We usually call them the *repelling sectors* for φ. See Figure 24.6.

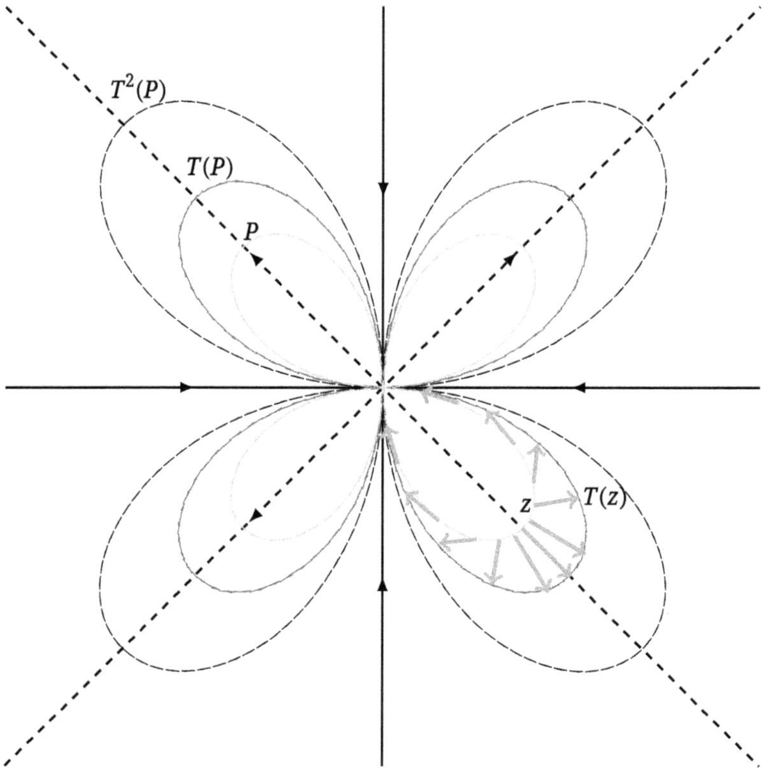

Figure 24.6: Four petals, case $p = 4$ (with $a = -1$). There are 4 attracting directions (indicated by rays and vectors pointing towards the parabolic fixed point) with respect to which the petals are symmetric and 4 repelling directions (indicated by dashed rays and vectors pointing away from the point). Some dynamics within one of the petals are depicted by the arrows. Reproduce them in the other three petals.

A straightforward consequence of these definitions is the following.

Lemma 24.2.16. *If φ is a locally holomorphic simple parabolic map at $\omega \in \mathbb{C}$, then for every $\alpha \in (0, \pi)$ both collections*

$$\{S_a^1(\omega,\alpha),\dots,S_a^p(\omega,\alpha)\} \quad \textit{and} \quad \{S_r^1(\omega,\alpha),\dots,S_r^p(\omega,\alpha)\}$$

consist of mutually disjoint sets.

Given the Fatou coordinates $\tilde{F}_\alpha$ defined through Theorem 24.2.14 for the map $\tilde{\varphi}_A$ associated with an attracting direction A, set

$$F_\alpha := \widetilde{F}_\alpha \circ (\rho_A \circ H)^{-1}. \tag{24.60}$$

The functions $\{F_\alpha\}_{\alpha \in (0,\pi)}$ are called the *Fatou coordinates* of the parabolic map φ.

As an immediate consequence of Lemma 24.2.4, Theorem 24.2.14 and Lemma 24.2.15, we obtain the following result.

Theorem 24.2.17. *If φ is a locally holomorphic simple parabolic map at $\omega \in \mathbb{C}$, then for every $\alpha \in (0,\pi)$ we have*

$$\varphi(S_a^j(\omega,\alpha)) \subseteq S_a^j(\omega,\alpha), \quad \forall j = 1,\dots,p, \tag{24.61}$$

and

$$F_\alpha \circ \varphi = F_\alpha + 1 \quad \text{on every } S_a^j(\omega,\alpha). \tag{24.62}$$

Moreover:

(a) *With appropriate normalizations (i. e., making the values of the F_α's coincide at one point), if $0 < \beta, \gamma < \pi$ then*

$$F_\gamma|_{S_a^j(\omega,\beta)\cap S_a^j(\omega,\gamma)} = F_\beta|_{S_a^j(\omega,\beta)\cap S_a^j(\omega,\gamma)}.$$

(b) *The map $F_\alpha|_{S_a^j(\omega,\alpha)}$ is univalent.*

(c) *If $\alpha \in [\pi/2,\pi)$, then there exists $s > 0$ such that*

$$F_\alpha(S_a^j(\omega,\alpha)) \supseteq \{\zeta \in \mathbb{C} : \operatorname{Re}(\zeta) > s\}.$$

24.2.2 Leau–Fatou flower petals

In this section, we continue our study of parabolic fixed and periodic points undertaken in the previous section. However, up to this point our considerations have had an entirely local character in the sense that the map T (denoted so far by φ) was defined only on a small neighborhood of its simple rationally indifferent (or parabolic) fixed point ω. Instead, let T be defined globally by letting

$$T : \widehat{\mathbb{C}} \to \widehat{\mathbb{C}}$$

be a rational function with $\deg(T) \geq 2$ that has a simple rationally indifferent fixed point ω. Let $\Omega(T)$ denote the set of rationally indifferent periodic points of T and let $\Omega_0(T) \subseteq \Omega(T)$ be the set of simple rationally indifferent fixed points of T, i. e.,

$$\Omega_0(T) = \{\omega \in \Omega(T) : T(\omega) = \omega \text{ and } T'(\omega) = 1\}. \tag{24.63}$$

Suppose that $\Omega_0(T) \neq \emptyset$ and $\omega \in \Omega_0(T)$. By Montel's criterion (Theorem 23.2.4) and (24.61), the attracting sector/flower petal $S^j_a(\omega, \alpha)$ is a subset of the Fatou set $\mathcal{F}(T)$ for every $j \in \{1, \ldots, p(\omega)\}$ and every $\alpha \in (0, \pi)$.

Let $\delta_T > 0$ be so small that the map $T|_{B(\omega, 2\delta_T)}$ is injective and that there exists

$$T_\omega^{-1} : B(\omega, \delta_T) \to \mathbb{C},$$

a unique holomorphic inverse branch of T that sends ω to ω.

Recall that the sets $S^j_r(\omega, \alpha)$, $j \in \{1, \ldots, p(\omega)\}$, are the attracting sectors for the map $T_\omega^{-1} : B(\omega, \delta_T) \to \mathbb{C}$.

For every $j \in \{1, \ldots, p(\omega)\}$, let $A_j^*(\omega)$ be the connected component of the Fatou set $\mathcal{F}(T)$ that contains $S^j_a(\omega, \alpha)$ for some (equivalently, for every) $\alpha \in (0, \pi)$. The domain $A_j^*(\omega)$ is called the *immediate basin of attraction of* ω *in the direction determined by* j. The sets $A_j^*(\omega)$, $j = 1, \ldots, p(\omega)$, are also called *Leau–Fatou petals* or *Leau domains*.

Further define

$$A_j(\omega) := \bigcup_{n=0}^{\infty} T^{-n}(A_j^*(\omega)).$$

The open set $A_j(\omega)$ is called the *basin of attraction of* ω *in the direction determined by* j. This basin is a subset of the Fatou set since $A_j^*(\omega) \subseteq \mathcal{F}(T)$ and the Fatou set is completely T-invariant according to Theorem 24.1.2.

Finally, the open set

$$A^*(\omega) := \bigcup_{j=1}^{p(\omega)} A_j^*(\omega)$$

is called the *immediate basin of attraction of* ω, whereas the open set

$$A(\omega) := \bigcup_{j=1}^{p(\omega)} A_j(\omega)$$

is said to be the *basin of attraction of* ω. These two sets are obviously contained in the Fatou set. Notice that

$$\omega \in \bigcap_{j=1}^{p(\omega)} \partial A_j^*(\omega) \cap \bigcap_{j=1}^{p(\omega)} \partial A_j(\omega) \cap \partial A^*(\omega) \cap \partial A(\omega).$$

The above terminologies make sense because of the following result, which is a somewhat direct consequence of the complete invariance of the Fatou set and Lemma 24.2.10.

Lemma 24.2.18. *Let* $T : \widehat{\mathbb{C}} \to \widehat{\mathbb{C}}$ *be a rational function with* $\deg(T) \geq 2$ *and with a simple rationally indifferent fixed point* ω, *i. e.,* $\omega \in \Omega_0(T)$. *Then*

$$T(A_j^*(\omega)) = A_j^*(\omega), \quad \forall j \in \{1, \dots, p(\omega)\} \quad \textit{and thus } T(A^*(\omega)) = A^*(\omega).$$

Moreover,

$$T^{-1}(A_j(\omega)) = A_j(\omega), \quad \forall j \in \{1, \dots, p(\omega)\} \quad \textit{and hence } T^{-1}(A(\omega)) = A(\omega).$$

Furthermore,

$$\lim_{n\to\infty} T^n(z) = \omega, \quad \forall z \in A(\omega).$$

The immediate basin of attraction of ω is a subset of the backward orbit of the attracting sectors/flower petals of any given angle α. The basin of attraction of ω coincides with the backward orbit of the immediate basin of attraction of ω as well as with the backward orbit of the attracting sectors/flower petals of any given angle α.

Lemma 24.2.19. *If $T : \widehat{\mathbb{C}} \to \widehat{\mathbb{C}}$ is a rational function with* $\deg(T) \geq 2$ *and $\omega \in \Omega_0(T)$, then for every $\alpha \in (0, \pi)$ we have that*

$$A^*(\omega) := \bigcup_{j=1}^{p(\omega)} A_j^*(\omega) \subseteq \bigcup_{n=0}^{\infty} T^{-n}\left(\bigcup_{j=1}^{p(\omega)} S_a^j(\omega, \alpha)\right)$$

and

$$A(\omega) = \bigcup_{n=0}^{\infty} T^{-n}(A^*(\omega)) = \bigcup_{n=0}^{\infty} T^{-n}\left(\bigcup_{j=1}^{p(\omega)} S_a^j(\omega, \alpha)\right).$$

Proof. As $A_j^*(\omega)$ is the Fatou component containing $S_a^j(\omega, \alpha)$, it is obvious that

$$\bigcup_{j=1}^{p(\omega)} S_a^j(\omega, \alpha) \subseteq \bigcup_{j=1}^{p(\omega)} A_j^*(\omega) =: A^*(\omega) \subseteq A(\omega).$$

Since $A(\omega)$ is completely T-invariant (see Lemma 24.2.18), one series of inclusions follows immediately.

For the opposite series, it suffices to show that

$$A^*(\omega) \subseteq \bigcup_{n=0}^{\infty} T^{-n}\left(\bigcup_{j=1}^{p(\omega)} S_a^j(\omega, \alpha)\right). \tag{24.64}$$

Indeed, if $\zeta \in A(\omega)$ then there is $1 \leq j \leq p(\omega)$ for which $\zeta \in A_j(\omega)$. This means that there is $k \geq 0$ such that $\zeta \in T^{-k}(A_j^*(\omega))$. So, $A(\omega) \subseteq \bigcup_{m=0}^{\infty} T^{-m}(A^*(\omega))$. If the above inclusion holds, then

$$A(\omega) \subseteq \bigcup_{m=0}^{\infty} T^{-m}(A^*(\omega)) \subseteq \bigcup_{m,n=0}^{\infty} T^{-(m+n)}\left(\bigcup_{j=1}^{p(\omega)} S_a^j(\omega, \alpha)\right) = \bigcup_{n=0}^{\infty} T^{-n}\left(\bigcup_{j=1}^{p(\omega)} S_a^j(\omega, \alpha)\right),$$

which is ultimately the series of inclusions sought.

Suppose, by way of contradiction, that (24.64) does not hold. This means that there exist some $i \in \{1, \ldots, p(\omega)\}$ and some

$$z \in A_i^*(\omega) \Big\backslash \bigcup_{n=0}^{\infty} T^{-n}\left(\bigcup_{j=1}^{p(\omega)} S_a^j(\omega, \alpha)\right). \tag{24.65}$$

Since $T(A_i^*(\omega)) = A_i^*(\omega)$ and since $\lim_{k\to\infty} T^k(z) = \omega$ by Lemma 24.2.18, passing to a sufficiently high iterate of z, we may assume without loss of generality that

$$T^k(z) \in B(\omega, \delta_T), \quad \forall k \geq 0. \tag{24.66}$$

As $\lim_{k\to\infty} T^k(z) = \omega$, we deduce from (24.65) that

$$T^n(z) \in \bigcup_{j=1}^{p(\omega)} S_r^j(\omega, \pi - \alpha) \tag{24.67}$$

for all $n \geq 0$ large enough. But, in light of Lemma 24.2.10, we know that

$$T_\omega^{-n}\left(\bigcup_{j=1}^{p(\omega)} S_r^j(\omega, \pi - \alpha)\right) \subseteq B(\omega, |z - \omega|/2) \tag{24.68}$$

for all $n \geq 0$ large enough. However, $z = T_\omega^{-k}(T^k(z))$ for all $k \geq 0$ by virtue of (24.66). We thus infer from (24.67) and (24.68) that $z \in B(\omega, |z - \omega|/2)$. This is impossible. □

Theorem 24.2.20. *If $T : \widehat{\mathbb{C}} \to \widehat{\mathbb{C}}$ is a rational function with $\deg(T) \geq 2$ and $\omega \in \Omega_0(T)$, then the Leau domains $A_j^*(\omega)$, $j = 1, \ldots, p(\omega)$, are mutually disjoint. Moreover, for every $j \in \{1, \ldots, p(\omega)\}$ and every $\alpha \in (0, \pi)$, we have that*

$$A_j^*(\omega) \subseteq \bigcup_{n=0}^{\infty} T^{-n}(S_a^j(\omega, \alpha)) \quad \text{and} \quad A_j(\omega) = \bigcup_{n=0}^{\infty} T^{-n}(S_a^j(\omega, \alpha)).$$

Proof. As $\bigcup_{n=0}^{\infty} T^{-n}(S_a^j(\omega, \alpha))$, $j = 1, \ldots, p(\omega)$, is a finite collection of mutually disjoint open sets and as the sets $A_i^*(\omega)$, $i = 1, \ldots, p(\omega)$, are each connected, Lemma 24.2.19 imposes that each set $A_i^*(\omega)$ be contained in exactly one set of the form $\bigcup_{n=0}^{\infty} T^{-n}(S_a^j(\omega, \alpha))$. Since $A_i^*(\omega) \supseteq S_a^i(\omega, \alpha)$, we deduce that

$$A_i^*(\omega) \subseteq \bigcup_{n=0}^{\infty} T^{-n}(S_a^i(\omega, \alpha)).$$

This not only implies that the Leau domains $A_i^*(\omega)$, $1 \leq i \leq p(\omega)$, are mutually disjoint but also that

$$A_i(\omega) := \bigcup_{m=0}^{\infty} T^{-m}(A_i^*(\omega)) \subseteq \bigcup_{m,n=0}^{\infty} T^{-(m+n)}(S_a^i(\omega, \alpha)) = \bigcup_{n=0}^{\infty} T^{-n}(S_a^i(\omega, \alpha)).$$

On the other hand,

$$S^i_a(\omega,\alpha) \subseteq A^*_i(\omega) \subseteq A_i(\omega).$$

Since $A_i(\omega)$ is completely T-invariant (see Lemma 24.2.18), we deduce that

$$\bigcup_{n=0}^{\infty} T^{-n}(S^i_a(\omega,\alpha)) \subseteq A_i(\omega).$$ □

Revisiting Theorem 24.2.17 in the context of a globally defined φ, i. e., for a rational function T, it turns out that all the maps F_α extend holomorphically to the entire Leau–Fatou petals $A^*_j(\omega)$.

Theorem 24.2.21. *If $T : \widehat{\mathbb{C}} \to \widehat{\mathbb{C}}$ is a rational function with $\deg(T) \geq 2$ and $\omega \in \Omega_0(T)$, then for every $j \in \{1,\ldots,p(\omega)\}$ there exists a holomorphic function $F : A^*_j(\omega) \to \mathbb{C}$ such that*

$$F \circ T = F + 1, \tag{24.69}$$

and for all $\alpha \in (0,\pi)$ the function $F|_{S^j_a(\omega,\alpha)}$ is univalent. In addition, if $\alpha \in [\pi/2,\pi)$ then there exists $s > 0$ such that

$$F(S^j_a(\omega,\alpha)) \supseteq \{\zeta \in \mathbb{C} : \operatorname{Re}(\zeta) > s\}. \tag{24.70}$$

Proof. Fix $z \in A^*_j(\omega)$. By Theorem 24.2.20, there exists $k \geq 0$ such that $T^k(z) \in S^j_a(\omega,\alpha)$. Define

$$F(z) := F_\alpha(T^k(z)) - k. \tag{24.71}$$

To see that this definition is independent of k, notice that Theorem 24.2.17 implies that

$$F(z) = F_\alpha(T^{k+j}(z)) - (k+j), \quad \forall j \geq 0.$$

So

$$F(z) = \lim_{n\to\infty} [F_\alpha(T^n(z)) - n] \tag{24.72}$$

is independent of k. It is also easy to show that (24.69) holds. Relation (24.70) is an immediate consequence of Theorem 24.2.17(c). The univalence of F ensues from Theorem 24.2.17(b). Finally, (24.71) directly implies that F is holomorphic. □

More generally, let $\omega \in \widehat{\mathbb{C}}$ be a rationally indifferent periodic point of T, i. e., $\omega \in \Omega(T)$. Then there is $\ell \in \mathbb{N}$ (ℓ is an appropriate multiple of the prime period of ω) such that ω is a simple rationally indifferent fixed point of T^ℓ, i. e., $\omega \in \Omega_0(T^\ell)$. Define $A^*(\omega)$ to be the immediate basin of attraction of ω under T^ℓ. Define also $A^*_j(\omega)$, $j = 1,\ldots,p(\omega)$, to be the Leau–Fatou petals of ω considered as a simple parabolic fixed point of T^ℓ. Further define

$$A(\omega) := \bigcup_{j=0}^{\infty} T^{-j}(A^*(\omega)),$$

and

$$A_p^*(\omega) := \bigcup_{j=0}^{\ell-1} A^*(T^j(\omega)), \quad A_p(\omega) := \bigcup_{j=0}^{\ell-1} A(T^j(\omega)).$$

Theorem 24.2.22. *If* $T : \widehat{\mathbb{C}} \to \widehat{\mathbb{C}}$ *is a rational function with* $\deg(T) \ge 2$ *and* $\omega \in \Omega(T)$, *then each Leau–Fatou petal of* ω *contains at least one critical point of* T.

Proof. The first part of the proof parallels the corresponding part of the proof of Theorem 24.1.20. However, the argument engendering a contradiction is more complicated. First, assume that ω is a simple rationally indifferent fixed point of T. Fix $j \in \{1, \dots, p(\omega)\}$. Suppose for a contradiction that $A_j^*(\omega)$ contains no critical point of T. We shall prove by induction that there exists an ascending sequence $(U_n)_{n=0}^{\infty}$ of open simply connected subsets of $A_j^*(\omega)$ such that

(a) $\omega \in \overline{U_0}$. Equivalently, $\omega \in \partial U_{n-1}$ for all $n \in \mathbb{N}$.
(b) For every $n \in \mathbb{N}$, there exists $T_n^{-1} : U_{n-1} \to U_n$, a unique surjective holomorphic inverse branch of T such that $T_n^{-1}(\omega) = \omega$.
(c) $T_{n+1}^{-1}|_{U_{n-1}} = T_n^{-1}$, $\forall n \in \mathbb{N}$.

Indeed, set

$$U_0 := S_a^j(\omega, \pi/2) \subseteq A_j^*(\omega).$$

It is clear that $\omega \in \partial U_0$. By relation (24.61) in Theorem 24.2.17, we know that

$$U_0 \subseteq T_\omega^{-1}(U_0) \subseteq A_j^*(\omega),$$

where T_ω^{-1} is the unique local holomorphic inverse branch of T that sends ω to ω. Set $T_1^{-1} := T_\omega^{-1}|_{U_0}$. As U_0 is open and simply connected, so is $U_1 := T_1^{-1}(U_0)$. Moreover, $U_1 \supseteq U_0$ by the above relation. The base step of the induction is complete.

For the inductive step, suppose that for some $k \in \mathbb{N}$ an ascending sequence of open simply connected sets $U_0 \subseteq U_1 \subseteq \cdots \subseteq U_k$ has been constructed such that properties (a) and (b) are satisfied for all $1 \le n \le k$ while (c) is satisfied for all $1 \le n < k$. Since $A_j^*(\omega) \cap T^{-1}(U_k)$ contains no critical point of T, since $T : A_j^*(\omega) \to A_j^*(\omega)$ is a holomorphic surjective map (per Lemma 24.2.18), and since $U_k \supseteq U_{k-1}$, there exists $T_{k+1}^{-1} : U_k \to A_j^*(\omega)$, a holomorphic inverse branch of T such that

$$T_{k+1}^{-1}|_{U_{k-1}} = T_k^{-1}. \tag{24.73}$$

So, (c) holds when $n = k$. Setting

$$U_{k+1} := T_{k+1}^{-1}(U_k), \tag{24.74}$$

property (b) holds when $n = k + 1$. Moreover, U_{k+1} is a simply connected open set since so is U_k and T_{k+1}^{-1} is holomorphic. We now check that $U_{k+1} \supseteq U_k$. Successively invoking (24.74), the assumption that $U_k \supseteq U_{k-1}$, (24.73) and property (b) with $n = k$, we infer that

$$U_{k+1} \supseteq T_{k+1}^{-1}(U_{k-1}) = T_k^{-1}(U_{k-1}) = U_k.$$

Finally, property (a) holds when $n = k + 1$ as it holds when $n = k$ and $U_{k+1} \supseteq U_k$ but $U_{k+1} \subseteq \mathcal{F}(T)$ while $\omega \in \mathcal{J}(T)$. The inductive step is thus finished.

Since $\bigcup_{n=0}^{\infty} U_n = A_j^*(\omega)$ by Theorem 24.2.20, conditions (a)–(c) along with ascendance of $(U_n)_{n=0}^{\infty}$ guarantee the existence of a holomorphic bijection $T_*^{-1} : A_j^*(\omega) \to A_j^*(\omega)$ such that $T_*^{-1}|_{U_{n-1}} = T_n^{-1}$ for every $n \in \mathbb{N}$. As $T \circ T_*^{-1} = \mathrm{Id}_{A_j^*(\omega)}$, the restriction $T : A_j^*(\omega) \to A_j^*(\omega)$ is one-to-one. Since we already know that it is onto, we conclude that $T : A_j^*(\omega) \to A_j^*(\omega)$ is bijective. Iterating formula (24.69) of Theorem 24.2.21 yields

$$F \circ T^n = F + n, \quad \forall n \geq 0. \tag{24.75}$$

Choosing distinct points $z, \zeta \in A_j^*(\omega)$, there is $n = n(z, \zeta) \geq 0$ so large that $T^n(z), T^n(\zeta) \in S_a^j(\omega, \pi/2)$. It follows from the bijectivity of $T^n : A_j^*(\omega) \to A_j^*(\omega)$, formula (24.75), and the univalence of $F|_{S_a^j(\omega,\pi/2)}$ in Theorem 24.2.21, that $F(z) \neq F(\zeta)$. In consequence, the holomorphic function $F : A_j^*(\omega) \to \mathbb{C}$ is univalent. Formula (24.75) further decrees that

$$F \circ T_*^{-n} = F - n, \quad \forall n \geq 0.$$

Therefore, it follows from inclusion (24.70) in Theorem 24.2.21 that

$$F(A_j^*(\omega)) \supseteq F(U_n) = F(T_*^{-n}(U_0)) = F(U_0) - n \supseteq \{z \in \mathbb{C} : \mathrm{Re}(z) > s - n\}.$$

Taking the union over all $n \in \mathbb{N}$, it ensues that

$$F(A_j^*(\omega)) = \mathbb{C}.$$

As the map $F : A_j^*(\omega) \to \mathbb{C}$ is univalent, its inverse $F^{-1} : \mathbb{C} \to A_j^*(\omega)$ is holomorphic. Because $A_j^*(\omega) \cap \mathcal{J}(T) = \emptyset$ and $\mathcal{J}(T)$ is a nonempty perfect set per Theorems 24.1.4 and 24.1.15, the function $F^{-1} : \mathbb{C} \to A_j^*(\omega)$ omits three distinct points and we conclude from Picard's theorem that F^{-1} is constant. This contradiction finishes the proof in the case where ω is a simple parabolic fixed point of T.

The general case, in which $\omega \in \Omega(T)$, follows by applying the case of a simple parabolic fixed point to an higher iterate T^ℓ such that $\omega \in \Omega_0(T^\ell)$. □

Since $\#\mathrm{Crit}(T) \le 2\deg(T) - 2$, the theorem just proved and Theorem 24.1.20 impose the following stronger constraint than Corollary 24.1.21.

Corollary 24.2.23. *If $T : \widehat{\mathbb{C}} \to \widehat{\mathbb{C}}$ is a rational function with $\deg(T) \ge 2$, then the total number of attracting and rationally indifferent periodic cycles of T does not exceed $2\deg(T) - 2$.*

24.2.3 Local and asymptotic behavior of rational functions around rationally indifferent periodic points: part II – Fatou's flower theorem and fundamental domains

In this section, $T : \widehat{\mathbb{C}} \to \widehat{\mathbb{C}}$ is once again a rational function with $\deg(T) \ge 2$ and $\omega \in \Omega_0(T)$. The results we present are of fairly classical nature and are scattered widely in the literature; for example, they can be found in [1, 8, 21, 37, 40, 94, 123].

As we have seen in Section 24.2.2, for a sufficiently small radius $\delta_T > 0$ there exists a unique holomorphic inverse branch $T_\omega^{-1} : B(\omega, \delta_T) \to \mathbb{C}$ which fixes ω. Thus, all the (local) results of Section 24.2.1 apply with both $\varphi = T$ and $\varphi = T_\omega^{-1}$. Fix $\alpha \in (0, \pi)$, $\kappa = 1/2$, and recall that

$$S_a^1(\omega, \alpha), \dots, S_a^{p(\omega)}(\omega, \alpha)$$

are the corresponding attracting sectors for T defined in (24.58) while

$$S_r^1(\omega, \alpha), \dots, S_r^{p(\omega)}(\omega, \alpha)$$

are the corresponding repelling sectors for T defined in (24.59). Equivalently, the repelling sectors for T are the attractive sectors for T_ω^{-1}. Since the family of iterates of T is not normal on any neighborhood of any point in the Julia set, we get the following celebrated classical result as a fairly immediate consequence of Lemma 24.2.10.

Theorem 24.2.24 (Fatou's flower theorem). *Let $T : \widehat{\mathbb{C}} \to \widehat{\mathbb{C}}$ be a rational function with $\deg(T) \ge 2$. If $\omega \in \Omega(T)$, i. e., if ω is a rationally indifferent periodic point of T, then for every $\alpha \in (0, \pi)$ there exists $\theta_\alpha(\omega) \in (0, \delta_T)$ such that*

$$\mathcal{J}(T) \cap B(\omega, \theta_\alpha(\omega)) \setminus \{\omega\} \subseteq \bigcup_{j=1}^{p(\omega)} S_r^j(\omega, \alpha).$$

Proof. Passing to a sufficiently high iterate of T, we may assume without loss of generality that ω is a simple parabolic fixed point of T, i. e., $\omega \in \Omega_0(T)$. Seeking a contradiction, suppose that there exists a sequence $(z_n)_{n=1}^\infty$ of points in $\mathcal{J}(T) \setminus \bigcup_{j=1}^{p(\omega)} S_r^j(\omega, \alpha)$ such that

$$\lim_{n\to\infty} |z_n - \omega| = 0 \quad \text{and} \quad z_n \neq \omega, \ \forall n \in \mathbb{N}.$$

Set $\beta \in (0,\alpha)$. There would then exist $i \in \{1,\dots,p(\omega)\}$ such that $z_n \in S_a^i(\omega,\pi-\beta)$ for every $n \in \mathbb{N}$ large enough. Fix one such n and i. By Lemma 24.2.10, the attracting petal $S_a^i(\omega,\pi-\beta)$ is a subset of $\mathcal{F}(T)$. Thus, $z_n \notin \mathcal{J}(T)$. This is a contradiction. □

Since the Julia set $\mathcal{J}(T)$ is completely invariant, we infer from Fatou's flower theorem (Theorem 24.2.24), Lemma 24.2.10, and (24.26) (actually the derivative of $\varphi_{A,0}$) that for every $\omega \in \Omega_0(T)$ there exists $\theta(\omega) \in (0,\theta_{\pi/4}(\omega))$ such that for every $0 < R \le \theta(\omega)$, we have

$$T_\omega^{-1}(\mathcal{J}(T)\cap B(\omega,R)) \subseteq \mathcal{J}(T)\cap B(\omega,R). \tag{24.76}$$

Thus, all the iterates

$$T_\omega^{-n} : \mathcal{J}(T)\cap B(\omega,R) \to \mathcal{J}(T)\cap B(\omega,R), \quad n \in \mathbb{N}, \tag{24.77}$$

are well-defined. For all $\alpha \in (0,\pi)$ and all $j \in \{1,2,\dots,p(\omega)\}$, we know by Lemma 24.2.10 that

$$T_\omega^{-1}(S_r^j(\omega,\alpha)) \subseteq S_r^j(\omega,\alpha). \tag{24.78}$$

Therefore, all the iterates

$$T_\omega^{-n} : S_r^j(\omega,\alpha) \to S_r^j(\omega,\alpha), \quad n \in \mathbb{N}, \tag{24.79}$$

are well-defined.

It ensues from the complete T-invariance of $\mathcal{J}(T)$ and (24.79) that

$$T_\omega^{-n}(\mathcal{J}(T)\cap S_r^j(\omega,\alpha)) \subseteq \mathcal{J}(T)\cap S_r^j(\omega,\alpha), \quad \forall n \in \mathbb{N}, \quad \forall 1 \le j \le p(\omega). \tag{24.80}$$

Furthermore, it follows from Lemma 24.2.10 that

$$\lim_{n\to\infty} T_\omega^{-n}(z) = \omega \quad \text{uniformly on } z \in S_r^j(\omega,\alpha), \tag{24.81}$$

while Lemma 24.2.11 states that

$$|T_\omega^{-1}(z)-\omega| < |z-\omega| \quad \text{and} \quad |(T_\omega^{-1})'(z)| < 1 \tag{24.82}$$

for every $z \in S_r^j(\omega,\alpha)\cap B(\omega,R_{\alpha,1/2}(\omega))\setminus\{\omega\}$ when $\alpha < \pi/p(\omega)$. Set

$$\theta = \theta(T) := \min_{\omega\in\Omega(T)} \min\{\theta(\omega), R_{\pi/4,1/2}(\omega)\}. \tag{24.83}$$

As an immediate consequence of Fatou's flower theorem (Theorem 24.2.24) and (24.77), we get the following.

Lemma 24.2.25. *Let $T : \widehat{\mathbb{C}} \to \widehat{\mathbb{C}}$ be a rational function with $\deg(T) \geq 2$. If $\tau > 0$ is sufficiently small (with $\theta > 0$ decreased if necessary), then for every $\omega \in \Omega_0(T)$ and every $z \in \mathcal{J}(T) \cap B(\omega,\theta)$ there exists $j \in \{1,\dots,p(\omega)\}$ such that*

$$B(z, 2\tau|z-\omega|) \subseteq B(\omega,\theta) \cap S_r^j(\omega,\pi/2).$$

In addition, all the local holomorphic inverse branches

$$T_\omega^{-n} : B(z, 2\tau|z-\omega|) \to \mathbb{C}$$

are well-defined for all $n \in \mathbb{N}$.

As a direct repercussion of Proposition 24.2.12 and Fatou's flower theorem (Theorem 24.2.24), we obtain the next result.

Lemma 24.2.26. *If $T : \widehat{\mathbb{C}} \to \widehat{\mathbb{C}}$ is a rational function with $\deg(T) \geq 2$, then for every $\omega \in \Omega_0(T)$, every $j \in \{1,\dots,p(\omega) =: p\}$, and every point $z \in S_r^j(\omega,\alpha)$ (including every $z \in \mathcal{J}(T) \cap B(\omega,\theta) \setminus \{\omega\}$), we have*

$$\lim_{n\to\infty} n^{\frac{p+1}{p}} |T_\omega^{-(n+1)}(z) - T_\omega^{-n}(z)| = (|a|p)^{-\frac{1}{p}} p^{-1}$$
$$\lim_{n\to\infty} n^{\frac{1}{p}} |T_\omega^{-n}(z) - \omega| = (|a|p)^{-\frac{1}{p}}.$$

In addition, in both limits, the convergence is uniform on the set $S_r^j(\omega,\alpha) \setminus \overline{B(\omega,R)}$ (and in particular on the set $\mathcal{J}(T) \cap B(\omega,\theta) \setminus \overline{B(\omega,R)}$) for every $R > 0$.

As a straightforward consequence of Proposition 24.2.13 and Fatou's flower theorem (Theorem 24.2.24), we observe the following fact.

Proposition 24.2.27. *If $T : \widehat{\mathbb{C}} \to \widehat{\mathbb{C}}$ is a rational function with $\deg(T) \geq 2$, then for every $\omega \in \Omega_0(T)$ and every $\alpha \in (0,\pi)$ there exists a holomorphic function*

$$(T_\omega^{-\infty})' : \bigcup_{j=1}^{p(\omega)} S_r^j(\omega,\alpha) =: S_r(\omega,\alpha) \longrightarrow \mathbb{C} \setminus \{0\}$$

with the following properties. Fix $j \in \{1,\dots,p(\omega) =: p\}$. For every $z \in S_r^j(\omega,\alpha)$ (including every $z \in \mathcal{J}(T) \cap B(\omega,\theta) \setminus \{\omega\}$), we have

$$\lim_{n\to\infty} n^{\frac{p+1}{p}} |(T_\omega^{-n})'(z)| = (|a|p)^{-\frac{p+1}{p}} |z-\omega|^{-(p+1)} |(T_\omega^{-\infty})'(z)| \tag{24.84}$$

and the convergence is uniform on $S_r^j(\omega,\alpha) \setminus \overline{B(\omega,R)}$ for every $R > 0$. If $\alpha \in (0,\pi/2)$, then

$$\lim_{S_r^j(\omega,\alpha) \ni z \to \omega} (T_\omega^{-\infty})'(z) = 1. \tag{24.85}$$

As the last result about the local behavior of a rational function around its parabolic points, we show the following.

Proposition 24.2.28 (Local expansiveness at rationally indifferent points). *Let $T : \widehat{\mathbb{C}} \to \widehat{\mathbb{C}}$ be a rational function with* $\deg(T) \ge 2$. *If $\omega \in \Omega_0(T)$, $z \in \mathcal{J}(T)$ and $T^n(z) \in B(\omega, \theta)$ for all $n \ge 0$, then $z = \omega$.*

Proof. Since $z \in \mathcal{J}(T)$, we know that $T^n(z) \in \mathcal{J}(T)$ for every $n \ge 0$. It follows from Fatou's flower theorem (Theorem 24.2.24) and the uniform convergence in (24.81) that

$$\omega = \lim_{n\to\infty} T_\omega^{-n}(T^n(z)) = \lim_{n\to\infty} z = z.$$

The uniform convergence is crucial because the point $T^n(z)$ varies with n in the sector involved in (24.81). □

We end this section with the following two lemmas, which are interesting on their own and constitute useful tools for applications.

Let

$$S_r(\omega, \alpha) := \bigcup_{j=1}^{p(\omega)} S_r^j(\omega, \alpha).$$

Lemma 24.2.29. *Let $T : \widehat{\mathbb{C}} \to \widehat{\mathbb{C}}$ be a rational function with* $\deg(T) \ge 2$. *If $\omega \in \Omega_0(T)$ and $\alpha \in (0, \pi)$, then the set $\mathcal{J}(T) \cap (S_r(\omega, \alpha) \setminus T_\omega^{-1}(S_r(\omega, \alpha)))$ is a fundamental domain for the map T_ω^{-1} acting on $\mathcal{J}(T)$ near ω. More precisely:*

(a) *If $(i, k), (j, l) \in \{1, \ldots, p(\omega)\} \times \{0, 1, 2, \ldots\}$ and $(i, k) \ne (j, l)$, then*

$$T_\omega^{-k}\Big(S_r^i(\omega, \alpha) \setminus T_\omega^{-1}(S_r(\omega, \alpha))\Big) \bigcap T_\omega^{-l}\Big(S_r^j(\omega, \alpha) \setminus T_\omega^{-1}(S_r(\omega, \alpha))\Big) = \emptyset.$$

(b)

$$\omega \in \mathrm{Int}_{\mathcal{J}(T)}\left(\mathcal{J}(T) \bigcap \left[\{\omega\} \cup \bigcup_{n=0}^{\infty} T_\omega^{-n}\Big(S_r(\omega, \alpha) \setminus T_\omega^{-1}(S_r(\omega, \alpha))\Big)\right]\right).$$

Proof. (a) Assume first that $i = j$ but $k < l$. Defining $\widetilde{T}_\omega^{-1} = (\rho_A \circ H)^{-1} \circ T_\omega^{-1}$ (where ρ_A is an attracting direction for T_ω^{-1}) and using (24.39) and (24.27), we get

$$\begin{aligned}
&T_\omega^{-k}\Big(S_r^i(\omega, \alpha) \setminus T_\omega^{-1}(S_r(\omega, \alpha))\Big) \bigcap T_\omega^{-l}\Big(S_r^i(\omega, \alpha) \setminus T_\omega^{-1}(S_r(\omega, \alpha))\Big) \\
&\quad = T_\omega^{-k}\Big(\rho_A \circ H(S(x(\alpha, \kappa), \alpha)) \setminus T_\omega^{-1}(\rho_A \circ H(S(x(\alpha, \kappa), \alpha)))\Big) \\
&\qquad \bigcap T_\omega^{-l}\Big(\rho_A \circ H(S(x(\alpha, \kappa), \alpha)) \setminus T_\omega^{-1}(\rho_A \circ H(S(x(\alpha, \kappa), \alpha)))\Big) \\
&\quad = T_\omega^{-k}\Big(\rho_A \circ H(S(x(\alpha, \kappa), \alpha)) \setminus \rho_A \circ H(\widetilde{T}_\omega^{-1}(S(x(\alpha, \kappa), \alpha)))\Big) \\
&\qquad \bigcap T_\omega^{-l}\Big(\rho_A \circ H(S(x(\alpha, \kappa), \alpha)) \setminus \rho_A \circ H(\widetilde{T}_\omega^{-1}(S(x(\alpha, \kappa), \alpha)))\Big)
\end{aligned}$$

$$
\begin{aligned}
&= T_\omega^{-k} \circ \rho_A \circ H\Big(S(x(\alpha,\kappa),\alpha) \setminus \widetilde{T}_\omega^{-1}(S(x(\alpha,\kappa),\alpha))\Big) \\
&\quad \bigcap T_\omega^{-l} \circ \rho_A \circ H\Big(S(x(\alpha,\kappa),\alpha) \setminus \widetilde{T}_\omega^{-1}(S(x(\alpha,\kappa),\alpha))\Big) \\
&= \rho_A \circ H \circ \widetilde{T}_\omega^{-k}\Big(S(x(\alpha,\kappa),\alpha) \setminus \widetilde{T}_\omega^{-1}(S(x(\alpha,\kappa),\alpha))\Big) \\
&\quad \bigcap \rho_A \circ H \circ \widetilde{T}_\omega^{-l}\Big(S(x(\alpha,\kappa),\alpha) \setminus \widetilde{T}_\omega^{-1}(S(x(\alpha,\kappa),\alpha))\Big) \\
&= \rho_A \circ H\Big(\widetilde{T}_\omega^{-k}\Big(S(x(\alpha,\kappa),\alpha) \setminus \widetilde{T}_\omega^{-1}(S(x(\alpha,\kappa),\alpha))\Big) \\
&\qquad\qquad \bigcap \widetilde{T}_\omega^{-l}\Big(S(x(\alpha,\kappa),\alpha) \setminus \widetilde{T}_\omega^{-1}(S(x(\alpha,\kappa),\alpha))\Big)\Big) \\
&= \rho_A \circ H \circ \widetilde{T}_\omega^{-k}\Big(\ \Big(S(x(\alpha,\kappa),\alpha) \setminus \widetilde{T}_\omega^{-1}(S(x(\alpha,\kappa),\alpha))\Big) \\
&\qquad\qquad \bigcap \widetilde{T}_\omega^{-(l-k)}\Big(S(x(\alpha,\kappa),\alpha) \setminus \widetilde{T}_\omega^{-1}(S(x(\alpha,\kappa),\alpha))\Big)\ \Big) \\
&\subseteq \rho_A \circ H \circ \widetilde{T}_\omega^{-k}\Big(\Big(S(x(\alpha,\kappa),\alpha) \setminus \widetilde{T}_\omega^{-1}(S(x(\alpha,\kappa),\alpha))\Big) \bigcap \widetilde{T}_\omega^{-(l-k)}(S(x(\alpha,\kappa),\alpha))\Big).
\end{aligned}
$$

As $l - k \geq 1$, Lemma 24.2.4 allows us to continue as follows:

$$
\begin{aligned}
&T_\omega^{-k}\Big(S_r^i(\omega,\alpha) \setminus T_\omega^{-1}(S_r(\omega,\alpha))\Big) \bigcap T_\omega^{-l}\Big(S_r^i(\omega,\alpha) \setminus T_\omega^{-1}(S_r(\omega,\alpha))\Big) \\
&\quad \subseteq \rho_A \circ H \circ \widetilde{T}_\omega^{-k}\Big(\Big(S(x(\alpha,\kappa),\alpha) \setminus \widetilde{T}_\omega^{-1}(S(x(\alpha,\kappa),\alpha))\Big) \bigcap \widetilde{T}_\omega^{-1}(S(x(\alpha,\kappa),\alpha))\Big) \\
&\quad = \rho_A \circ H \circ \widetilde{T}_\omega^{-k}(\emptyset) = \emptyset.
\end{aligned}
$$

We have thus ascertained statement (a) in the case $i = j$ but $k \neq l$.

If $i \neq j$, then we are immediately done thanks to Lemma 24.2.16 and relation (24.61) in Theorem 24.2.17.

Part (b) is an immediate consequence of (24.78) and Fatou's flower theorem (Theorem 24.2.24). □

Lemma 24.2.30. *Let $T : \widehat{\mathbb{C}} \to \widehat{\mathbb{C}}$ be a rational function with $\deg(T) \geq 2$. Let $\omega \in \Omega_0(T)$ and $\alpha \in (0,\pi)$. For every $u > 0$ and every $j \in \{1,2,\ldots,p(\omega)\}$, let*

$$\Delta_j(\omega,\alpha,u) := [S_r^j(\omega,\alpha) \setminus T_\omega^{-1}(S_r^j(\omega,\alpha))] \cap [\mathbb{C} \setminus B(\omega,u)]$$

and

$$\Delta(\omega,\alpha,u) := \bigcup_{j=1}^{p(\omega)} \Delta_j(\omega,\alpha,u).$$

Then there exists $u > 0$ such that the set $\Delta(\omega,\alpha,u)$ is a fundamental domain for the map T_ω^{-1} acting on $\mathcal{J}(T)$ near ω. More precisely:

(a) *If* $(i,k),(j,l)\in\{1,2,\dots,p(\omega)\}\times\{0,1,2,\dots\}$ *and* $(i,k)\neq(j,l)$, *then*

$$T_\omega^{-k}(\Delta_j(\omega,\alpha,u))\cap T_\omega^{-l}(\Delta_j(\omega,\alpha,u))=\emptyset.$$

(b) $\omega\in\mathrm{Int}_{\mathcal{J}(T)}\big(\mathcal{J}(T)\cap[\{\omega\}\cup\bigcup_{n=0}^{\infty}T_\omega^{-n}(\Delta(\omega,\alpha,u))]\big)$.

Proof. Part (a) is an immediate consequence of Lemma 24.2.29. We now prove (b). Since T_ω^{-1} is an open map (because it is holomorphic), there is $u\in(0,\theta_\alpha(\omega))$ such that

$$B(\omega,u)\subseteq T_\omega^{-1}(B(\omega,\theta_\alpha(\omega))).$$

It then follows from Fatou's flower theorem (Theorem 24.2.24) that

$$\mathcal{J}(T)\cap B(\omega,u)\subseteq\bigcup_{j=1}^{p(\omega)}T_\omega^{-1}(S_r^j(\omega,\alpha))=T_\omega^{-1}(S_r(\omega,\alpha)).\tag{24.86}$$

In light of Lemma 24.2.29(b), it suffices to show that

$$\mathcal{J}(T)\cap\left[\bigcup_{n=0}^{\infty}T_\omega^{-n}\big(S_r(\omega,\alpha)\setminus T_\omega^{-1}(S_r(\omega,\alpha))\big)\right]\subseteq\bigcup_{n=0}^{\infty}T_\omega^{-n}(\Delta(\omega,\alpha,u)).$$

Let z belong to the left-hand side of this claimed inclusion. Then there is $n\geq 0$ such that

$$z\in\mathcal{J}(T)\cap T_\omega^{-n}(S_r(\omega,\alpha)\setminus T_\omega^{-1}(S_r(\omega,\alpha))).$$

By (24.86), we get

$$\begin{aligned}T^n(z)&\in\mathcal{J}(T)\cap[S_r(\omega,\alpha)\setminus T_\omega^{-1}(S_r(\omega,\alpha))]\\&=\mathcal{J}(T)\cap[S_r(\omega,\alpha)\setminus T_\omega^{-1}(S_r(\omega,\alpha))]\cap[\mathbb{C}\setminus B(\omega,u)]\\&\subseteq\Delta(\omega,\alpha,u).\end{aligned}$$

Hence, $z\in T_\omega^{-n}(\Delta(\omega,\alpha,u))$. □

24.3 Nonattracting periodic points revisited: total number and denseness of repelling periodic points

We now show that the total number of nonrepelling cycles of a rational function does not exceed $6\deg(T)-6$. In fact, we first show the following somewhat stronger result.

Theorem 24.3.1. *Every rational function* $T:\widehat{\mathbb{C}}\to\widehat{\mathbb{C}}$ *with* $\deg(T)\geq 2$ *admits at most* $4\deg(T)-4$ *indifferent cycles with multipliers different from 1.*

Proof. Let $\omega_1, \dots, \omega_k$ be indifferent periodic points of T with respective multipliers $\lambda_1, \dots, \lambda_k$, all different from 1, and all belonging to distinct periodic cycles. By conjugation through a suitably chosen Möbius transformation, we may assume without loss of generality that all the periodic cycles of T containing the points $\omega_1, \dots, \omega_k$ do not visit the point ∞. Let $d = \deg(T)$. Write

$$T(z) = \frac{P(z)}{Q(z)},$$

where P and Q are relatively prime polynomials. Consider the one-parameter family of rational functions given by the formula

$$T_\xi(z) = \frac{(1-\xi)P(z) + \xi z^d}{(1-\xi)Q(z) + \xi}, \quad z, \xi \in \mathbb{C}. \tag{24.87}$$

Then

$$T_0 = T \quad \text{and} \quad T_1(z) = z^d.$$

For every $j \in \{1, \dots, k\}$, let $p_j \in \mathbb{N}$ be the prime period of ω_j under T. Consider the function

$$(z, \xi) \longmapsto F_j(z, \xi) := T_\xi^{p_j}(z) - z.$$

Then $F_j(\omega_j, 0) = T^{p_j}(\omega_j) - \omega_j = \omega_j - \omega_j = 0$ and

$$\frac{\partial F_j}{\partial z}(\omega_j, 0) = (T^{p_j})'(\omega_j) - 1 = \lambda_j - 1 \neq 0.$$

So, by the implicit function theorem (Theorem A.3.12), there exist $R_j > 0$ and a unique holomorphic function $H_j : B(0, R_j) \to \mathbb{C}$ such that

$$H_j(0) = \omega_j \quad \text{and} \quad F_j(H_j(\xi), \xi) = 0, \quad \forall \xi \in B(0, R_j).$$

This means that

$$T_\xi^{p_j}(H_j(\xi)) = H_j(\xi), \quad \forall \xi \in B(0, R_j). \tag{24.88}$$

We may further assume that the R_j's are such that the sets $H_j(B(0, R_j))$, $j = 1, \dots, k$, are mutually disjoint. Let $R = \min\{R_1, \dots, R_k\} > 0$. We now identify a finite "bad" set of parameters ξ whose relevance will soon become apparent. Let us declare that $\xi \in \text{Bad}$ if formula (24.87) is not well-defined for some $z \in \widehat{\mathbb{C}}$, which means that the numerator and the denominator of T_ξ have some common zero. For z to be a zero of the denominator of T_ξ means that

$$Q(z) = \frac{-\xi}{1-\xi} \tag{24.89}$$

whereas being a zero of the numerator means that

$$\frac{P(z)}{z^d} = \frac{-\xi}{1-\xi}. \tag{24.90}$$

So, if $\xi \in \text{Bad}$ then $P(z) - z^d Q(z) = 0$ for some $z \in \mathbb{C}$ (check that $z \neq \infty$). This polynomial equation has only finitely many (at most $2d$) solutions and, therefore, the corresponding Bad set of parameters ξ determined by (24.89) or (24.90) is finite.

Assume for a contradiction that some multiplier function $\lambda_j(\xi) := (T_\xi^{p_j})'(H_j(\xi))$ is constant on $B(0,R)$. Since $T_0 = T$ and $T_1(z) = z^d$, we know that $0, 1 \notin \text{Bad}$. Let $\gamma : [0,1] \to \mathbb{C}$ be a smooth curve from $\gamma(0) = 0$ to $\gamma(1) = 1$ that avoids the set Bad and never returns to $B(0,R)$ if it ever leaves it. We claim that H_j can be extended holomorphically from 0 to 1 along the curve γ. The function H_j is already defined for all $t \in [0,1]$ so small that $|\gamma(t)| < R$. Let s be the supremum of all those $t \in [0,1]$ for which an holomorphic extension of H_j exists. By continuous dependence on ξ, we see that $T_{\gamma(s)}^{p_j}(\omega) = \omega$, where $\omega \in \mathbb{C}$ is any limit point of a sequence $(H_j(t_k))_{k=1}^\infty$ such that $\lim_{k\to\infty} t_k = s$. Conjugating all the maps $T_\xi : \widehat{\mathbb{C}} \to \widehat{\mathbb{C}}$ by a common Möbius transformation if necessary, we may assume that the orbit of ω under $T_{\gamma(s)}$ lies in $\mathbb{C}$. As the function $\lambda_j(\xi)$ is constant, using once again the continuous dependence on ξ we infer that $(T_{\gamma(s)}^{p_j})'(\omega) = \lambda_j(0) = \lambda_j \neq 1$. We may therefore apply the implicit function theorem exactly as above and deduce that the function H_j extends uniquely and holomorphically to some open neighborhood of $\gamma(s)$ so that (24.88) holds. In particular, H_j extends along the curve γ beyond the point $\gamma(s)$. This implies that $s = 1$, as otherwise this would contradict the very definition of s. But $T_{\gamma(1)}(z) = T_1(z) = z^d$ and the only periodic points of T_1 are 0 (whose multiplier is zero) and points lying on the unit circle $\mathbb{S}^1 := \{z \in \mathbb{C} : |z| = 1\}$. The multipliers of these latter points all have moduli at least equal to $d > 1$. On the other hand, $T_{\gamma(0)} = T_0 = T$ and $H_j(0) = \omega_j$ has a multiplier of modulus 1. This contradicts the constantness of the multiplier function $\lambda_j(\xi)$ along γ. In conclusion, no multiplier function $\lambda_j : B(0,R) \to \mathbb{C}$ is constant. Thus, for every $1 \le j \le k$ there exists $n_j \in \mathbb{N}$ and a complex number $a_j \neq 0$ such that the Taylor series expansion of $\lambda_j/\lambda_j(0)$ about 0 takes the form

$$\frac{\lambda_j(\xi)}{\lambda_j(0)} = 1 + a_j \xi^{n_j} + O(\xi^{n_j+1}).$$

Since the Taylor series expansion of $\sqrt{1+x}$ about 0 is given by

$$\sqrt{1+x} = 1 + \frac{1}{2}x + O(x^2),$$

we obtain that

$$
\begin{aligned}
|\lambda_j(\xi)| = \left|\frac{\lambda_j(\xi)}{\lambda_j(0)}\right| &= \sqrt{[1+\mathrm{Re}(a_j\xi^{n_j})]^2 + [\mathrm{Im}(a_j\xi^{n_j})]^2 + O(\xi^{n_j+1})} \\
&= \sqrt{1 + 2\mathrm{Re}(a_j\xi^{n_j}) + (\mathrm{Re}(a_j\xi^{n_j}))^2 + (\mathrm{Im}(a_j\xi^{n_j}))^2 + O(\xi^{n_j+1})} \\
&= 1 + \mathrm{Re}(a_j\xi^{n_j}) + \frac{1}{2}(\mathrm{Re}(a_j\xi^{n_j}))^2 + \frac{1}{2}(\mathrm{Im}(a_j\xi^{n_j}))^2 + O(\xi^{n_j+1}) \\
&= 1 + \mathrm{Re}(a_j\xi^{n_j}) + O(\xi^{n_j+1}).
\end{aligned}
$$

Writing $\xi = re^{i\theta}$, let

$$s_j(\theta) := \mathrm{sgn}(\mathrm{Re}(a_j\xi^{n_j})).$$

If $s_j(\theta) = 1$, then it follows that $|\lambda_j(re^{i\theta})| > 1$ for all $r > 0$ small enough (depending on θ). While if $s_j(\theta) = -1$, then $|\lambda_j(re^{i\theta})| < 1$ for all $r > 0$ small enough (depending on θ).

Obviously, each s_j is a step function defined on the circle $\mathbb{R}/(2\pi\mathbb{Z})$, which takes on the values ± 1 everywhere except at $2n_j$ uniformly interspersed jump discontinuities where it is undefined. Hence,

$$\frac{1}{2\pi}\int_0^{2\pi} s_j(\theta)\,d\theta = 0. \tag{24.91}$$

Therefore,

$$\frac{1}{2\pi}\int_0^{2\pi} \sum_{j=1}^{k} s_j(\theta)\,d\theta = 0 \tag{24.92}$$

and, except at its finitely many discontinuities where it is undefined, the sum $\sum_{j=1}^{k} s_j(\theta)$ is an integer. For every $\theta \in \mathbb{R}/(2\pi\mathbb{Z})$, set

$$I_+(\theta) := \{1 \le j \le k \mid s_j(\theta) = +1\} \quad \text{and} \quad I_-(\theta) := \{1 \le j \le k \mid s_j(\theta) = -1\}.$$

If $\#I_+(\theta) > \#I_-(\theta)$ for some $\theta \in \mathbb{R}/(2\pi\mathbb{Z})$, then $\sum_{j=1}^{k} s_j(\theta) \ge 1$. Hence, if $\#I_+(\theta) > \#I_-(\theta)$ for all $\theta \in \mathbb{R}/(2\pi\mathbb{Z})$ (excepting discontinuities), then (24.92) would yield $0 \ge 1$. This contradiction produces some $\theta^* \in \mathbb{R}/(2\pi\mathbb{Z})$ (which is not a discontinuity) such that $\#I_+(\theta^*) \le \#I_-(\theta^*)$. Thus, $\#I_-(\theta^*) \ge k/2$. So, if $r > 0$ is small enough, then $T_{re^{i\theta^*}}$ has at least $\#I_-(\theta^*) \ge k/2$ attracting periodic points $H_j(re^{i\theta^*})$. In view of Corollary 24.1.21, this yields

$$\frac{k}{2} \le 2\deg(T_{re^{i\theta^*}}) - 2 \le 2d - 2.$$

Consequently, $k \le 4d - 4$. □

Combining the above result with Corollary 24.2.23 immediately gives the following result.

Theorem 24.3.2. *Every rational function $T : \widehat{\mathbb{C}} \to \widehat{\mathbb{C}}$ with* $\deg(T) \geq 2$ *has at most* $6\deg(T) - 6$ *nonrepelling periodic cycles (i. e., cycles with multipliers of moduli at most* 1*). In particular, there are finitely many nonrepelling periodic points.*

Remark 24.3.3. Using quasiconformal surgery, Shishikura proved in [119] that, in fact, the number of nonrepelling periodic cycles does not exceed $2\deg(T) - 2$ and this bound is sharp.

We are now ready to prove the following dynamical characterization of Julia sets.

Theorem 24.3.4. *The Julia set $\mathcal{J}(T)$ of a rational function $T : \widehat{\mathbb{C}} \to \widehat{\mathbb{C}}$ with* $\deg(T) \geq 2$ *is the closure of the set of its repelling periodic points.*

Proof. Let $\mathrm{Per}(T)$ denote the set of all periodic points of T and let $\mathrm{Per}_r(T)$ denote its subset of all repelling periodic points. As $\mathcal{J}(T)$ is closed, Theorem 24.1.22 yields that

$$\overline{\mathrm{Per}_r(T)} \subseteq \mathcal{J}(T).$$

Since the Julia set $\mathcal{J}(T)$ is perfect (by Theorem 24.1.15) and since the set of nonrepelling periodic points is finite (per Theorem 24.3.2), it suffices to show that

$$\mathcal{J}(T) \subseteq \overline{\mathrm{Per}(T)}.$$

To prove this inclusion, suppose by contradiction that there exists $w \in \mathcal{J}(T) \setminus \overline{\mathrm{Per}(T)}$. As $\mathcal{J}(T)$ is perfect, we may select $w \neq \infty$ not to be a critical value of T, i. e., not the image under T of a critical point of T. Then $T^{-1}(w)$ contains at least two distinct points, say a and b. Since neither a nor b are critical points of T, there are two holomorphic inverse branches $T_a^{-1} : D \to \mathbb{C}$ and $T_b^{-1} : D \to \mathbb{C}$, where D is some sufficiently small open disk centered at w, uniquely determined by the requirement that $T_a^{-1}(w) = a$ and $T_b^{-1}(w) = b$. Decreasing the disk D if necessary, we may assume that

$$D \cap \overline{\mathrm{Per}(T)} = \emptyset \tag{24.93}$$

and that

$$T_a^{-1}(D) \cap T_b^{-1}(D) = \emptyset. \tag{24.94}$$

Recall that given four complex numbers z_1, z_2, z_3, z_4, their *cross ratio* is defined by

$$[z_1, z_2, z_3, z_4] := \frac{(z_1 - z_3)(z_2 - z_4)}{(z_1 - z_4)(z_2 - z_3)}.$$

Consider the sequence of functions

$$\varphi_n := [T^n, T_a^{-1}, T_b^{-1}, \mathrm{Id}] : D \to \widehat{\mathbb{C}}.$$

That is,

$$\varphi_n(z) = \frac{(T^n(z) - T_b^{-1}(z))(T_a^{-1}(z) - z)}{(T^n(z) - z)(T_a^{-1}(z) - T_b^{-1}(z))}. \tag{24.95}$$

Because of (24.93)–(24.94), none of the four terms above ever vanishes (if $T^n(z) = \infty$, then $(T^n(z) - T_b^{-1}(z))/(T^n(z) - z) = 1$). Thus, the sequence $(\varphi_n)_{n=1}^{\infty}$ omits the number 0.

If it turned out that $\varphi_n(z) = 1$, then simple algebraic manipulations show that we would have $T^n(z) = T_a^{-1}(z)$ or $T_b^{-1}(z) = z$. This would imply that $T^{n+1}(z) = z$ or $T(z) = z$, contrary to (24.93). So, $(\varphi_n)_{n=1}^{\infty}$ also omits the number 1.

As neither of the terms in the denominator of (24.95) ever vanishes thanks to (24.93)–(24.94), if it happened that $\varphi_n(z) = \infty$ then this would mean that

$$T^n(z) - T_b^{-1}(z) = \infty \quad \text{or} \quad T_a^{-1}(z) - z = \infty.$$

The latter option does not hold since $T_a^{-1}(D) \subseteq \mathbb{C}$, which only leaves the possibility that $T^n(z) = \infty$ since $T_b^{-1}(D) \subseteq \mathbb{C}$. But then we would have that

$$\varphi_n(z) = \frac{T_a^{-1}(z) - z}{T_a^{-1}(z) - T_b^{-1}(z)} \neq \infty.$$

This contradiction shows that the sequence $(\varphi_n)_{n=1}^{\infty}$ omits the point ∞, too.

By Montel's theorem, we deduce that the sequence $(\varphi_n)_{n=1}^{\infty}$ is normal. Hence, the sequence $(\Phi_n(z))_{n=1}^{\infty}$ defined by

$$\Phi_n(z) := \frac{T^n(z) - T_b^{-1}(z)}{T^n(z) - z} = \varphi_n(z) \cdot \frac{T_a^{-1}(z) - T_b^{-1}(z)}{T_a^{-1}(z) - z}$$

is also normal. Finally, given that

$$T^n(z) = \frac{z\Phi_n(z) - T_b^{-1}(z)}{\Phi_n(z) - 1},$$

the sequence $(T^n|_D)_{n=1}^{\infty}$ is normal. So $w \in \mathcal{F}(T)$. This contradiction finishes the proof. □

We can now establish that the restriction of T to $\mathcal{J}(T)$ is topologically exact.

Corollary 24.3.5. *Let $T : \widehat{\mathbb{C}} \to \widehat{\mathbb{C}}$ be a rational function with $\deg(T) \geq 2$. Then $T : \mathcal{J}(T) \to \mathcal{J}(T)$ is topologically exact.*

Proof. By Theorem 24.1.4, we know that $\mathcal{J}(T) \neq \emptyset$. Let $U \subseteq \widehat{\mathbb{C}}$ be an open set such that $U \cap \mathcal{J}(T) \neq \emptyset$. We need to show that there is $N = N(U) \in \mathbb{N}$ such that $T^N(U) \supseteq \mathcal{J}(T)$. By Theorem 24.3.4, the map T has at least one repelling periodic point in $U \cap \mathcal{J}(T)$; call it z.

Let $q \in \mathbb{N}$ be a period of z. Since z is repelling, we know that $|(T^q)'(z)| > 1$ and the mean value inequality applied to the local inverse branch of T^q that maps z to itself entails the existence of $R > 0$ such that

$$T^q(B(z,r)) \supseteq B(z,r), \quad \forall r \in (0,R]. \tag{24.96}$$

Fix $r \in (0,R]$ so small that

$$U \supseteq B(z,r). \tag{24.97}$$

Applying Corollary 24.1.10 to T^q and $B(z,r)$, there is $n = n(r) \in \mathbb{N}$ such that

$$\bigcup_{k=0}^{n} T^{kq}(B(z,r)) \supseteq \mathcal{J}(T^q).$$

Along with (24.96)–(24.97) and the fact that $\mathcal{J}(T^q) = \mathcal{J}(T)$ per Theorem 24.1.5, this implies that

$$T^{nq}(U) \supseteq T^{nq}(B(z,r)) = \bigcup_{k=0}^{n} T^{kq}(B(z,r)) \supseteq \mathcal{J}(T^q) = \mathcal{J}(T). \qquad \square$$

As an immediate consequence of this corollary we get the following two ramifications. The first one ensues from the fact that any topologically exact system is topologically transitive and that topological transitivity can be characterized by the absence of any proper subsystem with nonempty interior (see Theorem 1.5.11).

Corollary 24.3.6. *If $T : \widehat{\mathbb{C}} \to \widehat{\mathbb{C}}$ is a rational function with $\deg(T) \geq 2$, then either $\mathcal{J}(T)$ is a nowhere dense subset of $\widehat{\mathbb{C}}$ or $\mathcal{J}(T) = \widehat{\mathbb{C}}$.*

The second repercussion follows from the fact that any topologically exact system is strongly transitive and from the very definition of strong transitivity (see Definition 1.5.14).

Corollary 24.3.7. *If $T : \widehat{\mathbb{C}} \to \widehat{\mathbb{C}}$ is a rational function with $\deg(T) \geq 2$, then for each $z \in \mathcal{J}(T)$ we have*

$$\overline{\bigcup_{n=0}^{\infty} T^{-n}(z)} = \mathcal{J}(T)$$

and $\mathcal{J}(T)$ is the only nonempty closed backward T-invariant subset of $\mathcal{J}(T)$, i. e., the only nonempty closed subset F of $\mathcal{J}(T)$ satisfying

$$T^{-1}(F) \subseteq F.$$

We end this section with a useful technical fact.

Proposition 24.3.8. *If* $T : \widehat{\mathbb{C}} \to \widehat{\mathbb{C}}$ *is a rational function with* $\deg(T) \geq 2$, *then* $z \in \mathcal{F}(T)$ *if and only if there exist an open neighborhood* $U_z \subseteq \widehat{\mathbb{C}}$ *of* z *and a strictly increasing sequence* $(n_k)_{k=1}^{\infty}$ *of positive integers such that the family* $\{T^{n_k}|_{U_z}\}_{k=1}^{\infty}$ *is normal.*

Proof. The direct implication is trivial. For the converse, let $\zeta \in \mathcal{J}(T)$. By Lemma 24.1.6, we may assume without loss of generality that $\infty \notin \mathcal{J}(T)$. It suffices to show that for any open neighborhood U of ζ and any strictly increasing sequence $(n_k)_{k=1}^{\infty}$ of positive integers, the sequence $(T^{n_k}|_U)_{k=1}^{\infty}$ is not normal. Suppose for a contradiction that such a sequence is normal. Passing to a subsequence if necessary, there exist $R > 0$ and an analytic function $f : B(\zeta, R) \to \widehat{\mathbb{C}}$ such that $B(\zeta, R) \subseteq U$ and the sequence $(T^{n_k}|_{B(\zeta,R)})_{k=1}^{\infty}$ converges uniformly to f. Since f is continuous, there exists $r \in (0, R)$ such that

$$f(B(\zeta, r)) \subseteq B\Big(\zeta, \frac{1}{4}\operatorname{diam}(\mathcal{J}(T))\Big). \tag{24.98}$$

But $T^{n_k}(B(\zeta, r)) \supseteq \mathcal{J}(T)$ for all $k \in \mathbb{N}$ large enough, per Corollary 24.3.5. Consequently, $f(B(\zeta, r)) \supseteq \mathcal{J}(T)$. This contradicts (24.98). □

24.4 The structure of the Fatou set

In this section, we provide a complete description of the Fatou set of any rational function $T : \widehat{\mathbb{C}} \to \widehat{\mathbb{C}}$ with $\deg(T) \geq 2$. We first classify the connected components of $\mathcal{F}(T)$ that are fixed (i. e., forward invariant) under T. From this, we easily derive the classification of the periodic components of $\mathcal{F}(T)$. Next, by using this and the Riemannn–Hurwitz formula we show that every periodic component of $\mathcal{F}(T)$ is either a Herman ring or is simply connected or infinitely connected. Then we will cite Sullivan's nonwandering theorem, which asserts that every connected component of the Fatou set is eventually periodic.

Several deep and beautiful theorems from complex analysis and from the theory of Riemann surfaces will be crucial in the proofs that follow, including Koebe's uniformization theorem (indirectly) and Koebe's distortion theorem discussed in the first two chapters of this volume.

24.4.1 Forward invariant components of the Fatou set

The primary goal in front of us is to classify the forward invariant connected components of the Fatou set. This theorem is essentially due to Cremer [29] and Fatou [49]. Fatou [49, Section 56, p. 249] proved that given a rational function T with $\deg(T) \geq 2$, if $\{T^n|_V\}$ has only constant limit functions, then V is an immediate attractive basin or a Leau–Fatou domain. In the case of transcendental functions, his proof shows that the only further possibility is that of a Baker domain. Cremer [29, p. 317] proved that if $\{T^n|_V\}$

has nonconstant limit functions, then V is a Siegel disk or a Herman ring. Neither Fatou nor Cremer stated the full classification theorem but Töpfer's remarks in [135, p. 211] came fairly close to it.

The reader may want to reminisce about Subsection 24.1.2.2 prior to reading onward.

Theorem 24.4.1. *Let $T : \widehat{\mathbb{C}} \to \widehat{\mathbb{C}}$ be a rational function with $\deg(T) \geq 2$ and let V be a forward T-invariant connected component of the Fatou set $\mathcal{F}(T)$, i. e., a component such that $T(V) \subseteq V$. There are four possibilities:*

(a) *V is the immediate basin of attraction to some attracting fixed point of T.*
(b) *V is the immediate basin of attraction to some simple rationally indifferent fixed point of T lying in ∂V.*
(c) *V is a topological disk (i. e., is homeomorphic to the open unit disk $\mathbb{D}$), and the map $T|_V : V \to V$ is analytically conjugate to an irrational rotation $R_{2\pi\alpha} : \mathbb{D} \to \mathbb{D}$. The domain V is then referred to as a* Siegel disk.
(d) *There exists an annulus $A_r(0) := \{z \in \mathbb{C} : r < |z| < 1\}$ with $r \in (0,1)$ such that the map $T|_V : V \to V$ is analytically conjugate to an irrational rotation $R_{2\pi\alpha} : A_r(0) \to A_r(0)$. The domain V is then called a* Herman ring.

In either case, $T(V) = V$.

The proof of this theorem is long and will be broken down into several steps. Let us begin with the following lemma, which is in part a direct proof of Theorem 24.1.8(a).

Lemma 24.4.2. *If V is a connected component of the Fatou set $\mathcal{F}(T)$, then so is $T(V)$. In addition, V is a connected component of $T^{-1}(T(V))$.*

Proof. Recall that the completely T-invariant Fatou set $\mathcal{F}(T)$ is open in $\widehat{\mathbb{C}}$, and hence each of its connected components, such as V, is open not only in $\mathcal{F}(T)$ but also in $\widehat{\mathbb{C}}$. Recall also that T is an open map, and thus $T(V)$ is open. Since $T(V)$ is a connected subset of $\mathcal{F}(T)$, there exists a unique connected component W of $\mathcal{F}(T)$ such that $T(V) \subseteq W$. We shall show that in fact

$$T(V) = W. \tag{24.99}$$

By way of contradiction, suppose that this equality does not hold. As W is connected, we know that $\partial_W(T(V)) \neq \emptyset$, where $\partial_W(T(V))$ is the boundary of the set $T(V)$ with respect to the domain W. Fix some $y \in \partial_W(T(V))$. Then $y \in W \setminus T(V)$ and there exists a sequence $(y_n)_{n=1}^{\infty}$ with $y_n \in T(V)$ for all $n \in \mathbb{N}$ such that $\lim_{n\to\infty} y_n = y$. So, for every $n \in \mathbb{N}$ there exists $x_n \in V$ such that $T(x_n) = y_n$. Passing to a subsequence if necessary, we may assume that the sequence $(x_n)_{n=1}^{\infty}$ converges to some point $x \in \widehat{\mathbb{C}}$. Of course, $x \in \overline{V}$. If it turned out that $x \in V$, then we would have that

$$y = \lim_{n\to\infty} y_n = \lim_{n\to\infty} T(x_n) = T(x) \in T(V),$$

and this contradiction shows that $y \in \overline{V} \setminus V = \partial V \subseteq \mathcal{J}(T)$. But then $y \in \mathcal{J}(T) \cap W \subseteq \mathcal{J}(T) \cap \mathcal{F}(T) = \emptyset$. This contradiction establishes (24.99).

Since $V \subseteq T^{-1}(T(V)) \subseteq \mathcal{F}(T)$ and V is a connected component of $\mathcal{F}(T)$, we conclude that V is also a connected component of $T^{-1}(T(V))$. □

Corollary 24.4.3. *If V is a forward T-invariant connected component of $\mathcal{F}(T)$, then $T(V) = V$.*

We now describe the conformal automorphisms of an annulus $A_r(0)$.

Proposition 24.4.4. *Let $r \in (0,1)$. Every conformal automorphism of the annulus $A_r(0)$ is either a rotation $R_{2\pi\alpha}(z) = e^{2\pi i\alpha}z$ or a composition $S \circ R_{2\pi\alpha}$, where $\alpha \in [0,1]$ and $S : A_r(0) \to A_r(0)$ is defined by $S(z) = r/z$.*

Proof. For every $t > 0$, let $\mathbb{S}_t = \{z \in \mathbb{C} : |z| = t\}$, and for every $0 < i < o$ let $A(0; i, o)$ denote the open annulus centered at the origin with inner radius i and outer radius o. Let $R : A_r(0) \to A_r(0)$ be a conformal automorphism, where we recall that $A_r(0) = A(0; r, 1)$. As R is a homeomorphism, we have that:

(a) Either: For every $u \in \mathbb{S}_r$ and every $v \in \mathbb{S}_1$, if $\lim_{n\to\infty} u_n = u$ and $\lim_{n\to\infty} v_n = v$, where $u_n, v_n \in A_r(0)$ for all $n \in \mathbb{N}$, then

$$\lim_{n\to\infty} R(u_n) = r \quad \text{and} \quad \lim_{n\to\infty} R(v_n) = 1,$$

(b) Or: For every $u \in \mathbb{S}_r$ and every $v \in \mathbb{S}_1$, if $\lim_{n\to\infty} u_n = u$ and $\lim_{n\to\infty} v_n = v$, where $u_n, v_n \in A_r(0)$ for all $n \in \mathbb{N}$, then

$$\lim_{n\to\infty} R(u_n) = 1 \quad \text{and} \quad \lim_{n\to\infty} R(v_n) = r.$$

If (a) holds, set $R^* := R$, whereas if (b) holds, set $R^* := S \circ R$, where we recall that $S(z) := r/z$. Then (a) holds in both cases if R is replaced by R^*. With this in mind, by applying the Schwarz reflection principle (Theorem A.3.6) and replacing the complex conjugacy $z \mapsto \overline{z}$ by the inversion $z \mapsto r^2/\overline{z}$ in the circle $\mathbb{S}_r$, we can extend R^* to a holomorphic map (denoted by the same symbol R^*) from $A_{r^2}(0)$ to $A_{r^2}(0)$ such that (a) holds with: (1) R replaced by the extended R^*; (2) r replaced by r^2; and (3) the annulus $A_r(0)$ replaced by the annulus $A_{r^2}(0)$. So, proceeding by induction, we construct a holomorphic map $R^*_\infty : A_0(0) \to A_0(0)$ such that:

(c) $R^*_\infty|_{A_r(0)} = R^*$;

(d) $R^*_\infty(A(0; r^{2^{n+1}}, r^{2^n})) = A(0; r^{2^{n+1}}, r^{2^n})$ for all $n \geq 0$; and

(e) $R^*_\infty(\mathbb{S}_{r^{2^n}}) = \mathbb{S}_{r^{2^n}}$.

It follows from (d) that $\lim_{z\to 0} R^*_\infty(z) = 0$ and therefore R^*_∞ extends to a holomorphic map from $\mathbb{D}$ to $\mathbb{D}$ such that $R^*_\infty(0) = 0$. Because of this and (e), Schwarz lemma (Lemma A.3.5) implies that R^*_∞ is a rotation. Thus, either $R = R^* = R_{2\pi\alpha}|_{A_r(0)}$ or $R = S^{-1} \circ R^* = S \circ R^*_\infty|_{A_r(0)} = S \circ R_{2\pi\alpha}|_{A_r(0)}$. □

As an almost immediate consequence of the above proposition, we obtain the following.

Corollary 24.4.5. *Let $r \in (0,1)$. Every conformal automorphism $R : A_r(0) \to A_r(0)$ of infinite order (i. e., such that $R^n \neq \mathrm{Id}$ for all $n \in \mathbb{N}$) is an irrational rotation of the annulus $A_r(0)$.*

Proof. Given the notation from Proposition 24.4.4, observe that $S \circ R_{2\pi\alpha} = R_{2\pi\alpha}^{-1} \circ S$. If $R = S \circ R_{2\pi\alpha}$, then $R^2 = S^2 = \mathrm{Id}$. In particular, R is of finite order. It therefore follows from Proposition 24.4.4 that $R = R_{2\pi\alpha}$. Since all rational rotations are of finite order, α must be irrational. □

Let V be a forward T-invariant connected component of $\mathcal{F}(T)$. In order to determine the dynamics of the restriction $T|_V : V \to V$ and ultimately prove Theorem 24.4.1, we first consider two cases characterized by the existence/nonexistence of nonconstant limit functions of the sequence $(T^n|_V)_{n=0}^{\infty}$.

Lemma 24.4.6. *Let V be a forward T-invariant connected component of $\mathcal{F}(T)$. If the sequence $(T^n|_V)_{n=0}^{\infty}$ has a nonconstant limit function, then either part* (c) *or part* (d) *of Theorem* 24.4.1 *holds.*

Proof. By hypothesis, there exist a nonconstant meromorphic function $R : V \to \widehat{\mathbb{C}}$ and an increasing sequence $(n_k)_{k=1}^{\infty}$ such that the sequence $(T^{n_k}|_V)_{k=1}^{\infty}$ converges to R uniformly on compact subsets of V and such that $\lim_{k\to\infty}(n_{k+1} - n_k) = \infty$. Given that $T^n(V) \subseteq V$ for all $n \in \mathbb{N}$, we deduce that $R(V) \subseteq \overline{V}$. In fact, $R(V) \subseteq V$ as $R(V)$ is open. Let $m_k := n_{k+1} - n_k$. Passing to a subsequence if necessary, we may assume that the sequence $(T^{m_k}|_V)_{k=1}^{\infty}$ converges uniformly on compact subsets of V to some meromorphic function $S : V \to \widehat{\mathbb{C}}$. Since $R(V) \subseteq V$, since images of compact sets under R are compact, and since $T^{n_{k+1}} = T^{m_k} \circ T^{n_k}$, we deduce that

$$R = S \circ R.$$

So, $S|_{R(V)}$ is the identity function. As $R(V) \subseteq V$ is open, it follows that $S = \mathrm{Id}|_V$. Passing to a subsequence, we may assume without loss of generality that $(T^{m_k-1}|_V)_{k=1}^{\infty}$ converges to some meromorphic function $M : V \to \widehat{\mathbb{C}}$. But then

$$\mathrm{Id}|_V = \lim_{k\to\infty} T^{m_k}|_V = \lim_{k\to\infty} T^{m_k-1}|_V \circ T|_V = M \circ T|_V. \tag{24.100}$$

Therefore, $M(T(V)) = V$. Since V is forward T-invariant, Corollary 24.4.3 asserts that $T(V) = V$. Thus, $M(V) \subseteq V$, and hence

$$\mathrm{Id}|_V = S = \lim_{k\to\infty} T \circ T^{m_k-1}|_V = T \circ M = T|_V \circ M.$$

This and (24.100) show that $T|_V : V \to V$ is a conformal automorphism and $(T|_V)^{-1} = M$. As $\mathrm{Id}|_V = \lim_{k\to\infty} T^{m_k}|_V$, we thereby see that the topological group of conformal automorphisms of the Riemann surface V does not have the discrete topology. Invoking also the fact that the complement of V contains at least three points, we infer (see [136]) that V is conformally equivalent either to the open unit disk $\mathbb{D}$ or to an annulus $A_r(0)$ for some $r \in [0,1)$.

Let us first consider the case where V is conformally equivalent to $\mathbb{D}$. Recalling that $\mathbb{D}$ is conformally equivalent to the upper-half plane $\mathbb{H} := \{z \in \mathbb{C} \mid \operatorname{Im}(z) > 0\}$, let $\varphi : V \to \mathbb{H}$ be a conformal isomorphism. Then $\varphi \circ T \circ \varphi^{-1} : \mathbb{H} \to \mathbb{H}$ is a conformal automorphism of $\mathbb{H}$, and hence takes the form of a Möbius transformation $\mathbb{H} \ni z \mapsto \frac{az+b}{cz+d} \in \mathbb{H}$. By the same formula, the map $\varphi \circ T \circ \varphi^{-1}$ extends to a Möbius map from $\widehat{\mathbb{C}}$ to $\widehat{\mathbb{C}}$ and $\varphi \circ T \circ \varphi^{-1}(\overline{\mathbb{H}}) = \overline{\mathbb{H}}$. So, $\varphi \circ T \circ \varphi^{-1} : \overline{\mathbb{H}} \to \overline{\mathbb{H}}$ has a fixed point according to Brouwer's fixed-point theorem ($\overline{\mathbb{H}}$ is understood as a subset of $\widehat{\mathbb{C}}$ and is therefore homeomorphic to the closed unit disk $\overline{\mathbb{D}}$).

Suppose for a contradiction that $\varphi \circ T \circ \varphi^{-1}$ has no fixed point in $\mathbb{H}$. Then it has either exactly one or exactly two fixed points in $\partial\overline{\mathbb{H}}$ satisfying the equation $z = (az+b)/(cz+d)$. In the former case, conjugating $\varphi \circ T \circ \varphi^{-1}$ by a Möbius transformation which preserves $\overline{\mathbb{H}}$, we may assume without loss of generality that the sole fixed point is ∞. But then $\varphi \circ T \circ \varphi^{-1}(z) = z + b$ for some $b \in \mathbb{R} \setminus \{0\}$. Thus, $\lim_{n\to\infty} \varphi \circ T^n \circ \varphi^{-1}(z) = \lim_{n\to\infty}(z+nb) = \infty$ for all $z \in \mathbb{H}$. However, this implies that the sequence $(T^n(\zeta))_{n=1}^{\infty}$ converges to ∂V for every $\zeta \in V$, which contradicts the assumption that the sequence $(T^n|_V)_{n=1}^{\infty}$ has a nonconstant limit. So, $\varphi \circ T \circ \varphi^{-1}$ cannot have only one fixed point in $\partial\overline{\mathbb{H}}$. Assume then that $\varphi \circ T \circ \varphi^{-1}$ has exactly two fixed points in $\partial\overline{\mathbb{H}}$. In that case, conjugating $\varphi \circ T \circ \varphi^{-1}$ by a Möbius transformation that preserves $\mathbb{H}$, we may assume without loss of generality that these two fixed points are 0 and ∞. But then $\varphi \circ T \circ \varphi^{-1}(z) = az$ for some $a > 0$ because $\varphi \circ T \circ \varphi^{-1}$ preserves $\mathbb{H}$. Moreover, $a \neq 1$ because $\varphi \circ T \circ \varphi^{-1}$ has no fixed point in $\mathbb{H}$ by assumption. If $a > 1$, then we once again get $\lim_{n\to\infty} \varphi \circ T^n \circ \varphi^{-1}(z) = \lim_{n\to\infty} a^n z = \infty$ for all $z \in \mathbb{H}$, which leads to a contradiction in precisely the same way as in the previous case. A similar contradiction is reached if $0 < a < 1$. In conclusion, the map $\varphi \circ T \circ \varphi^{-1} : \mathbb{H} \to \mathbb{H}$ must have a fixed point and, therefore, $T : V \to V$ has a fixed point, say w.

Let $\psi : \mathbb{D} \to V$ be a conformal isomorphism such that $\psi(0) = w$. Then $\psi^{-1} \circ T \circ \psi(0) = 0$. It follows from the classification of conformal automorphisms of the unit disk $\mathbb{D}$ that $\psi^{-1} \circ T \circ \psi : \mathbb{D} \to \mathbb{D}$ is a rotation through some angle α, i. e., $\psi^{-1} \circ T \circ \psi(z) = e^{2\pi i\alpha} z$. Observe that α must be irrational, as otherwise $w \in V \subseteq \mathcal{F}(T)$ would be a rationally indifferent periodic point of T and would thereby belong to $\mathcal{J}(T)$, an impossibility. So V is a Siegel disk, i. e., part (c) of Theorem 24.4.1 holds in this case.

Finally, we consider the remaining case of V being conformally isomorphic to an annulus $A_r(0)$ for some $0 \le r < 1$. Let us first rule out the possibility that $r = 0$. Indeed, a conformal isomorphism $\varphi : A_r(0) \to V$ would then extend continuously to a map $\overline{\varphi} : \mathbb{D} \to V \cup \{\overline{\varphi}(0)\} \subseteq \overline{V}$ and $\overline{\varphi}(0) \in \partial V$ would be an isolated point of $\mathcal{J}(T)$, contrary to the perfectness of Julia sets. So, $r > 0$. Note that $T|_V$ is not of finite order, i. e., there is no $k \in \mathbb{N}$ such that $T^k|_V = \mathrm{Id}$. Otherwise, we would have that

$T^k = \mathrm{Id}_{\widehat{\mathbb{C}}}$, whence $1 = \deg(\mathrm{Id}_{\widehat{\mathbb{C}}}) = (\deg(T))^k > 1$. Thus, $T|_V$ is of infinite order. By Corollary 24.4.5, we conclude that V is a Herman ring, i. e., part (d) of Theorem 24.4.1 holds in this case. □

Lemma 24.4.6 affirms that in the case where the sequence $(T^n|_V)_{n=0}^\infty$ has a nonconstant limit function, the forward T-invariant component V is either a Siegel disk or a Herman ring. What can be said when the sequence $(T^n|_V)_{n=0}^\infty$ has only constant limit functions? We need to split this case into several subcases, starting with the following.

Lemma 24.4.7. *Let V be a forward T-invariant connected component of $\mathcal{F}(T)$. If the sequence $(T^n|_V)_{n=0}^\infty$ has a constant limit function whose value belongs to V, then part* (a) *of Theorem* 24.4.1 *holds.*

Proof. The hypothesis means that there exists an increasing sequence $(n_k)_{k=1}^\infty$ of positive integers such that the sequence $(T^{n_k}|_V)_{k=1}^\infty$ converges to some point $w \in V$ uniformly on compact subsets of V. But then, for every $z \in V$ we know that $T(z) \in V$, and hence

$$w = \lim_{k\to\infty} T^{n_k}(T(z)) = T\Big(\lim_{k\to\infty} T^{n_k}(z)\Big) = T(z) \tag{24.101}$$

and $\lim_{k\to\infty}(T^{n_k})'(z) = 0$ uniformly on compact subsets of V. In particular, $w = T(w)$ and $\lim_{k\to\infty}(T'(w))^{n_k} = 0$, which implies that $|T'(w)| < 1$. In other words, w is an attracting fixed point of T and $w \in V \cap A^*(w)$, where $A^*(w)$ is the immediate basin of attraction of w. Hence, V and $A^*(w)$ are connected components of the Fatou set $\mathcal{F}(T)$ with nonempty intersection. Thus, $V = A^*(w)$. □

The remaining case to consider is when the sequence $(T^n|_V)_{n=0}^\infty$ has only constant limit functions whose values belong to the boundary $\partial V = \overline{V} \setminus V$ of the domain V. To deal with this case, we shall first prove another helpful lemma.

Lemma 24.4.8. *Let V be a forward T-invariant connected component of $\mathcal{F}(T)$. If all the limit functions of the sequence $(T^n|_V)_{n=0}^\infty$ are constant and their values belong to ∂V, then T has a fixed point $w \in \partial V$ such that $|T'(w)| = 1$ and the sequence $(T^n|_V)_{n=0}^\infty$ converges to w uniformly on compact subsets of V.*

Proof. Suppose that $(n_k)_{k=1}^\infty$ is an increasing sequence of positive integers such that the sequence $(T^{n_k}|_V)_{k=1}^\infty$ converges to some point $w \in \partial V$ uniformly on compact subsets of V. The same calculation as in (24.101) shows that $T(z) = w$ for all $z \in V$ and $T(w) = w$ thanks to the continuity of T. Fix $R > 0$ so small that $B(w, 2R)$ contains no fixed point of T other than w. Fix also an arbitrary point $\xi \in V$. The set

$$A := \{n \geq 0 \mid T^n(\xi) \in B(w, R)\} \tag{24.102}$$

is infinite as it contains a tail of the sequence $(n_k)_{k=1}^\infty$. If $(m_k)_{k=1}^\infty$ is any strictly increasing sequence of elements in A such that the sequence $(T^{m_k}|_V)_{k=1}^\infty$ converges uniformly on compact subsets of V, then the limit of $(T^{m_k}|_V)_{k=1}^\infty$ is some constant function ζ, and as above, $T(\zeta) = \zeta$. It follows from (24.102) that $\rho(\zeta, w) \le R < 2R$, where ρ is the spherical metric on $\widehat{\mathbb{C}}$ and, therefore, $\zeta = w$. If $\mathbb{N} \setminus A$ is infinite, then there is a strictly increasing sequence $(s_k)_{k=1}^{\infty}$ in $\mathbb{N} \setminus A$ such that $s_k - 1 \in A$ for all $k \in \mathbb{N}$. Passing to a subsequence if necessary, we may assume without loss of generality that the sequence $(T^{s_k}|_V)_{k=1}^{\infty}$ converges uniformly on compact subsets of V. By hypothesis, its limit is a constant function $c \in \partial V$. By (24.102), we know that $\rho(c, w) \ge R$ and, in particular, $c \ne w$. On the other hand, as $s_k - 1 \in A$, with any $z \in V$ we obtain that

$$c = \lim_{k\to\infty} T^{s_k}(z) = T\Big(\lim_{k\to\infty} T^{s_k - 1}(z)\Big) = T(w) = w.$$

This contradiction demonstrates that $\mathbb{N} \setminus A$ must be finite and, therefore, the sequence $(T^n|_V)_{n=0}^{\infty}$ converges to w uniformly on compact subsets of V. This proves the second assertion of the lemma.

To prove the first assertion, note that if $|T'(w)| < 1$, then $V \cup A^*(w)$ would be a connected component of the Fatou set $\mathcal{F}(T)$ properly containing V. But this would contradict the fact that V is a connected component of $\mathcal{F}(T)$. Hence, $|T'(w)| \ge 1$. If $|T'(w)| > 1$, then for every point $z \in V$ we would deduce from the convergence of $(T^n(z))_{n=1}^{\infty}$ to w that $T^n(z) = w$ for some $n \in \mathbb{N}$, with n depending on z. But then $T^{-n}(w)$ would be uncountable for some $n \in \mathbb{N}$, yielding $T^n \equiv w$. This contradiction implies that $|T'(w)| = 1$. □

The final piece of the puzzle is contained in the following lemma.

Lemma 24.4.9. *Let V be a forward T-invariant connected component of $\mathcal{F}(T)$. If the sequence $(T^n|_V)_{n=0}^{\infty}$ has a constant limit function whose value $w \in \partial V$ is a fixed point for T, then $T'(w) = 1$.*

Proof. We may assume without loss of generality that $w = 0$. Let $\Delta \subseteq \overline{\Delta} \subseteq V$ be an open connected set. Since V is an open connected subset of $\widehat{\mathbb{C}}$ and $\Delta \cup T(\Delta) \subseteq V$, there exists a smooth arc $\gamma \subseteq V$ joining Δ and $T(\Delta)$. Taking $W \subseteq \overline{W} \subseteq V$ an open connected neighborhood of γ, the set

$$D := \Delta \cup W \cup T(\Delta)$$

satisfies $\overline{D} \subseteq V$ and has the following iterative properties for every $n \ge 0$:

(a) The set $T^n(D) \subseteq V$ is open and connected.

(b) $T^{n+1}(D) \cap T^n(D) \ne \emptyset$.

(c) $0 \notin T^n(D)$.

For every $n \ge 0$, let

$$D_n := \bigcup_{k=n}^{\infty} T^k(D).$$

In light of (a), (b), (c) and Lemma 24.4.8, the sets $(D_n)_{n=0}^{\infty}$ have the following properties:

(d) The set $D_n \subseteq V$ is open and connected.
(e) $\lim_{n\to\infty} \max\{\rho(z,0) : z \in \overline{D_n}\} = 0$.
(f) $T(D_n) = D_{n+1} \subseteq D_n$.
(g) $0 \notin D_n$.

Since $|T'(0)| = 1 \neq 0$ by Lemma 24.4.8, there exists $s > 0$ such that $T|_{B(0,s)}$ is an injection. By (e), we can fix $k \geq 0$ such that $\overline{D_k} \subseteq B(0,s) \subseteq \mathbb{C}$. From this and (f), we see that all the iterates $T^n|_{\overline{D_k}}$, $n \geq 0$, are injective. Choose arbitrarily $\widehat{\xi} \in D_k$ and per (e)+(g) fix $q > k$ such that $\widehat{\xi} \notin D_q$. For every $n \in \mathbb{N}$, define the meromorphic function $\widehat{g}_n : D_q \to \widehat{\mathbb{C}}$ by

$$\widehat{g}_n(z) := \frac{T^n(z)}{T^n(\widehat{\xi})}.$$

The sequence of functions $(\widehat{g}_n)_{n=1}^{\infty}$ does not take the value ∞, as $T^n(z) \in D_q \subseteq B(0,s) \subseteq \mathbb{C}$ and $T^n(\widehat{\xi}) \in D_k \subseteq V \subseteq \mathbb{C} \setminus \{0\}$. It also omits 0, as $T^n(z) \in D_q \subseteq V \subseteq \mathbb{C} \setminus \{0\}$ and $T^n(\widehat{\xi}) \in \mathbb{C}$. And it avoids 1, as $T^n|_{D_q}$ is one-to-one. Thus, by Montel's theorem, the sequence of functions $(\widehat{g}_n)_{n=1}^{\infty}$ is normal. As $T(0) = 0$ and $\lambda := T'(0) \neq 0$, the Taylor series expansion of T around 0 can be written as

$$T(z) = \lambda z + az^{p+1} + O(z^{p+2}),$$

with $p \in \mathbb{N}$ and $a \in \mathbb{C} \setminus \{0\}$. Then

$$\frac{T(z)}{z} = \lambda + az^p + O(z^{p+1}).$$

Therefore,

$$\lim_{z\to 0} \frac{T(z)}{z} = \lambda \quad \text{uniformly on } B(0,s) \tag{24.103}$$

provided that $s > 0$ is sufficiently small. Consequently,

$$\lim_{z\to 0} \frac{T^j(z)}{z} = \lambda^j \quad \text{uniformly on } D_q \tag{24.104}$$

for every $j \in \mathbb{N}$. Set $\xi := T^{q-k}(\widehat{\xi}) \in D_q$. For every $n \in \mathbb{N}$, define the meromorphic function $g_n : D_q \to \widehat{\mathbb{C}}$ by

$$g_n(z) := \frac{T^n(z)}{T^n(\xi)}.$$

Then

$$g_n(z) = \frac{T^n(z)}{T^n(\widehat{\xi})} \cdot \frac{T^n(\widehat{\xi})}{T^n(\xi)} = \widehat{g}_n(z) \cdot \frac{T^n(\widehat{\xi})}{T^{q-k}(T^n(\widehat{\xi}))}.$$

It follows from (24.104) that

$$\lim_{n\to\infty} \frac{g_n(z)}{\widehat{g}_n(z)} = \lambda^{k-q} \quad \text{uniformly on } D_q.$$

So, the sequence $(g_n)_{n=1}^{\infty}$ is, like $(\widehat{g}_n)_{n=1}^{\infty}$, normal. Fix a strictly increasing sequence $(n_j)_{j=1}^{\infty}$ such that $(g_{n_j})_{j=1}^{\infty}$ converges to some analytic function $g : D_q \to \widehat{\mathbb{C}}$ uniformly on compact subsets of D_q. Then

$$\begin{aligned}
g_{n_j+1}(z) &= \frac{T^{n_j+1}(z)}{T^{n_j+1}(\xi)} = \frac{T(T^{n_j}(z))}{T(T^{n_j}(\xi))} \\
&= \frac{\lambda T^{n_j}(z) + a(T^{n_j}(z))^{p+1} + O((T^{n_j}(z))^{p+2})}{\lambda T^{n_j}(\xi) + a(T^{n_j}(\xi))^{p+1} + O((T^{n_j}(\xi))^{p+2})} \\
&= \frac{\lambda T^{n_j}(z)[1 + a\lambda^{-1}(T^{n_j}(z))^p + O((T^{n_j}(z))^{p+1})]}{\lambda T^{n_j}(\xi)[1 + a\lambda^{-1}(T^{n_j}(\xi))^p + O((T^{n_j}(\xi))^{p+1})]} \\
&= g_{n_j}(z) \cdot \frac{1 + a\lambda^{-1}(T^{n_j}(z))^p + O((T^{n_j}(z))^{p+1})}{1 + a\lambda^{-1}(T^{n_j}(\xi))^p + O((T^{n_j}(\xi))^{p+1})} \\
&\to g(z) \quad \text{as } j \to \infty
\end{aligned}$$

while, by (24.103),

$$\begin{aligned}
g_{n_j+1}(z) &= \frac{T^{n_j}(T(z))}{T^{n_j}(T(\xi))} = \frac{T^{n_j}(T(z))}{T^{n_j}(\xi)} \cdot \frac{T^{n_j}(\xi)}{T(T^{n_j}(\xi))} \\
&\to g(T(z)) \cdot \lambda^{-1} \quad \text{as } j \to \infty.
\end{aligned}$$

Reconciling these two limit functions yields

$$g \circ T = \lambda g \quad \text{on } D_q. \tag{24.105}$$

As the functions $(T^n|_{D_q})_{n=1}^{\infty}$ are univalent, so are the functions $(g_n)_{n=1}^{\infty}$. It follows from Hurwitz's theorem (Theorem A.3.8) that either $g : D_q \to \widehat{\mathbb{C}}$ is univalent or is constant. Assume first that g is univalent. Iterating (24.105) gives

$$g \circ T^n = \lambda^n g, \quad \forall n \in \mathbb{N}. \tag{24.106}$$

In particular,

$$g \circ T^n(\xi) = \lambda^n g(\xi) = \lambda^n, \quad \forall n \in \mathbb{N}.$$

If λ were a root of unity, say $\lambda^p = 1$ for some $p \in \mathbb{N}$, then $g(T^{pn}(\xi)) = 1$ for all $n \in \mathbb{N}$, contrary to the facts that the set $\{T^{pn}(\xi) : n \in \mathbb{N}\}$ is infinite and the function g is one-to-one. So, λ is not a root of unity. Consequently, the set $\{\lambda^n : n \in \mathbb{N}\}$ is dense in the unit circle $\mathbb{S}^1$. Since $T^n|_V$ tends to 0 uniformly on compact subsets of V and $0 \notin D_q$, there exists $j \in \mathbb{N}$ so large that

$$T^n(T^q(D)) \cap T^q(D) = \emptyset, \quad \forall n \geq j. \tag{24.107}$$

Let $N \geq j$ be such that λ^N is close enough to 1 that

$$\lambda^N g(T^q(D)) \cap g(T^q(D)) \neq \emptyset.$$

It follows from this and (24.106) that

$$g(T^N(T^q(D))) \cap g(T^q(D)) \neq \emptyset.$$

In conjunction with (24.107), this implies that $g : D_q \to \widehat{\mathbb{C}}$ is not one-to-one. This contradiction rules out the case of $g : D_q \to \widehat{\mathbb{C}}$ being univalent. As the only other outcome of Hurwitz's theorem, the function $g : D_q \to \widehat{\mathbb{C}}$ must be constant. Since $g(\xi) = 1$, we must have $g \equiv 1$. But then (24.105) implies that $\lambda = 1$. □

We are finally in a position to establish Theorem 24.4.1.

Proof of Theorem 24.4.1. Three of the possibilities, namely (a), (c) and (d), have already been ascertained through Lemmas 24.4.6–24.4.7. The remaining case to consider is when the sequence $(T^n|_V)_{n=0}^{\infty}$ has only constant limit functions whose values belong to ∂V. In light of Lemmas 24.4.8–24.4.9, the map T has a simple rationally indifferent fixed point $w \in \partial V$ and $V \cap A_j^*(w) \neq \emptyset$ for some $j \in \{1, \ldots, p(w)\}$. Since both V and $A_j^*(w)$ are connected components of $\mathcal{F}(T)$, it follows that $V = A_j^*(w)$ and part (b) describes the fourth possibility. □

24.4.2 Periodic components of the Fatou set

As a direct consequence of Theorem 24.4.1, we obtain an analogous statement for the periodic components of the Fatou set.

Theorem 24.4.10 (Classification of the periodic components of the Fatou set). *Let $T : \widehat{\mathbb{C}} \to \widehat{\mathbb{C}}$ be a rational function with* $\deg(T) \geq 2$ *and let V be a periodic component of the Fatou set $\mathcal{F}(T)$, i. e., such that $T^q(V) \subseteq V$ for some $q \in \mathbb{N}$. Then $T^q(V) = V$ and one of the following four alternatives holds:*

(a) *V is the immediate basin of attraction to some attracting periodic point of period q for T.*
(b) *V is the immediate basin of attraction to some rationally indifferent periodic point of period q lying in ∂V.*

(c) *V is a topological disk (i. e., is homeomorphic to the open unit disk $\mathbb{D}$), and the map $T^q|_V : V \to V$ is analytically conjugate to an irrational rotation $R_{2\pi\alpha} : \mathbb{D} \to \mathbb{D}$. The domain V is then referred to as a* Siegel disk.
(d) *There exists an annulus $A_r(0)$ with $r \in (0,1)$ such that the map $T^q|_V : V \to V$ is analytically conjugate to an irrational rotation $R_{2\pi\alpha} : A_r(0) \to A_r(0)$. The domain V is then called a* Herman ring.

Each such type of periodic component exists for some rational functions. The existence of rational functions having type (a) or type (b) components is directly evidenced through the examination of the fixed point 0 for the one-parameter family of quadratic polynomials $z \mapsto z^2 + \lambda z$. The existence of Siegel disks and Herman rings is subtler. As we have already mentioned in Subsection 24.1.2.2, Carl Ludwig Siegel [121] demonstrated the existence of such disks (which is also of profound importance in the KAM theory); the proof can also be found in [21]. The study of Siegel disks, particularly the nature of the rotational numbers α for which they exist, has been the topic of extensive research led by many mathematicians. With Herman rings, the situation is even more delicate. They cannot exist for polynomials (see Theorem 24.7.3 below). Their existence was shown by Michel Herman [58] in 1984 only. His argument is based on Arnold's theorem. A proof due to Shishikura [119] can also be found in [21].

We now delimit the connectivity of the periodic components of the Fatou set.

Theorem 24.4.11. *If $T : \widehat{\mathbb{C}} \to \widehat{\mathbb{C}}$ is a rational function with* $\deg(T) \geq 2$, *then each periodic connected component of $\mathcal{F}(T)$ is either simply connected, doubly connected (in which case it is a Herman ring), or infinitely connected.*

Proof. Let V be a periodic component of the Fatou set $\mathcal{F}(T)$, i. e., a component such that $T^q(V) \subseteq V$ for some $q \in \mathbb{N}$. Replacing T^q by T, we may assume that $T(V) \subseteq V$.

Suppose that V is finitely connected and denote its connectivity by k. It then follows from formula (22.23) that the Euler characteristic $\chi(V)$ is finite; more precisely $\chi(V) = 2 - k$. In view of Lemma 24.1.7, we may apply the general Riemann–Hurwitz formula (Theorem 22.5.1) to the map $T|_V : V \to V$ to get

$$\chi(V) + \delta_T(V) = \deg(T|_V) \cdot \chi(V).$$

Equivalently,

$$(2 - k) + \delta_T(V) = \deg(T|_V) \cdot (2 - k). \tag{24.108}$$

If it turned out that $3 \leq k < \infty$, then (24.108) and the fact that $\delta_T(V) \geq 0$ would yield $\deg(T|_V) \cdot (2 - k) \geq (2 - k)$. So $\deg(T|_V) = 1$. But then $\delta_T(V) = 0$ per (24.108). This would mean that T has no critical point in V. But according to Theorems 24.1.20 and 24.2.22, this eliminates possibilities (a) and (b) in Theorem 24.4.10. And the remaining possibilities (c) and (d) imply that $k = 1$ and $k = 2$. This contradiction rules out the possibility that $3 \le k < \infty$.

If $k = 2$, then V obviously cannot be a Siegel disk. Moreover, (24.108) reduces to $\delta_T(V) = 0$, i. e., T has no critical point in V. Therefore, by virtue of Theorems 24.1.20 and 24.2.22, V cannot be the immediate basin of attraction of an attracting periodic point or of a rationally indifferent periodic point. It thus follows from Theorem 24.4.10 that V must be a Herman ring.

The only other possibilities are $k = 1$ and $k = \infty$. □

We now state without proof Sullivan's nonwandering theorem ([131]) which, along with the classification theorem 24.4.10, gives a complete description of the structure of Fatou components. Sullivan's nonwandering theorem constituted a breakthrough in the development of the theory of iteration of rational functions. In fact, it brought the theory to its modern state. Proofs of Sullivan's nonwandering theorem and its various generalizations, particularly to the dynamics of transcendental meromorphic functions, can be found in many sources, one of the clearest appearing in Steinmetz [123].

Theorem 24.4.12 (Sullivan's nonwandering theorem). *Let $T : \widehat{\mathbb{C}} \to \widehat{\mathbb{C}}$ be a rational function with* $\deg(T) \ge 2$. *Each connected component V of the Fatou set $\mathcal{F}(T)$ is eventually periodic, i. e., there are integers $0 \le p < q$ (depending on V) such that $T^{q-p}(T^p(V)) = T^p(V)$.*

As an immediate consequence of the last two theorems we get the following result.

Corollary 24.4.13. *Let $T : \widehat{\mathbb{C}} \to \widehat{\mathbb{C}}$ be a rational function with* $\deg(T) \ge 2$. *If K is a compact subset of the Fatou set $\mathcal{F}(T)$ which is disjoint from the basins of attraction of all rationally indifferent periodic points, then*

$$\overline{\bigcup_{n=0}^{\infty} T^n(K)} \subseteq \mathcal{F}(T)$$

and

$$\mathcal{J}(T) \cap \overline{\bigcup_{n=0}^{\infty} T^n(\mathrm{Crit}(T) \cap \mathcal{F}(T))} \subseteq \Omega(T),$$

where $\Omega(T)$ is the (finite) set of all rationally indifferent periodic points of T.

24.4.3 The postcritical set

Recall that the finite set

$$\mathrm{Crit}(T) := \{z \in \widehat{\mathbb{C}} \mid T'(z) = 0\}$$

is naturally called the *critical set* of T while its image $T(\mathrm{Crit}(T))$ is the *set of critical values* of T. These two sets play important roles for several reasons. One of those reasons is the subject of the following remark.

Remark 24.4.14. Every point z outside of $T(\mathrm{Crit}(T))$ admits a neighborhood N_z on which T^{-1} consists of finitely many (more precisely, $\deg(T)$ many) holomorphic inverse branches of T. For instance, $N_z = B(z, \delta_z)$ where $0 < \delta_z \le \mathrm{dist}(z, T(\mathrm{Crit}(T)))$ (see Corollary 22.5.6).

Dynamically, we are spontaneously led to examine the union of all iterates of the critical set.

Definition 24.4.15. Let $T : \widehat{\mathbb{C}} \to \widehat{\mathbb{C}}$ be a rational function. The countable set

$$\mathrm{PC}(T) := \bigcup_{n=1}^{\infty} T^n(\mathrm{Crit}(T)) = \bigcup_{n=0}^{\infty} T^n(T(\mathrm{Crit}(T)))$$

is instinctively named the *postcritical set* of T.

Given any $N \in \mathbb{N}$, we readily notice that

$$\mathrm{PC}(T^N) = \bigcup_{n=1}^{\infty} T^{Nn}(\mathrm{Crit}(T^N)) = \bigcup_{k=1}^{\infty} T^k(\mathrm{Crit}(T)) = \mathrm{PC}(T). \tag{24.109}$$

For a slew of reasons, the postcritical set plays an important role in complex dynamics. The following remark constitutes one of them.

Remark 24.4.16.
(a) Every point z outside of $\mathrm{PC}(T)$ is such that $(T^n)^{-1} = T^{-n}$ consists of finitely many (more precisely, $\deg(T^n) = (\deg(T))^n$ many) holomorphic inverse branches of T^n on some neighborhood of z, and this for every $n \in \mathbb{N}$. That neighborhood may depend on n. Given that $\mathrm{Crit}(T^n) = \bigcup_{k=0}^{n-1} T^{-k}(\mathrm{Crit}(T))$, and thus $T^n(\mathrm{Crit}(T^n)) = \bigcup_{j=1}^{n} T^j(\mathrm{Crit}(T)) \subseteq \mathrm{PC}(T)$, we can take $N_{z,n} = B(z, \delta_{z,n})$ where $0 < \delta_{z,n} \le \mathrm{dist}(z, T^n(\mathrm{Crit}(T^n)))$.
(b) Every point z outside of $\overline{\mathrm{PC}(T)}$ is such that every $(T^n)^{-1} = T^{-n}$ consists of finitely many (more precisely, $\deg(T^n) = (\deg(T))^n$ many) holomorphic inverse branches of T^n on some neighborhood of z. This time, though, a neighborhood independent of n exists. Set $N_z = B(z, \delta_z)$ where $0 < \delta_z \le \mathrm{dist}(z, \overline{\mathrm{PC}(T)})$ (see Corollary 22.5.6).

It is easy to see that $\mathrm{PC}(T) = [\mathrm{PC}(T) \cap \mathcal{F}(T)] \cup [\mathrm{PC}(T) \cap \mathcal{J}(T)]$ and

$$\mathrm{PC}(T) \cap \mathcal{F}(T) = \bigcup_{n=1}^{\infty} T^n(\mathrm{Crit}(T) \cap \mathcal{F}(T)) \quad \text{and} \quad \mathrm{PC}(T) \cap \mathcal{J}(T) = \bigcup_{n=1}^{\infty} T^n(\mathrm{Crit}(T) \cap \mathcal{J}(T)).$$

Obviously, $\overline{\mathrm{PC}(T)} = \overline{\mathrm{PC}(T) \cap \mathcal{F}(T)} \cup \overline{\mathrm{PC}(T) \cap \mathcal{J}(T)}$. Note also that

$$\overline{\mathrm{PC}(T)} = \mathrm{PC}(T) \cup \omega(\mathrm{Crit}(T)).$$

Moreover, $\mathrm{Int}(\mathrm{PC}(T)) = \emptyset$ since $\mathrm{PC}(T)$ is countable. The second part of Corollary 24.4.13 thereby reduces to the following.

Corollary 24.4.17. *If $T : \widehat{\mathbb{C}} \to \widehat{\mathbb{C}}$ is a rational function with* $\deg(T) \geq 2$, *then*

$$\mathcal{J}(T) \cap \overline{\mathrm{PC}(T) \cap \mathcal{F}(T)} = \mathcal{J}(T) \cap \partial(\mathrm{PC}(T) \cap \mathcal{F}(T)) \subseteq \Omega(T).$$

Finally, the sets $\mathrm{PC}(T)$, $\mathrm{PC}(T) \cap \mathcal{F}(T)$ and $\mathrm{PC}(T) \cap \mathcal{J}(T)$ are forward T-invariant. So are their closures, due to the continuity of T.

24.5 Cremer points, boundary of Siegel disks and Herman rings

In this section, we prove that the boundary of each Siegel disk and Herman ring is contained in the closure of the forward orbit of critical points. We will thereafter prove the same for Cremer points.

Theorem 24.5.1. *Let $T : \widehat{\mathbb{C}} \to \widehat{\mathbb{C}}$ be a rational function with* $\deg(T) \geq 2$. *If G is a Siegel disk or a Herman ring for T, then*

$$\partial G \subseteq \omega(\mathrm{Crit}(T) \cap \mathcal{J}(T)).$$

Proof. Seeking a contradiction, suppose that $\partial G \nsubseteq \omega(\mathrm{Crit}(T) \cap \mathcal{J}(T))$. Pick a point $y \in \partial G \setminus \omega(\mathrm{Crit}(T) \cap \mathcal{J}(T))$. Then there is $\delta > 0$ such that

$$B(y, 2\delta) \cap \overline{\mathrm{PC}(T) \cap \mathcal{J}(T)} \subseteq \{y\}. \tag{24.110}$$

Since ∂G consists either of one (if G is a Siegel disk) or two (if G is a Herman ring) connected components, none of which is a singleton, the boundary ∂G is perfect. Therefore, there exists a point

$$\xi \in (\partial G \cap B(y, \delta)) \setminus (\{y\} \cup \Omega(T)),$$

where $\Omega(T)$ is the (finite) set of all rationally indifferent periodic points of T. It then ensues from (24.110), Corollary 24.4.17, and the fact that $\partial G \subseteq \mathcal{J}(T)$, that there exists $\kappa > 0$ such that

$$B(\xi, 3\kappa) \cap \overline{\mathrm{PC}(T)} = \emptyset.$$

Pick a point

$$w \in B(\xi, \kappa) \cap G. \tag{24.111}$$

Then

$$B(w, 2\kappa) \cap \overline{\mathrm{PC}(T)} = \emptyset. \tag{24.112}$$

Let

$$\Delta := \mathbb{D} = B(0, 1)$$

if G is a Siegel disk and

$$\Delta := A(0; r, 1) = A_r(0)$$

with $r \in (0, 1)$ coming from Theorem 24.4.10(d) if G is a Herman ring. Let

$$H : \Delta \to G$$

be the analytic conjugacy arising from Theorem 24.4.10(c)/(d) in the Siegel/Herman case, respectively. Let

$$F := H(\{z \in \Delta : |z| = |H^{-1}(w)|\}). \tag{24.113}$$

Note that F is a compact set (homeomorphic to a circle) and $F \subseteq G$. Since $w \in G$, for every $n \in \mathbb{N}$ the intersection $G \cap T^{-n}(w)$ must be a singleton (otherwise, this would contradict Theorem 24.4.10(c)/(d)). Denoting its only element by w_n, we can readily check that

$$w_n \in F. \tag{24.114}$$

By (24.112) and Remark 24.4.16, for every $n \in \mathbb{N}$ there is a unique holomorphic branch

$$T_{w_n}^{-n} : B(w, 2\kappa) \to \widehat{\mathbb{C}} \tag{24.115}$$

of T^{-n} sending w to w_n. As $\xi \in \partial G \subseteq \mathcal{J}(T)$, we know that $T_{w_n}^{-n}(\xi) \in \mathcal{J}(T)$ for every $n \geq 0$. Thus, $T_{w_n}^{-n}(\xi) \notin G$ for all $n \geq 0$. Using (24.111) and (24.114), we conclude that

$$\operatorname{diam}(T_{w_n}^{-n}(B(\xi, \kappa))) \geq \operatorname{dist}(w_n, T_{w_n}^{-n}(\xi)) \geq \operatorname{dist}(F, \widehat{\mathbb{C}} \setminus G) > 0, \quad \forall n \geq 0.$$

Consequently,

$$\liminf_{n \to \infty} \operatorname{diam}(T_{w_n}^{-n}(B(\xi, \kappa))) \geq \operatorname{dist}(F, \widehat{\mathbb{C}} \setminus G) > 0,$$

contrary to Lemma 24.1.13. □

Theorem 24.5.2. *Let $T : \widehat{\mathbb{C}} \to \widehat{\mathbb{C}}$ be a rational function with $\deg(T) \geq 2$. If ξ is a Cremer periodic point of T, then $\xi \in \overline{\mathrm{PC}(T)} \cap \mathcal{J}(T)$.*

Proof. Let $p \in \mathbb{N}$ be the prime period of ξ. Seeking a contradiction, suppose that $\xi \notin \overline{\mathrm{PC}(T) \cap \mathcal{J}(T)}$. As $\xi \in \mathcal{J}(T) \backslash \Omega(T)$, combining the above assumption with Corollary 24.4.17 allows us to infer that $\xi \notin \overline{\mathrm{PC}(T)}$. By Remark 24.4.16, there exists $r > 0$ (e. g., choose $r = \frac{1}{2}\mathrm{dist}(\xi, \overline{\mathrm{PC}(T)})$) such that for every $n \in \mathbb{N}$ there exists a unique holomorphic branch

$$T_\xi^{-pn} : B(\xi, 2r) \to \widehat{\mathbb{C}} \tag{24.116}$$

of T^{-pn} that fixes ξ. But because $|(T^{pn})'(\xi)| = 1$ for all $n \in \mathbb{N}$ and $\xi \in \mathcal{J}(T)$, this contradicts Lemma 24.1.13. □

24.6 Continuity of Julia sets

If X is a set and (Y, ρ) is a bounded metric space, we consider the metric space (Y^X, ρ_∞) of all functions from X to Y with the metric of uniform convergence ρ_∞ given by

$$\rho_\infty(f, g) := \sup\{\rho(f(x), g(x)) : x \in X\}. \tag{24.117}$$

If ρ and d are two bounded metrics on Y, then the topologies induced by ρ_∞ and d_∞ on Y^X are the same whenever ρ and d are uniformly equivalent, i. e., when the identity map $\mathrm{Id}_Y : (Y, d) \to (Y, \rho)$ is uniformly bicontinuous. In particular, if Y is a compact metrizable space, then Y^X is (canonically) topologized by any metric on Y compatible with the topology of Y. In this section, $\widehat{\mathbb{C}}^{\widehat{\mathbb{C}}}$ is understood as a compact metrizable space topologized in that way. Moreover, we consider the set $\mathcal{R}$ of all rational functions on $\widehat{\mathbb{C}}$ as a (metrizable) subspace of $\widehat{\mathbb{C}}^{\widehat{\mathbb{C}}}$. We also consider the metric space $\mathcal{K}(\widehat{\mathbb{C}})$ of all compact subsets of $\widehat{\mathbb{C}}$ endowed with the Hausdorff metric ρ_H induced by the spherical metric ρ on $\widehat{\mathbb{C}}$. We further note that any metric on $\widehat{\mathbb{C}}$ compatible with its standard topology induces the same topology as ρ_H on $\mathcal{K}(\widehat{\mathbb{C}})$.

Theorem 24.6.1. *The function $\mathcal{R} \ni T \longmapsto \mathcal{J}(T) \in \mathcal{K}(\widehat{\mathbb{C}})$ is lower semicontinuous. More precisely, for every $S \in \mathcal{R}$ and every $\varepsilon > 0$ there exists an open neighborhood U of S in $\mathcal{R}$ such that*

$$T \in U \quad \Longrightarrow \quad \mathcal{J}(S) \subseteq B_\rho(\mathcal{J}(T), \varepsilon).$$

In addition, if the Fatou set $\mathcal{F}(S)$ consists only of basins of attraction of attracting periodic points, then S is a continuity point of the function $\mathcal{R} \ni T \mapsto \mathcal{J}(T) \in \mathcal{K}(\widehat{\mathbb{C}})$.

Proof. Concerning the lower semicontinuity, it follows from Theorem 24.3.4 that there is a finite subset E of the set of repelling periodic points of S in $\mathbb{C}$ such that

$$\bigcup_{w \in E} B_\rho(w, \varepsilon/4) \supseteq \mathcal{J}(S). \tag{24.118}$$

Let $p \in \mathbb{N}$ be the product of the prime periods of all points in E. Then

$$S^p(w) = w \quad \text{and} \quad |(S^p)'(w)| > \lambda \tag{24.119}$$

for all $w \in E$ and some $\lambda \in (1, \infty)$. Since the function $\mathcal{R} \ni T \mapsto |(T^p)'(w)| \in \mathbb{R}$ is continuous for each $w \in E$, an application of Hurwitz's theorem (Theorem A.3.8) confirms the existence of $\delta > 0$ and $\theta \in (0, 3\varepsilon/4]$ such that

$$T \in B_{\rho_\infty}(S, \delta) \quad \Longrightarrow \quad |(T^p)'(z)| > \sqrt{\lambda}, \quad \forall z \in \bigcup_{w \in E} B(w, 4\theta) \tag{24.120}$$

and such that for every $w \in E$ the map $T^p|_{B(w,4\theta)}$ is one-to-one and satisfies

$$|T^p(w) - w| < \theta(\sqrt{\lambda} - 1). \tag{24.121}$$

For each $w \in E$, let

$$T_w^{-p} := (T^p|_{B(w,2\theta)})^{-1} : T^p(B(w, 2\theta)) \to B(w, 2\theta).$$

For each $w \in E$, we infer from (24.120) that

$$|(T_w^{-p})'(z)| < (\sqrt{\lambda})^{-1} < 1, \quad \forall z \in \overline{B(w, \theta)}.$$

It ensues from the mean value inequality that

$$T_w^{-p}(\overline{B(w, \theta)}) \subseteq \overline{B(w, \theta)}, \quad \forall w \in E.$$

The Banach fixed-point theorem then asserts that for every $w \in E$ there is $\xi_w \in \overline{B(w, \theta)}$ such that $T_w^{-p}(\xi_w) = \xi_w$. Hence,

$$T^p(\xi_w) = \xi_w \quad \text{and} \quad |(T^p)'(\xi_w)| > \sqrt{\lambda} > 1.$$

Therefore, $\xi_w \in \mathcal{J}(T)$. If $z \in \mathcal{J}(S)$, then in view of (24.118) there is $w \in E$ such that $\rho(z, w) < \varepsilon/4$. Consequently,

$$\rho(z, \mathcal{J}(T)) \le \rho(z, \xi_w) \le \rho(z, w) + \rho(w, \xi_w) < \varepsilon/4 + \theta \le \varepsilon,$$

and the lower semicontinuity holds.

Regarding the continuity points, assume that $\varepsilon > 0$ is so small that $B_\rho(\mathcal{J}(S), 4\varepsilon)$ contains no attracting periodic points of S. Given that $\widehat{\mathbb{C}} \backslash B_\rho(\mathcal{J}(S), \varepsilon)$ is a compact set and, by hypothesis, $\widehat{\mathbb{C}} \backslash B_\rho(\mathcal{J}(S), \varepsilon)$ is contained in the union of the basins of attraction of all (finitely many) attracting periodic points of S, there exists $q \in \mathbb{N}$ such that

$$S^q(\widehat{\mathbb{C}} \setminus B_\rho(\mathcal{J}(S), \varepsilon)) \subseteq \widehat{\mathbb{C}} \setminus B_\rho(\mathcal{J}(S), 2\varepsilon).$$

Hence, there exists $\eta > 0$ such that

$$T \in B_{\rho_\infty}(S,\eta) \quad \Longrightarrow \quad T^q(\widehat{\mathbb{C}} \setminus B_\rho(\mathcal{J}(S),\varepsilon)) \subseteq \widehat{\mathbb{C}} \setminus B_\rho(\mathcal{J}(S),\varepsilon).$$

Therefore, $T^{qn}(\widehat{\mathbb{C}} \setminus B_\rho(\mathcal{J}(S),\varepsilon)) \subseteq \widehat{\mathbb{C}} \setminus B_\rho(\mathcal{J}(S),\varepsilon)$ for every $n \geq 0$. Thus, the family of maps $\{T^{qn}\}_{n=0}^\infty$, restricted to the open set $\widehat{\mathbb{C}} \setminus \overline{B_\rho(\mathcal{J}(S),\varepsilon)}$, is normal. It follows from Proposition 24.3.8 that

$$\widehat{\mathbb{C}} \setminus \overline{B_\rho(\mathcal{J}(S),\varepsilon)} \subseteq \mathcal{F}(T).$$

Equivalently,

$$\mathcal{J}(T) \subseteq \overline{B_\rho(\mathcal{J}(S),\varepsilon)}.$$

If $T \in B_{\rho_\infty}(S,\min\{\eta,\delta\})$, where $\delta > 0$ arises from the lower semicontinuity part of the proof, then

$$\mathcal{J}(S) \subseteq \overline{B_\rho(\mathcal{J}(T),\varepsilon)} \quad \text{and} \quad \mathcal{J}(T) \subseteq \overline{B_\rho(\mathcal{J}(S),\varepsilon)}.$$

Consequently,

$$\rho_H(\mathcal{J}(S),\mathcal{J}(T)) \leq \varepsilon.$$

This proves that S is a continuity point. □

In order to make this theorem more concrete and effective, we now show a result which, together with Theorem 24.6.1, entails (semi)continuity of Julia sets in terms of their dependence on the coefficients of the underlying rational functions. Fix an integer $d \geq 2$ and consider the space

$$\mathcal{RP}_d := \left\{ a = (a_0, a_1, \ldots, a_d, a_{d+1}, \ldots, a_{2d}, a_{2d+1}) \in \mathbb{C}^{2d+2} : \sum_{k=d+1}^{2d+1} |a_k| \neq 0 \right\},$$

with the standard metric inherited from $\mathbb{C}^{2d+2}$. Consider the map $\mathcal{RP}_d \ni a \longmapsto R_a \in \mathcal{R}$ defined by

$$R_a(z) := \frac{\sum_{j=0}^d a_j z^j}{\sum_{j=d+1}^{2d+1} a_j z^{j-(d+1)}} \in \widehat{\mathbb{C}}.$$

Lemma 24.6.2. *For every integer $d \geq 2$, the function $R : \mathcal{RP}_d \to \mathcal{R}$ is continuous.*

Proof. Fix $b \in \mathcal{RP}_d$. Fix $\varepsilon > 0$. For every $z \in \widehat{\mathbb{C}}$ there exist $\delta_z > 0$ and $r_z > 0$ such that

$$\rho(R_a(w), R_b(w)) < \varepsilon$$

for all $a \in B_{\mathcal{RP}_d}(b,\delta_z)$ and all $w \in B_\rho(z,r_z)$. Since the space $\widehat{\mathbb{C}}$ is compact and $\{B_\rho(z,r_z)\}_{z\in\widehat{\mathbb{C}}}$ is an open cover of $\widehat{\mathbb{C}}$, there is a finite set $D \subseteq \widehat{\mathbb{C}}$ such that

$$\bigcup_{z\in D} B_\rho(z, r_z) = \widehat{\mathbb{C}}.$$

Let

$$\delta := \min\{\delta_z : z \in D\}.$$

We then have

$$\rho(R_a(\xi), R_b(\xi)) < \varepsilon, \quad \forall \xi \in \widehat{\mathbb{C}}.$$

So $\rho_\infty(R_a, R_b) \le \varepsilon$. □

More about continuity of Julia sets can be found in [92], for instance.

24.7 Polynomials

In this section, we prove selected facts about the Fatou and Julia sets of polynomials. Most proofs provided here are inspired by Beardon's approach in [8]. The first easy but fundamental fact about polynomials P concerns their common fixed point: ∞. Its basin of attraction is $A(\infty) = \{z \in \widehat{\mathbb{C}} : \lim_{n\to\infty} P^n(z) = \infty\}$. Its immediate basin of attraction $A^*(\infty)$ is the connected component of $A(\infty)$ that comprises ∞. In fact, these two basins coincide. Moreover, the boundary of that basin is the Julia set.

Theorem 24.7.1. *If* $P : \widehat{\mathbb{C}} \to \widehat{\mathbb{C}}$ *is a polynomial with* $\deg(P) \ge 2$*, then* $A(\infty) = A^*(\infty)$*, i. e.,* $P^{-1}(A^*(\infty)) = A^*(\infty)$*. Moreover,* $\mathcal{J}(P) = \partial A^*(\infty)$*.*

Proof. Since $P^{-1}(\{\infty\}) = \{\infty\}$, the equalities $A(\infty) = A^*(\infty)$ and $P^{-1}(A^*(\infty)) = A^*(\infty)$ follow from Theorem 24.1.8. So, $\overline{A^*(\infty)} = \overline{P^{-1}(A^*(\infty))}$.

On one hand, the continuity of P and the definition of closure imply that

$$\overline{A^*(\infty)} = \overline{P^{-1}(A^*(\infty))} \subseteq P^{-1}(\overline{A^*(\infty)}). \tag{24.122}$$

On the other hand,

$$P^{-1}(A^*(\infty)) \subseteq \overline{P^{-1}(A^*(\infty))} = \overline{A^*(\infty)}. \tag{24.123}$$

Let $z \in \partial A^*(\infty)$. Suppose for a contradiction that $P^{-1}(z) \setminus \overline{A^*(\infty)} \neq \emptyset$. Then $z \in P(\widehat{\mathbb{C}} \setminus \overline{A^*(\infty)}) \cap \partial A^*(\infty)$. The openness of P implies that $P(\widehat{\mathbb{C}} \setminus \overline{A^*(\infty)}) \cap A^*(\infty) \neq \emptyset$. So $(\widehat{\mathbb{C}} \setminus \overline{A^*(\infty)}) \cap P^{-1}(A^*(\infty)) \neq \emptyset$. This contradicts (24.123), and thus we deduce that $P^{-1}(z) \subseteq \overline{A^*(\infty)}$ for any $z \in \partial A^*(\infty)$. This means that

$$P^{-1}(\partial A^*(\infty)) \subseteq \overline{A^*(\infty)}. \tag{24.124}$$

Altogether (24.122)–(24.124) yield

$$P^{-1}(\overline{A^*(\infty)}) = \overline{P^{-1}(A^*(\infty))} = \overline{A^*(\infty)}.$$

But as $\overline{A^*(\infty)} = A^*(\infty) \cup \partial A^*(\infty)$ and $P^{-1}(\partial A^*(\infty)) \cap A^*(\infty) = \emptyset$, we conclude that

$$P^{-1}(\partial A^*(\infty)) \subseteq \partial A^*(\infty).$$

Hence, we deduce from Corollary 24.3.7 that $\mathcal{J}(P) = \partial A^*(\infty)$. □

We now describe the connectivity of the (immediate) basin of attraction of ∞.

Theorem 24.7.2. *If $P : \widehat{\mathbb{C}} \to \widehat{\mathbb{C}}$ is a polynomial with $\deg(P) \geq 2$, then:*
(a) *$A(\infty) = A^*(\infty)$ is either simply connected or infinitely connected.*
(b) *$A(\infty) = A^*(\infty)$ is simply connected if and only if the Julia set $\mathcal{J}(P)$ is connected.*
(c) *All (necessarily bounded) connected components of $\widehat{\mathbb{C}} \setminus \mathcal{J}(P)$ different from $A^*(\infty)$ are simply connected.*

Proof. As ∞ is a superattracting fixed point of P, the basin $A(\infty)$ cannot be doubly connected and item (a) is an immediate consequence of Theorem 24.4.11. Since an open connected subset of $\widehat{\mathbb{C}}$ is simply connected if and only if its boundary is connected, item (b) follows directly from Theorem 24.7.1. All connected components of $\widehat{\mathbb{C}} \setminus \mathcal{J}(P)$ different from $A^*(\infty)$ are exactly the connected components of $\widehat{\mathbb{C}} \setminus \overline{A^*(\infty)}$. But since the closure $\overline{A^*(\infty)}$ is connected (as $A^*(\infty)$ is), it is a general fact from topology of the plane (see [98], e. g.) that all connected components of its complement are simply connected, meaning that item (c) holds. □

A straightforward repercussion of this theorem is the following fact.

Theorem 24.7.3. *No polynomial $P : \widehat{\mathbb{C}} \to \widehat{\mathbb{C}}$ admits a Herman ring.*

We now fully characterize the cases where the basin $A(\infty) = A^*(\infty)$ is simply connected or infinitely connected.

Theorem 24.7.4. *For a polynomial $P : \widehat{\mathbb{C}} \to \widehat{\mathbb{C}}$ with $\deg(P) \geq 2$, the following statements are equivalent:*
(a) *The basin $A(\infty) = A^*(\infty)$ is simply connected.*
(b) *The Julia set $\mathcal{J}(P)$ is connected.*
(c) *$A(\infty) = A^*(\infty)$ contains no finite (i. e., different from ∞) critical point of P.*

Proof. The equivalence of items (a) and (b) is just Theorem 24.7.2(b). Denote $d = \deg(P)$. We now show that (a) entails (c). Suppose that $A^*(\infty)$ is simply connected. Then the Euler characteristic of that basin is $\chi(A^*(\infty)) = 1$ and it follows from Lemma 24.1.7 that the general Riemann–Hurwitz formula (Theorem 22.5.1) applies to the restriction $P : A^*(\infty) \to A^*(\infty)$ and yields

$$1 + \delta_P(A^*(\infty)) = d.$$

Equivalently, $\delta_P(A^*(\infty)) = d - 1$. But given that $\deg_\infty(P) = d$, this implies that P cannot have any other critical point in $A^*(\infty)$, whence item (c) holds.

We now show the converse, i. e., that (c) entails (a). Assume that ∞ is the only critical point in $A(\infty)$. Since $P^{-1}(\{\infty\}) = \{\infty\}$ and $P^{-1}(A(\infty)) = A(\infty)$, it ensues that for every $n \in \mathbb{N}$ the polynomial P^n has no finite critical point in $A(\infty)$. Since ∞ is a (super)attracting fixed point of P, there is an open disk D centered at ∞ such that

$$\infty \in P(D) \subseteq D \subseteq A(\infty).$$

Then $\{\infty\} = P^{-1}(\{\infty\}) \subseteq D \subseteq P^{-1}(P(D)) \subseteq P^{-1}(D) \subseteq P^{-1}(A(\infty)) = A(\infty)$ and by induction

$$\infty \in P^{-(n-1)}(D) \subseteq P^{-n}(D) \subseteq A(\infty), \quad \forall n \in \mathbb{N}. \tag{24.125}$$

Hence, $D \subseteq P^{-n}(D)$ for every $n \in \mathbb{N}$. Let U_n be the connected component of $P^{-n}(D)$ containing D. Then the map $P^n : U_n \to D$ is proper by Lemma 24.1.7 and

$$\deg(P^n|_{U_n}) \leq \deg(P^n) = d^n.$$

On the other hand, since $\deg_\infty(P) = d$ and $U_n \supseteq D \ni \infty$, it follows that

$$\deg(P^n|_{U_n}) \geq d^n.$$

Thus, $\deg(P^n|_{U_n}) = d^n$. Consequently, U_n is the only connected component of $P^{-n}(D)$, i. e., $P^{-n}(D)$ is connected for every $n \in \mathbb{N}$. It ensues from Lemma 24.1.7 that the general Riemann–Hurwitz formula (Theorem 22.5.1) applies to each map $P^n : P^{-n}(D) \to D$ to give

$$\chi(P^{-n}(D)) + (d^n - 1) = \chi(P^{-n}(D)) + \delta_{P^n}(P^{-n}(D)) = d^n \chi(D).$$

As D is simply connected, $\chi(D) = 1$ and we deduce that $\chi(P^{-n}(D)) = 1$. So, every $P^{-n}(D)$ is simply connected. According to (24.125), the sequence $(P^{-n}(D))_{n=0}^{\infty}$ is ascending. By the definition of D and as ∞ is an attracting fixed point of P, we infer that

$$\bigcup_{n=0}^{\infty} P^{-n}(D) = A(\infty).$$

So, $A(\infty)$ is the union of an ascending sequence of simply connected sets and thereby is likewise simply connected, meaning that (a) holds. □

Theorem 24.7.4 implies that if $A(\infty)$ contains a finite critical point, then the Julia set $\mathcal{J}(P)$ is not connected, i. e., it has at least two connected components. Assuming that all the critical points of P are in $A(\infty)$, we can prove much more.

Theorem 24.7.5. *If* $P : \widehat{\mathbb{C}} \to \widehat{\mathbb{C}}$ *is a polynomial with* $\deg(P) \geq 2$ *such that* $\mathrm{Crit}(P) \subseteq A(\infty)$, *then the Julia set* $\mathcal{J}(P)$ *is homeomorphic to the middle-third Cantor set, i. e., it is perfect, metrizable and totally disconnected.*

Proof. Let $\overline{D}$ be a compact disk in $A(\infty)$ centered at ∞. Since $\mathrm{Crit}(P) \subseteq A(\infty)$, the set

$$A := \mathrm{PC}(P) \setminus \overline{D} = \bigcup_{n=1}^{\infty} P^n(\mathrm{Crit}(P)) \setminus \overline{D}$$

is finite. Join each point $a \in A$ with $\partial\overline{D}$ via a simple (i. e., without self-intersections) closed polygonal arc γ_a in $A(\infty)$ such that the collection $\{\gamma_a\}_{a\in A}$ consists of mutually disjoint arcs. Then for some $\varepsilon > 0$ small enough, the set

$$\overline{D} \cup \bigcup_{a\in A} \overline{B(\gamma_a, \varepsilon)}$$

is a closed topological disk (i. e., a set homeomorphic to $\overline{\mathbb{D}}$) contained in $A(\infty)$. Denote its boundary by Γ and its interior by W. Then $\overline{D} = W \cup \Gamma$ and, by the Jordan–Schoenflies theorem, the complement $V := \widehat{\mathbb{C}} \setminus (W \cup \Gamma)$ is an open topological disk. Observe that

$$(V \cup \Gamma) \cap \mathrm{PC}(P) = \emptyset. \tag{24.126}$$

Since $W \cup \Gamma$ is a compact set contained in $A(\infty)$, there exists $N \in \mathbb{N}$ such that

$$P^N(W \cup \Gamma) \subseteq W \tag{24.127}$$

and

$$P^{-N}(V \cup \Gamma) \subseteq V. \tag{24.128}$$

We deduce from (24.126) and (24.109) that

$$(V \cup \Gamma) \cap \mathrm{PC}(P^N) = \emptyset. \tag{24.129}$$

Given that $\mathcal{J}(P^N) = \mathcal{J}(P)$, replacing P^N by P we get from (24.127)–(24.129) that

$$P(W \cup \Gamma) \subseteq W, \quad P^{-1}(V \cup \Gamma) \subseteq V, \quad (V \cup \Gamma) \cap \mathrm{PC}(P) = \emptyset, \quad \mathcal{J}(P) \subseteq V.$$

Because $\mathcal{J}(P)$ is compact, V is open and $\mathcal{J}(P) \subseteq V$, there exists $\delta > 0$ such that

$$\mathcal{J}(P) \subseteq V \setminus \overline{B(\partial V, \delta)} \tag{24.130}$$

and

$$P^{-1}(V \cup \Gamma) \subseteq V \setminus \overline{B(\partial V, \delta)}.$$

If U is a connected component of $P^{-1}(V)$, then it follows from Lemma 24.1.7, Corollary 22.5.4, and the fact that V is an open topological disk in $\widehat{\mathbb{C}}$, that the map $P|_U : U \to V$ is a conformal homeomorphism. So, there are $d := \deg(P)$ connected components of $P^{-1}(V)$. Label them by $U_1, U_2, \ldots, U_d$, and for each $j = 1, 2, \ldots, d$ denote

$$\varphi_j := P|_{U_j}^{-1} : V \to U_j \subseteq V \setminus \overline{B(\partial V, \delta)}.$$

The maps φ_j, $1 \le j \le d$, form a conformal iterated function system (CIFS) Φ in the sense of Chapter 19 once we show that

$$\|(\varphi_\omega)'\|_\infty < 1$$

for some $k \in \mathbb{N}$ and all $\omega \in \{1, 2, \ldots, d\}^k$. This directly follows from Lemma 24.1.13. By construction, the CIFS Φ satisfies the Strong Separation Condition (SSC; also called the Boundary Separation Condition (BSC); see Definitions 19.7.5–19.7.6). It follows from Theorem 19.7.8 that the limit set J_Φ of the system Φ is homeomorphic to the middle-third Cantor set.

It thus suffices to demonstrate that $\mathcal{J}(P) = J_\Phi$. Since the Julia set $\mathcal{J}(P)$ is completely P-invariant, it ensues from (24.130) that

$$\mathcal{J}(P) \subseteq \bigcap_{\omega \in \{1,2,\ldots,d\}^k} \varphi_\omega(V \setminus \overline{B(\partial V, \delta)})$$

for every $k \in \mathbb{N}$. Therefore,

$$\mathcal{J}(P) \subseteq \bigcap_{k=1}^{\infty} \bigcap_{\omega \in \{1,2,\ldots,d\}^k} \varphi_\omega(V \setminus \overline{B(\partial V, \delta)}) = J_\Phi. \tag{24.131}$$

On the other hand, for every $k \in \mathbb{N}$ and every $\omega \in \{1, 2, \ldots, d\}^k$ denote by x_ω the only fixed point of the map $\varphi_\omega : V \setminus \overline{B(\partial V, \delta)} \to V \setminus \overline{B(\partial V, \delta)}$. As

$$J_\Phi = \overline{\bigcup_{k=1}^{\infty} \bigcup_{\omega \in \{1,2,\ldots,d\}^k} \{x_\omega\}}$$

and since the points x_ω are all repelling periodic points of P, we deduce from Theorem 24.1.22 that $J_\Phi \subseteq \mathcal{J}(P)$. Along with (24.131), this means that $J_\Phi = \mathcal{J}(P)$. So, $\mathcal{J}(P) = J_\Phi$ is homeomorphic to the middle-third Cantor set. □

The following is an immediate consequence of this theorem, Theorem 24.7.4 and the fact that each quadratic polynomial has exactly one finite critical point.

Corollary 24.7.6. *If $P : \widehat{\mathbb{C}} \to \widehat{\mathbb{C}}$ is a quadratic polynomial, then the following statements are equivalent:*
(a) *The Julia set $\mathcal{J}(P)$ is not connected.*

(b) *The Julia set $\mathcal{J}(P)$ is totally disconnected, i. e., it is homeomorphic to the middle-third Cantor set.*
(c) *The finite critical point of P belongs to $A(\infty)$.*
(d) *If c is the finite critical point of P, then $\lim_{n\to\infty} P^n(c) = \infty$.*

24.8 Exercises

Exercise 24.8.1. Prove that a map from the Riemann sphere into iself is an analytic endomorphism if and only if it is a rational function.

Exercise 24.8.2. Find a rational function T (even a polynomial) of degree one such that $\mathcal{J}(T) = \emptyset$. Provide a complete description of the Julia sets of rational functions of degree one.

Exercise 24.8.3. Show that any rational function T of degree at least two such that $T^{-1}(\{\infty\}) = \{\infty\}$ is a polynomial.

Exercise 24.8.4. Let $a \neq 0$ and $d \geq 2$. Show that the Julia set of the polynomial $P(z) = az^d$ is the circle $C(0, |a|^{-1/(d-1)}) := \partial B(0, |a|^{-1/(d-1)})$. Show also that $E(P) = \{0, \infty\}$.

Exercise 24.8.5. Suppose that an holomorphic function f has a Maclaurin series expansion of the form

$$f(z) = az + b_1 z^{p+1} + O(z^2) \quad \text{where} \quad ab_1 \neq 0 \text{ and } p \in \mathbb{N}.$$

(a) Prove that

$$f^n(z) = a^n z + b_n z^{p+1} + O(z^{p+2}) \quad \text{with} \quad b_n = a^{n-1} b_1 \sum_{k=0}^{n-1} a^k, \quad \forall n \in \mathbb{N}.$$

(b) Infer that $b_n = nb_1$ when $a = 1$.
(c) Deduce that $b_n = 0$ if and only if $a^p \neq 1$ and $a^{np} = 1$.
(d) Show that if f^n has a single fixed point at 0, then so does f.
(e) Prove that f^n has at least as many fixed points at 0 as f. Moreover, if f^n has more then $a \neq 1$ but $a^n = 1$.

Exercise 24.8.6. This exercise concerns infinite products and series.
(a) Show that $\log x \leq x - 1$ for all $x > 0$ and that the equality prevails only when $x = 1$.
(b) Let $a > 1$. Prove that $a(x - 1) \leq \log x$ for all $x \in [1/a, 1]$ and that the equality prevails precisely when $x = 1$.
(c) Let $a > 1$ and $a_n \in [1/a, 1]$ for all $n \in \mathbb{N}$. Deduce from (a) and (b) that

$$\exp\Big(-a \sum_{n=1}^{\infty} (1 - a_n)\Big) \leq \prod_{n=1}^{\infty} a_n \leq \exp\Big(-\sum_{n=1}^{\infty} (1 - a_n)\Big).$$

(d) Suppose that $0 < a_n < 1$ for all $n \in \mathbb{N}$. Show that $\prod_{n=1}^{\infty} a_n > 0 \iff \sum_{n=1}^{\infty}(1 - a_n) < \infty$.
(e) Let $b_n > 1$ for all $n \in \mathbb{N}$. Deduce from (d) that $\prod_{n=1}^{\infty} b_n < \infty \iff \sum_{n=1}^{\infty}(1 - \frac{1}{b_n}) < \infty$.

Exercise 24.8.7. Prove relation (24.76).

Exercise 24.8.8 (Cf. Exercise 4.2.2, p. 73 in [8]). Suppose that C is a backward T-invariant circle (i. e., such that $T^{-1}(C) \subseteq C$). Show that either $\mathcal{J}(T) = C$ or $\mathcal{J}(T)$ is a totally disconnected subset of C. This applies to the case when T is a Blaschke product (a finite product of Möbius transformations, each leaving the unit disc $\mathbb{D} = B(0,1)$ invariant).

Exercise 24.8.9 (Cf. Exercise 6.1.5, p. 101 in [8]). This exercise shows that given any $N \in \mathbb{N}$, there is a polynomial with an attracting cycle of prime period at least N. We may assume that N is prime. For each $c \in \mathbb{C}$, define the quadratic polynomial

$$Q_c(z) = z^2 + c.$$

Fix $n \in \mathbb{N}$. As the coefficients of $Q_c^n(z)$ are polynomials in c, the map

$$\begin{array}{rccc} P_n & : & \mathbb{C} & \to & \mathbb{C} \\ & & c & \mapsto & Q_c^n(c) \end{array}$$

is a polynomial in c. Moreover, as $Q_c^{n+1}(z) = (Q_c^n(z))^2 + c$, the map P satisfies

$$P_{n+1}(c) = (P_n(c))^2 + c.$$

Prove that

$$P_n(0) = 0, \quad P_n'(0) = 1, \quad P_n(-2) = 2, \quad \forall n \in \mathbb{N}.$$

Deduce that for each $n \in \mathbb{N}$ there is some number $\gamma = \gamma(n) \in (-2, 0)$ such that $P_n(\gamma) = 0$.

Show that $Q_c^{n-1}(0) = P_n(c)$: thus there exists some $\gamma \in (-2, 0)$ with $Q_\gamma^N(0) = 0$. This shows that the origin has period k for Q_γ, where k divides N. As N is prime, we infer that $k \in \{1, N\}$, and $k \neq 1$ as 0 is not fixed by Q_γ; thus, 0 belongs to a superattracting cycle of period N.

Exercise 24.8.10 (Cf. Exercise 6.1.6, p. 101 in [8]). Prove that the fixed points of a polynomial cannot all be attracting.

Exercise 24.8.11. Prove that the Julia set of a rational function is either connected or has uncountably many connected components.

Exercise 24.8.12. Show that there exists a compact, connected, perfect set in $\widehat{\mathbb{C}}$ which is not the Julia set of any rational function.

Exercise 24.8.13. Show that there exists a topological Cantor set (i. e., a compact, totally disconnected, perfect set) in $\widehat{\mathbb{C}}$ which is not the Julia set of any rational function.

Exercise 24.8.14. With the notation and terminology of Section 24.6, show that the subset of $\mathcal{K}(\widehat{\mathbb{C}})$ consisting of all compact sets that are not Julia sets of any rational function is dense in $\mathcal{K}(\widehat{\mathbb{C}})$.

Exercise 24.8.15. Show that there are two rational functions of degree 2 whose Julia sets are homeomorphic but whose restrictions to their Julia sets are not topologically conjugate.

Exercise 24.8.16. Show that there are two polynomials whose Julia sets are homeomorphic but whose restrictions to their Julia sets are not topologically conjugate.

Exercise 24.8.17 (Cf. Exercise 6.2.1, p. 103 in [8]). Show that every quadratic polynomial P is conjugate via a Möbius transformation to one and only one polynomial $Q_c(z) = z^2 + c$. Show also that $z^2 - z$ is conjugate to $Q_{-3/4}$. Explain why $Q_c(z) - z$ divides $P^2(z) - z$. Using this, show that if Q_c has no periodic point of period 2, then $c = -3/4$.

Exercise 24.8.18 (Cf. Exercise 6.2.3, p. 103 in [8]). For $a \neq 1$, let

$$T_a(z) = \frac{z^2 - z}{1 + az}.$$

Show that:

(a) T_a has 0, ∞ and $\alpha = 2/(1 - a)$ for fixed points.
(b) T_a^2 has three fixed points at the origin (i. e., the origin is a fixed point with multiplicity 3).
(c) The five fixed points of T_a^2 are $0, 0, 0, \infty$ and α.
(d) T_a has no two-cycle.

Prove that T_a is conjugate through a Möbius transformation to T_b if and only if $b = a$ or $b = (3-a)/(1+a)$. [Hint: If $g \circ T_a \circ g^{-1} = T_b$, then T_b must fix $g(0)$, $g(\infty)$ and $g(\alpha)$, so these three points must coincide with $0, \infty, \beta$, where $\beta = 2/(1 - b)$. Further, the multipliers at corresponding points must coincide.]

Exercise 24.8.19 (Cf. Exercise 6.2.5, p. 103 in [8]). Let T be a rational function with $\deg(T) \geq 2$, and let

$$G = \{g : g \text{ Möbius}, \ g \circ T \circ g^{-1} = T\}.$$

Use the fact that T has periodic points of period p for some prime number p to show that G is a finite group. Select a rational function of your choice and attempt to find G explicitly.

Exercise 24.8.20 (Cf. Exercise 6.2.6, p. 104 in [8]). Prove that every quadratic polynomial has points of period 3.

Exercise 24.8.21 (Cf. Exercise 6.4.2, p. 110 in [8]). Let ζ be a repelling point of T. Show that if $T^n(z) \to \zeta$ as $n \to \infty$, then $T^n(z) = \zeta$ for some $n \geq 0$, and so $z \in \mathcal{J}(T)$.

Show, however, that it is possible to have $z \in \mathcal{F}(T)$ and $T^n(z) \to \xi$ for some fixed point $\xi \in \mathcal{J}(T)$. This happens, e. g., when $T(z) = z/(1+z^2)$ and $\xi = 0$.

Exercise 24.8.22 (Cf. Exercise 6.5.2, p. 132 in [8]). Let $P(z) = -z + z^{p+1}$, where $p \in \mathbb{N}$. How many petals does P have at the origin?

Exercise 24.8.23 (Cf. Exercise 7.1.1, p. 161 in [8]). Let $R(z) = z(z-a)/(1-\bar{a}z)$, where $0 < |a| < 1$. Show that $B(0,1)$ is a simply connected attracting component of $\mathcal{F}(R)$. What can be said of $\widehat{\mathbb{C}} \backslash \overline{B(0,1)}$?

Exercise 24.8.24 (Cf. Exercise 7.1.2, p. 161 in [8]). Let $S(z) = z/(2-z^2)$. Show that $\mathcal{F}(S)$ has an attracting component of infinite connectivity.

Exercise 24.8.25 (Cf. Exercise 7.1.4, p. 161, in [8]). Let $P(z) = 6z(1-z)$. Draw the graph of $P(x)$ for $x \in [0,1]$ and show that $\bigcup_{n=0}^{\infty} P^{-n}([0,1])$ is a Cantor set C. Prove that $\mathcal{J}(P) = C$, and deduce that P has a superattracting component of infinite connectivity.

Exercise 24.8.26 (Cf. Exercise 9.5.1, p. 209 in [8]). Let $P(z) = z^2 - 1$. For each $w \in \mathcal{F}(P)$, let F_w be the component of $\mathcal{F}(P)$ containing w. Show that the only finite critical point of P is in an attracting two-cycle of P. Deduce that:

(i) $\mathcal{F}(P)$ has infinitely many components.
(ii) Every component of $\mathcal{F}(P)$ is simply connected.
(iii) P has no Siegel disk.
(iv) $\{F_\infty\}$ and $\{F_0, F_{-1}\}$ are the only periodic cycles of components of $\mathcal{F}(P)$.
(v) Every bounded component B of $\mathcal{F}(P)$ is mapped by some P^n into F_0; thus, the sequence $(P^n(B))_{n=0}^{\infty}$ is eventually alternating between F_0 and F_{-1}.

Find the multiplicity of P as a map acting between the given components in each of the following cases:

$$F_\infty \to F_\infty, \quad F_0 \to F_{-1}, \quad F_{-1} \to F_0.$$

Exercise 24.8.27. Show that if two rational functions commute, then they have the same Julia set.

Exercise 24.8.28 (Cf. Exercise 9.5.8, p. 210 in [8]). Let $P(z) = z(1+z^2)$ and $Q(z) = -P(z)$. Show that $\mathcal{J}(P) = \mathcal{J}(Q)$ but that P and Q are not conjugate.

Exercise 24.8.29. Let $T : X \to X$ be a self-map of a metric space (X, d). A point $w \in X$ is called *Lyapunov T-stable* if for every $\varepsilon > 0$ there exists $\delta > 0$ such that

$$d(z, w) < \delta \quad \Longrightarrow \quad \sup_{n \in \mathbb{N}} d(T^n(z), T^n(w)) < \varepsilon.$$

Let $T : \widehat{\mathbb{C}} \to \widehat{\mathbb{C}}$ be a rational function. Prove that the Fatou set $\mathcal{F}(T)$ is equal to the set of all Lyapunov T-stable points.

Exercise 24.8.30 (Cf. Exercise 5.4.1, p. 89 in [8]). Find examples of domains U and V, and a rational function R mapping U onto V in such a way that ∂V is a proper subset of $R(\partial U)$.

Exercise 24.8.31 (Cf. Exercise 5.4.2, p. 89 in [8]). Suppose that $R(z) = z^d$ and that V is a Euclidean disk with $0 \notin \partial V$. Show that $R^{-1}(V)$ is either simply connected (and R is a d-fold map of it onto V) or it consists of d simply connected domains (each homeomorphic to V under R). What is $R^{-1}(V)$ when $V = B(1,1)$?

Exercise 24.8.32. Let $T : \widehat{\mathbb{C}} \to \widehat{\mathbb{C}}$ be a rational function with $\deg(T) \geq 2$, and let $k \in \mathbb{N}$. Prove the following statements:

(a) $\mathrm{Crit}(T^k) = \bigcup_{n=0}^{k-1} T^{-n}(\mathrm{Crit}(T)) \supseteq \mathrm{Crit}(T)$.

(b) $\mathrm{PC}(T^k) = \mathrm{PC}(T)$.

(c) $\Omega(T^k) = \Omega(T)$.

25 Selected technical properties of rational functions

25.1 Passing near critical points: results and applications

Let $D \subseteq \mathbb{C}$ be an open set and let $f : D \to \mathbb{C}$ be a nonconstant analytic function. For every $w \in D$, let $\deg_w(f)$ be the local degree of f at w, sometimes called the multiplicity of f at w; more precisely, $\deg_w(f)$ is the unique $d \in \mathbb{N}$ such that the Taylor series expansion of f around w has the form

$$f(z) = f(w) + \sum_{k=d}^{\infty} a_k (z - w)^k, \tag{25.1}$$

with $a_d := \frac{1}{d!} f^{(d)}(w) \neq 0$. So, $\deg_w(f) \geq 2$ if and only if $w \in \mathrm{Crit}(f)$. Equivalently,

$$w \notin \mathrm{Crit}(f) \quad \Longleftrightarrow \quad \deg_w(f) = 1. \tag{25.2}$$

Let $c \in \mathrm{Crit}(f)$. Then there exist $\widetilde{r}(f, c) > 0$ and $A = A(f, c) \geq 1$ such that

$$A^{-1}|z - c|^{\deg_c(f)} \leq |f(z) - f(c)| \leq A|z - c|^{\deg_c(f)}, \quad \forall z \in B(c, \widetilde{r}(f, c)) \tag{25.3}$$

and

$$A^{-1}|z - c|^{\deg_c(f)-1} \leq |f'(z)| \leq A|z - c|^{\deg_c(f)-1}, \quad \forall z \in B(c, \widetilde{r}(f, c)). \tag{25.4}$$

We now prove the following description, well known in complex analysis, of an analytic function's local behavior. In the vicinity of a function's critical point, this description is frequently more convenient (even more powerful) and easier to apply than formulas (25.1), (25.3) and (25.4).

Theorem 25.1.1. *Let $D \subseteq \mathbb{C}$ be an open set and $f : D \to \mathbb{C}$ a nonconstant analytic function. Then for every point $w \in D$ the map f restricted to some sufficiently small neighborhood of w is biholomorphically conjugate to the map $\xi \mapsto \xi^{\deg_w(f)}$ on some neighborhood of 0. More precisely, there is a biholomorphic map $\varphi_w : U_w \to V_w$ whose domain U_w is some neighborhood of w, whose codomain V_w is some neighborhood of 0 and such that $\varphi_w(w) = 0$ and the following diagram commutes:*

$$\begin{array}{ccc} U_w & \xrightarrow[f]{} & f(U_w) \\ \varphi_w \big\downarrow & & \big\downarrow \mathrm{Tr}_{f(w)} \\ V_w & \xrightarrow{\xi \mapsto \xi^{\deg_w(f)}} & f(U_w) - f(w) = (V_w)^{\deg_w(f)} \end{array}$$

where $\mathrm{Tr}_{f(w)}$ is the translation map $\mathrm{Tr}_{f(w)}(z) = z - f(w)$ and $(V_w)^{\deg_w(f)} = \{z^{\deg_w(f)} : z \in V_w\}$.

https://doi.org/10.1515/9783110769876-025

Proof. Let $\mathrm{Tr}_w : \mathbb{C} \to \mathbb{C}$ be the translation $\mathrm{Tr}_w(z) = z - w$. Consider the analytic map

$$\hat{f} := \mathrm{Tr}_{f(w)} \circ f \circ \mathrm{Tr}_w^{-1} : \mathrm{Tr}_w(D) \to \mathrm{Tr}_{f(w)}(f(D)).$$

Since f is nonconstant, so is $\hat{f}$. Moreover,

$$\hat{f}(0) = \mathrm{Tr}_{f(w)} \circ f(\mathrm{Tr}_w^{-1}(0)) = \mathrm{Tr}_{f(w)}(f(w)) = 0$$

and

$$\deg_0(\hat{f}) = \deg_{f(w)}(\mathrm{Tr}_{f(w)}) \cdot \deg_w(f) \cdot \deg_0(\mathrm{Tr}_w^{-1}) = 1 \cdot \deg_w(f) \cdot 1 = \deg_w(f).$$

Alternatively, observe that $\hat{f}(z) = \sum_{k=d}^{\infty} a_k z^k = z^d \sum_{j=0}^{\infty} a_{j+d} z^j$. Denote $d := \deg_w(f)$. Applying (25.1) to $\hat{f}$, we deduce that there exist an open simply connected neighborhood W_1 of 0 and a bounded holomorphic function $g : W_1 \to \mathbb{C}$ such that $g(0) \neq 0$ and

$$\hat{f}(z) = z^d g(z), \quad \forall z \in W_1.$$

Hence, there exists a bounded holomorphic function $h : W_1 \to \mathbb{C}$ such that

$$g(z) = (h(z))^d, \quad \forall z \in W_1.$$

Consider the map $\psi : W_1 \to \mathbb{C}$ defined by

$$\psi(z) = zh(z).$$

Then $\psi'(z) = h(z) + zh'(z)$ and, therefore, $\psi'(0) = h(0) \neq 0$. Thus, there exists an open simply connected set $W_2 \subseteq W_1$ containing 0 such that the map $\psi|_{W_2}$ is one-to-one and so is biholomorphic. Let $U_w := \mathrm{Tr}_w^{-1}(W_2)$ and

$$\varphi_w := \psi \circ \mathrm{Tr}_w : U_w \to \mathbb{C}.$$

Then φ_w is a biholomorphic map, $\varphi_w(w) = \psi(\mathrm{Tr}_w(w)) = \psi(0) = 0$ and

$$\begin{aligned}(\varphi_w(z))^d &= (\psi \circ \mathrm{Tr}_w(z))^d = (\mathrm{Tr}_w(z)h(\mathrm{Tr}_w(z)))^d = (\mathrm{Tr}_w(z))^d g(\mathrm{Tr}_w(z)) \\ &= \hat{f}(\mathrm{Tr}_w(z)) = \hat{f} \circ \mathrm{Tr}_w(z) = \mathrm{Tr}_{f(w)} \circ f(z).\end{aligned}$$

□

As a direct consequence of this theorem, we get a characterization of the local degree of an analytic function.

Theorem 25.1.2. *If $D \subseteq \mathbb{C}$ is an open set and $f : D \to \mathbb{C}$ is a nonconstant analytic function, then for every point $w \in D$ the local degree $\deg_w(f)$ is the unique $d \in \mathbb{N}$ for which there exists a local base $\{U_n(w)\}_{n=1}^{\infty}$ of topology at w consisting of open sets contained in D such that the restriction*

$$f|_{U_n(w)\setminus\{w\}} : U_n(w) \setminus \{w\} \to \mathbb{C}$$

*is d-to-*1.

We now prove several more specific, but interesting on their own, consequences of relations (25.1), (25.3)–(25.4) and Theorem 25.1.1 for rational functions. The first one is obvious: At any positive distance ε away from the critical set, the function is one-to-one on balls of a certain radius $\alpha(\varepsilon)$ and the inverse branches are defined on balls of some radius $\beta(\varepsilon)$.

Lemma 25.1.3. *If $T : \widehat{\mathbb{C}} \to \widehat{\mathbb{C}}$ is a rational function, then for every $\varepsilon > 0$ there exist $0 < \alpha(\varepsilon) < \varepsilon$ and $\beta(\varepsilon) > 0$ such that for every $z \in \widehat{\mathbb{C}} \setminus B(\mathrm{Crit}(T), \varepsilon)$ the restriction $T|_{B(z,\alpha(\varepsilon))}$ is injective and there is a unique holomorphic inverse branch $T_z^{-1} : B(T(z), \beta(\varepsilon)) \to \widehat{\mathbb{C}}$ of T such that $T_z^{-1}(T(z)) = z$.*

The next result concerns (1) the local preservation of simple connectedness for preimages' components; (2) the control on/uniform boundedness of the components' diameters; and, finally, (3) the conformal homeomorphy of those components that are disjoint from the critical set.

Lemma 25.1.4. *If $T : \widehat{\mathbb{C}} \to \widehat{\mathbb{C}}$ is a rational function, then for every $\varepsilon > 0$ there exists $\delta > 0$ such that if $V \subseteq \widehat{\mathbb{C}}$ is an open simply connected set with* $\mathrm{diam}(V) < \delta$, *then each connected component U of $T^{-1}(V)$ is simply connected and* $\mathrm{diam}(U) < \varepsilon$. *In addition, if $U \cap \mathrm{Crit}(T) = \emptyset$ then $T|_U : U \to V$ is a conformal homeomorphism.*

Proof. Write $d = \deg(T)$. As the map T is open and continuous and $\widehat{\mathbb{C}}$ is compact, it follows from Lemma 4.2.2 and Remark 4.2.3 (in Volume 1) that there is $\delta > 0$ such that

$$T\left(B\left(x, \frac{\varepsilon}{2d}\right)\right) \supseteq B(T(x), \delta), \quad \forall x \in \widehat{\mathbb{C}}. \tag{25.5}$$

We shall first prove that

$$T^{-1}(B(w, \delta)) \subseteq \bigcup_{z \in T^{-1}(w)} B\left(z, \frac{\varepsilon}{2d}\right), \quad \forall w \in \widehat{\mathbb{C}}. \tag{25.6}$$

Indeed, fix $w \in \widehat{\mathbb{C}}$ and let $\xi \in T^{-1}(B(w, \delta))$. This means that $T(\xi) \in B(w, \delta)$; equivalently, $w \in B(T(\xi), \delta)$. By virtue of (25.5),

$$T\left(B\left(\xi, \frac{\varepsilon}{2d}\right)\right) \supseteq B(T(\xi), \delta) \ni w.$$

So, there exists $z \in B(\xi, \frac{\varepsilon}{2d})$ with $T(z) = w$. As $\xi \in B(z, \frac{\varepsilon}{2d})$, the proof of (25.6) is done.

It ensues from (25.6) that for each connected component W of $T^{-1}(B(w, \delta))$ there are points $z_1, z_2, \ldots, z_q$, $1 \le q \le d$, in $T^{-1}(w)$, such that

$$W \subseteq \bigcup_{j=1}^{q} B\Big(z_j, \frac{\varepsilon}{2d}\Big) \quad \text{and} \quad B\Big(z_i, \frac{\varepsilon}{2d}\Big) \cap B\Big(z_{i+1}, \frac{\varepsilon}{2d}\Big) \neq \emptyset, \quad \forall 1 \le i < q.$$

Hence,

$$\operatorname{diam}(W) \le q \cdot \frac{\varepsilon}{d} \le \varepsilon. \tag{25.7}$$

If $V \subseteq \widehat{\mathbb{C}}$ is a nonempty set with $\operatorname{diam}(V) < \delta$, then fixing any $w \in V$ we know that $V \subseteq B(w,\delta)$. Therefore, each connected component U of $T^{-1}(V)$ is contained in a (unique) connected component W of $T^{-1}(B(w,\delta))$. It follows from (25.7) that

$$\operatorname{diam}(U) \le \operatorname{diam}(W) \le \varepsilon. \tag{25.8}$$

It just remains to prove that U is simply connected. Let

$$\eta = \min\{|c_1 - c_2| : c_1, c_2 \in \operatorname{Crit}(T),\ c_1 \neq c_2\}.$$

By the already proven part of the lemma, there is $\theta > 0$ such that $\operatorname{diam}(U) < \eta$ for every open simply connected set $V \subseteq \widehat{\mathbb{C}}$ with $\operatorname{diam}(V) < \theta$ and for each connected component U of $T^{-1}(V)$. But then U contains at most one critical point of T, and the simple connectedness of U directly follows from Corollaries 22.5.4–22.5.5. Finally, if U contains no critical point, then $T|_U$ is a conformal homeomorphism according to Corollary 22.5.4. □

As a straightforward consequence of this lemma, we get its extension to a finite number of iterates of a rational function.

Lemma 25.1.5. *Let $T : \widehat{\mathbb{C}} \to \widehat{\mathbb{C}}$ be a rational function. For every $n \in \mathbb{N}$ and every $\varepsilon > 0$, there exists $\delta > 0$ such that if $V \subseteq \widehat{\mathbb{C}}$ is an open simply connected set with $\operatorname{diam}(V) < \delta$, then for every $0 \le k \le n$ each connected component U of $T^{-k}(V)$ is simply connected and $\operatorname{diam}(U) < \varepsilon$. In addition, if $U \cap \operatorname{Crit}(T^k) = \emptyset$ then $T^k|_U : U \to V$ is a conformal homeomorphism.*

In particular, it follows from Lemma 25.1.4 and (25.3)–(25.4) that there exist $R = R(T,c) > 0$ and $A = A(T,c) \ge 1$ such that

$$A^{-1}|z-c|^{\deg_c(T)} \le |T(z) - T(c)| \le A|z-c|^{\deg_c(T)} \tag{25.9}$$

and

$$A^{-1}|z-c|^{\deg_c(T)-1} \le |T'(z)| \le A|z-c|^{\deg_c(T)-1} \tag{25.10}$$

for every $z \in \operatorname{Comp}(c,T,R)$, where $\operatorname{Comp}(c,T,R)$ denotes the connected component of $T^{-1}(B(T(c),R))$ containing c. Furthermore,

$$T(\operatorname{Comp}(c,T,R)) = B(T(c),R)$$

and there exists $0 < r(T,c) \le \tilde{r}(T,c)$ such that

$$B(c, r(T,c)) \subseteq \operatorname{Comp}(c, T, R(T,c)). \tag{25.11}$$

Notably, relations (25.9)–(25.10) hold for all $z \in B(c, r(T,c))$.

The next result provides, at any positive distance R away from the critical set, a uniform bound on the diameters of the components of the preimages of balls. In addition to depending on R, that bound depends linearly on the radius r of the balls.

Lemma 25.1.6. *If $T : \widehat{\mathbb{C}} \to \widehat{\mathbb{C}}$ is a rational function, then for every $R > 0$ there exists a constant $K_R \ge 1$ such that for every $z \in \widehat{\mathbb{C}} \setminus B(\operatorname{Crit}(T), R)$ we have for every $r > 0$ that*

$$\operatorname{diam}(U_{z,r}) < K_R r,$$

where $U_{z,r}$ is the connected component of $T^{-1}(B(T(z), r))$ containing z.

Proof. Let $R > 0$ and let $z \in \widehat{\mathbb{C}} \setminus B(\operatorname{Crit}(T), R)$. Using Lemma 25.1.4 with $\varepsilon = R/2$ and defining $\widehat{R} = \delta/2$, we know that $\operatorname{diam}(U) < R/2$ for every $w \in \widehat{\mathbb{C}}$ and every connected component U of $T^{-1}(B(w, \widehat{R}))$. In particular,

$$\operatorname{diam}(U_{z,\widehat{R}}) < R/2.$$

Since $z \in U_{z,\widehat{R}} \cap [\widehat{\mathbb{C}} \setminus B(\operatorname{Crit}(T), R)]$, we deduce that

$$U_{z,\widehat{R}} \subseteq \widehat{\mathbb{C}} \setminus B(\operatorname{Crit}(T), R/2).$$

Assume first that $r \ge \widehat{R}$. Then

$$\operatorname{diam}(U_{z,r}) \le \operatorname{diam}(\widehat{\mathbb{C}}) \le \frac{\operatorname{diam}(\widehat{\mathbb{C}})}{\widehat{R}} r.$$

Now assume that $r \le \widehat{R}$. Given that $U_{z,r} \subseteq U_{z,\widehat{R}} \subseteq \widehat{\mathbb{C}} \setminus B(\operatorname{Crit}(T), R/2)$, it ensues that

$$\inf\{|T'(x)| : x \in U_{z,r}\} \ge \|T'\|_R := \inf\{|T'(y)| : y \in \widehat{\mathbb{C}} \setminus B(\operatorname{Crit}(T), R/2)\} > 0$$

and $U_{z,r} \cap \operatorname{Crit}(T) = \emptyset$. By Lemma 25.1.4, the restriction $T|_{U_{z,r}} : U_{z,r} \to B(T(z), r)$ is a biholomorphic map. Applying the mean value inequality to $T|_{U_{z,r}}^{-1}$, we deduce that

$$\begin{aligned}
\operatorname{diam}(U_{z,r}) &\le \sup\{|(T|_{U_{z,r}}^{-1})'(w)| : w \in B(T(z), r)\} \cdot 2r \\
&= \inf\{|T'(\zeta)|^{-1} : \zeta \in U_{z,r}\} \cdot 2r \\
&\le 2\|T'\|_R^{-1} r.
\end{aligned}$$

Set

$$K_R := \max\left\{\frac{\operatorname{diam}(\widehat{\mathbb{C}})}{\widehat{R}}, 2\|T'\|_R^{-1}\right\}. \qquad \square$$

The following lemma in some way generalizes the previous one. It offers a uniform bound on the diameters of the components of the preimages of balls. That bound evidently depends on the radius r of the balls though not linearly this time.

Lemma 25.1.7. *If $T : \widehat{\mathbb{C}} \to \widehat{\mathbb{C}}$ is a rational function, then for every $r \in (0,1/2]$ small enough and all $w \in \widehat{\mathbb{C}}$ we have that*

$$\operatorname{diam}(U) \le r^{\frac{1}{2d_T}}$$

for every connected component U of $T^{-1}(B(w,r))$, where

$$d_T := \max\{\deg_c(T) : c \in \operatorname{Crit}(T)\}.$$

Proof. Pick $w \in \widehat{\mathbb{C}}$. Recalling the definition of $r(T,c)$ from (25.11), let

$$\rho := \frac{1}{2}\min\{r(T,c) : c \in \operatorname{Crit}(T)\} > 0. \tag{25.12}$$

Using Lemma 25.1.4, assume that $r \in (0,1/2]$ is so small that

$$\operatorname{diam}(V) < \rho/2 \tag{25.13}$$

for every $\zeta \in \widehat{\mathbb{C}}$ and every connected component V of $T^{-1}(B(\zeta,r))$.

Fix a point $z \in U \cap T^{-1}(w)$. If $\operatorname{dist}(z,\operatorname{Crit}(T)) \ge \rho/2$, then Lemma 25.1.6 asserts that

$$\operatorname{diam}(U) \le K_{\rho/2} r \le r^{\frac{1}{2d_T}}, \tag{25.14}$$

provided that $r \le 1/2$ is small enough.

So, assume that $\operatorname{dist}(z,\operatorname{Crit}(T)) < \rho/2$. There exists $c \in \operatorname{Crit}(T)$ such that $|z-c| < \rho/2$. Using (25.13) and (25.11), we get that

$$U \subseteq B\Big(c, \frac{\rho}{2} + \operatorname{diam}(U)\Big) \subseteq B(c,\rho) \subseteq \operatorname{Comp}(c,T,R(T,c)). \tag{25.15}$$

Suppose first that

$$r \le \frac{1}{2}|w - T(c)|. \tag{25.16}$$

Denote $d_c := \deg_c(T)$. With $A = \max\{A(T,x) : x \in \operatorname{Crit}(T)\}$, it follows from (25.9)–(25.10) and (25.15) that for every point $\xi \in U$,

$$|T'(\xi)| \ge A^{-1}|\xi - c|^{d_c - 1} \ge A^{-(1+\frac{1}{d_c})}|T(\xi) - T(c)|^{1-\frac{1}{d_c}}$$

$$\geq A^{-(1+\frac{1}{d_c})}(|w - T(c)| - |T(\xi) - w|)^{1-\frac{1}{d_c}} \geq A^{-(1+\frac{1}{d_c})}(2r - r)^{1-\frac{1}{d_c}} = A^{-(1+\frac{1}{d_c})}r^{1-\frac{1}{d_c}}.$$

By the mean value inequality, we deduce that

$$\operatorname{diam}(U) \leq A^{1+\frac{1}{d_c}} r^{\frac{1}{d_c}-1} \cdot (2r) = 2A^{1+\frac{1}{d_c}} r^{\frac{1}{d_c}} \leq r^{\frac{1}{2d_c}} \leq r^{\frac{1}{2d_T}}, \tag{25.17}$$

where the second inequality sign was written assuming that $r \leq 1/2$ is small enough. Suppose now that $r > \frac{1}{2}|w - T(c)|$. Then

$$B(w, r) \subseteq B(T(c), 3r). \tag{25.18}$$

There exists a point $y \in U$ such that $|y - c| \geq \frac{1}{3}\operatorname{diam}(U)$ (otherwise, $U \subseteq B(c, \frac{1}{3}\operatorname{diam}(U))$, which implies that $\operatorname{diam}(U) \leq 2 \cdot \frac{1}{3}\operatorname{diam}(U)$, an impossibility). Using (25.9) and (25.17)–(25.18), we infer that

$$3r \geq |T(y) - T(c)| \geq A^{-1}|y - c|^{d_c} \geq 3^{-d_c}A^{-1}(\operatorname{diam}(U))^{d_c}.$$

Thus,

$$\operatorname{diam}(U) \leq 3^{1+\frac{1}{d_c}} A^{\frac{1}{d_c}} r^{\frac{1}{d_c}} \leq r^{\frac{1}{2d_T}},$$

where the second inequality sign was written assuming that $r \leq 1/2$ is small enough. This inequality along with (25.14) and (25.17) complete the proof. □

The upcoming two results apply to any analytic function. The first one pertains to their injectivity on balls disjoint from the critical set. Unlike in previous results, the balls do not stay uniformly away from the critical points.

Lemma 25.1.8. *Let $D \subseteq \mathbb{C}$ be an open set and $f : D \to \mathbb{C}$ a nonconstant analytic function. For every $c \in \operatorname{Crit}(f)$, there exist two numbers $r_1(f, c) \in (0, \frac{1}{2}\tilde{r}(f, c))$ and $s(f, c) \in (0, \frac{1}{2}]$, this latter depending only on $\deg_c(f)$, such that for any $w \in B(c, r_1(f, c))$ the restriction*

$$f|_{B(w,s(f,c)|w-c|)} : B(w, s(f, c)|w - c|) \to \mathbb{C}$$

is one-to-one. (*Note: $\tilde{r}(f, c)$ arises from* (25.3)–(25.4).)

Proof. We define $s^* = s^*(f, c)$ to be the largest number in $(0, \frac{1}{2}]$ such that

$$\angle(1 + s^* i, 0, 1 - s^* i) \leq \frac{\pi}{\deg_c(f)}.$$

Denote

$$d_c := \deg_c(f).$$

For every $\xi \in \mathbb{C}$, the map $\mathbb{C} \ni z \mapsto E_{d_c}(z) := z^{d_c} \in \mathbb{C}$ restricted to the ball $B(\xi, s^*|\xi|)$ is one-to-one. Decreasing the neighborhood U_c of c produced in Theorem 25.1.1 if necessary, we may assume that

$$0 < m_c := \inf\{|\varphi_c'(\xi)| : \xi \in U_c\} \leq \sup\{|\varphi_c'(\xi)| : \xi \in U_c\} =: M_c < \infty.$$

Let

$$s = m_c M_c^{-1} s^*$$

and let $r_1(f,c) \in (0, \frac{1}{2}\tilde{r}(f,c))$ be so small that

$$B(c, 2r_1(f,c)) \subseteq U_c.$$

Let $w \in B(c, r_1(f,c))$. By the mean value inequality, we get for every $z \in B(w, s|w-c|)$ that

$$|\varphi_c(z) - \varphi_c(w)| \leq M_c|z-w| \leq M_c s|w-c|$$

and

$$|w-c| = |\varphi_c^{-1}(\varphi_c(w)) - \varphi_c^{-1}(\varphi_c(c))| \leq m_c^{-1}|\varphi_c(w) - \varphi_c(c)| = m_c^{-1}|\varphi_c(w)|.$$

Hence,

$$|\varphi_c(z) - \varphi_c(w)| \leq M_c s m_c^{-1}|\varphi_c(w)| = s^*|\varphi_c(w)|.$$

Therefore,

$$\varphi_c(B(w, s|w-c|)) \subseteq B(\varphi_c(w), s^*|\varphi_c(w)|).$$

So, the map $E_{d_c}|_{\varphi_c(B(w,s|w-c|))}$ is one-to-one. Consequently, by Theorem 25.1.1 the map

$$f|_{B(w,s|w-c|)} = \mathrm{Tr}_{f(c)}^{-1} \circ E_{d_c} \circ \varphi_c|_{B(w,s|w-c|)}$$

is one-to-one, too. Set $s(f,c) = s$. □

As a fairly easy consequence of this lemma, we get an upper bound on the ratio of the derivatives of the function on balls disjoint from the critical points.

Lemma 25.1.9. *Let $D \subseteq \mathbb{C}$ be an open set and $f : D \to \mathbb{C}$ a nonconstant analytic function. For any $\eta > 0$ and $c \in \mathrm{Crit}(f)$, there is $r_\eta(f,c) \in (0, \frac{1}{2}r_1(f,c))$ such that if $w \in B(c, r_\eta(f,c))$ then*

$$\frac{|f'(z)|}{|f'(w)|} \leq (1+\eta)(1+s)^{\deg_c(f)-1}$$

for every $s \in (0, s(f,c)]$ *and every* $z \in B(w, s|w-c|)$. *(Note:* $r_1(f,c)$ *and* $s(f,c)$ *come from Lemma* 25.1.8.)

Proof. A variant of this lemma, which would frequently be sufficient in this book, would be an immediate consequence of Lemma 25.1.8 and Koebe's distortion theorem. We rather provide an elementary proof of Lemma 25.1.9, i. e., a proof that neither relies on Lemma 25.1.8 nor on Koebe's distortion theorem but merely on the Taylor series expansion formula (25.1). As before, denote

$$d_c := \deg_c(f).$$

By (25.1), there exist $\widehat{r} = \widehat{r}(f,c) \in (0, \frac{1}{2}\widetilde{r}(f,c)]$ and a bounded holomorphic function $\widehat{f} : B(c,\widehat{r}) \to \mathbb{C}$ with bounded derivative such that

$$f(z) = f(c) + a_{d_c}(z-c)^{d_c} + (z-c)^{d_c+1}\widehat{f}(z), \quad \forall z \in B(c,\widehat{r}).$$

Differentiating this formula, we get

$$f'(z) = a_{d_c}d_c(z-c)^{d_c-1} + (z-c)^{d_c}\big[(d_c+1)\widehat{f}(z) + (z-c)\widehat{f}'(z)\big], \quad \forall z \in B(c,\widehat{r}). \tag{25.19}$$

Define the function $\widetilde{f} : B(c,\widehat{r}) \to \mathbb{C}$ by

$$\widetilde{f}(z) := (a_{d_c}d_c)^{-1}\big[(d_c+1)\widehat{f}(z) + (z-c)\widehat{f}'(z)\big]. \tag{25.20}$$

Let $r_\eta \in (0, \frac{\widehat{r}}{1+s}]$ be so small that

$$\frac{1 + 2\|\widetilde{f}\|_\infty r_\eta}{1 - \|\widetilde{f}\|_\infty r_\eta} < 1 + \eta. \tag{25.21}$$

Let $w \in B(c, r_\eta)$ and $z \in B(w, s|w-c|)$. Then $|z-c| < (1+s)|w-c| < (1+s)r_\eta \le \widehat{r}$. Using (25.19)–(25.21), we get that

$$\begin{aligned}
\frac{|f'(z)|}{|f'(w)|} &= \frac{|a_{d_c}d_c(z-c)^{d_c-1} + (z-c)^{d_c}\cdot a_{d_c}d_c\widetilde{f}(z)|}{|a_{d_c}d_c(w-c)^{d_c-1} + (w-c)^{d_c}\cdot a_{d_c}d_c\widetilde{f}(w)|} \\
&= \left|\frac{z-c}{w-c}\right|^{d_c-1}\frac{|1+\widetilde{f}(z)(z-c)|}{|1+\widetilde{f}(w)(w-c)|} \le (1+s)^{d_c-1}\frac{1 + 2\|\widetilde{f}\|_\infty r_\eta}{1 - \|\widetilde{f}\|_\infty r_\eta} \\
&\le (1+s)^{d_c-1}(1+\eta).
\end{aligned}$$

□

We now prove a global version of the previous lemma for rational functions. This time, we do rely on Koebe's distortion theorem.

Lemma 25.1.10. *Let $T : \widehat{\mathbb{C}} \to \widehat{\mathbb{C}}$ be a rational function. For every $\varepsilon > 0$ there is $s_\varepsilon \in (0, \frac{1}{2}]$ such that for any $w \in \widehat{\mathbb{C}}$ and $z \in B(w, s_\varepsilon \operatorname{dist}(w, \operatorname{Crit}(T)))$ we have*

$$\frac{|T'(z)|}{|T'(w)|} \le 1 + \varepsilon.$$

Proof. We stick to the notation adopted in Lemmas 25.1.8–25.1.9 and in their proofs. Fix $\varepsilon > 0$. Let

$$d_T := \max\{\deg_c(T) \mid c \in \operatorname{Crit}(T)\} \quad \text{and} \quad s(T) := \min\{s(T,c) \mid c \in \operatorname{Crit}(T)\}.$$

Given $0 < \eta < \varepsilon$, let $\widehat{s} \in (0, s(T))$ be so small that

$$(1+\eta)(1+\widehat{s})^{d_T-1} \le 1+\varepsilon. \tag{25.22}$$

For every $c \in \operatorname{Crit}(T)$, let $r_\eta(T,c) \in (0, \frac{\widehat{r}(T,c)}{1+s(T,c)}]$ be so small that

$$\frac{1 + 2\|\widetilde{T}\|_\infty r_\eta(T,c)}{1 - \|\widetilde{T}\|_\infty r_\eta(T,c)} < 1 + \eta.$$

Thereafter, let

$$r_\eta(T) := \min\{r_\eta(T,c) \mid c \in \operatorname{Crit}(T)\}.$$

Let

$$X := \widehat{\mathbb{C}} \setminus B(\operatorname{Crit}(T), r_\eta(T)).$$

Then $T'(\xi) \neq 0$ for all $\xi \in X$. Thus, for every $\xi \in X$ there exists $\rho_\xi > 0$ such that the map $T|_{B(\xi,\rho_\xi)}$ is one-to-one. Since X is compact, the open cover $\{B(\xi, \rho_\xi) : \xi \in X\}$ of X has a Lebesgue number $\rho > 0$. If $x \in X$, then there exists $\widehat{x} \in X$ such that $B(x, \rho/2) \subseteq B(\widehat{x}, \rho_{\widehat{x}})$. Hence, the map $T|_{B(x,\rho/2)}$ is one-to-one. So, by virtue of Koebe's distortion theorem (Theorem 23.1.7), there exists $\widetilde{s} \in (0, \widehat{s})$ such that

$$\frac{|T'(x)|}{|T'(\xi)|} \le 1 + \varepsilon \tag{25.23}$$

whenever $\xi \in X$ and $x \in B(\xi, \widetilde{s}\rho/2)$. Set

$$s_\varepsilon := \min\{\widetilde{s}, \operatorname{diam}^{-1}(\widehat{\mathbb{C}})\widetilde{s}\rho/2\}.$$

Take $w \in \widehat{\mathbb{C}}$ and $z \in B(w, s_\varepsilon \operatorname{dist}(w, \operatorname{Crit}(T)))$. If $w \in B(\operatorname{Crit}(T), r_\eta(T))$, then there exists $c \in \operatorname{Crit}(T)$ such that $w \in B(c, r_\eta(T,c))$. Moreover, $z \in B(w, \widehat{s}|w-c|)$. It follows from Lemma 25.1.9 and (25.22) that

$$\frac{|T'(z)|}{|T'(w)|} \leq 1 + \varepsilon.$$

On the other hand, if $w \in X$ then

$$|z - w| < s_\varepsilon \operatorname{dist}(w, \operatorname{Crit}(T)) \leq s_\varepsilon \operatorname{diam}(\widehat{\mathbb{C}}) \leq \tilde{s}\rho/2,$$

whence

$$\frac{|T'(z)|}{|T'(w)|} \leq 1 + \varepsilon$$

by virtue of (25.23). □

We now show that, whenever the iterates of a point stay away from the critical set for some time and the modulus of the map's derivative exceeds one at all these iterates, there exists a unique inverse branch on a small enough ball.

Lemma 25.1.11. *Let $T : \widehat{\mathbb{C}} \to \widehat{\mathbb{C}}$ be a rational function with $\deg(T) \geq 2$. For every $\varepsilon > 0$ and every $\lambda > 1$, there exists $\theta = \theta(\varepsilon, \lambda) > 0$ such that if for some $n \in \mathbb{N}$ and some point $z \in \widehat{\mathbb{C}}$ we have*

$$\{z, T(z), \dots, T^{n-1}(z)\} \subseteq \widehat{\mathbb{C}} \setminus B(\operatorname{Crit}(T), \varepsilon) \quad \text{and} \quad |T'(T^j(z))| \geq \lambda, \quad \forall 0 \leq j < n,$$

then there is a unique holomorphic branch $T_z^{-n} : B(T^n(z), 2\theta(\varepsilon, \lambda)) \to \widehat{\mathbb{C}}$ of T^{-n} sending $T^n(z)$ to z.

Proof. Without losing generality, we may assume that $0 < \varepsilon \leq \pi/4$. Let $\delta > 0$ be ascribed to $\varepsilon = \pi/4$ per Lemma 25.1.4. Let also $\beta(\varepsilon) > 0$ arise from Lemma 25.1.3. Set $\beta'(\varepsilon) := \min\{\delta, \beta(\varepsilon)\}$. Take $t \in (0, 1)$ so small that $\lambda^{-1} k_{2R,s}(t) < 1$, where $k_{2R,s}(t)$ is the function produced in Theorem 23.1.9 (Koebe's distortion theorem, spherical version) with $R = \beta'(\varepsilon)$ and $s = \pi/4$. It follows from Lemma 25.1.3 that for every $j = 1, 2, \dots, n$ there exists a unique holomorphic inverse branch $T_j^{-1} : B(T^j(z), \beta'(\varepsilon)) \to \widehat{\mathbb{C}}$ satisfying $T_j^{-1}(T^j(z)) = T^{j-1}(z)$. Hence, in view of Theorem 23.1.9, we get

$$T_j^{-1}(B(T^j(z), t\beta'(\varepsilon))) \subseteq B(T^{j-1}(z), \lambda^{-1} k_{2R,s}(t) t\beta'(\varepsilon)) \subseteq B(T^{j-1}(z), t\beta'(\varepsilon)).$$

Therefore, the composition $T_1^{-1} \circ T_2^{-1} \circ \dots \circ T_n^{-1}$ is well-defined on $B(T^n(z), t\beta'(\varepsilon))$ and $T_1^{-1} \circ T_2^{-1} \circ \dots \circ T_n^{-1}(T^n(z)) = z$. Set $\theta = t\beta'(\varepsilon)/2$ and $T_z^{-n} = T_1^{-1} \circ T_2^{-1} \circ \dots \circ T_n^{-1}$. □

In a similar vein, we establish the following fact about inverse branches.

Lemma 25.1.12. *If $T : \widehat{\mathbb{C}} \to \widehat{\mathbb{C}}$ is a rational function with $\deg(T) \geq 2$, then there exists $\tau \in (0, 1/2)$ such that for every $c \in \operatorname{Crit}(T)$ and every*

$$z \in B\Big(c, \frac{1}{2} \min\{|c_1 - c_2| : c_1, c_2 \in \operatorname{Crit}(T), c_1 \neq c_2\}\Big)$$

there exists a unique holomorphic inverse branch $T_z^{-1} : B(T(z), 2\tau|T(z) - T(c)|) \to \widehat{\mathbb{C}}$ *such that* $T_z^{-1}(T(z)) = z$.

Proof. Recall that $\mathrm{Comp}(\zeta, T, r)$ denotes the connected component of $T^{-1}(B(T(\zeta), r))$ containing ζ. By virtue of Lemma 25.1.4, there exists $\tau \in (0, 1/2)$ so small that

$$\mathrm{diam}(\,\mathrm{Comp}(w, T, 2\tau|T(w) - T(c)|)\,) < \frac{1}{2}\min\{|c_1 - c_2| : c_1, c_2 \in \mathrm{Crit}(T), c_1 \neq c_2\}$$

for all $w \in \widehat{\mathbb{C}}$. Hence, for every z as in the statement of the lemma, we know that

$$\mathrm{Crit}(T) \cap \mathrm{Comp}(z, T, 2\tau|T(z) - T(c)|) = \emptyset.$$

Thus, in view of Corollary 22.5.4, the restriction of T to $\mathrm{Comp}(z, T, 2\tau|T(z) - T(c)|)$ is a conformal homeomorphism onto $B(T(z), 2\tau|T(z) - T(c)|)$. Its inverse is the required map T_z^{-1}. □

Now we identify symmetric lower and upper bounds on the ratio of the diameters of a connected set and its image. These bounds depend on the minimal "distance" separating the set from the critical points.

Lemma 25.1.13. *If* $T : \widehat{\mathbb{C}} \to \widehat{\mathbb{C}}$ *is a rational function with* $\deg(T) \geq 2$, *then there exists a constant* $E \geq 1$ *such that*

$$E^{-1} \leq \frac{\mathrm{diam}(T(F))}{\mathrm{diam}(F)\min\{(\mathrm{Dist}(c, F))^{\deg_c(T)-1} : c \in \mathrm{Crit}(T)\}} \leq E$$

for every connected set $F \subseteq \widehat{\mathbb{C}}$ *containing at least two points, where*

$$\mathrm{Dist}(A, B) := \sup\{|a - b| : a \in A, b \in B\}.$$

Proof. Denote $d_c(T) := \deg_c(T)$ and let

$$d_T := \max\{d_c(T) : c \in \mathrm{Crit}(T)\} \quad \text{and} \quad d_T^* := \min\{d_c(T) : c \in \mathrm{Crit}(T)\}.$$

Pick $k \geq 2$ such that $R := \frac{1}{k}\min\{|c_1 - c_2| : c_1, c_2 \in \mathrm{Crit}(T), c_1 \neq c_2\} < 1$ is as small as required in (25.3)–(25.4), in Theorem 25.1.1, and that satisfies

$$(R/4)^{d_T^*-1} \leq [(4k-1)(R/4)]^{d_T-1}. \tag{25.24}$$

Case 1: Assume first that

$$F \cap [\widehat{\mathbb{C}} \setminus B(\mathrm{Crit}(T), R/4)] \neq \emptyset. \tag{25.25}$$

Then

$$R/4 \le \min\{\mathrm{Dist}(c,F) : c \in \mathrm{Crit}(T)\} \le \mathrm{diam}(\widehat{\mathbb{C}}). \tag{25.26}$$

Subcase 1a: Suppose that

$$\mathrm{diam}(F) \le \eta := \frac{1}{4}\min\{\alpha(R/4), K^{-1}\|T'\|_\infty^{-1}\beta(R/4)\}, \tag{25.27}$$

where $\alpha(R/4)$ and $\beta(R/4)$ originate from Lemma 25.1.3 and K from Koebe's distortion theorem (Theorem 23.1.9 applied to the map $T : B(\xi, 2\mathrm{diam}(F)) \to \widehat{\mathbb{C}}$ with $K = k_{r,s}(1/2)$, where $r = \frac{1}{4}\alpha(R/4)$ and $s = \frac{1}{2}[\mathrm{diam}(\widehat{\mathbb{C}}) - \beta(R/4)]$).

For any $\xi \in F \cap [\widehat{\mathbb{C}} \setminus B(\mathrm{Crit}(T), R/4)]$, we get by Lemma 25.1.3 and Theorem 23.1.9 that

$$\mathrm{diam}(T(F)) \le \mathrm{diam}(T(B(\xi, \mathrm{diam}(F)))) \le K|T'(\xi)|\mathrm{diam}(F) \le \frac{1}{4}\beta(R/4) \tag{25.28}$$

and

$$\mathrm{diam}(F) \le \mathrm{diam}(T_\xi^{-1}(B(T(\xi), \mathrm{diam}(T(F))))) \le K|T'(\xi)|^{-1}\mathrm{diam}(T(F)). \tag{25.29}$$

Using (25.26) and (25.28), we obtain an upper bound on the ratio

$$\frac{\mathrm{diam}(T(F))}{\mathrm{diam}(F)\min\{(\mathrm{Dist}(c,F))^{d_c(T)-1} : c \in \mathrm{Crit}(T)\}} \le \frac{K\|T'\|_\infty}{(R/4)^{d_T-1}},$$

On the other hand, using (25.26) and (25.29) we get a lower bound on the ratio

$$\frac{\mathrm{diam}(T(F))}{\mathrm{diam}(F)\min\{(\mathrm{Dist}(c,F))^{d_c(T)-1} : c \in \mathrm{Crit}(T)\}} \ge \frac{K^{-1}\inf\{|T'(z)| : z \in \widehat{\mathbb{C}} \setminus B(\mathrm{Crit}(T), R/4)\}}{(\mathrm{diam}(\widehat{\mathbb{C}}))^{d_T-1}}.$$

So, the result holds in Subcase 1*a*.

Subcase 1b: Suppose that

$$\mathrm{diam}(F) > \eta := \frac{1}{4}\min\{\alpha(R/4), K^{-1}\|T'\|_\infty^{-1}\beta(R/4)\}. \tag{25.30}$$

Employing (25.26) and (25.30), we obtain an upper bound on the ratio

$$\frac{\mathrm{diam}(T(F))}{\mathrm{diam}(F)\min\{(\mathrm{Dist}(c,F))^{d_c(T)-1} : c \in \mathrm{Crit}(T)\}} \le \frac{\mathrm{diam}(\widehat{\mathbb{C}})}{\eta(R/4)^{d_T-1}}.$$

On the other hand, for any $\xi \in F \cap [\widehat{\mathbb{C}} \setminus B(\mathrm{Crit}(T), R/4)]$ we know that

$$\eta/2 < \mathrm{diam}(F \cap B(\xi, \eta/2)) \le \eta.$$

Using this and (25.29) with F replaced by $F \cap B(\xi, \eta/2)$, we get a lower bound:

$$\begin{aligned}
&\frac{\operatorname{diam}(T(F))}{\operatorname{diam}(F)\min\{(\operatorname{Dist}(c,F))^{d_c(T)-1} : c\in \operatorname{Crit}(T)\}}\\
&\qquad\ge \frac{\operatorname{diam}(T(F\cap B(\xi,\eta/2)))}{\operatorname{diam}(\widehat{\mathbb{C}})\cdot(\operatorname{diam}(\widehat{\mathbb{C}}))^{d_T-1}}\\
&\qquad\ge \frac{K^{-1}|T'(\xi)|\operatorname{diam}(F\cap B(\xi,\eta/2))}{(\operatorname{diam}(\widehat{\mathbb{C}}))^{d_T}}\\
&\qquad\ge \frac{K^{-1}\inf\{|T'(z)| : z\in\widehat{\mathbb{C}}\setminus B(\operatorname{Crit}(T),R/4)\}\cdot\eta/2}{(\operatorname{diam}(\widehat{\mathbb{C}}))^{d_T}}.
\end{aligned}$$

Thus, the result holds in Subcase 1b, too, which means that it holds in Case 1 altogether.

Case 2: Assume now that

$$F\cap[\widehat{\mathbb{C}}\setminus B(\operatorname{Crit}(T),R/4)]=\emptyset. \tag{25.31}$$

This means that $F\subseteq B(\operatorname{Crit}(T),R/4)$. As F is connected, the definition of R imposes the existence of a unique $c_F\in\operatorname{Crit}(T)$ such that

$$F\subseteq B(c_F,R/4).$$

Then

$$D:=\operatorname{Dist}(c_F,F)\le R/4.$$

Given that $F\subseteq\overline{B(c_F,D)}$, it ensues from the right inequality in (25.4) that

$$\operatorname{diam}(T(F))\le\operatorname{diam}(F)\sup\{|T'(z)| : z\in\overline{B(c_F,D)}\}\le AD^{d_{c_F}(T)-1}\operatorname{diam}(F).$$

If $c\neq c_F$, then $\operatorname{Dist}(c,F)\ge\max\{D,R/4\}$ and thus

$$\frac{D^{d_{c_F}(T)-1}}{\operatorname{Dist}(c,F)^{d_c(T)-1}}\le\frac{\operatorname{Dist}(c,F)^{d_{c_F}(T)-1}}{\operatorname{Dist}(c,F)^{d_c(T)-1}}=\operatorname{Dist}(c,F)^{d_{c_F}(T)-d_c(T)}\le\max\{\pi^{d_T},(R/4)^{-d_T}\}.$$

So

$$\frac{\operatorname{diam}(T(F))}{\operatorname{diam}(F)\operatorname{Dist}(c,F)^{d_c(T)-1}}=\frac{\operatorname{diam}(T(F))}{\operatorname{diam}(F)D^{d_{c_F}(T)-1}}\frac{D^{d_{c_F}(T)-1}}{\operatorname{Dist}(c,F)^{d_c(T)-1}}\le A\max\{\pi^{d_T},(R/4)^{-d_T}\}.$$

This results in an upper bound for the entire case:

$$\frac{\operatorname{diam}(T(F))}{\operatorname{diam}(F)\min\{(\operatorname{Dist}(c,F))^{d_c(T)-1} : c\in\operatorname{Crit}(T)\}}\le A\max\{\pi^{d_T},(R/4)^{-d_T}\}. \tag{25.32}$$

For a lower bound, we will soon split the case into subcases. Since $\overline{F}$ is connected, $\operatorname{diam}(\overline{F}) = \operatorname{diam}(F)$ and $\operatorname{diam}(T(\overline{F})) = \operatorname{diam}(\overline{T(F)}) = \operatorname{diam}(T(F))$, we may assume without loss of generality that F is a closed set, whence compact. Let $\xi \in F$ be such that

$$|\xi - c_F| = D. \tag{25.33}$$

Let $\tau \in (0, 1/2)$ come from Lemma 25.1.12.

Subcase 2a: Suppose first that

$$\operatorname{diam}(T(F)) \le \frac{1}{2}\tau |T(\xi) - T(c_F)|.$$

Then

$$F = \bigcup_{z \in T^{-1}(T(\xi)) \cap \overline{B(c_F, D)}} T_z^{-1}(T(F)).$$

As the set F is connected and all the sets $T_z^{-1}(T(F))$, $z \in T^{-1}(T(\xi))$, are closed and mutually disjoint, there exists a unique $z \in T^{-1}(T(\xi))$ such that $F \subseteq T_z^{-1}(T(F))$. In fact, $z = \xi$ since $\xi \in F \cap T^{-1}(T(\xi)) \cap \overline{B(c_F, D)}$. It then follows from Koebe's distortion theorem and the left inequality in (25.4) that

$$\begin{aligned}
\operatorname{diam}(F) &\le \operatorname{diam}(T_z^{-1}(T(F))) \le \operatorname{diam}(T_z^{-1}(\overline{B(T(z), \operatorname{diam}(T(F)))})) \\
&\le K|T'(z)|^{-1} \cdot 2\operatorname{diam}(T(F)) \le K\big[A^{-1} D^{d_{c_F}(T)-1}\big]^{-1} \cdot 2\operatorname{diam}(T(F)) \\
&= 2KAD^{-(d_{c_F}(T)-1)} \operatorname{diam}(T(F)),
\end{aligned}$$

hence leading to a lower bound:

$$\frac{\operatorname{diam}(T(F))}{\operatorname{diam}(F) \min\{(\operatorname{Dist}(c, F))^{d_c(T)-1} : c \in \operatorname{Crit}(T)\}} = \frac{\operatorname{diam}(T(F))}{\operatorname{diam}(F)\, D^{d_{c_F}(T)-1}} \ge (2KA)^{-1}.$$

We are done in Subcase 2a.

Subcase 2b: Suppose now that

$$\operatorname{diam}(T(F)) \ge \frac{1}{2}\tau |T(\xi) - T(c_F)|.$$

By definition of ξ and the left inequality in (25.3), we have that

$$\operatorname{diam}(T(F)) \ge \frac{1}{2}\tau A^{-1} |\xi - c_F|^{d_{c_F}(T)} = \frac{1}{2}\tau A^{-1} D^{d_{c_F}(T)}.$$

As $F \subseteq \overline{B(c_F, D)}$, we know that $\operatorname{diam}(F) \le 2D$ and, therefore,

$$\frac{\operatorname{diam}(T(F))}{\operatorname{diam}(F)} \geq \frac{\frac{1}{2}\tau A^{-1}D^{d_{c_F}(T)}}{2D} = \frac{1}{4}\tau A^{-1}D^{d_{c_F}(T)-1},$$

and we are done in Subcase 2b, too. So, Case 2 holds overall and the proof of Lemma 25.1.13 is complete. □

Finally, we establish the integrability of some logarithmic functions on the Julia set based on the integrability of the logarithm of the modulus of the derivative of the map on that set.

Lemma 25.1.14. *Let $T : \widehat{\mathbb{C}} \to \widehat{\mathbb{C}}$ be a rational function with $\deg(T) \geq 2$. If μ is a finite Borel measure on $\mathcal{J}(T)$ and if the function $\log|T'|$ is μ-integrable, then the function*

$$\mathcal{J}(T) \ni z \longmapsto \log|z - c|$$

is μ-integrable for every $c \in \operatorname{Crit}(T) \cap \mathcal{J}(T)$. If in addition μ is T-invariant, then the function

$$\mathcal{J}(T) \ni z \longmapsto \log|z - T(c)|$$

is μ-integrable.

Proof. The first assertion follows immediately from (25.4) and since outside of any neighborhood of $\operatorname{Crit}(T)$, the function $|T'|$ is uniformly bounded away from zero and infinity.

In order to prove the second part of the lemma, consider a closed curve R emanating from $T(c)$ such that $\mu(R) = 0$ and a disk $B(T(c), r)$ with $r \in (0, R(T, c))$ so small that a holomorphic inverse branch of T,

$$T_c^{-1} : B(T(c), r) \setminus R \to \mathbb{C}$$

with the property that $\lim_{B(T(c),r)\ni z \to T(c)} T_c^{-1}(z) = c$, is well-defined. Let

$$D := B(T(c), r) \setminus R.$$

By virtue of (25.9), we have

$$|z - T(c)| \asymp |T_c^{-1}(z) - c|^{\deg_c(T)}, \quad \forall z \in T_c^{-1}(B(T(c), r)),$$

where $\asymp$ is the usual symbol of multiplicative comparability:

$$A \asymp B \text{ means that } C^{-1}A \leq B \leq CA \text{ for some constant } C \geq 1. \tag{25.34}$$

It suffices to show that

$$\int_D \log|z - T(c)|\, d\mu(z)$$

is finite. Using the T-invariance of μ, we get

$$\begin{aligned}\int_D \log|z - T(c)|\,d\mu(z) &= \int_{\mathcal{J}(T)} \mathbb{1}_D(z)\log|z - T(c)|\,d\mu(z)\\ &\asymp_+ \int_{\mathcal{J}(T)} \mathbb{1}_D(z)\log|T_c^{-1}(z) - c|^{\deg_c(T)}\,d\mu(z)\\ &= \int_{\mathcal{J}(T)} (\mathbb{1}_D \circ T)(z)\log|z - c|^{\deg_c(T)}\,d\mu(z)\\ &= \deg_c(T)\int_{\mathcal{J}(T)} \mathbb{1}_{T^{-1}(D)}(z)\log|z - c|\,d\mu(z).\end{aligned}$$

Notice that the functions

$$z \mapsto \mathbb{1}_D(z)\log|z - T(c)| \quad \text{and} \quad z \mapsto \mathbb{1}_D(z)\log|T_c^{-1}(z) - c|^{\deg_c(T)}$$

are well-defined on $\mathcal{J}(T)$ and that the sign $\asymp_+$ stands for additive comparability. The finiteness of the last integral follows from the first part of the lemma. $\square$

25.2 Two rules for critical points

In this section, we first prove a lower bound on the “first-return time” of any ball centered on a critical point that lies in the Julia set. This “first-return time” obviously depends on the radius of the ball: it tends to infinity as the radius decreases to zero. We call it Rule I. Note that this result generally does not hold for critical points in the Fatou set; for instance, $c = 0$ is a superattractive fixed point of $z \mapsto z^2$. Rule I was proved in [107]. Later we prove Rule II. It is kind of a strengthening of Rule I and was proved in [33]. We also derive consequences of these two rules.

Lemma 25.2.1 (Rule I). *For any rational function $T : \widehat{\mathbb{C}} \to \widehat{\mathbb{C}}$ there is a constant $C_1 > 0$ such that if $T^n(B(c,\varepsilon)) \cap B(c,\varepsilon) \neq \emptyset$ for some $c \in \mathrm{Crit}(T) \cap \mathcal{J}(T)$, $\varepsilon > 0$ and $n \in \mathbb{N}$, then*

$$n \geq -C_1 \log\varepsilon.$$

Proof. As $\#\mathrm{Crit}(T) < \infty$, by Taylor’s theorem (see (25.3)) there are $\delta \in (0, 1/2)$ and $\Gamma > 0$ such that

$$\operatorname{diam}(T(B(c,2\varepsilon))) \leq \Gamma\varepsilon^2, \quad \forall \varepsilon \in (0,\delta],\ \forall c \in \mathrm{Crit}(T).$$

Using the mean value inequality, we then get

$$\operatorname{diam}(T^k(B(c,2\varepsilon))) \leq \Gamma\varepsilon^2\|T'\|_\infty^{k-1}, \quad \forall k \in \mathbb{N},\ \forall \varepsilon \in (0,\delta],\ \forall c \in \mathrm{Crit}(T).$$

Increasing Γ and decreasing δ if necessary, we may assume that

$$\Gamma \geq \|T'\|_\infty \quad \text{and} \quad C := \frac{1}{\log \|T'\|_\infty}\left[1 + \frac{\log \Gamma - \log \|T'\|_\infty}{\log \delta}\right] > 0. \tag{25.35}$$

Suppose that $T^n(B(c,\varepsilon)) \cap B(c,\varepsilon) \neq \emptyset$ for some $c \in \mathrm{Crit}(T) \cap \mathcal{J}(T)$, some $\varepsilon > 0$ and some $n \in \mathbb{N}$. If $\varepsilon \in (0,\delta]$ and it turned out that $\Gamma\varepsilon^2\|T'\|_\infty^{n-1} < \varepsilon$, then we would deduce from the above that $T^n(B(c,2\varepsilon)) \subseteq B(c,2\varepsilon)$. By induction, we would have

$$T^{nk}(B(c,2\varepsilon)) \subseteq B(c,2\varepsilon), \quad \forall k \in \mathbb{N}.$$

Hence, the family of iterates $\{T^{nk}|_{B(c,2\varepsilon)}\}_{k=1}^\infty$ would be normal, whence $c \in \mathcal{F}(T^n) = \mathcal{F}(T)$, which would contradict the hypothesis that $c \in \mathcal{J}(T)$. Therefore, $\Gamma\varepsilon^2\|T'\|_\infty^{n-1} \geq \varepsilon$. Then

$$n \geq \frac{-\log \varepsilon - \log \Gamma}{\log \|T'\|_\infty} + 1 = \frac{-\log \varepsilon}{\log \|T'\|_\infty}\left[1 + \frac{\log \Gamma - \log \|T'\|_\infty}{\log \varepsilon}\right] \geq -C \log \varepsilon.$$

On the other hand, if $\varepsilon > \delta$ then

$$n \geq 1 = \frac{-\log \delta}{-\log \delta} \geq \frac{1}{-\log \delta}(-\log \varepsilon).$$

Set

$$C_1 := \min\left\{C, \frac{1}{-\log \delta}\right\}.$$

□

We will not really need the next lemma, but it is interesting in itself and is a fairly immediate consequence of Lemma 25.2.1. For every $n \in \mathbb{N}$, denote

$$\mathrm{Critv}(T^n) := T^n(\mathrm{Crit}(T^n)) = \bigcup_{k=1}^{n} T^k(\mathrm{Crit}(T)),$$

i. e., $\mathrm{Critv}(T^n)$ is the set of all critical values of T^n. Then

$$\#\mathrm{Critv}(T^n) \leq \#\mathrm{Crit}(T) \cdot n \leq [2\deg(T) - 2]n.$$

Lemma 25.2.2 (Finiteness lemma). *If $T : \widehat{\mathbb{C}} \to \widehat{\mathbb{C}}$ is a rational function, then there exists $M \in \mathbb{N}$ such that*

$$\#[B(\mathrm{Crit}(T) \cap \mathcal{J}(T), e^{-n}) \cap \mathrm{Critv}(T^n)] \leq M, \quad \forall n \in \mathbb{N}.$$

Proof. Suppose that $T^k(\widetilde{c}), T^l(\widetilde{c}) \in B(c, e^{-n})$ for some $c \in \mathrm{Crit}(T) \cap \mathcal{J}(T)$, $n \in \mathbb{N}$, $l > k \geq 1$ and some $\widetilde{c} \in \mathrm{Crit}(T)$. It then follows from Lemma 25.2.1 that $k - l \geq -C_1 \log(e^{-n}) = C_1 n$. So, the iterate T^n has at most $(C_1^{-1} + 1)\#\mathrm{Crit}(T)$ critical values in $B(c, e^{-n})$. Take $M := (C_1^{-1} + 1)(\#\mathrm{Crit}(T))^2$. □

Let $T : \widehat{\mathbb{C}} \to \widehat{\mathbb{C}}$ be a rational function. Fix any $a \in (0,1)$ such that

$$B(\mathrm{Crit}(T) \cap \mathcal{J}(T), a) \cap \mathrm{Crit}(T) = \mathrm{Crit}(T) \cap \mathcal{J}(T). \tag{25.36}$$

For each $c \in \mathrm{Crit}(T) \cap \mathcal{J}(T)$, define the function $k_c : \widehat{\mathbb{C}} \to \mathbb{N}_0 \cup \{\infty\} = \{0,1,2,\dots\} \cup \{\infty\}$ by the formula

$$k_c(x) := \inf\{k \in \mathbb{N}_0 : x \notin B(c, ae^{-(k+1)})\}, \quad (\inf(\emptyset) = \infty).$$

For every $x \in B(c,a) \setminus \{c\}$, this is equivalent to $k_c(x)$ being the unique $k \in \mathbb{N}_0$ such that

$$x \in B(c, ae^{-k}) \setminus B(c, ae^{-(k+1)}).$$

Obviously, $k_c(c) = \infty$ and $k_c(x) = 0$ for all $x \notin B(c,a)$. For any x, set also

$$k(x) := \max\{k_c(x) : c \in \mathrm{Crit}(T) \cap \mathcal{J}(T)\}.$$

As a rather straightforward repercussion of Lemma 25.2.1, we prove the following.

Lemma 25.2.3 (Rule I'). *If $T : \widehat{\mathbb{C}} \to \widehat{\mathbb{C}}$ is a rational function, then for every $c \in \mathrm{Crit}(T) \cap \mathcal{J}(T)$, every $x \in \widehat{\mathbb{C}}$ and every $n \in \mathbb{N}$, we have that*

$$\min\{k_c(x), k_c(T^n(x))\} \le C_2 n,$$

where $C_2 = (1 + C_1)/C_1$ and $C_1 > 0$ is the constant produced in Lemma 25.2.1.

Proof. By way of contradiction, suppose that $\min\{k_c(x), k_c(T^n(x))\} > C_2 n$ for some $c \in \mathrm{Crit}(T) \cap \mathcal{J}(T)$, $x \in \widehat{\mathbb{C}}$ and $n \in \mathbb{N}$. Then $x, T^n(x) \in B(c, ae^{-(C_2 n+1)})$, where a comes from (25.36). So,

$$T^n(x) \in T^n(B(c, ae^{-(C_2 n+1)})) \cap B(c, ae^{-(C_2 n+1)}).$$

In particular, $T^n(B(c, ae^{-(C_2 n+1)})) \cap B(c, ae^{-(C_2 n+1)}) \neq \emptyset$. It follows from Lemma 25.2.1 that

$$\begin{aligned} n &\ge -C_1 \log(ae^{-(C_2 n+1)}) = -C_1 \log a + C_1(C_2 n + 1) \ge C_1 C_2 n + C_1(1 - \log a) \\ &> C_1 \frac{1 + C_1}{C_1} n + C_1(1 - \log a) = (1 + C_1)n + C_1(1 - \log a). \end{aligned}$$

Hence, $C_1 n < -C_1(1 - \log a)$. So $n < -(1 - \log a) < 0$, a contradiction. □

As a direct consequence of Taylor's theorem (see (25.4)), we record the following.

Observation 25.2.4. If $T : \widehat{\mathbb{C}} \to \widehat{\mathbb{C}}$ is a rational function, then there exists $\theta > 0$ (depending on a) such that

$$\exp(-\theta - (\deg_c(T) - 1)k_c(x)) \le |T'(x)| \le \exp(\theta - (\deg_c(T) - 1)k_c(x)) \tag{25.37}$$

for every $c \in \mathrm{Crit}(T)$ and every point $x \in \widehat{\mathbb{C}}$.

Lemma 25.2.5 (Rule II). *For any rational function $T : \widehat{\mathbb{C}} \to \widehat{\mathbb{C}}$ there exists a constant $H > 0$ such that if $c \in \mathrm{Crit}(T) \cap \mathcal{J}(T)$, $n \in \mathbb{N}$ and $x \in \widehat{\mathbb{C}}$ satisfy*

$$k_c(T^j(x)) \le k_c(T^n(x)), \quad \forall 1 \le j < n, \tag{25.38}$$

then

$$\min\{k_c(x), k_c(T^n(x))\} + \sum_{j=1}^{n-1} k_c(T^j(x)) \le Hn. \tag{25.39}$$

Proof. The proof is by induction with respect to $n \in \mathbb{N}$. For $n = 1$, the statement is Lemma 25.2.3. The procedure for the inductive step will be as follows:

Given x and $n \in \mathbb{N}$ satisfying (25.38) we shall decompose the string (trajectory) $x, T(x), \dots, T^n(x)$ into two blocks:

(a) $x, T(x), \dots, T^m(x), m \le n$, for which we shall prove (25.39);
(b) $T^m(x), \dots, T^n(x)$ for which we can apply the inductive hypothesis.

Summing up these two estimates will establish (25.39) for the string $x, T(x), \dots, T^n(x)$. We will demonstrate that it suffices to take

$$H := \alpha^{-1}\big(m(\theta + \alpha) + \log 2 + 1 - \log(e - 1)\big), \tag{25.40}$$

where $\alpha := \deg_c(T) - 1$ and θ arises from Observation 25.2.4. Denote

$$k' := \min\{k_c(x), k_c(T^n(x))\} \quad \text{and} \quad B := B(c, ae^{-(k'-1)}),$$

where a comes from (25.36).

If $k' \ge 1$, denote by m the least positive integer such that

$$k_c(T^m(x)) > \min\{k_c(T^m(z)) : z \in B\} + 1 \tag{25.41}$$

or

$$k_c(T^m(x)) \ge k'. \tag{25.42}$$

In that case, observe that $1 \le m \le n$ and $\{x, T^n(x)\} \subseteq B(c, ae^{-k'}) \subseteq B$.

If $k' = 0$, set

$$m = 1. \tag{25.43}$$

In all these cases, the sequence $y = T^m(x), T(y), \dots, T^{n-m}(y)$ satisfies (25.38) and thereby

$$k_c(y) = \min\{k_c(y), k_c(T^{n-m}(y))\}.$$

Hence, by the inductive hypothesis,

$$\sum_{j=m}^{n-1} k_c(T^j(x)) \leq H(n-m). \tag{25.44}$$

Assume first that $k' \geq 1$. By definition of m, neither (25.41) nor (25.42) holds for any $1 \leq j < m$. Therefore, for every $1 \leq j < m$ and for every $z \in B$ we have that

$$k_c(T^j(x)) \leq k_c(T^j(z)) + 1.$$

Hence, by Observation 25.2.4,

$$\begin{aligned}|(T^{m-1})'(T(z))| &= \prod_{k=0}^{m-2} |T'(T^{k+1}(z))| \leq \prod_{k=0}^{m-2} \exp(\theta - (\deg_c(T)-1)k_c(T^{k+1}(z))) \\ &\leq \exp\Big((m-1)\theta - \alpha \sum_{j=1}^{m-1} [k_c(T^j(x)) - 1]\Big), \quad \forall z \in B.\end{aligned}$$

By definition of B, we also know that $k_c(z) \geq k' - 1$ for every $z \in B$. By the same Observation 25.2.4, we then get

$$|T'(z)| \leq \exp(\theta - (\deg_c(T)-1)k_c(z)) \leq \exp(\theta - \alpha(k'-1)), \quad \forall z \in B.$$

Thus,

$$\frac{\operatorname{diam}(T^m(B))}{\operatorname{diam}(B)} \leq \sup_{z \in B} |(T^m)'(z)| \leq \exp\Big(m\theta + m\alpha - \alpha\Big[k' + \sum_{j=1}^{m-1} k_c(T^j(x))\Big]\Big). \tag{25.45}$$

Moreover, for every $z \in B$ we have

$$T^m(z) \in B(c, ae^{-k_c(T^m(z))}) \setminus B(c, ae^{-(k_c(T^m(z))+1)}).$$

In particular,

$$T^m(x) \in B(c, ae^{-k_c(T^m(x))}) \setminus B(c, ae^{-(k_c(T^m(x))+1)}).$$

If (25.41) holds but (25.42) does not, then there is $z \in B$ for which $k_c(T^m(x)) > k_c(T^m(z)) + 1$ and we deduce that

$$\begin{aligned}\operatorname{diam}(T^m(B)) &\geq |T^m(z) - T^m(x)| \\ &\geq ae^{-(k_c(T^m(z))+1)} - ae^{-k_c(T^m(x))} \\ &\geq a(e^{-(k_c(T^m(x))-1)} - e^{-k_c(T^m(x))}) \\ &= ae^{-k_c(T^m(x))}(e-1)\end{aligned}$$

$$\geq ae^{-k'}(e-1).$$

Combining this with (25.45) gives

$$\frac{e-1}{2e} \leq \exp\left(m(\theta+\alpha) - \alpha\left[k' + \sum_{j=1}^{m-1} k_c(T^j(x))\right]\right).$$

Consequently,

$$k' + \sum_{j=1}^{m-1} k_c(T^j(x)) \leq \alpha^{-1}(m(\theta+\alpha) + \log 2 + 1 - \log(e-1)) = H. \tag{25.46}$$

If (25.42) holds, we obtain

$$\operatorname{diam}(T^m(B)) \geq ae^{-k'}(e-1). \tag{25.47}$$

Otherwise, given any $w \in T^m(B)$ and having already asserted that $x \in B$, we would have

$$|w - T^m(x)| \leq \operatorname{diam}(T^m(B)) < ae^{-k'}(e-1).$$

Furthermore, as $k_c(T^m(x)) \geq k'$ by (25.42), the definition of k_c would yield that

$$T^m(x) \in B(c, ae^{-k_c(T^m(x))}) \subseteq B(c, ae^{-k'}),$$

i. e., $|T^m(x) - c| < ae^{-k'}$. Then we would deduce that

$$|w-c| \leq |w - T^m(x)| + |T^m(x) - c| < ae^{-k'}(e-1) + ae^{-k'} = ae^{-(k'-1)},$$

i. e., $w \in B(c, ae^{-(k'-1)}) =: B$. So, we would infer that $T^m(B) \subseteq B$. Since $\operatorname{diam}(B) \leq 2a < \operatorname{diam}(\widehat{\mathbb{C}})$, the set $\widehat{\mathbb{C}} \setminus B$ contains at least 3 points. Therefore, the family of iterates $(T^{mj}|_B : B \to \widehat{\mathbb{C}})_{j=1}^{\infty}$ would be normal according to Montel's theorem, which would contradict the hypothesis that $c \in \mathcal{J}(T)$. Consequently, (25.47) holds, and hence (25.46) is valid in this case, too.

Finally, if $k' = 0$ then (25.43) holds and (25.46) is trivially satisfied. The combination of (25.44) and (25.46) means that (25.39) holds with H defined by (25.40). □

We now prove the following consequence of this lemma, i. e., of Rule II.

Lemma 25.2.6. *For every rational function $T : \widehat{\mathbb{C}} \to \widehat{\mathbb{C}}$, there exist constants $\kappa \in (0,1]$ and $\beta > 0$ such that if $x \in \mathcal{J}(T)$ and $n > 2\#(\mathrm{Crit}(T) \cap \mathcal{J}(T))$, then there are integers j and l such that $0 \leq j < j + l \leq n$, $l \geq n\kappa$, and*

$$|(T^l)'(T^j(x))| \geq e^{-\beta n}.$$

Proof. Temporarily fix $c \in \mathrm{Crit}(T) \cap \mathcal{J}(T)$. Set $i_0 = 0$. Let $q(c) = i_1$ denote an integer in $\{0, 1, \ldots, n\}$ for which $k_c(T^i(x))$ attains its maximum; recall that even $k_c(T^i(x)) = \infty$ is possible if $c = T^i(x)$, but there exists at most one such i. Define inductively i_l to be the integer in $\{i_{l-1} + 1, i_{l-1} + 2, \ldots, n\}$ where $k_c(T^i(x))$ attains its maximum. This procedure terminates after finitely many steps, say u, and of course $i_u = n$.

Decompose the trajectory $x, T(x), \ldots, T^n(x)$ into the following pieces (with overlapping ends):

$$(x, \ldots, T^{i_1}(x)), (T^{i_1}(x), \ldots, T^{i_2}(x)), \ \ldots, \ (T^{i_{u-1}}(x), \ldots, T^{i_u}(x)).$$

Observe that each of these pieces satisfies the hypotheses of Lemma 25.2.5 and

$$k_c(T^{i_1}(x)) \geq k_c(T^{i_2}(x)) \geq \cdots \geq k_c(T^{i_{u-1}}(x)) \geq k_c(T^{i_u}(x)) = k_c(T^n(x)).$$

Applying Lemma 25.2.5 to all these pieces and remembering that $i_1 = q(c)$, we obtain

$$\min\{k_c(T^{i_k}(x)), k_c(T^{i_{k+1}}(x))\} + \sum_{j=i_k+1}^{i_{k+1}-1} k_c(T^j(x)) \leq H(i_{k+1} - i_k)$$

for all $0 \leq k < u$ and thereby

$$\begin{aligned} \sum_{j=0}^{i_1-1} k_c(T^j(x)) + \sum_{k=1}^{u-1} \sum_{j=i_k+1}^{i_{k+1}} k_c(T^j(x)) &\leq \sum_{k=0}^{u-1} H(i_{k+1} - i_k) \\ -k_c(T^{i_1}(x)) + \sum_{j=i_0}^{i_u} k_c(T^j(x)) &\leq H(i_u - i_0) \\ -k_c(T^{q(c)}(x)) + \sum_{j=0}^{n} k_c(T^j(x)) &\leq Hn. \end{aligned} \tag{25.48}$$

Denote

$$N := \#(\mathrm{Crit}(T) \cap \mathcal{J}(T)) \quad \text{and} \quad Z_n := \{0, 1, \ldots, n\} \setminus \{q(c) : c \in \mathrm{Crit}(T) \cap \mathcal{J}(T)\}.$$

By hypothesis, $n > 2N$. Recalling that

$$k(x) = \max\{k_c(x) : c \in \mathrm{Crit}(T) \cap \mathcal{J}(T)\},$$

we get from (25.48) that

$$\sum_{j \in Z_n} k(T^j(x)) \leq NHn. \tag{25.49}$$

As the set Z_n is a union of at most $N + 1$ blocks, there is at least one such block whose length is at least

$$\frac{n-N+1}{N+1}.$$

Denote this block's first element by j and the last one by $j+l$. Invoking (25.49), we have

$$0 \le j < j+l \le n, \quad l \ge \frac{n-N+1}{N+1} \ge \frac{n/2}{N+1} = \frac{1}{2(N+1)}n \quad \text{and} \quad \sum_{i=j}^{j+l-1} k(T^j(x)) \le NHn.$$

From the chain rule and Observation 25.2.4, we know that

$$|(T^l)'(T^j(x))|^{-1} \le \exp\left(\theta l + (\deg(T)-1)\sum_{i=j}^{j+l-1} k(T^j(x))\right).$$

We therefore obtain that

$$|(T^l)'(T^j(x))|^{-1} \le \exp\left(\theta l + (\deg(T)-1)NHn\right) \le \exp\left([\theta + (\deg(T)-1)NH]n\right).$$

Hence,

$$|(T^l)'(T^j(x))| \ge \exp\left(-[\theta + (\deg(T)-1)NH]n\right).$$

Take

$$\kappa := \frac{1}{2(N+1)} \quad \text{and} \quad \beta := \theta + (\deg(T)-1)NH. \qquad \square$$

25.3 Expanding subsets of Julia sets

Definition 25.3.1. Let $T : \widehat{\mathbb{C}} \to \widehat{\mathbb{C}}$ be a rational function. A forward T-invariant closed (so compact) subset X of the Julia set $\mathcal{J}(T)$ is called *expanding* (*or hyperbolic*) if there exists $q \in \mathbb{N}$ such that $|(T^q)'(z)| > 1$ for all $z \in X$.

The proof of the following characterizations of expanding sets is left to the reader.

Theorem 25.3.2. *Let $T : \widehat{\mathbb{C}} \to \widehat{\mathbb{C}}$ be a rational function. If X is a forward T-invariant closed subset of the Julia set $\mathcal{J}(T)$, then the following statements are equivalent:*
(a) *The set X is expanding.*
(b) *There exist $\lambda > 1$ and $n \in \mathbb{N}$ such that $|(T^n)'(z)| \ge \lambda$ for all $z \in X$.*
(c) *For each $\lambda > 1$, there exists $n \in \mathbb{N}$ such that $|(T^n)'(z)| \ge \lambda$ for all $z \in X$.*
(d) *There exist $\lambda > 1$ and $c > 0$ such that $|(T^n)'(z)| \ge c\lambda^n$ for all $z \in X$ and all $n \in \mathbb{N}$.*
(e) *For each $\lambda > 1$, there exists $c > 0$ such that $|(T^n)'(z)| \ge c\lambda^n$ for all $z \in X$ and all $n \in \mathbb{N}$.*

Definition 25.3.3. A rational function $T : \widehat{\mathbb{C}} \to \widehat{\mathbb{C}}$ is called *expanding* (*or hyperbolic*) if the Julia set $\mathcal{J}(T)$ is an expanding set for T, i. e., if there exists $q \in \mathbb{N}$ such that $|(T^q)'(z)| > 1$ for all $z \in \mathcal{J}(T)$.

In this section, we will prove selected results about expanding subsets of Julia sets and identify significant subsets which are not expanding. We first provide basic examples of expanding sets.

Proposition 25.3.4. *Let $T : \widehat{\mathbb{C}} \to \widehat{\mathbb{C}}$ be a rational function. Every finite collection of repelling periodic orbits is an expanding set. Moreover, if $X \subseteq \mathcal{J}(T)$ is an expanding set, then any forward T-invariant closed set $Y \subseteq X$ is expanding as well. In particular, if T is expanding then every forward T-invariant closed subset of $\mathcal{J}(T)$ is expanding.*

We now record two obvious observations, the latter of which is actually a reformulation of Proposition 25.3.4.

Observation 25.3.5. Let $T : \widehat{\mathbb{C}} \to \widehat{\mathbb{C}}$ be a rational function. If $X \subseteq \mathcal{J}(T)$ is an expanding set for T, then X is an expanding set for every higher iterate T^k, $k \in \mathbb{N}$.

Observation 25.3.6. Let $T : \widehat{\mathbb{C}} \to \widehat{\mathbb{C}}$ be a rational function. If a forward T-invariant closed set $X \subseteq \mathcal{J}(T)$ is not expanding, then no forward T-invariant closed set $X \subseteq Y \subseteq \mathcal{J}(T)$ is expanding.

Next, we prove the following result.

Theorem 25.3.7. *If $T : \widehat{\mathbb{C}} \to \widehat{\mathbb{C}}$ is a rational function with $\deg(T) \geq 2$, then none of the following forward T-invariant closed subsets of the Julia set $\mathcal{J}(T)$ is expanding:*
(a) *Rationally indifferent periodic orbits.*
(b) *Boundaries of cycles of Siegel disks.*
(c) *Boundaries of cycles of Herman rings.*

Proof. Item (a) is obvious. We shall prove items (b) and (c) simultaneously. The proof is similar to that of Theorem 24.5.1 but this time the holomorphic inverse branches originate from Lemma 25.1.11. Let G be either a Siegel disk or a Herman ring for T. Denote its period by p. Seeking a contradiction, suppose that $\partial(\bigcup_{i=0}^{p-1} T^i(G))$ is an expanding set for T. By Observation 25.3.5, it is also an expanding set for T^p. Since $T^p(\partial G) \subseteq \partial G$, Proposition 25.3.4 asserts that ∂G is an expanding set for T^p. Replacing T by T^{ip} with a sufficiently large integer $i \in \mathbb{N}$, we may assume that

$$T(\partial G) \subseteq \partial G \quad \text{and} \quad |T'(z)| \geq 2, \quad \forall z \in \partial G. \tag{25.50}$$

Then $\mathrm{Crit}(T) \cap \partial G = \emptyset$, and let

$$\varepsilon = \mathrm{dist}(\mathrm{Crit}(T), \partial G). \tag{25.51}$$

Let

$$\Delta = B(0,1) \quad \text{if } G \text{ is a Siegel disk}$$

and

$$\Delta = A(0; r, 1) \quad \text{if } G \text{ is a Herman ring,}$$

with $r \in (0,1)$ coming from Theorem 24.4.10(d). Let

$$H : \Delta \to G$$

be the analytic conjugacy resulting from Theorem 24.4.10(c)/(d) in the Siegel/Herman case, respectively. Fix $w \in G$ such that

$$\operatorname{dist}(w, \partial G) < \frac{1}{2}\theta(\varepsilon, 2), \tag{25.52}$$

where $\theta(\varepsilon, 2) > 0$ comes from Lemma 25.1.11. Fix a point $\xi \in \partial G$ such that

$$|w - \xi| = \operatorname{dist}(w, \partial G). \tag{25.53}$$

Let $\gamma : [0,1] \to \mathbb{C}$ be the function

$$\gamma(t) = w + t(\xi - w).$$

Then $\gamma(0) = w,\ \gamma(1) = \xi,\ \gamma([0,1])$ is the closed line segment from w to ξ, and

$$\gamma([0,1)) \subseteq G.$$

Let

$$F := H(\{z \in \Delta : |z| = |H^{-1}(w)|\}). \tag{25.54}$$

Note that F is a compact set (homeomorphic to a circle) and $F \subseteq G$. Therefore,

$$\operatorname{dist}(F, \partial G) > 0. \tag{25.55}$$

Since the map $T|_G : G \to G$ is a holomorphic homeomorphism, its inverse $T|_G^{-1} : G \to G$ is also a holomorphic homeomorphism. Fix $n \in \mathbb{N}$. By way of contradiction, suppose that $\overline{T|_G^{-n}(\gamma([0,1)))} \subseteq G$. As $T^n(\overline{T|_G^{-n}(\gamma([0,1)))})$ is a compact set, it is closed and we would then have

$$\xi \in \gamma([0,1]) = T^n(\overline{T|_G^{-n}(\gamma([0,1)))}) \subseteq T^n(G) = G.$$

This contradiction shows that $\overline{T|_G^{-n}(\gamma([0,1)))} \nsubseteq G$. Hence, there exists a point

$$\xi_n \in \left[\overline{T|_G^{-n}(\gamma([0,1)))} \setminus T|_G^{-n}(\gamma([0,1)))\right] \cap \partial G.$$

Then

$$T^n(\xi_n) \in T^n\big(\overline{T|_G^{-n}(\gamma([0,1)))}\big) \setminus G = \gamma([0,1]) \setminus G = \{\xi\}.$$

So, $T^n(\xi_n) = \xi$. By (25.50)–(25.51) and Lemma 25.1.11, there is a unique holomorphic branch $T_{\xi_n}^{-n} : B(\xi, 2\theta(\varepsilon,2)) \to \widehat{\mathbb{C}}$ of T^{-n} sending ξ to ξ_n. By definition of ξ_n, there exists a sequence $(t_k)_{k=1}^{\infty}$ in $[0,1)$ such that

$$\lim_{k\to\infty} t_k = 1 \quad \text{and} \quad \lim_{k\to\infty} T|_G^{-n}(\gamma(t_k)) = \xi_n.$$

Consequently, for every $k \in \mathbb{N}$ large enough we have

$$T|_G^{-n}(\gamma(t_k)) \in T_{\xi_n}^{-n}\Big(B\Big(\xi, \frac{1}{2}\theta(\varepsilon,2)\Big)\Big).$$

Hence,

$$T_{\xi_n}^{-n}(\gamma(t_k)) = T|_G^{-n}(\gamma(t_k)) \tag{25.56}$$

for all k large enough. It follows from (25.52)–(25.53) that

$$\gamma([0,1)) \subseteq B(w, |w-\xi|) \subseteq B(\xi, \theta(\varepsilon,2)) \cap G.$$

Thus, both holomorphic branches $T|_G^{-n}$ and $T_{\xi_n}^{-n}$ of T^{-n} are defined on $B(w, |w-\xi|)$. As both of them, restricted to $B(w, |w-\xi|)$, are uniquely determined by their value at any point in $B(w, |w-\xi|)$, we infer from (25.56) that

$$(T_{\xi_n}^{-n})|_{B(w,|w-\xi|)} = (T|_G^{-n})|_{B(w,|w-\xi|)}.$$

In particular,

$$T_{\xi_n}^{-n}(w) = T|_G^{-n}(w). \tag{25.57}$$

Using also the fact that $T|_G^{-n}(w) \in F$, we deduce that

$$\operatorname{diam}(T_{\xi_n}^{-n}(B(\xi, \theta(\varepsilon,2)))) \geq |\xi_n - T_{\xi_n}^{-n}(w)| = |\xi_n - T|_G^{-n}(w)| \geq \operatorname{dist}(\partial G, F).$$

By (25.55), we conclude that

$$\liminf_{n\to\infty} \operatorname{diam}(T_{\xi_n}^{-n}(B(\xi, \theta(\varepsilon,2)))) \geq \operatorname{dist}(\partial G, F) > 0,$$

contrary to Lemma 24.1.13 (recall that $\xi \in \partial G \subseteq \mathcal{J}(T)$). □

We now prove the following result in a “positive” direction, i. e., about expanding sets. The proof stems from [110].

Theorem 25.3.8. *Let $T : \widehat{\mathbb{C}} \to \widehat{\mathbb{C}}$ be a rational function with $\deg(T) \geq 2$. If $X \subseteq \mathcal{J}(T)$ is a forward T-invariant closed set, homeomorphic to the Cantor set, such that the map $T|_X : X \to X$ is expanding, then for every $\eta > 0$ there exists a forward T-invariant closed set $X \subseteq X_\eta \subseteq B(X, \eta)$, homeomorphic to the Cantor set, such that the map $T|_{X_\eta} : X_\eta \to X_\eta$ is open and expanding. In addition, if $T|_X$ is topologically transitive, then the map $T|_{X_\eta}$ is likewise topologically transitive.*

Proof. Since $X \subseteq \mathcal{J}(T)$, we may assume that $X \subseteq \mathbb{C}$ by performing a Möbius change of coordinates. As the set X is expanding, there exists $q \in \mathbb{N}$ such that

$$|(T^q)'(x)| \geq 3, \quad \forall x \in X.$$

Denote by $(\langle \cdot, \cdot \rangle_x)_{x \in \mathbb{C}}$ the standard Euclidean Riemannian metric on $\mathbb{C}$ and define a new Riemannian metric $(\langle \cdot, \cdot \rangle_x^*)_{x \in \mathbb{C}}$ by

$$\langle v, w \rangle_x^* := \sum_{j=0}^{q-1} \langle (T^j)'(x)v, (T^j)'(x)w \rangle_{T^j(x)} = \sum_{j=0}^{q-1} |(T^j)'(x)|^2 \langle v, w \rangle_x.$$

Claim 1. *There exist $R > 0$ and $\lambda > 1$ such that*

$$|T'(z)|^* \geq \lambda, \quad \forall z \in B^*(X, R),$$

where $|T'(z)|^*$ is the modulus of the derivative (scaling factor) given by

$$|T'(z)|^* = \frac{\|T'(z)v\|_{T(z)}}{\|v\|_z}.$$

Proof of Claim 1. As X is compact, a continuity argument yields the existence of some $R > 0$ such that

$$|(T^q)'(z)| \geq 2, \quad \forall z \in B^*(X, 2R). \tag{25.58}$$

For every point $z \in \mathbb{C}$ and every tangent vector $v \in T_z\mathbb{C}$, we have

$$\|v\|_z^{*2} = \|v\|_z^2 \sum_{j=0}^{q-1} |(T^j)'(z)|^2. \tag{25.59}$$

Therefore, for every $z \in B^*(X, R)$,

$$\begin{aligned}\|T'(z)v\|_{T(z)}^{*2} &= \|T'(z)v\|_{T(z)}^2 \sum_{j=0}^{q-1} |(T^j)'(T(z))|^2 = |T'(z)|^2 \|v\|_z^2 \sum_{j=0}^{q-1} |(T^j)'(T(z))|^2 \\ &= \|v\|_z^2 \sum_{j=0}^{q-1} |(T^{j+1})'(z)|^2 = \|v\|_z^2 \left(\sum_{j=0}^{q-1} |(T^j)'(z)|^2 - 1 + |(T^q)'(z)|^2 \right)\end{aligned}$$

$$= \|v\|_z^{*2} + \|v\|_z^2 \cdot \left(\left|(T^q)'(z)\right|^2 - 1\right) \geq \|v\|_z^{*2} + 3\|v\|_z^2. \tag{25.60}$$

It ensues from (25.59) that

$$\|v\|_z^{*2} \leq \|v\|_z^2 \sum_{j=0}^{q-1} \|T'\|_\infty^{2j} = \frac{\|T'\|_\infty^{2q} - 1}{\|T'\|_\infty^2 - 1} \cdot \|v\|_z^2,$$

where

$$\|T'\|_\infty := \sup\{|T'(w)| : w \in \overline{B^*(X,R)}\} < \infty.$$

Inserting this into (25.60), we get

$$\|T'(z)v\|_{T(z)}^{*2} \geq \left(1 + 3\frac{\|T'\|_\infty^2 - 1}{\|T'\|_\infty^{2q} - 1}\right)\|v\|_z^{*2}.$$

Taking

$$\lambda = \left(1 + 3\frac{\|T'\|_\infty^2 - 1}{\|T'\|_\infty^{2q} - 1}\right)^{1/2} > 1$$

completes the proof of Claim 1. ■

Denote by $|w - z|^*$ the distance between any two points $w, z \in \mathbb{C}$ generated by the Riemannian metric $(\langle\cdot,\cdot\rangle_x^*)_{x\in\mathbb{C}}$.

Fix any $r \in (0, R]$ and define a relation $\sim_r$ on $B^*(X, r)$ by declaring that $z \sim_r w$ if and only if there are finitely many points $x_1, x_2, \ldots, x_j$ in X such that $z \in B^*(x_1, r)$, $w \in B^*(x_j, r)$, and $B^*(x_i, r) \cap B^*(x_{i+1}, r) \neq \emptyset$ for all $i = 1, 2, \ldots, j-1$. It is immediate that $\sim_r$ is an equivalence relation on X.

Claim 2. *Each equivalence class of the relation $\sim_r$ is open in $\mathbb{C}$ and intersects X. Moreover, it is a union of open *balls of radius r centered at points in X.*

Proof of Claim 2. The latter assertion is obvious. For the former one, suppose that $z \in B^*(X, r)$ and $[z]_r$ is the equivalence class of z. Let $w \in [z]_r$. This means that $z \sim_r w$. Let $x_1, \ldots, x_k \in X$ be points witnessing this relation between z and w. Then $w \in B^*(x_k, r)$ and each element of $B^*(x_k, r)$ is in relation with z through the same points $x_1, \ldots, x_k$. So, $x_k \in B^*(x_k, r) \subseteq [z]_r$. The proof of Claim 2 is complete. ■

It follows from this claim and the compactness of X that the number $N(r)$ of equivalence classes of $\sim_r$ is finite. It is obvious that if $0 < r \leq s \leq R$ and $z \sim_r w$ then $z \sim_s w$, whence each equivalence class of the relation $\sim_r$ is contained in a unique equivalence class of the relation $\sim_s$. Thus, $N(s) \leq N(r)$. This means that the function

$$(0, R] \ni r \mapsto N(r)$$

is monotone decreasing. Since it takes only positive integer values, the set Y of its discontinuity points, or equivalently its points of nonconstantness, is a countable set whose sole accumulation point may be 0.

Claim 3. *If $r \in (0,R]\backslash Y$, then the closures of any two different equivalence classes of the relation $\sim_r$ are disjoint.*

Proof of Claim 3. Let $\delta > 0$ be so small that

$$B(r, 2\delta) \cap Y = \emptyset.$$

Then

$$N(r+\delta) = N(r). \tag{25.61}$$

Seeking a contradiction, suppose that A and B are two distinct classes of the relation $\sim_r$ such that $\overline{A} \cap \overline{B} \neq \emptyset$. Let $z \in \overline{A} \cap \overline{B}$. Then $A \cap B^*(z,\delta) \neq \emptyset$ and $B \cap B^*(z,\delta) \neq \emptyset$. Let

$$x \in A \cap B^*(z,\delta) \quad \text{and} \quad y \in B \cap B^*(z,\delta).$$

By Claim 2, there are two points $x', y' \in X$ such that

$$x \in B^*(x', r) \quad \text{and} \quad y \in B^*(y', r).$$

Hence, $x' \in A, y' \in B$ and $z \in B^*(x', r+\delta) \cap B^*(y', r+\delta) \neq \emptyset$. Thus, $x' \sim_{r+\delta} y'$. Therefore, A and B are contained in the same class of the relation $\sim_{r+\delta}$. So, $N(r+\delta) \leq N(r) - 1$, contrary to (25.61). The proof of Claim 3 is complete. ■

By set Y's structure, there exists a strictly decreasing sequence $(r_k)_{k=1}^{\infty}$ converging to 0 such that $r_k \in (0,R]\backslash Y$ for all $k \in \mathbb{N}$. Denote by E_k the collection of all equivalence classes of the relation $\sim_{r_k}$.

Claim 4. $\lim\limits_{k\to\infty} \max\{\text{diam}^*(A) : A \in E_k\} = 0.$

Proof of Claim 4. Fix $\varepsilon > 0$. Since X is a topological Cantor set, there exists a finite partition α of X consisting of closed sets such that

$$\text{diam}^*(A) < \varepsilon, \quad \forall A \in \alpha. \tag{25.62}$$

As $\lim\limits_{k\to\infty} r_k = 0$, there is $N \in \mathbb{N}$ such that

$$r_k < \frac{1}{2}\min\{\text{dist}^*(A,B) : A, B \in \alpha \text{ and } A \neq B\} \tag{25.63}$$

for all $k \geq N$. Fix any such k. Then each atom $A \in \alpha$ is contained in a unique class $\widehat{A} \in E_k$ and

$$\widehat{B} \cap \widehat{C} = \emptyset, \quad \forall B, C \in \alpha,\ B \neq C. \tag{25.64}$$

We will prove that

$$\max\{\mathrm{diam}^*(D) : D \in E_k\} \leq \varepsilon. \tag{25.65}$$

Seeking a contradiction, suppose that $\mathrm{diam}^*(D) > \varepsilon$ for some $D \in E_k$. Then there are two points $z, w \in D$ such that $|z - w|^* > \varepsilon$. In view of (25.62), there are then two different atoms $B, C \in \alpha$ such that $z \in B$ and $w \in C$. As $B \cap D \neq \emptyset$ and $C \cap D \neq \emptyset$, we also have $\widehat{B} \cap D \neq \emptyset$ and $\widehat{C} \cap D \neq \emptyset$. Hence, $\widehat{B} = D = \widehat{C}$, contrary to (25.64). Thus, (25.65) holds for all $k \geq N$. This finishes the proof of Claim 4. ■

Now, it follows from (25.58) that $T'(x) \neq 0$ for all $x \in X$. So it ensues from the inverse function theorem (Theorem A.3.13) along with a standard compactness argument that there exists $R_1 \in (0, R]$ such that for every $x \in X$ there is a unique holomorphic branch $T_x^{-1} : B^*(T(x), R_1) \to \mathbb{C}$ of T^{-1} sending $T(x)$ to x. Decreasing R_1 if necessary, we have that

$$T_x^{-1}(B^*(T(x), R_1)) \subseteq B^*(x, R_1). \tag{25.66}$$

By Claim 4, there exists $q \in \mathbb{N}$ such that

$$\max\{r_q, \mathrm{diam}^*(A)\} < \min\{\eta^*, R_1/4\}, \quad \forall A \in E_q,$$

where $\eta^* \in (0, \eta)$ is so small that $B^*(X, \eta^*) \subseteq B(X, \eta)$. Given $A \in E_q$, fix an element $y_A \in A \cap X$. Pick also any $x \in X \cap T^{-1}(y_A)$. If $z, w \in A$, then $z \sim_{r_q} w$ and Claim 1 affirms that $T_x^{-1}(z) \sim_{\lambda^{-1} r_q} T_x^{-1}(w)$. Hence, $T_x^{-1}(z) \sim_{r_q} T_x^{-1}(w)$. Thus, there is a unique element $A_x \in E_q$ such that

$$T_x^{-1}(A) \subseteq A_x. \tag{25.67}$$

Take $V := E_q$ for (finite) vertex set. For each vertex $v = A \in E_q = V$, define $X_v := \overline{A}$ and $W_v := B^*(y_A, R_1)$. Take $E := \{(A, x) : A \in E_q,\ x \in X \cap T^{-1}(y_A)\}$ for (finite) alphabet (i. e., set of edges). For each letter/edge $e = (A, x) \in E$, respectively define the initial and terminal vertices of e by $i(e) = A_x$ and $t(e) = A$, where A_x arises from (25.67). Then associate the map (generator) $\varphi_e = \varphi_{A,x} := T_x^{-1} : B^*(y_A, R_1) \to \mathbb{C}$. In light of Claim 1 and (25.66)–(25.67), the family $\Phi := \{\varphi_e\}_{e \in E}$ forms a CGDS (i. e., a CGDMS whose matrix solely reflects the system's underlying graph; for more information on CGDMSs, see Chapter 19). By Claim 3, this CGDMS satisfies the Strong Separation Condition (SSC; also called the Boundary Separation Condition (BSC); see Definitions 19.7.5–19.7.6). Consequently, the limit set $\mathcal{J}$ of the system Φ is compact and the corresponding coding map $\pi : E_G^\infty \to \mathcal{J}$ is a homeomorphism, where G is the incidence matrix of the system Φ. We will prove several claims about $\mathcal{J}$. The first is immediate.

Claim 5. $\mathcal{J} \subseteq B^*(X, \eta^*) \subseteq B(X, \eta)$.

Claim 6. $X \subseteq \mathcal{J}$.

Proof of Claim 6. Let $x \in X$. For every $n \geq 0$, consider the holomorphic inverse branch $T^{-1}_{T^n(x)} : B^*(T^{n+1}(x), R_1) \to \mathbb{C}$ and let $x_n := T^{-1}_{T^n(x)}(y_{A_{T^{n+1}(x)}})$. By Claim 1, we conclude that

$$\pi\big(((A_{T^{n+1}(x)}, x_n))_{n=0}^{\infty}\big) = x.$$

So $x \in \mathcal{J}$. Thus, $X \subseteq \mathcal{J}$ and Claim 6 is proved. ■

Claim 7. $T(\mathcal{J}) \subseteq \mathcal{J}$.

Proof of Claim 7. Let $x \in \mathcal{J}$. Then there exists a (unique) sequence $\omega \in E_G^\infty$ such that

$$x = \pi(\omega) = \varphi_{\omega_1}(\pi(\sigma(\omega))).$$

Therefore, $T(x) = T \circ \varphi_{\omega_1}(\pi(\sigma(\omega))) = \pi(\sigma(\omega)) \in \mathcal{J}$. Claim 7 is established. ■

As an immediate consequence of Claims 1 and 7, we get the following.

Claim 8. *The map* $T : \mathcal{J} \to \mathcal{J}$ *is expanding.*

Claim 9. *The maps* $T : \mathcal{J} \to \mathcal{J}$ *and* $\sigma : E_G^\infty \to E_G^\infty$ *are topologically conjugate. More precisely, the map* $\pi : E_G^\infty \to \mathcal{J}$ *is a topological conjugacy, i. e., a homeomorphism such that the following diagram commutes:*

$$\begin{array}{ccc} E_G^\infty & \xrightarrow{\ \sigma\ } & E_G^\infty \\ \pi \downarrow & & \downarrow \pi \\ \mathcal{J} & \xrightarrow[\ T\ ]{} & \mathcal{J} \end{array}$$

i. e., $\pi \circ \sigma = T \circ \pi$.

Proof of Claim 9. We have already observed that π is a homeomorphism. Let $\omega \in E_G^\infty$. Then $T \circ \pi(\omega) = T \circ \varphi_{\omega_1}(\pi(\sigma(\omega))) = \pi(\sigma(\omega))$. So $T \circ \pi = \pi \circ \sigma$, and Claim 9 is proved. ■

As the map $\sigma : E_G^\infty \to E_G^\infty$ is open, Claim 9 entails the following.

Claim 10. *The map* $T : \mathcal{J} \to \mathcal{J}$ *is open.*

If the map $T : X \to X$ is transitive, then the incidence matrix G is irreducible, whence the shift map $\sigma : E_G^\infty \to E_G^\infty$ is transitive. Thus, invoking Claim 9 again, we obtain the following.

Claim 11. *If the map* $T : X \to X$ *is transitive, then the map* $T : \mathcal{J} \to \mathcal{J}$ *is also transitive.*

Setting $X_\eta := \mathcal{J}$ completes the proof of Theorem 25.3.8. □

In Definition 16.2.1, we adopted a definition a little too narrow. In fact, all results derived about conformal repellers in Volume 2 are valid when part (d) of their definition is replaced by the following weaker condition:

(d') There exists $\lambda > 1$ and $q \in \mathbb{N}$ such that $|(T^q)'(x)| \geq \lambda$ for all $x \in X$.

As a complementary result to Theorem 25.3.8, we shall prove the following.

Theorem 25.3.9. *If $T : \widehat{\mathbb{C}} \to \widehat{\mathbb{C}}$ is a rational function with $\deg(T) \geq 2$ and $X \subseteq \mathcal{J}(T)$ is an expanding set such that the restriction $T|_X : X \to X$ is open and transitive, then there exists an open set $X \subseteq U \subseteq \widehat{\mathbb{C}}$ for which the triple (X, U, T) is a conformal expanding repeller.*

Proof. Since X is expanding, there are $\lambda > 1$ and $q \in \mathbb{N}$ such that

$$|(T^q)'(x)| \geq \lambda, \quad \forall x \in X. \tag{25.68}$$

By standard compactness arguments, there exists $\delta > 0$ such that

$$T^q|_{B(x,4\delta)} \text{ is 1-to-1}, \quad \forall x \in X \tag{25.69}$$

and

$$|(T^q)'(z)| \geq \sqrt{\lambda}, \quad \forall z \in B(X, 4\delta) := \bigcup_{x \in X} B(x, 4\delta). \tag{25.70}$$

Consequently, for every $x \in X$ there is a unique holomorphic inverse branch $T_x^{-q} : B(T^q(x), 4\delta) \to \widehat{\mathbb{C}}$ such that $T_x^{-q}(T^q(x)) = x$. By (25.70), we have that

$$T^q(B(x,r)) \supseteq B(T^q(x), \sqrt{\lambda}\, r), \quad |(T_x^{-q})'(z)| \leq \lambda^{-1/2}, \quad \forall r \in [0, 4\delta], \ \forall z \in B(x, 4\delta), \forall x \in X \tag{25.71}$$

and

$$\rho(T_x^{-q}(w), T_x^{-q}(z)) \leq \lambda^{-1/2} \rho(w, z), \quad \forall w, z \in B(x, 4\delta), \ \forall x \in X. \tag{25.72}$$

Applying Corollary 4.3.6 to the open distance expanding map $T^q : X \to X$, there exists $\alpha \in (0, \delta)$ such that every $(1 + \mathrm{Lip}(T^q))\alpha$-pseudo-orbit for $T^q|_X$ is δ-shadowed by a point in X, where $\mathrm{Lip}(T^q)$ is a Lipschitz constant for T^q. Let

$$U = B(X, \alpha).$$

All we need to do for the proof of this theorem is to show that $\bigcap_{n=0}^{\infty} T^{-n}(U) \subseteq X$. Of course, it is sufficient to establish that

$$\bigcap_{n=0}^{\infty} T^{-qn}(U) \subseteq X.$$

So, let

$$z \in \bigcap_{n=0}^{\infty} T^{-qn}(U).$$

Then, for every $n \geq 0$ there exists $x_n \in X$ such that

$$T^{qn}(z) \in B(x_n, \alpha).$$

Therefore

$$\begin{aligned}\rho(x_{n+1}, T^q(x_n)) &\leq \rho(x_{n+1}, T^{q(n+1)}(z)) + \rho(T^{q(n+1)}(z), T^q(x_n)) < \alpha + \rho(T^q(T^{qn}(z)), T^q(x_n))\\ &\leq \alpha + \mathrm{Lip}(T^q)\,\rho(T^{qn}(z), x_n) < [1 + \mathrm{Lip}(T^q)]\alpha.\end{aligned}$$

This means that $(x_n)_{n=0}^{\infty}$ is a $(1 + \mathrm{Lip}(T^q))\alpha$-pseudo-orbit for $T^q|_X$. Thus, it is δ-shadowed by a point $x \in X$. Hence, for all $n \geq 0$,

$$\rho(T^{qn}(z), T^{qn}(x)) \leq \rho(T^{qn}(z), x_n) + \rho(x_n, T^{qn}(x)) < \alpha + \delta < 2\delta. \tag{25.73}$$

It follows from (25.69) and (25.72) that

$$T^{qn}(z) = T^{-q}_{T^{qn}(x)}(T^{q(n+1)}(z)).$$

Hence, by applying (25.72) again, we get

$$\rho(T^{qn}(z), T^{qn}(x)) \leq \lambda^{-1/2}\rho(T^{q(n+1)}(z), T^{q(n+1)}(x)).$$

So, by induction and (25.73), we obtain that

$$\rho(z, x) \leq 2\delta\lambda^{-n/2}, \quad \forall n \geq 0.$$

Thus $\rho(z, x) = 0$, whence $z = x \in X$. □

We will also need the following general result about topological pressure.

Theorem 25.3.10. *Let $T : X \to X$ be a topological dynamical system, i. e. T is a continuous self-map of a compact metrizable space X. For any (continuous) potential $\varphi : X \to \mathbb{R}$,*

$$\mathrm{P}(T, \varphi) = \sup\{\mathrm{P}(T|_F, \varphi|_F) : F \subseteq X \text{ closed, forward } T\text{-invariant and } T|_F \text{ transitive}\}.$$

Proof. Obviously, $\mathrm{P}(T, \varphi) \geq \sup\{\mathrm{P}(T|_F, \varphi|_F)\}$. We now prove the opposite inequality. If μ is an ergodic T-invariant Borel probability measure on X, then $\mathrm{supp}(\mu)$ is a closed, forward T-invariant subset of X. Moreover, according to Theorem 8.2.27, the subsystem $T|_{\mathrm{supp}(\mu)}$ is topologically transitive. By the variational principle (Theorem 12.1.1), we further know that

$$\mathrm{P}(T|_{\mathrm{supp}(\mu)}, \varphi|_{\mathrm{supp}(\mu)}) \geq \mathrm{h}_\mu(T|_{\mathrm{supp}(\mu)}) + \int\limits_{\mathrm{supp}(\mu)} \varphi\, d\mu = \mathrm{h}_\mu(T) + \int\limits_X \varphi\, d\mu.$$

It therefore follows from this same variational principle and Remark 12.1.2 that

$$\mathrm{P}(T,\varphi) = \sup_\mu \Big\{\mathrm{h}_\mu(T) + \int\limits_X \varphi\, d\mu\Big\} \leq \sup_\mu\{\mathrm{P}(T|_{\mathrm{supp}(\mu)}, \varphi|_{\mathrm{supp}(\mu)})\} \leq \sup_F\{\mathrm{P}(T|_F, \varphi|_F)\},$$

where the leftmost two suprema are taken over all ergodic T-invariant Borel probability measures μ on X while the rightmost supremum is taken over all closed, forward T-invariant subsets F of X such that the subsystem $T|_F$ is transitive. □

We now introduce some quantities that are offshoots of the topological pressure function.

Definition 25.3.11. Let $T : \widehat{\mathbb{C}} \to \widehat{\mathbb{C}}$ be a rational function with $\deg(T) \geq 2$. For every $t \in \mathbb{R}$, we denote

$$\mathrm{P}_{\exp}(t) := \sup\{\mathrm{P}(T|_X, -t\log|T'|) \mid X \subseteq \mathcal{J}(T) \text{ expanding forward } T\text{-invariant closed set}\}$$

$$\mathrm{P}^*_{\exp}(t) := \sup\left\{\mathrm{P}(T|_X, -t\log|T'|) \;\middle|\; \begin{array}{l} X \subseteq \mathcal{J}(T) \text{ expanding forward } T\text{-invariant closed set} \\ \text{and } T|_X : X \to X \text{ transitive} \end{array}\right\}$$

$$\mathrm{P}^{**}_{\exp}(t) := \sup\left\{\mathrm{P}(T|_X, -t\log|T'|) \;\middle|\; \begin{array}{l} X \subseteq \mathcal{J}(T) \text{ expanding forward } T\text{-invariant closed set} \\ \text{and } T|_X : X \to X \text{ open and transitive} \end{array}\right\}$$

As a direct consequence of Theorems 25.3.8–25.3.10, we get the following assertion on the pressure function.

Proposition 25.3.12. *If $T : \widehat{\mathbb{C}} \to \widehat{\mathbb{C}}$ is a rational function with $\deg(T) \geq 2$, then for every $t \in \mathbb{R}$ we have*

$$\mathrm{P}_{\exp}(t) = \mathrm{P}^*_{\exp}(t) = \mathrm{P}^{**}_{\exp}(t).$$

25.4 Lyubich's geometric lemma

Let (X,ρ) be a metric space and let $Z \subseteq X$ be a finite set. Fix $R > \varepsilon > 0$ and $\beta \in (0,1)$. Let

$$\#[Z; R, \varepsilon, \beta]$$

be the minimum of the cardinalities of all sets $F \subseteq X$ such that

$$B(Z,R) \setminus B(Z,\varepsilon) \subseteq \bigcup_{x\in F} B(x, \beta\rho(x,Z)).$$

The next lemma's proof and formulation are adapted from [79].

Lemma 25.4.1. *Let $Z \subset \mathbb{C}$ be a nonempty finite set. If $R > \varepsilon > 0$ and $\beta \in (0,1)$, then*

$$\#[Z;R,\varepsilon,\beta] \le A(\beta)(\#Z)^2 \log(R/\varepsilon),$$

where $A(\beta)$ is a constant which depends only on β (and thus is independent of Z, R and ε).

Proof. Fix $r > 0$. For every $w \in Z$, we have

$$\partial B(w,r) \setminus B(Z,r) = \partial B(w,r) \setminus B(Z\backslash\{w\},r) = \bigcup_{L\in\mathcal{A}_w(r)} L,$$

where $\mathcal{A}_w(r)$ is a finite collection of mutually disjoint closed topological arcs in $\partial B(w,r)$. Observe that $\#\mathcal{A}_w(r) \le \#Z - 1$. Consider an arbitrary arc $L \in \mathcal{A}_w(r)$. Denote its endpoints by a_L and b_L. Let $u_1, u_2, \ldots, u_{s_L}$, where $s_L \in \mathbb{N}$, be a unique sequence of points in L such that $u_1 = a_L$, $|\angle(u_j, w, u_{j+1})| = \beta$ for all $1 \le j < s_L$ and $0 < |\angle(u_{s_L}, w, b_L)| \le \beta$. So $(s_L - 1)\beta + |\angle(u_{s_L}, w, b_L)| = |\angle(a_L, w, b_L)|$. Then

$$s_L < \frac{|\angle(a_L, w, b_L)|}{\beta} + 1.$$

Thus,

$$\sum_{L\in\mathcal{A}_w(r)} s_L \le \frac{2\pi}{\beta} + \#\mathcal{A}_w(r) \le \frac{2\pi}{\beta} + \#Z - 1 \le \frac{2\pi\#Z}{\beta}.$$

Hence,

$$\sum_{w\in Z} \sum_{L\in\mathcal{A}_w(r)} s_L \le \frac{2\pi}{\beta}(\#Z)^2. \tag{25.74}$$

As $\partial B(Z,r) \subseteq \bigcup_{w\in Z}[\partial B(w,r) \setminus B(Z,r)]$, it follows from our construction that

$$\partial B(Z,r) \subseteq \bigcup_{w\in Z} \bigcup_{L\in\mathcal{A}_w(r)} \bigcup_{j=1}^{s_L} B(u_j, \beta r).$$

But since the set on the left is open while the set on the right is compact, there exists $\eta = \eta(\beta) > 0$ such that

$$\bigcup_{w\in Z} \bigcup_{L\in\mathcal{A}_w(r)} \bigcup_{j=1}^{s_L} B(u_j, \beta r) \supseteq B(\partial B(Z,r), \eta r) \supseteq B(Z, (1+\eta)r) \setminus B(Z,r). \tag{25.75}$$

Obviously, $1+\eta$ is independent of R and ε. We can clearly have it independent of Z and, by a similarity argument, independent of r.

Let N be the largest integer such that $(1+\eta)^N\varepsilon < R$. Then

$$N < \frac{\log(R/\varepsilon)}{\log(1+\eta)}. \tag{25.76}$$

Using (25.75), we obtain

$$\bigcup_{i=0}^{N}\bigcup_{w\in Z}\bigcup_{L\in\mathcal{A}_w((1+\eta)^i\varepsilon)}\bigcup_{j=1}^{s_L} B(u_j,\beta(1+\eta)^i\varepsilon) \supseteq B(Z,(1+\eta)^{N+1}\varepsilon)\setminus B(Z,\varepsilon) \supseteq B(Z,R)\setminus B(Z,\varepsilon).$$

But given that for each ball $B(u_j,\beta(1+\eta)^i\varepsilon)$ appearing in this union, we have

$$B(u_j,\beta(1+\eta)^i\varepsilon) \subseteq B(u_j,\beta\rho(u_j,Z)),$$

we get that

$$\sum_{i=0}^{N}\sum_{w\in Z}\sum_{L\in\mathcal{A}_w((1+\eta)^i\varepsilon)} s_L \le \frac{2\pi}{\beta}(\#Z)^2\frac{\log(R/\varepsilon)}{\log(1+\eta)}.$$

Choose $A(\beta) = 2\pi/[\beta\log(1+\eta(\beta))]$. □

25.5 Two auxiliary partitions

For the purpose of this section, denote the Lebesgue measure of any Lebesgue measurable set $A \subseteq \mathbb{R}$ by $\mathrm{Leb}(A)$.

25.5.1 Boundary partition

The next three results have been proved in Volume 2. We restate them for convenience.

Lemma 25.5.1. *Every increasing function $k : I \to \mathbb{R}$ defined on a compact interval $I \subset \mathbb{R}$ is Lipschitz continuous at Lebesgue almost every point in I. More precisely, for every $\varepsilon > 0$ there exist a number $L > 0$ and a Lebesgue measurable set $A \subseteq I$ such that* $\mathrm{Leb}(I \setminus A) < \varepsilon$ *and the function $k : I \to \mathbb{R}$ is Lipschitz continuous at each point of A with Lipschitz constant not exceeding L.*

Proof. See Lemma 16.5.4. □

Corollary 25.5.2. *For every Borel probability measure ν on a compact metric space (X,ρ) and for every $r > 0$ there exists a finite Borel partition $\mathcal{P} = \{P_j\}_{j=1}^M$ of X with the following three properties:*

(a)
$$\nu(P) > 0, \quad \forall P \in \mathcal{P}. \tag{25.77}$$

(b)
$$\operatorname{diam}(\mathcal{P}) < r. \tag{25.78}$$

(c) *There exists a constant $C > 0$ such that for every $\eta > 0$ it holds that*
$$\nu(\partial_\eta \mathcal{P}) \le C\eta, \quad \textit{where} \quad \partial_\eta \mathcal{P} := \bigcap_{j=1}^{M} \bigcup_{k\neq j} B(P_k, \eta). \tag{25.79}$$

Proof. See Corollary 16.5.5. □

25.5.2 Exponentially large partition

Corollary 25.5.3. *Let ν be a Borel probability measure on a compact metric space (X, ρ) and let $T : X \to X$ be a Borel endomorphism of X preserving the measure ν. Then for every $r > 0$ there exists a finite partition $\mathcal{P}$ of X into Borel sets of positive ν-measure with* $\operatorname{diam}(\mathcal{P}) < r$ *such that for every $\delta > 0$ and ν–a. e. $x \in X$ there exists an integer $N(x) \ge 0$ for which*
$$B(T^n(x), e^{-n\delta}) \subseteq \mathcal{P}(T^n(x)), \quad \forall n \ge N(x), \tag{25.80}$$
where $\mathcal{P}(y)$ denotes the atom $P \in \mathcal{P}$ such that $y \in P$.

Proof. See Corollary 16.5.7. □

25.5.3 Mañé's partition

In this subsection, basically following Mañé's book [83], we construct the so-called Mañé partition, which will play an important role in the proof of part of the volume lemma given in the next section. We begin with the following elementary fact.

Lemma 25.5.4. *If $x_n \in (0,1)$ for every $n \in \mathbb{N}$ and $\sum_{n=1}^{\infty} nx_n < \infty$, then*
$$\sum_{n=1}^{\infty} -x_n \log x_n < \infty.$$

Proof. Let $S := \{n \in \mathbb{N} \mid -\log x_n \ge n\}$. Then
$$\sum_{n=1}^{\infty} -x_n \log x_n = \sum_{n\notin S} -x_n \log x_n + \sum_{n\in S} -x_n \log x_n \le \sum_{n=1}^{\infty} nx_n + \sum_{n\in S} -x_n \log x_n.$$

Since $n \in S$ means that $x_n \le e^{-n}$ and since $\log t \le 2\sqrt{t}$ for all $t \ge 1$, we have

$$\sum_{n\in S} x_n \log \frac{1}{x_n} \le 2 \sum_{n=1}^{\infty} x_n \sqrt{\frac{1}{x_n}} \le 2 \sum_{n=1}^{\infty} e^{-n/2} < \infty. \qquad \square$$

The next lemma is the main, and simultaneously the last, result of this subsection.

Lemma 25.5.5 (Mañé's partition). *If μ is a Borel probability measure supported on a bounded subset M of a (finite-dimensional) Euclidean space and $\rho : M \to (0,1]$ is a Borel measurable function such that $\log \rho$ is integrable with respect to μ (i. e., $\int_M \log \rho \, d\mu > -\infty$), then there exists a countable Borel partition $\mathcal{P}$ of M such that*

$$\mathrm{H}_\mu(\mathcal{P}) < \infty \quad \textit{and} \quad \mathrm{diam}(\mathcal{P}(x)) \le \rho(x) \textit{ for } \mu\textit{-a.e. } x \in M,$$

where $\mathrm{H}_\mu(\mathcal{P})$ is the measure-theoretic entropy of $\mathcal{P}$ with respect to μ (cf. Definition 9.4.1).

Proof. Let d be the dimension of the Euclidean space containing M. Since M is bounded, there is a constant $C > 0$ such that for every $0 < r < 1$ there exists a Borel partition $\mathcal{P}_r$ of M with $\mathrm{diam}(\mathcal{P}_r) \le r$ and $\#\mathcal{P}_r \le Cr^{-d}$. For every $n \ge 0$, set

$$U_n := \{x \in M \mid e^{-(n+1)} < \rho(x) \le e^{-n}\}.$$

The family $\mathcal{U} := \{U_n\}_{n=0}^{\infty}$ is clearly a Borel partition of M. Since $\log \rho$ is a nonpositive integrable function, we have

$$\sum_{n=1}^{\infty} -n\mu(U_n) \ge \sum_{n=1}^{\infty} \int_{U_n} \log \rho \, d\mu = \int_M \log \rho \, d\mu > -\infty, \tag{25.81}$$

so that

$$\sum_{n=1}^{\infty} n\mu(U_n) < \infty. \tag{25.82}$$

For every $n \ge 0$, let $r_n := e^{-(n+1)}$ and $\mathcal{Q}_n := \mathcal{P}_{r_n}|_{U_n}$ be the Borel partition of U_n whose atoms are of the form $Q \cap U_n$, where $Q \in \mathcal{P}_{r_n}$. Thereafter, define the Borel partition $\mathcal{P} := \bigcup_{n=0}^{\infty} \mathcal{Q}_n$. Then

$$\mathrm{H}_\mu(\mathcal{P}) := -\sum_{P\in\mathcal{P}} \mu(P) \log \mu(P) = \sum_{n=0}^{\infty} \Big(-\sum_{P\in\mathcal{Q}_n} \mu(P) \log \mu(P) \Big).$$

For every $n \ge 0$, let μ_{U_n} be the conditional measure of μ on U_n. Then

$$-\sum_{P\in\mathcal{Q}_n} \mu(P) \log \mu(P) = \mu(U_n) \sum_{P\in\mathcal{Q}_n} -\frac{\mu(P)}{\mu(U_n)} \log \frac{\mu(P)}{\mu(U_n)} - \mu(U_n) \sum_{P\in\mathcal{Q}_n} \frac{\mu(P)}{\mu(U_n)} \log \mu(U_n)$$

$$
\begin{aligned}
&= \mu(U_n)\mathrm{H}_{\mu_{U_n}}(\mathcal{Q}_n) - \mu(U_n)\log\mu(U_n)\\
&\le \mu(U_n)\log\#\mathcal{Q}_n - \mu(U_n)\log\mu(U_n)\\
&\le \mu(U_n)\log\#\mathcal{P}_{r_n} - \mu(U_n)\log\mu(U_n)\\
&\le \mu(U_n)(\log C - d\log r_n) - \mu(U_n)\log\mu(U_n)\\
&= \mu(U_n)\log C + d(n+1)\mu(U_n) - \mu(U_n)\log\mu(U_n).
\end{aligned}
$$

Summing over all $n \ge 0$, we obtain

$$
\mathrm{H}_\mu(\mathcal{P}) \le \log C + d + d\sum_{n=0}^{\infty} n\mu(U_n) + \sum_{n=0}^{\infty} -\mu(U_n)\log\mu(U_n).
$$

In light of (25.82) and Lemma 25.5.4, we conclude that $\mathrm{H}_\mu(\mathcal{P})$ is finite. Moreover, if $x \in U_n$ then the atom $\mathcal{P}(x)$ is contained in some atom of $\mathcal{P}_{r_n}$. Therefore,

$$
\mathrm{diam}(\mathcal{P}(x)) \le r_n = e^{-(n+1)} < \rho(x).
$$

As $\bigcup_{n=0}^{\infty} U_n = M$, the relation $\mathrm{diam}(\mathcal{P}(x)) < \rho(x)$ holds for all $x \in M$. □

25.6 Miscellaneous facts

25.6.1 Miscellaneous general facts

Lemma 25.6.1. *Let $d \in \mathbb{N}$ and $a \in \mathbb{R}^d$. If μ is a finite Borel measure on $\mathbb{R}^d$ such that the function $\mathbb{R}^d \ni z \longmapsto \log\|z-a\|$ is μ-integrable, then for every $C > 0$ and every $0 < t < 1$ we have*

$$
\sum_{n=1}^{\infty} \mu(B(a, Ct^n)) < \infty.
$$

Proof. As μ is finite and as for any $0 < t < s < 1$ there exists $N \in \mathbb{N}$ such that $Ct^n \le s^n$ for all $n \ge N$, we may assume that $C = 1$ without losing any generality. Given $b \in \mathbb{R}^d$ and two numbers $0 \le r < R$, consider the “half-closed, half-open” annulus

$$
\underline{A}(b; r, R) := \{z \in \mathbb{R}^d : r \le \|z-b\| < R\}.
$$

Since $-\log(t^n) \le -\log\|z-a\|$ for every $z \in B(a, t^n)$, we get

$$
\begin{aligned}
\sum_{n=1}^{\infty} \mu(B(a,t^n)) &= \sum_{n=1}^{\infty} \mu(\underline{A}(a; t^{n+1}, t^n)) \le \frac{-1}{\log t}\sum_{n=1}^{\infty} -\log(t^n)\mu(\underline{A}(a; t^{n+1}, t^n))\\
&\le \frac{-1}{\log t}\int_{B(a,t)} -\log\|z-a\|\,d\mu(z) < \infty.
\end{aligned}
$$

□

The following lemma is obvious and we state it without proof.

Lemma 25.6.2. *Let (X,ρ) be a metric space and let ν be a Borel probability measure on X. Fix $x \in X$ and $\alpha > 0$. Suppose that there exist $C_1, C_2 \geq 1$ and $r_0 > 0$ with the following property: For every $0 < r < r_0$ there are $0 < r_1 \leq C_1 r$ and $r_2 \geq C_2^{-1} r$ for which*

$$\nu(B(x,r_1)) \geq C_1^{-1} r^\alpha \quad \text{and} \quad \nu(B(x,r_2)) \leq C_2 r^\alpha.$$

Then there exists $C_3 \geq 1$ such that

$$C_3^{-1} \leq \frac{\nu(B(x,r))}{r^\alpha} \leq C_3, \quad \forall 0 < r \leq 1,$$

i. e., ν is a geometric measure with exponent α (cf. Definition 15.6.12).

25.6.2 Miscellaneous facts about rational functions

The following fact is well known for all topologically exact, open, distance expanding systems, including $T : \mathcal{J}(T) \to \mathcal{J}(T)$ (cf. Proposition 13.8.6). We repeat its short proof in this complex-analytic setting.

Proposition 25.6.3. *If $T : \widehat{\mathbb{C}} \to \widehat{\mathbb{C}}$ is a rational function with $\deg(T) \geq 2$, then*

$$\lim_{n\to\infty} \sup\{\mathrm{dist}(z, T^{-n}(\zeta)) : z, \zeta \in \mathcal{J}(T)\} = 0.$$

Proof. Fix $\varepsilon > 0$. As $\mathcal{J}(T)$ is compact, there is a finite set $J \subseteq \mathcal{J}(T)$ such that

$$\bigcup_{w\in J} B(w, \varepsilon/2) \supseteq \mathcal{J}(T).$$

Since the map $T : \mathcal{J}(T) \to \mathcal{J}(T)$ is topologically exact (per Corollary 24.3.5), for every $w \in J$ there exists $N_w \geq 0$ such that $T^{N_w}(B(w,\varepsilon/2)) \supseteq \mathcal{J}(T)$. Setting $N_\varepsilon = \max\{N_w : w \in J\}$ yields

$$T^n(B(w,\varepsilon/2)) \supseteq \mathcal{J}(T), \quad \forall w \in J,\ \forall n \geq N_\varepsilon.$$

Take $z, \zeta \in \mathcal{J}(T)$. Then $z \in B(w,\varepsilon/2)$ for some $w \in J$. For every $n \geq N_\varepsilon$, there exists $\zeta_n \in B(w,\varepsilon/2)$ such that $T^n(\zeta_n) = \zeta$. Thus,

$$\mathrm{dist}(z, T^{-n}(\zeta)) \leq |z - \zeta_n| \leq |z - w| + |w - \zeta_n| < \varepsilon/2 + \varepsilon/2 = \varepsilon. \qquad \square$$

For every rational function, we now establish the existence of finite Borel partitions of the Riemann sphere of arbitrarily small diameters on the atoms of which the function is injective.

Lemma 25.6.4. *Let $T : \widehat{\mathbb{C}} \to \widehat{\mathbb{C}}$ be a rational function with* $\deg(T) \geq 2$. *For every $\delta > 0$, there exists a finite Borel partition $\alpha(\delta)$ of $\widehat{\mathbb{C}}$ such that* $\operatorname{diam}(\alpha(\delta)) < \delta$ *and the restriction $T|_A : A \to \widehat{\mathbb{C}}$ is injective on every $A \in \alpha(\delta)$.*

Proof. Fix $\delta > 0$. Let $c \in \operatorname{Crit}(T)$. It follows from Theorem 25.1.1 that there exist $\eta_c \in (0, \delta)$ and $\deg_c(T)$ mutually disjoint Borel sets $\{B_k(c)\}_{k=1}^{\deg_c(T)}$ such that

$$\bigcup_{k=1}^{\deg_c(T)} B_k(c) = B(c, \eta_c) \quad \text{and} \quad T|_{B_k(c)} \text{ is injective for every } 1 \leq k \leq \deg_c(T).$$

On the other hand, if $z \in \widehat{\mathbb{C}} \backslash \operatorname{Crit}(T)$ then there exists $\eta_z \in (0, \delta)$ such that the map $T|_{B(z,\eta_z)}$ is injective.

Since $\widehat{\mathbb{C}}$ is compact, there is a finite subcover of the open cover $\{B(w, \eta_w)\}_{w \in \widehat{\mathbb{C}}}$. Therefore, there is a finite subcover $\mathcal{B} = \{B_1, \ldots, B_n\}$ of the Borel cover

$$\{B(z, \eta_z) : z \in \widehat{\mathbb{C}} \backslash \operatorname{Crit}(T)\} \cup \{B_k(c) : c \in \operatorname{Crit}(T), 1 \leq k \leq \deg_c(T)\}.$$

We form the sought-after Borel partition $\alpha(\delta)$ of $\widehat{\mathbb{C}}$ from the cover $\mathcal{B}$ in the following standard way:

$$A_1 := B_1 \quad \text{and} \quad A_j := B_j \setminus \bigcup_{i<j} B_i, \quad \forall 2 \leq j \leq n. \qquad \square$$

26 Expanding (or hyperbolic), subexpanding, and parabolic rational functions: topological outlook

In this chapter, we study three special classes of rational functions: expanding, subexpanding, and parabolic rational functions. We also define in a natural way the class of expansive rational functions and show that it consists exactly in the union of the expanding and parabolic classes of functions.

Expanding rational functions are also frequently called hyperbolic. They form the most "regular" class of rational functions and their properties are best known and most transparent. We start with them.

26.1 Expanding rational functions

Recall from Definition 25.3.3 that a rational function $T : \widehat{\mathbb{C}} \to \widehat{\mathbb{C}}$ is called *expanding* (or *hyperbolic*) if the Julia set $\mathcal{J}(T)$ is an expanding set for T, i. e., if there exists $q \in \mathbb{N}$ such that $|(T^q)'(z)| > 1$ for all $z \in \mathcal{J}(T)$.

As a direct consequence of Theorem 25.3.2, we get the following.

Theorem 26.1.1. *If $T : \widehat{\mathbb{C}} \to \widehat{\mathbb{C}}$ is a rational function, then the following statements are equivalent:*
(a) *The map $T : \widehat{\mathbb{C}} \to \widehat{\mathbb{C}}$ is expanding.*
(b) *There exist $\lambda > 1$ and $n \in \mathbb{N}$ such that $|(T^n)'(z)| \geq \lambda$ for all $z \in \mathcal{J}(T)$.*
(c) *For each $\lambda > 1$, there exists $n \in \mathbb{N}$ such that $|(T^n)'(z)| \geq \lambda$ for all $z \in \mathcal{J}(T)$.*
(d) *There exist $\lambda > 1$ and $c > 0$ such that $|(T^n)'(z)| \geq c\lambda^n$ for all $z \in \mathcal{J}(T)$ and all $n \in \mathbb{N}$.*
(e) *For each $\lambda > 1$, there exists $c > 0$ such that $|(T^n)'(z)| \geq c\lambda^n$ for all $z \in \mathcal{J}(T)$ and all $n \in \mathbb{N}$.*

As a straightforward repercussion, we have the following.

Proposition 26.1.2. *If $T : \widehat{\mathbb{C}} \to \widehat{\mathbb{C}}$ is a rational function, then the following statements are equivalent:*
(a) *The map $T : \widehat{\mathbb{C}} \to \widehat{\mathbb{C}}$ is expanding.*
(b) *There exists $n \in \mathbb{N}$ such that the iterate $T^n : \widehat{\mathbb{C}} \to \widehat{\mathbb{C}}$ is expanding.*
(c) *For every $n \in \mathbb{N}$, the iterate $T^n : \widehat{\mathbb{C}} \to \widehat{\mathbb{C}}$ is expanding.*

Apart from Theorem 26.1.1 itself, this proposition adds the decisive argument to coin the concerned functions as being expanding.

Recall that the finite set

$$\mathrm{Crit}(T) := \{z \in \widehat{\mathbb{C}} \mid T'(z) = 0\}$$

is the critical set of T while the countable set

https://doi.org/10.1515/9783110769876-026

$$\mathrm{PC}(T) := \bigcup_{n=1}^{\infty} T^n(\mathrm{Crit}(T))$$

is the postcritical set of T. Note that

$$\overline{\mathrm{PC}(T)} = \mathrm{PC}(T) \cup \omega(\mathrm{Crit}(T)).$$

We now characterize expandingness as the disjointness of the Julia set from the closure of the postcritical set. This provides further insight into the role played by the critical points and their orbits.

Theorem 26.1.3. *A rational function* $T : \widehat{\mathbb{C}} \to \widehat{\mathbb{C}}$ *with* $\deg(T) \geq 2$ *is expanding if and only if*

$$\mathcal{J}(T) \cap \overline{\mathrm{PC}(T)} = \emptyset.$$

Proof. Assume first that the map $T : \widehat{\mathbb{C}} \to \widehat{\mathbb{C}}$ is expanding. The very definition of this property implies that $\mathcal{J}(T) \cap \mathrm{Crit}(T) = \emptyset$. Since $\mathcal{J}(T)$ is completely T-invariant, it follows that $\mathcal{J}(T) \cap \mathrm{PC}(T) = \emptyset$. Suppose for a contradiction that $\mathcal{J}(T) \cap \overline{\mathrm{PC}(T)} \neq \emptyset$. As $\overline{\mathrm{PC}(T)} = \mathrm{PC}(T) \cup \omega(\mathrm{Crit}(T))$, this means that $\mathcal{J}(T) \cap \omega(\mathrm{Crit}(T)) \neq \emptyset$. The finiteness of $\mathrm{Crit}(T)$ implies that there exists a point $c \in \mathrm{Crit}(T) \subseteq \mathcal{F}(T)$ such that

$$\mathcal{J}(T) \cap \omega(c) \neq \emptyset. \tag{26.1}$$

By virtue of Sullivan's nonwandering theorem (Theorem 24.4.12), there then exist an integer $n \geq 0$ and a periodic connected component V of $\mathcal{F}(T)$ such that $T^n(c) \in V$. The classification theorem 24.4.10 then affirms that the only possibility for (26.1) to hold is that V be the immediate basin of attraction to some simple rationally indifferent periodic point $\zeta \in \partial V \subseteq \mathcal{J}(T)$. But then statement (e) in Theorem 26.1.1 fails for $z = \zeta$. This contradiction shows that $\mathcal{J}(T) \cap \overline{\mathrm{PC}(T)} = \emptyset$. This proves the direct implication.

For the converse, assume that $\mathcal{J}(T) \cap \overline{\mathrm{PC}(T)} = \emptyset$. The compactness of the sets $\mathcal{J}(T)$ and $\overline{\mathrm{PC}(T)}$ implies that there is $\delta > 0$ for which

$$B(\mathcal{J}(T), 2\delta) \cap \overline{\mathrm{PC}(T)} = \emptyset. \tag{26.2}$$

By way of contradiction, suppose that the map $T : \widehat{\mathbb{C}} \to \widehat{\mathbb{C}}$ is not expanding. This means that there exist a strictly increasing sequence $(n_k)_{k=1}^{\infty}$ of positive integers and a sequence $(\xi_k)_{k=1}^{\infty}$ of points in $\mathcal{J}(T)$ such that

$$|(T^{n_k})'(\xi_k)| < 2, \quad \forall k \in \mathbb{N}. \tag{26.3}$$

Passing to a subsequence if necessary, we may assume that both of the sequences $(\xi_k)_{k=1}^{\infty}$ and $(T^{n_k}(\xi_k))_{k=1}^{\infty}$ in $\mathcal{J}(T)$ respectively converge to, say, $\xi \in \mathcal{J}(T)$ and $w \in \mathcal{J}(T)$. Disregarding finitely many terms, we may further assume that

$$|\xi_k - \xi| < \delta/16, \quad \forall k \in \mathbb{N} \tag{26.4}$$

and that

$$\left|T^{n_k}(\xi_k) - w\right| < \delta, \quad \forall k \in \mathbb{N}. \tag{26.5}$$

By (26.2), we also know that

$$B(w, 2\delta) \cap \overline{\mathrm{PC}(T)} = \emptyset. \tag{26.6}$$

Per Remark 24.4.16 and (26.5)–(26.6), there exists for every $k \in \mathbb{N}$ a unique analytic branch $T_{\xi_k}^{-n_k} : B(w, 2\delta) \to \widehat{\mathbb{C}}$ of T^{-n_k} that sends $T^{n_k}(\xi_k)$ to ξ_k. By Koebe's $\frac{1}{4}$–theorem (Theorem 23.1.3) and (26.3)–(26.5), we get

$$\begin{aligned}
T_{\xi_k}^{-n_k}(B(w, 2\delta)) &\supseteq T_{\xi_k}^{-n_k}(B(T^{n_k}(\xi_k), \delta)) \\
&\supseteq B\left(T_{\xi_k}^{-n_k}(T^{n_k}(\xi_k)), \frac{1}{4}|(T_{\xi_k}^{-n_k})'(T^{n_k}(\xi_k))|\delta\right) \\
&= B\left(\xi_k, \frac{1}{4}|(T^{n_k})'(\xi_k)|^{-1}\delta\right) \\
&\supseteq B(\xi_k, \delta/8) \\
&\supseteq B(\xi, \delta/16), \quad \forall k \in \mathbb{N}.
\end{aligned}$$

Consequently, $T^{n_k}(B(\xi, \delta/16)) \subseteq B(w, 2\delta)$ for all $k \in \mathbb{N}$. Therefore, the sequence $(T^{n_k}|_{B(\xi,\delta/16)})_{k=1}^{\infty}$ is normal. By virtue of Proposition 24.3.8, this tells us that $\xi \in \mathcal{F}(T)$. On the other hand, $\xi = \lim_{k\to\infty} \xi_k \in \mathcal{J}(T)$. This contradiction confirms the converse implication. □

This theorem, when taken in conjunction with the classification theorem 24.4.10 and Sullivan's nonwandering theorem (Theorem 24.4.12), leads to the following result.

Theorem 26.1.4. *A rational function $T : \widehat{\mathbb{C}} \to \widehat{\mathbb{C}}$ with $\deg(T) \geq 2$ is expanding if and only if its critical set $\mathrm{Crit}(T)$ is contained in a (finite) union of connected components of basins of attraction to attracting periodic points of T.*

As belonging to a connected component of the basin of attraction to an attracting periodic point is an open condition and as, by Corollary 24.1.21, there are at most $2d-2$ (a finite number depending only on d) attracting periodic cycles for any rational function of degree d, an immediate ramification of Theorem 26.1.4 is the following.

Corollary 26.1.5. *Expanding rational functions of a given degree $d \geq 2$ form an open subset of the space of all rational functions of degree d.*

Theorem 26.1.6. *An expanding rational function $T : \widehat{\mathbb{C}} \to \widehat{\mathbb{C}}$ with $\deg(T) \geq 2$ admits no Cremer periodic point, rationally indifferent periodic point, Siegel disk or Herman ring. In* *particular, the Fatou set $\mathcal{F}(T)$ consists only of basins of attraction to attracting periodic points of T.*

Proof. The presence of rationally indifferent periodic points is ruled out by Definition 25.3.3. The rest of the first assertion is an immediate consequence of Theorems 26.1.3, 24.5.2 and 24.5.1. The last assertion immediately follows from the first one and Theorem 24.4.10. □

With the terminology and notation of Section 24.6 and as an immediate consequence of the previous theorem and Theorem 24.6.1, we get the following (cf. [92]).

Theorem 26.1.7. *Every expanding rational function $T : \widehat{\mathbb{C}} \to \widehat{\mathbb{C}}$ with $\deg(T) \geq 2$ is a continuity point of the function $\mathcal{R} \ni S \longmapsto \mathcal{J}(S) \subseteq \mathcal{K}(\widehat{\mathbb{C}})$.*

We can actually say more: expanding rational functions are structurally stable on their Julia sets. To see this, we start with the following repercussion of Theorem 26.1.4 and of the very definition of expandingness.

Observation 26.1.8. If $T : \widehat{\mathbb{C}} \to \widehat{\mathbb{C}}$ is an expanding rational function with $\deg(T) \geq 2$, then for every open set $U \subseteq \widehat{\mathbb{C}}$ containing the Julia set $\mathcal{J}(T)$ and whose closure $\overline{U}$ contains no attracting periodic point of T, the triple $(\mathcal{J}(T), U, T)$ is a topologically exact conformal expanding repeller in the sense of Definition 16.2.1. Consequently, it is a distance expanding map. (For more information, see Corollary 24.3.5 and Chapters 16 and 4.)

We can now prove the following remarkable result.

Theorem 26.1.9. *Let $T : \widehat{\mathbb{C}} \to \widehat{\mathbb{C}}$ be an expanding rational function with $\deg(T) \geq 2$ and let Λ be an open subset of $\mathbb{C}$. If $(T_\lambda : \widehat{\mathbb{C}} \to \widehat{\mathbb{C}})_{\lambda\in\Lambda}$ is a collection of rational functions of the same degree such that $T_{\lambda_0} = T$ for some $\lambda_0 \in \Lambda$ and if for every $z \in \widehat{\mathbb{C}}$ the map*

$$\Lambda \ni \lambda \longmapsto T_\lambda(z) \in \widehat{\mathbb{C}}$$

is analytic (in fact, it suffices to know this for every z in some neighborhood of $\mathcal{J}(T)$), then there exist an open neighborhood $\widehat{\Lambda} \subseteq \Lambda$ of λ_0 and an open neighborhood $\widehat{U}$ of $\mathcal{J}(T)$ such that for every $\lambda \in \widehat{\Lambda}$ the following hold:

(a) *The rational function $T_\lambda : \widehat{\mathbb{C}} \to \widehat{\mathbb{C}}$ is expanding and $(\mathcal{J}(T_\lambda), T_\lambda, \widehat{U})$ is a conformal expanding repeller.*
(b) *The maps $T_\lambda : \mathcal{J}(T_\lambda) \to \mathcal{J}(T_\lambda)$ and $T : \mathcal{J}(T) \to \mathcal{J}(T)$ are topologically conjugate. More precisely, there exists a homeomorphism $\tau_\lambda : \mathcal{J}(T) \to \mathcal{J}(T_\lambda)$ with the following properties:*
    (i) *For every $z \in \mathcal{J}(T)$, the map*

$$\widehat{\Lambda} \ni \lambda \longmapsto \tau_\lambda(z) \in \widehat{\mathbb{C}}$$

is holomorphic.

(ii) *The map* τ_{λ_0} *is the identity map and the map*

$$\mathcal{J}(T) \ni z \longmapsto \tau_\lambda(z) \in \mathcal{J}(T_\lambda)$$

is quasiconformal and Hölder continuous with exponent converging to 1 *uniformly as* $\lambda \to \lambda_0$.

(iii) *The homeomorphism* τ_λ *conjugates the maps* T *and* T_λ, *i. e.*,

$$T_\lambda \circ \tau_\lambda = \tau_\lambda \circ T|_{\mathcal{J}(T)}.$$

(iv) *The map* $\widehat{\Lambda} \times \mathcal{J}(T) \ni (\lambda, z) \longmapsto \tau_\lambda(z) \in \widehat{U}$ *is continuous.*

Proof. Having Observation 26.1.8 and noticing that the family $(T_\lambda : \widehat{\mathbb{C}} \to \widehat{\mathbb{C}})_{\lambda\in\Lambda}$ induces an analytic perturbation for every set U appearing in this observation (cf. Definition 16.4.1), our theorem almost entirely follows from Theorem 16.4.7. All we need to show is that

$$\tau_\lambda(\mathcal{J}(T)) = \mathcal{J}(T_\lambda), \quad \forall \lambda \in \widehat{\Lambda}. \tag{26.7}$$

In view of Theorem 16.4.7, each set $\tau_\lambda(\mathcal{J}(T))$ is a conformal expanding repeller for the map $T_\lambda : \widehat{\mathbb{C}} \to \widehat{\mathbb{C}}$. Therefore, $\tau_\lambda(\mathcal{J}(T))$ is the closure of a set of repelling periodic points of T_λ (cf. Definition 16.2.1 and Corollary 4.3.10). Thus,

$$\tau_\lambda(\mathcal{J}(T)) \subseteq \mathcal{J}(T_\lambda).$$

Since the maps $T|_{\mathcal{J}(T)}$ and $T_\lambda|_{\tau_\lambda(\mathcal{J}(T))}$ are conjugate (by virtue of Theorem 16.4.7), we have that

$$\deg(T_\lambda|_{\tau_\lambda(\mathcal{J}(T))}) = \deg(T|_{\mathcal{J}(T)}) = \deg(T).$$

But by hypothesis, $\deg(T) = \deg(T_\lambda)$, whence we get

$$\deg(T_\lambda|_{\tau_\lambda(\mathcal{J}(T))}) = \deg(T_\lambda).$$

This implies that

$$T_\lambda^{-1}(\tau_\lambda(\mathcal{J}(T))) \subseteq \tau_\lambda(\mathcal{J}(T)).$$

Thus, (26.7) directly follows from Corollary 24.3.7 and the proof is over.

Note: In fact, by item (iii) we also know that

$$T_\lambda(\tau_\lambda(\mathcal{J}(T))) = \tau_\lambda(\mathcal{J}(T)) \subseteq T_\lambda^{-1}(\tau_\lambda(\mathcal{J}(T))).$$

So, $T_\lambda(\tau_\lambda(\mathcal{J}(T))) = \tau_\lambda(\mathcal{J}(T)) = T_\lambda^{-1}(\tau_\lambda(\mathcal{J}(T)))$. ☐

26.1.1 The Mandelbrot set

For every $c \in \mathbb{C}$, let $Q_c : \widehat{\mathbb{C}} \to \widehat{\mathbb{C}}$ be the quadratic polynomial

$$Q_c(z) := z^2 + c.$$

The Mandelbrot set (see Figure 26.1) is possibly the most famous fractal. It is commonly denoted by $\mathcal{M}$ and is defined as the set of all $c \in \mathbb{C}$ such the set

$$\{Q_c^n(0) : n \geq 0\}$$

is bounded or, equivalently,

$$\liminf_{n\to\infty} |Q_c^n(0)| < \infty.$$

The Mandelbrot set was defined and studied numerically by Benoit Mandelbrot [81]. Then came Douady and Hubbard's pioneering work [43, 44] on the theoretical understanding of the topological and conformal structures of this set. This was followed by a huge quantity of research aiming to prove various properties of the Mandelbrot set, including typicality of expandingness and local connectedness of the boundary of $\mathcal{M}$. We only mention here the breakthrough concept of Yoccoz puzzles and Shishikura's work [120] on the Hausdorff dimension of the boundary of the Mandelbrot set, which both had enormous influence on the direction of development of the research on that set.

Denote by $\mathcal{M}_E$ the set of all parameters $c \in \mathbb{C}$ for which the map Q_c is expanding. Obviously, $0 \in \mathcal{M}_E$. In view of Corollary 26.1.5, the set $\mathcal{M}_E$ is open, and hence $c \in \mathcal{M}_E$ for every $c \in \mathbb{C}$ sufficiently small in modulus. The main cardioid $\mathcal{M}_0$ of the Mandelbrot set $\mathcal{M}$ is the connected component of $\mathcal{M}_E$ containing 0. So, the main cardioid $\mathcal{M}_0$ of the Mandelbrot set $\mathcal{M}$ contains a neighborhood of 0.

As a nearly immediate consequence of Theorem 26.1.9, we can prove the following.

Theorem 26.1.10. *If c belongs to the main cardioid $\mathcal{M}_0$ of the Mandelbrot set $\mathcal{M}$, then the Julia set J_c of the quadratic polynomial Q_c is a Jordan (i. e., a simple closed) curve. Furthermore, J_c is a quasicircle, i. e., it is a quasiconformal image of the unit circle.*

Proof. By Theorem 26.1.9, for every $s \in \mathcal{M}_0$ there exists $r_s > 0$ such that for every $c \in B(s, r_s)$ the maps $Q_c|_{J_c}$ and $Q_s|_{J_s}$ are topologically conjugate via a quasiconformal homeomorphism. Since $\mathcal{M}_0$ is a connected set, it follows that all maps $Q_c|_{J_c}$ are topologically conjugate to the map $Q_0|_{J_0}$ via a quasiconformal homeomorphism. As J_0 is the unit circle $\mathbb{S}^1$, the proof of Theorem 26.1.10 is complete. □

Remark 26.1.11. An analogous theorem holds for the one-parameter family of polynomials $z \mapsto z^d + c$ for any integer $d \geq 2$.

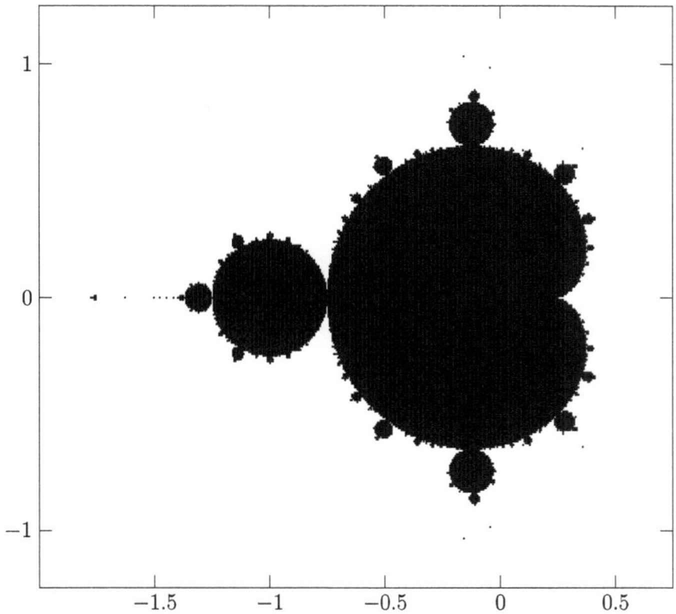

Figure 26.1: The Mandelbrot set.

26.2 Expansive and parabolic rational functions

Definition 26.2.1. A rational function $T : \widehat{\mathbb{C}} \to \widehat{\mathbb{C}}$ is called *expansive* if the map $T|_{\mathcal{J}(T)} : \mathcal{J}(T) \to \mathcal{J}(T)$ is expansive.

Clearly, all expanding rational functions are expansive but this latter class is bigger and their difference will be the primary consideration in this section and, most notably, in Chapter 31. The current section can be partly viewed as a preparation for that chapter.

We now characterize expansive rational functions in terms that are purely topological and less dynamical than general expansiveness. The statement is as simple as those of Theorems 26.1.3–26.1.4. The proof, however, is somewhat more complex. It was first provided in [37].

We begin with an intermediate result. Recall that $\Omega(T)$ denotes the set of all rationally indifferent periodic points of T, which is finite according to Theorem 24.3.2.

Proposition 26.2.2. *If $T : \widehat{\mathbb{C}} \to \widehat{\mathbb{C}}$ is a rational function with $\deg(T) \geq 2$ and such that $\mathcal{J}(T) \cap \mathrm{Crit}(T) = \emptyset$, then for every $\eta > 0$ there exists $\theta_\eta \in (0, \eta)$ for which*

$$B(\mathcal{J}(T) \backslash B(\Omega(T), \eta), 2\theta_\eta) \cap \overline{\mathrm{PC}(T)} = \emptyset.$$

In particular, given any $z \in \mathcal{J}(T) \backslash B(\Omega(T), \eta)$, the analytic inverse branches of T^n are well-defined on $B(z, 2\theta_\eta)$ for every $n \in \mathbb{N}$.

Proof. Since $\mathcal{J}(T) \cap \mathrm{Crit}(T) = \emptyset$, we know that $\mathcal{J}(T) \cap \mathrm{PC}(T) = \emptyset$. It also follows from the classification theorem 24.4.10 and Sullivan's nonwandering theorem (Theorem 24.4.12) that $\omega(\mathrm{Crit}(T)) \cap \mathcal{J}(T) \subseteq \Omega(T)$. These two facts taken together imply that $[\mathcal{J}(T)\backslash B(\Omega(T),\eta)] \cap \overline{\mathrm{PC}(T)} = \emptyset$. Since the sets $\mathcal{J}(T)\backslash B(\Omega(T),\eta)$ and $\overline{\mathrm{PC}(T)}$ are compact, the existence of the required $\theta_\eta > 0$ ensues. □

We can now derive the sought characterization of expansiveness.

Theorem 26.2.3. *If $T : \widehat{\mathbb{C}} \to \widehat{\mathbb{C}}$ is a rational function with $\deg(T) \geq 2$, then the map $T|_{\mathcal{J}(T)} : \mathcal{J}(T) \to \mathcal{J}(T)$ is expansive if and only if $\mathcal{J}(T) \cap \mathrm{Crit}(T) = \emptyset$.*

Proof. Suppose that $\mathcal{J}(T) \cap \mathrm{Crit}(T) \neq \emptyset$ and let $c \in \mathcal{J}(T) \cap \mathrm{Crit}(T)$. By Theorem 24.1.15, the Julia set $\mathcal{J}(T)$ is perfect. Therefore, there exists a sequence $(\xi_n)_{n=1}^{\infty}$ of points in $\mathcal{J}(T)\backslash\{c\}$ converging to c. Since $c \in \mathrm{Crit}(T)$, for all $n \in \mathbb{N}$ large enough (removing finitely many ξ_n's, we get that for all $n \in \mathbb{N}$) there are points $\xi_n' \in \widehat{\mathbb{C}}$ such that $\xi_n' \neq \xi_n$, $T(\xi_n') = T(\xi_n)$, and $\lim_{n\to\infty} \xi_n' = c$. But then $\xi_n' \in \mathcal{J}(T)$, and we hence constructed a sequence of pairs of arbitrarily close but distinct points ξ_n and ξ_n' in $\mathcal{J}(T)$ that are mapped to the same points. Thus, $T|_{\mathcal{J}(T)}$ fails to be expansive. The direct implication has been established.

To prove the converse, note first that $\mathcal{J}(T) \cap \mathrm{Crit}(T) = \emptyset$ entails the existence of a constant $\chi > 0$ such that for every $z \in \mathcal{J}(T)$ there is a unique holomorphic branch $T_z^{-1} : B(T(z),\chi) \to \mathbb{C}$ of T^{-1} mapping $T(z)$ to z. A standard argument (cf. Lemma 4.2.2) shows that

$$\delta_1 := \sup\{r > 0 : B(z,r) \subseteq T_z^{-1}(B(T(z),\chi)),\ \forall z \in \mathcal{J}(T)\} > 0.$$

Let δ_2 be the θ from (24.83) and Proposition 24.2.28. Subsequently, let $\theta_{\delta_2/2}$ arise from Proposition 26.2.2. Finally, let $\delta = \frac{1}{2}\min\{\chi, \delta_1, \delta_2, \theta_{\delta_2/2}\}$. We shall prove that δ is an expansive constant for $T : \mathcal{J}(T) \to \mathcal{J}(T)$. Assume that $x, y \in \mathcal{J}(T)$ and $|T^n(x) - T^n(y)| < \delta$ for all $n \geq 0$. As $\delta < \delta_2/2$, it follows from Proposition 24.2.28 and the finiteness of $\Omega(T)$ that the set

$$Z := \{n \geq 0 \,|\, \{T^n(x), T^n(y)\} \not\subseteq B(\Omega(T), \delta_2/2)\}$$

is infinite. This means that at least one of the sets $Z_x := \{n \geq 0 \mid T^n(x) \notin B(\Omega(T), \delta_2/2)\}$ or $Z_y := \{n \geq 0 \mid T^n(y) \notin B(\Omega(T), \delta_2/2)\}$ is infinite. Assume without loss of generality that Z_x is infinite. By virtue of Proposition 26.2.2, for every $n \in Z_x$ there exists a unique analytic branch $T_x^{-n} : B(T^n(x), \theta_{\delta_2/2}) \to \widehat{\mathbb{C}}$ of T^{-n} sending $T^n(x)$ to x.

Claim. *If $n \in Z_x$, then $T_x^{-n}(T^n(y)) = y$.*

Proof of Claim. Note that $T^n(y) \in B(T^n(x), \delta) \subseteq B(T^n(x), \frac{1}{2}\theta_{\delta_2/2})$, and thus $T_x^{-n}(T^n(y))$ is well-defined. Suppose for a contradiction that $n \in Z_x$ and $T_x^{-n}(T^n(y)) \neq y$. Let $k \in \{0, 1, \ldots, n\}$ be the largest integer such that $T^k(y) \neq T^k(T_x^{-n}(T^n(y)))$. Then $k < n$ and

$$T^{k+1}(y) = T^{k+1}(T_x^{-n}(T^n(y))). \tag{26.8}$$

In addition, from the definitions of δ and δ_1, we know that

$$T^k(y) \in B(T^k(x),\delta) \subseteq B(T^k(x),\delta_1) \subseteq T^{-1}_{T^k(x)}(B(T^{k+1}(x),\chi)). \tag{26.9}$$

Finally, from (26.8) notice that $T^k(T_x^{-n}(T^n(y))) \in T^{-1}(T^{k+1}(y))$. As $|T^j(x) - T^j(y)| < \delta$ for all $j \geq 0$ (and in particular for $j = k+1$), there exists $z \in T^{-1}(T^{k+1}(x))$ for which $T^k(T_x^{-n}(T^n(y))) = T_z^{-1}(T^{k+1}(y))$. But then $T^k(x) = T_z^{-1}(T^{k+1}(x)) = z$. Therefore,

$$\begin{aligned} T^k(T_x^{-n}(T^n(y))) = T^{-1}_{T^k(x)}(T^{k+1}(y)) &\subseteq T^{-1}_{T^k(x)}(B(T^{k+1}(x),\delta)) \\ &\subseteq T^{-1}_{T^k(x)}(B(T^{k+1}(x),\chi)). \end{aligned} \tag{26.10}$$

In sum, (26.9), (26.10) and (26.8) assert that $T^k(y)$ and $T^k(T_x^{-n}(T^n(y)))$ both lie in $T^{-1}_{T^k(x)}(B(T^{k+1}(x),\chi))$ and have the same image under T. The injectivity of T on $T^{-1}_{T^k(x)}(B(T^{k+1}(x),\chi))$ imposes that $T^k(y) = T^k(T_x^{-n}(T^n(y)))$. This contradicts the definition of k and concludes the proof of the claim. ■

Write Z_x as a strictly increasing sequence $(n_k)_{k=1}^{\infty}$. By the above claim, $x,y \in T_x^{-n_k}(B(T^{n_k}(x),\frac{1}{2}\theta_{\delta_2/2}))$ for all $k \in \mathbb{N}$. However, since $x \in \mathcal{J}(T)$, it follows from Lemma 24.1.13 that

$$\lim_{k\to\infty} \operatorname{diam}\left(T_x^{-n_k}\left(B\left(T^{n_k}(x),\frac{1}{2}\theta_{\delta_2/2}\right)\right)\right) = 0.$$

So, $\lfloor y - x\rfloor = 0$. This means that $x = y$. □

We now describe interesting ramifications of expansiveness.

Theorem 26.2.4. *An expansive rational function $T : \widehat{\mathbb{C}} \to \widehat{\mathbb{C}}$ admits no Cremer periodic point, Siegel disk or Herman ring. In particular, the Fatou set $\mathcal{F}(T)$ consists only of basins of attraction to attracting periodic points of T and basins of attraction to rationally indifferent periodic points.*

Proof. The first assertion is an immediate consequence of Theorems 24.5.1, 24.5.2 and 26.2.3. The last assertion immediately follows from the first one and Theorem 24.4.10. □

Let

$$\Omega_\infty^-(T) := \bigcup_{n=0}^{\infty} T^{-n}(\Omega(T))$$

be the completely T-invariant backward orbit of $\Omega(T)$.

Lemma 26.2.5. *Let $T : \widehat{\mathbb{C}} \to \widehat{\mathbb{C}}$ be an expansive rational function with* $\deg(T) \geq 2$. *If $z \in \mathcal{J}(T)$, then*

$$\limsup_{n\to\infty}|(T^n)'(z)| > 0.$$

More precisely:

(a) *If* $z \in \Omega_\infty^-(T)$, *then*

$$0 < \liminf_{n\to\infty}|(T^n)'(z)| \le \limsup_{n\to\infty}|(T^n)'(z)| < \infty.$$

(b) *If* $z \notin \Omega_\infty^-(T)$, *then*

$$\limsup_{n\to\infty}|(T^n)'(z)| = \infty.$$

In fact, if $\delta_T > 0$ *is an expansive constant for* $T : \mathcal{J}(T) \to \mathcal{J}(T)$ *so small that* $[T(B(\omega, \delta_T))] \cap [B(\Omega(T)\backslash\{T(\omega)\}, \delta_T)] = \emptyset$ *for all* $\omega \in \Omega(T)$, *then*

$$\lim_{k\to\infty}|(T^{n_k(z)})'(z)| = \infty,$$

where $(n_k(z))_{k=1}^\infty$ *is the sequence of all* $n \in \mathbb{N}$ *such that* $T^n(z) \notin B(\Omega(T), \delta_T)$.

Proof. Assertion (a) is obvious since $\mathcal{J}(T)\cap\mathrm{Crit}(T) = \emptyset$ by Theorem 26.2.3, and $T^q(\omega) = \omega$ and $(T^q)'(\omega) = 1$ for some $q \in \mathbb{N}$ and all $\omega \in \Omega(T)$. For (b), let δ_T and $(n_k(z))_{k=1}^\infty$ be as in the statement (observe that $(n_k(z))_{k=1}^\infty$ does exist). By Proposition 26.2.2, for every $k \in \mathbb{N}$ there exists a unique analytic branch $T_z^{-n_k(z)} : B(T^{n_k(z)}(z), 2\theta_{\delta_T}) \to \widehat{\mathbb{C}}$ of $T^{-n_k(z)}$ sending $T^{n_k(z)}(z)$ to z. Lemma 24.1.13 then entails that $\lim_{k\to\infty}|(T^{n_k(z)})'(z)| = \infty$. □

Definition 26.2.6. A rational function $T : \widehat{\mathbb{C}} \to \widehat{\mathbb{C}}$ is called *parabolic* if it is expansive but not expanding.

Together, Theorems 26.2.3, 26.1.3, 24.4.12 and 24.4.10 provide a characterization of parabolic rational functions.

Theorem 26.2.7. *A rational function* $T : \widehat{\mathbb{C}} \to \widehat{\mathbb{C}}$ *with* $\deg(T) \ge 2$ *is parabolic if and only if*

$$\mathcal{J}(T) \cap \mathrm{Crit}(T) = \emptyset \quad \textit{and} \quad \Omega(T) \neq \emptyset.$$

It is worth noticing that an iterate T^k of a rational function T is parabolic if and only if the map T is parabolic too (see Exercise 24.8.32).

Parabolic rational functions abound. Perhaps the most famous one is the quadratic polynomial $Q_{1/4}(z) = z^2 + 1/4$ whose Julia set is commonly called the cauliflower. In fact, every quadratic polynomial of the form $z \mapsto \lambda z(1 + z)$, where λ is a root of unity, is a parabolic rational function.

26.3 Subexpanding rational functions

Let $R : \widehat{\mathbb{C}} \to \widehat{\mathbb{C}}$ be a rational function. Define the compact set

$$\omega(R) := \mathcal{J}(R) \cap \omega(\mathrm{Crit}(R)),$$

where $\omega(\mathrm{Crit}(R)) = \bigcup_{\zeta \in \mathrm{Crit}(R)} \omega(\zeta)$ is the set of limit points of the forward orbits of the critical points of R.

The set $\omega(R)$ is forward R-invariant.

Proposition 26.3.1. *If $R : \widehat{\mathbb{C}} \to \widehat{\mathbb{C}}$ is a rational function, then $\omega(R) \subseteq R^{-1}(\omega(R))$. Equivalently, $R(\omega(R)) \subseteq \omega(R)$.*

This forward invariance ensures that the following concept is well-defined.

Definition 26.3.2. A rational function $R : \widehat{\mathbb{C}} \to \widehat{\mathbb{C}}$ is said to be *subexpanding* if the map $R|_{\omega(R)} : \omega(R) \to \omega(R)$ is expanding, i. e., if there exists $s \in \mathbb{N}$ such that

$$|(R^s)'(z)| > 1, \quad \forall z \in \omega(R). \tag{26.11}$$

Given that the (immediate) basin of attraction of any rationally indifferent periodic point would contain a critical point, a subexpanding rational function cannot admit such a periodic point. The following statement is then a simple consequence of Theorem 24.2.22 and Corollary 24.4.17.

Proposition 26.3.3. *If $R : \widehat{\mathbb{C}} \to \widehat{\mathbb{C}}$ is a subexpanding rational function with $\deg(R) \geq 2$, then $\Omega(R) = \emptyset$ and*

$$\mathcal{J}(R) \cap \omega(\mathrm{Crit}(R) \cap \mathcal{F}(R)) = \emptyset, \quad \textit{i. e.,} \quad \omega(R) = \omega(\mathrm{Crit}(R) \cap \mathcal{J}(R)).$$

Observe also that any subexpanding rational function R satisfies

$$\mathrm{Crit}(R) \cap \omega(R) = \emptyset. \tag{26.12}$$

In fact, relation (26.12) is equivalent to the subexpanding property. This easily follows from Mañé's theorem [85].

Of course, all expanding rational functions are subexpanding (cf. Definition 25.3.3) but the latter class is bigger. Among others, let us draw the attention of the reader to the family of critically finite rational functions. A rational function R is called *critically finite* if R has no rationally indifferent periodic point and the forward orbit of each critical point in the Julia set of R is finite. In this case, Mañé's theorem is not needed; a direct argument establishes that each critically finite rational function is subexpanding. Indeed, $\omega(R)$ then consists of finitely many repelling periodic orbits, whence it is an expanding set. Any critically finite rational function whose Julia set contains a critical point is thus subexpanding but not expanding. Examples of such maps include the quadratic polynomials $Q_i(z) = z^2 + i$, $Q_{-2}(z) = z^2 - 2$, Lattés examples from [74] and especially from John Milnor's article [95].

In [97], Michal Misiurewicz proved the existence of uncountably (in fact, continuum) many real numbers c such that the quadratic polynomial $Q_c(z) = z^2 + c$ is subexpanding. For such c's, he showed that $c \notin \omega(c)$ and the Julia set of Q_c is a compact subset of $\mathbb{R}$ (see [28] and [93] for more on this).

The difference between the classes of expanding and subexpanding rational functions will be the main concern in this section and, most notably, in Chapter 32. The present section can be partially considered as a preparation for that chapter. Both this section and Chapter 32 are substantial alterations and developments of the paper [39].

We begin with a topological property of $\omega(R)$ as a subset of $\mathcal{J}(R)$.

Proposition 26.3.4. *If $R : \widehat{\mathbb{C}} \to \widehat{\mathbb{C}}$ is a subexpanding rational function with* $\deg(R) \geq 2$, *then the compact set $\omega(R)$ is nowhere dense in $\mathcal{J}(R)$.*

Proof. If $\mathrm{Crit}(R) \cap \mathcal{J}(R) = \emptyset$ then $\omega(R) = \emptyset$ by Proposition 26.3.3, and the result is trivially true. Otherwise, by subexpandingness (see (26.11)), the compact set $\omega(R)$ is a proper subset of $\mathcal{J}(R)$. Hence, the result is an immediate consequence of the forward invariance of $\omega(R)$ (see Proposition 26.3.1) and the topological exactness of the map $R : \mathcal{J}(R) \to \mathcal{J}(R)$ (see Corollary 24.3.5). □

As a repercussion of Theorems 24.5.1 and 24.4.10, we deduce the next property, which subexpanding rational functions share with expanding ones (cf. Theorem 26.1.6).

Theorem 26.3.5. *A subexpanding rational function $R : \widehat{\mathbb{C}} \to \widehat{\mathbb{C}}$ admits no Cremer periodic point, rationally indifferent periodic point, Siegel disk or Herman ring. In particular, the Fatou set $\mathcal{F}(R)$ consists only of basins of attraction to attracting periodic points of R.*

With the terminology and notation from Section 24.6 and as an immediate consequence of Theorems 26.3.5 and 24.6.1, we identify another property that extends to the class of subexpanding rational functions (cf. Theorem 26.1.7).

Theorem 26.3.6. *Every subexpanding rational function $R : \widehat{\mathbb{C}} \to \widehat{\mathbb{C}}$ is a continuity point of the function $\mathcal{R} \ni S \longmapsto \mathcal{J}(S) \subseteq \mathcal{K}(\widehat{\mathbb{C}})$.*

We now turn our attention toward the existence of, and the domain and range of, holomorphic inverse branches of subexpanding rational functions in the vicinity of various points.

Let $R : \widehat{\mathbb{C}} \to \widehat{\mathbb{C}}$ be a subexpanding rational function. Set

$$T := R^s, \tag{26.13}$$

where s comes from (26.11). Since $\omega(T) \subseteq \omega(R)$, we have

$$|T'(z)| > 1, \quad \forall z \in \omega(T).$$

So T is also subexpanding, i. e., $\mathrm{Crit}(T) \cap \omega(T) = \emptyset$. As the map T' is continuous and the set $\omega(T)$ is compact, it follows from Proposition 26.3.3 that there exist

$$\eta \in (0, 2^{-6}\mathrm{dist}(\mathcal{J}(T), \overline{[\mathrm{Crit}(T) \cup \mathrm{PC}(T)] \cap \mathcal{F}(T)}) \quad \text{and} \quad \lambda > 1 \tag{26.14}$$

such that

$$|T'(z)| \geq \lambda, \quad \forall z \in B(\omega(T), 2\eta). \tag{26.15}$$

By definition of $\omega(T)$, there exists $p \geq 2$ such that

$$T^j(\mathrm{Crit}(T) \cup \mathrm{PC}(T)) \subseteq B(\omega(T), \eta/2) \cup [\widehat{\mathbb{C}} \setminus B(\mathcal{J}(T), 2^6\eta)], \quad \forall j \geq p-1. \tag{26.16}$$

Therefore, denoting $\|T'\|_\infty := \sup\{|T'(z)| : z \in \widehat{\mathbb{C}}\}$ and setting

$$\gamma := \frac{\eta}{4\|T'\|_\infty^{p-1}}, \tag{26.17}$$

we infer the following about holomorphic inverse branches near some points of the Julia set.

Lemma 26.3.7. *Let $R : \widehat{\mathbb{C}} \to \widehat{\mathbb{C}}$ be a subexpanding rational function with* $\deg(R) \geq 2$. *If $z \in \mathcal{J}(T)$, $n \geq p-1$ and $T^n(z) \notin B(\omega(T), \eta)$ then*

$$T^{n-p+1}(z) \notin B(\mathrm{PC}(T), 2\gamma),$$

where T, p, η and γ are defined in (26.13)–(26.17). *Thus, there is a unique holomorphic branch $T_z^{-(n-p+1)} : B(T^{n-p+1}(z), 2\gamma) \to \widehat{\mathbb{C}}$ of $T^{-(n-p+1)}$ such that $T_z^{-(n-p+1)}(T^{n-p+1}(z)) = z$.*

Proof. Otherwise, there are $z \in \mathcal{J}(T)$, $n \geq p-1$ and $T^n(z) \notin B(\omega(T), \eta)$ for which

$$T^{n-p+1}(z) \in B(\mathrm{PC}(T), 2\gamma).$$

Then

$$T^n(z) = T^{p-1}(T^{n-p+1}(z)) \in B(\omega(T), \eta) \quad \text{by (26.16)–(26.17).}$$

This is a contradiction. □

By virtue of (26.15) and (25.3)–(25.4), there exists $\varepsilon \in (0, \eta)$ so small that conditions (26.18)–(26.21) are satisfied:

$$B(\mathrm{Crit}(T), \varepsilon) \cap B(\omega(T), 2\eta) = \emptyset; \tag{26.18}$$

there exists $A \geq 1$ such that for every $c \in \mathrm{Crit}(T^p)$ and every $z \in B(c, \varepsilon)$, we have

$$A^{-1}|z-c|^{\deg_c(T^p)} \leq |T^p(z) - T^p(c)| \leq A|z-c|^{\deg_c(T^p)} \tag{26.19}$$

$$A^{-1}|z-c|^{\deg_c(T^p)-1} \le |(T^p)'(z)| \le A|z-c|^{\deg_c(T^p)-1}; \tag{26.20}$$

and finally

$$B(c_1,2\varepsilon)\cap B(c_2,2\varepsilon)=\emptyset, \quad \forall c_1,c_2\in \mathrm{Crit}(T^p),\ c_1\neq c_2. \tag{26.21}$$

We continue with a few more results about analytic inverse branches. As a direct consequence of Lemmas 25.1.12 and 25.1.5, we obtain the following fact.

Lemma 26.3.8. *If $R:\widehat{\mathbb{C}}\to\widehat{\mathbb{C}}$ is a subexpanding rational function with $\deg(R)\ge 2$, then there is $\tau\in(0,1/2)$ such that for each $c\in \mathrm{Crit}(T^p)$ and each $z\in B(c,\varepsilon)$ there exists a unique holomorphic inverse branch $T_z^{-p}: B(T^p(z),2\tau|T^p(z)-T^p(c)|)\to B(z,\gamma)$ such that $T_z^{-p}(T^p(z))=z$, where T, p and γ are defined in (26.13)–(26.17).*

Proof. By virtue of Lemmas 25.1.12 and 25.1.5, there exists $\tau\in(0,1/2)$ so small that

$$\mathrm{diam}(\,\mathrm{Comp}(w,T^p,2\tau|T^p(w)-T^p(c)|)\,)<\min\{\gamma,\varepsilon\}$$

for all $w\in\widehat{\mathbb{C}}$, where the reader will recall that $\mathrm{Comp}(w,T^p,2\tau|T^p(w)-T^p(c)|)$ is the connected component of $T^{-p}(B(T^p(w),2\tau|T^p(w)-T^p(c)|))$ containing w. It follows from (26.21) that

$$\mathrm{Crit}(T^p)\cap \mathrm{Comp}(z,T^p,2\tau|T^p(z)-T^p(c)|)=\emptyset.$$

Thus, in view of Corollary 22.5.4, the restriction of T^p to $\mathrm{Comp}(z,T^p,2\tau|T^p(z)-T^p(c)|)$ is a conformal homeomorphism onto $B(T^p(z),2\tau|T^p(z)-T^p(c)|)$. Its inverse is the required map T_z^{-p}. □

Lemma 26.3.9. *If $R:\widehat{\mathbb{C}}\to\widehat{\mathbb{C}}$ is a subexpanding rational function with $\deg(R)\ge 2$, then there exists $\Gamma_R>0$ such that*

$$\limsup_{n\to\infty}\sup_{(z,V)}\mathrm{diam}(V)=0,$$

where the supremum is taken over all $z\in\mathcal{J}(R)$ and all connected components V of $R^{-n}(B(z,4\Gamma_R))$.

Proof. Because of the uniform continuity of R (and all its iterates) it suffices to prove this lemma with R replaced by T. Suppose first that

$$z\in\mathcal{J}(T)\setminus B(\omega(T),\eta).$$

It follows from Lemma 26.3.7 that for any $w\in T^{-(p-1)}(z)$, any $n\ge 0$ and any $x\in T^{-n}(w)$, there exists a unique holomorphic branch $T_x^{-n}: B(w,2\gamma)\to\widehat{\mathbb{C}}$ of T^{-n} sending w to x. It then ensues from Lemma 24.1.13 that

$$\lim_{n\to\infty} \sup\{\operatorname{diam}(T_x^{-n}(B(w,\gamma))) : z \in \mathcal{J}(T)\backslash B(\omega(T),\eta),\ w \in T^{-(p-1)}(z),\ x \in T^{-n}(w)\} = 0.$$

On the other hand, Lemma 25.1.4 implies that there exists $\Delta_1 > 0$ such that for every $z \in \mathcal{J}(T) \setminus B(\omega(T),\eta)$ and every $w \in T^{-(p-1)}(z)$, the connected component of $T^{-(p-1)}(B(z,\Delta_1))$ containing w is contained in $B(w,\gamma)$. Therefore, the lemma is proved for the supremum taken over all points $z \in \mathcal{J}(T) \setminus B(\omega(T),\eta)$.

Let's now look at $\mathcal{J}(T) \cap B(\omega(T),\eta)$. Fixing $\delta > 0$, there exists $N_\delta \geq p$ such that

$$\operatorname{diam}(U) < \delta \tag{26.22}$$

for every $y \in \mathcal{J}(T) \setminus B(\omega(T),\eta)$, every $n \geq N_\delta$ and every connected component U of $T^{-n}(B(z,\Delta_1))$. Now take

$$\Delta_2 \in (0, \min\{\Delta_1, \gamma, \theta(\varepsilon,\lambda)\}) \tag{26.23}$$

so small (see Lemma 25.1.4) that

$$\operatorname{diam}(D) < \Delta_1 \tag{26.24}$$

for every $y \in \mathcal{J}(T)$, every $0 \leq q \leq N$ and every connected component D of $T^{-q}(B(y,\Delta_2))$, and take $N \in \mathbb{N}$ so large that

$$\theta(\varepsilon,\lambda)\lambda^{-N} < \Delta_1. \tag{26.25}$$

Also, note that because of Lemma 25.1.4 there exists $\tilde{\delta} \in (0,\delta]$ such that

$$\operatorname{diam}(W) < \delta \tag{26.26}$$

for every $y \in \mathcal{J}(T)$, every $0 \leq j \leq N_\delta$ and every connected component W of $T^{-j}(B(y,\tilde{\delta}))$.

Finally, take $M \in \mathbb{N}$ so large that

$$\theta(\varepsilon,\lambda)\lambda^{-M} < \tilde{\delta}. \tag{26.27}$$

Fix any

$$n \geq N + N_\delta + M \tag{26.28}$$

and suppose that

$$z \in \mathcal{J}(T) \cap B(\omega(T),\eta).$$

Set $w \in T^{-n}(z)$. If $T^j(w) \in B(\omega(T),\eta)$ for all $j = 0,1,\ldots,n$, then it follows from (26.18) and Lemma 25.1.11 that the diameter of the connected component G of $T^{-n}(B(z,\theta(\varepsilon,\lambda)))$ containing w satisfies

$$\operatorname{diam}(G) \le \theta(\varepsilon,\lambda)\lambda^{-n} < \delta, \tag{26.29}$$

where the last inequality holds thanks to (26.28), (26.27) and the fact that $\tilde{\delta} \le \delta$. Otherwise, let k be the largest integer in $\{0, 1, \ldots, n-1\}$ such that

$$T^k(w) \notin B(\omega(T), \eta).$$

Let V be the connected component of $T^{-n}(B(z, \Delta_2))$ containing w. Consider several cases.

Case 1: $k \ge n - N$. Then $n - k \le N$ and it ensues from (26.24) that $\operatorname{diam}(T^k(V)) < \Delta_1$. Since $k \ge n - N \ge N_\delta$ by (26.28), it follows from (26.22) that

$$\operatorname{diam}(V) < \delta. \tag{26.30}$$

Case 2: $k < n - N$. Consider two subcases:

Subcase 2a: $k \ge N_\delta$. Since $n - k > N$, we infer from (26.25), (26.23) and (26.29) that $\operatorname{diam}(T^k(V)) < \Delta_1$. As $k \ge N_\delta$, it ensues from (26.22) that

$$\operatorname{diam}(V) < \delta. \tag{26.31}$$

Subcase 2b: $k < N_\delta$. Then $n - (k+1) \ge n - N_\delta \ge M$, whence by (26.29) and (26.27),

$$\operatorname{diam}(T^{k+1}(V)) \le \theta(\varepsilon,\lambda)\lambda^{-[n-(k+1)]} \le \theta(\varepsilon,\lambda)\lambda^{-M} < \tilde{\delta}.$$

As $k + 1 \le N_\delta$, it follows from (26.26) that $\operatorname{diam}(V) < \delta$. This, along with (26.29)–(26.31), shows that the lemma holds if the supremum is taken over all $z \in \mathcal{J}(T) \cap B(\omega(T), \eta)$. □

As an immediate reverberation of this lemma, of Lemma 25.1.5, and of the fact that $c \notin \omega(c)$ for every $c \in \operatorname{Crit}(R) \cap \mathcal{J}(R)$ per (26.12), we deduce the following.

Corollary 26.3.10. *If $R : \widehat{\mathbb{C}} \to \widehat{\mathbb{C}}$ is a subexpanding rational function with $\deg(R) \ge 2$, then for every $z \in \mathcal{J}(R)$, every $n \ge 0$, and every connected component V of $R^{-n}(B(z, 4\Gamma_R))$, the number of integers $0 \le k \le n$ such that*

$$R^k(V) \cap \operatorname{Crit}(R) \ne \emptyset$$

is bounded above by $\#\operatorname{Crit}(R)$ and each such nonempty intersection is a singleton.

Lemma 26.3.11. *If $R : \widehat{\mathbb{C}} \to \widehat{\mathbb{C}}$ is a subexpanding rational function with $\deg(R) \ge 2$, then there exist $\kappa \in (0,1]$ and $\vartheta \in (0,1]$ such that the following holds: Let $z \in \mathcal{J}(R)$ and let $n \ge 0$ be an integer. Suppose that $Q^{(1)} \subseteq Q^{(2)} \subseteq B(R^n(z), \kappa\Gamma_R)$ are connected sets. Let $Q_n^{(1)}$ be a connected component of $R^{-n}(Q^{(1)})$ contained in the connected component $V_n(z, \kappa\Gamma_R)$ of $R^{-n}(B(R^n(z), \kappa\Gamma_R))$ containing z and let $Q_n^{(2)}$ be the connected component of $R^{-n}(Q^{(2)})$ containing $Q_n^{(1)}$. Then*

$$\frac{\operatorname{diam}(Q_n^{(1)})}{\operatorname{diam}(Q_n^{(2)})} \geq \vartheta \frac{\operatorname{diam}(Q^{(1)})}{\operatorname{diam}(Q^{(2)})}. \tag{26.32}$$

Proof. Since $\mathcal{J}(R)$ contains no superattracting periodic point of R, there exists $q \in \mathbb{N}$ such that

$$\operatorname{Crit}(R) \cap R^{q-1}(\operatorname{PC}(R) \cap \mathcal{J}(R)) = \emptyset. \tag{26.33}$$

Otherwise, there would exist a strictly increasing sequence $(n_k)_{k=1}^{\infty}$ such that $\operatorname{Crit}(R) \cap R^{n_k}(\operatorname{Crit}(R) \cap \mathcal{J}(R)) \neq \emptyset$. Due to the finiteness of $\operatorname{Crit}(R)$, there would then be $c_1 \in \operatorname{Crit}(R)$, $c_2 \in \operatorname{Crit}(R) \cap \mathcal{J}(R)$ and $k < l \in \mathbb{N}$ for which $c_1 = R^{n_k}(c_2) = R^{n_l}(c_2)$. Then we would have $R^{n_l-n_k}(c_1) = R^{n_l-n_k}(R^{n_k}(c_2)) = R^{n_l}(c_2) = c_1$. Moreover, the complete invariance of the Julia set would impose that $c_1 \in \mathcal{J}(R)$, just like c_2. This would contradict the fact that $\mathcal{J}(R)$ contains no superattracting periodic point of R.

As $\operatorname{Crit}(R) \cap \omega(\operatorname{Crit}(R) \cap \mathcal{J}(R)) = \operatorname{Crit}(R) \cap \omega(R) = \emptyset$, there is $0 < \Delta < \pi/2$ such that

$$B(\operatorname{Crit}(R), \Delta) \cap R^{q-1}(\operatorname{PC}(R) \cap \mathcal{J}(R)) = \emptyset \tag{26.34}$$

and

$$B(c, \Delta) \cap (\operatorname{Crit}(R) \cup \operatorname{PC}(R)) = \{c\}, \quad \forall c \in \operatorname{Crit}(R). \tag{26.35}$$

Fix $n \geq 0$ and $z \in \mathcal{J}(R)$. For every $r \in (0, 4\Gamma_R)$, let $V_n(z,r) := \operatorname{Comp}(z, R^n, r)$ be the connected component of $R^{-n}(B(R^n(z), r))$ containing z. In view of Lemma 25.1.5 and Lemma 26.3.9, there exists $\tau \in (0,1]$ such that

$$\operatorname{diam}(R^k(V_n(z, 4\tau\Gamma_R))) < \Delta, \quad \forall 0 \leq k \leq n. \tag{26.36}$$

By virtue of Theorem 25.1.1, there exist $\Delta_1 \in (0, \Delta]$ and $A \in [1, \infty)$ such that

$$B(c, A^{-1} r^{\frac{1}{d_c(R^k)}}) \subseteq \operatorname{Comp}(c, R^k, r) \subseteq B(c, A r^{\frac{1}{d_c(R^k)}}) \tag{26.37}$$

for every $1 \leq k \leq q+1$, every $c \in \operatorname{Crit}(R^k)$ and every $r \in (0, \Delta_1]$, where $d_c(R^k) = \deg_c(R^k)$.

Finally, let $\kappa := \rho\tau$, where $\rho \in (0,1]$ is so small that

$$2A(2K\rho)^{\frac{1}{d_w(R^{q+1})}} \leq A^{-1}(K^{-1} - K\rho)^{\frac{1}{d_w(R^{q+1})}}, \quad \forall w \in \operatorname{Crit}(R^{q+1}) \tag{26.38}$$

and $K := k_{2\tau\Gamma_R, \pi/2}(1/2)$ is the Koebe constant from Theorem 23.1.9.

We will consider several cases.

Case 1: For every $0 \leq k < n$,

$$R^k(V_n(z, 4\tau\Gamma_R)) \cap \operatorname{Crit}(R) = \emptyset.$$

It then follows from Koebe's distortion theorem that

$$\frac{\operatorname{diam}(Q_n^{(1)})}{\operatorname{diam}(Q_n^{(2)})} \geq K^{-2}\frac{\operatorname{diam}(Q^{(1)})}{\operatorname{diam}(Q^{(2)})},$$

and we are done with Case 1.

Case 2: There exists $0 \leq u < n$ such that

$$R^u(V_n(z, 4\tau\Gamma_R)) \cap \operatorname{Crit}(R) \neq \emptyset \tag{26.39}$$

and assume that u is the largest number with that property. It then ensues from Theorem 23.1.9 that

$$\frac{\operatorname{diam}(R^{u+1}(Q_n^{(1)}))}{\operatorname{diam}(R^{u+1}(Q_n^{(2)}))} \geq K^{-2}\frac{\operatorname{diam}(Q^{(1)})}{\operatorname{diam}(Q^{(2)})}. \tag{26.40}$$

In light of Corollary 26.3.10, there exists $c \in \widehat{\mathbb{C}}$ such that

$$R^u(V_n(z, 4\tau\Gamma_R)) \cap \operatorname{Crit}(R) = \{c\}. \tag{26.41}$$

Subcase 2a: $c \notin R^u(V_n(z, 2\rho\tau\Gamma_R))$.

In light of (26.41), this implies that

$$R^u(V_n(z, 2\rho\tau\Gamma_R)) \cap \operatorname{Crit}(R) = \emptyset. \tag{26.42}$$

Suppose for a contradiction that

$$R^u(V_n(z, 2\rho\tau\Gamma_R)) \cap \operatorname{PC}(R) \neq \emptyset. \tag{26.43}$$

As $\operatorname{diam}(R^u(V_n(z, 4\tau\Gamma_R))) < \Delta$ by (26.36) and as $c \in R^u(V_n(z, 4\tau\Gamma_R))$ by (26.41), we infer that

$$B(c, \Delta) \supseteq R^u(V_n(z, 4\tau\Gamma_R)) \supseteq R^u(V_n(z, 2\rho\tau\Gamma_R)). \tag{26.44}$$

It follows from (26.43)–(26.44) that

$$B(c, \Delta) \cap \operatorname{PC}(R) \neq \emptyset. \tag{26.45}$$

We deduce from (26.35) that

$$B(c, \Delta) \cap \operatorname{PC}(R) = \{c\}.$$

It ensues from this and (26.43)–(26.44) that

$$R^u(V_n(z, 2\rho\tau\Gamma_R)) \cap \operatorname{PC}(R) = \{c\}.$$

In particular, $c \in R^u(V_n(z, 2\rho\tau\Gamma_R))$. But this is not the case in Subcase 2a. So, (26.43) is impossible in that subcase, and thus

$$R^u(V_n(z, 2\rho\tau\Gamma_R)) \cap \mathrm{PC}(R) = \emptyset. \tag{26.46}$$

Now, if there were some $0 \le j < u$ such that $R^j(V_n(z, 2\rho\tau\Gamma_R)) \cap \mathrm{Crit}(R) \neq \emptyset$, then (26.46) would be contradicted. Hence,

$$R^j(V_n(z, 2\rho\tau\Gamma_R)) \cap \mathrm{Crit}(R) = \emptyset, \quad \forall 0 \le j < u. \tag{26.47}$$

It follows from (26.47), (26.42) and Koebe's distortion theorem (Theorem 23.1.9) that

$$\frac{\mathrm{diam}(Q_n^{(1)})}{\mathrm{diam}(Q_n^{(2)})} \ge K^{-2} \frac{\mathrm{diam}(R^{u+1}(Q_n^{(1)}))}{\mathrm{diam}(R^{u+1}(Q_n^{(2)}))}. \tag{26.48}$$

We deduce from (26.40) and (26.48) that

$$\frac{\mathrm{diam}(Q_n^{(1)})}{\mathrm{diam}(Q_n^{(2)})} \ge K^{-4} \frac{\mathrm{diam}(Q^{(1)})}{\mathrm{diam}(Q^{(2)})},$$

and we are done in Subcase 2a.

Subcase 2b: $c \in R^u(V_n(z, 2\rho\tau\Gamma_R))$. Let

$$k := \min\{q+1, u+1\}.$$

Since both sets $R^{u+1-k}(Q_n^{(1)})$ and $R^{u+1-k}(Q_n^{(2)})$ are connected, we get for $j = 1, 2$ from Lemma 25.1.13 that

$$E^{-1} \le \frac{\mathrm{diam}(R^{u+1}(Q_n^{(j)}))}{\mathrm{diam}(R^{u+1-k}(Q_n^{(j)})) \cdot [\mathrm{Dist}(c, R^{u+1-k}(Q_n^{(j)}))]^{d_c(R^k)-1}} \le E.$$

where

$$E := \max\{E_{R^{k'}} : 1 \le k' \le q+1\}$$

with $E_{R^{k'}} \in [1, \infty)$, $1 \le k' \le q+1$, being the constants produced in Lemma 25.1.13. Hence,

$$\begin{aligned}\frac{\mathrm{diam}(R^{u+1-k}(Q_n^{(1)}))}{\mathrm{diam}(R^{u+1-k}(Q_n^{(2)}))} &\ge E^{-2} \frac{\mathrm{diam}(R^{u+1}(Q_n^{(1)}))}{[\mathrm{Dist}(c, R^{u+1-k}(Q_n^{(1)}))]^{d_c(R^k)-1}} \cdot \frac{[\mathrm{Dist}(c, R^{u+1-k}(Q_n^{(2)}))]^{d_c(R^k)-1}}{\mathrm{diam}(R^{u+1}(Q_n^{(2)}))} \\ &\ge E^{-2} \frac{\mathrm{diam}(R^{u+1}(Q_n^{(1)}))}{\mathrm{diam}(R^{u+1}(Q_n^{(2)}))}.\end{aligned}$$

Invoking (26.40), we thereby get that

$$\frac{\operatorname{diam}(R^{u+1-k}(Q_n^{(1)}))}{\operatorname{diam}(R^{u+1-k}(Q_n^{(2)}))} \geq (EK)^{-2}\frac{\operatorname{diam}(Q^{(1)})}{\operatorname{diam}(Q^{(2)})}. \tag{26.49}$$

Sub-subcase 2b1: $k = u + 1$. Then (26.49) yields

$$\frac{\operatorname{diam}(Q_n^{(1)})}{\operatorname{diam}(Q_n^{(2)})} \geq (EK)^{-2},$$

and we are done in this sub-subcase.

Sub-subcase 2b2: $k = q + 1$. Recall that $\operatorname{Comp}(\zeta, f, r)$ denotes the connected component of $f^{-1}(B(f(\zeta), r))$ containing ζ. Remember also that $V_n(z, 2\rho\tau\Gamma_R) := \operatorname{Comp}(z, R^n, 2\rho\tau\Gamma_R)$. In Subcase 2$b$, by hypothesis $c \in R^u(V_n(z, 2\rho\tau\Gamma_R))$. By definition,

$$R^{u+1}(V_n(z, 2\rho\tau\Gamma_R)) \subseteq R^{u+1} \circ R_z^{-n}(B(R^n(z), 2\rho\tau\Gamma_R)).$$

Applying Koebe's distortion theorem (Theorem 23.1.9) to the right-hand side yields

$$R^{u+1}(V_n(z, 2\rho\tau\Gamma_R)) \subseteq B(R^{u+1}(z), 2\rho\tau\Gamma_R K|(R^{n-(u+1)})'(R^{u+1}(z))|^{-1}).$$

Thus,

$$\operatorname{diam}(\, R^{u+1}(V_n(z, 2\rho\tau\Gamma_R))\,) \leq 4\rho\tau\Gamma_R K|(R^{n-(u+1)})'(R^{u+1}(z))|^{-1}.$$

Since $R(c) \in R^{u+1}(V_n(z, 2\rho\tau\Gamma_R))$, we have

$$\begin{aligned} R^{u+1}(V_n(z, 2\rho\tau\Gamma_R)) &\subseteq B(R(c), \operatorname{diam}(R^{u+1}(V_n(z, 2\rho\tau\Gamma_R)))) \\ &\subseteq B(R(c), 4\rho\tau\Gamma_R K|(R^{n-(u+1)})'(R^{u+1}(z))|^{-1}). \end{aligned}$$

Fix a point

$$c^* \in R^{(u+1)-(q+1)}(V_n(z, 2\rho\tau\Gamma_R)) \cap R^{-q}(c). \tag{26.50}$$

Then $(R^{q+1})'(c^*) = 0$, i. e., $c^* \in \operatorname{Crit}(R^{q+1})$. Moreover,

$$R^{u+1}(V_n(z, 2\rho\tau\Gamma_R)) \subseteq B(R^{q+1}(c^*), 4\rho\tau\Gamma_R K|(R^{n-(u+1)})'(R^{u+1}(z))|^{-1}).$$

It follows from this, (26.50) and (26.37) that

$$\begin{aligned} R^{(u+1)-(q+1)}(V_n(z, 2\rho\tau\Gamma_R)) &= \operatorname{Comp}(c^*, R^{q+1}, R^{u+1}(V_n(z, 2\rho\tau\Gamma_R))) \\ &\subseteq B\Big(c^*, A[4K\rho\tau\Gamma_R|(R^{n-(u+1)})'(R^{u+1}(z))|^{-1}]^{\frac{1}{d_{c^*}(R^{q+1})}}\Big). \end{aligned} \tag{26.51}$$

On the other hand, by the definition of u and Koebe's distortion theorem we have

$$R^{u+1}(V_n(z, 2\tau\Gamma_R)) \supseteq B(R^{u+1}(z), 2\tau\Gamma_R K^{-1}|(R^{n-(u+1)})'(R^{u+1}(z))|^{-1})$$

$$\begin{aligned}
&\supseteq B(R(c), 2\tau\Gamma_R K^{-1} |(R^{n-(u+1)})'(R^{u+1}(z))|^{-1} - |R(c) - R^{u+1}(z)|) \\
&\supseteq B(R(c), 2\tau\Gamma_R K^{-1} |(R^{n-(u+1)})'(R^{u+1}(z))|^{-1} - \operatorname{diam}(V_n(z, 2\tau\Gamma_R))) \\
&\supseteq B(R(c), 2\tau\Gamma_R (K^{-1} - K\rho) |(R^{n-(u+1)})'(R^{u+1}(z))|^{-1}). \qquad (26.52)
\end{aligned}$$

Relations (26.37) and (26.52) imply that

$$R^{(u+1)-(q+1)}(V_n(z, 2\tau\Gamma_R)) \supseteq B\Big(c^*, A^{-1}[2\tau\Gamma_R(K^{-1} - K\rho)|(R^{n-(u+1)})'(R^{u+1}(z))|^{-1}]^{\frac{1}{d_{c^*}(R^{q+1})}}\Big). \qquad (26.53)$$

It follows from (26.38) and (26.53) that

$$R^{(u+1)-(q+1)}(V_n(z, 2\tau\Gamma_R)) \supseteq B\Big(c^*, 2A[4K\rho\tau\Gamma_R|(R^{n-(u+1)})'(R^{u+1}(z))|^{-1}]^{\frac{1}{d_{c^*}(R^{q+1})}}\Big). \qquad (26.54)$$

Seeking a contradiction, suppose that

$$R^{(u+1)-(q+1)}(V_n(z, 2\tau\Gamma_R)) \cap \operatorname{PC}(R) \neq \emptyset.$$

Applying R^q, we obtain that

$$R^u(V_n(z, 2\tau\Gamma_R)) \cap R^q(\operatorname{PC}(R)) \neq \emptyset.$$

But $c \in R^u(V_n(z, 2\tau\Gamma_R))$, as we are in Subcase 2*b*. Using (26.36), we get

$$B(c, \Delta) \cap R^q(\operatorname{PC}(R)) \neq \emptyset.$$

However, this contradicts (26.34) and establishes that

$$R^{(u+1)-(q+1)}(V_n(z, 2\tau\Gamma_R)) \cap \operatorname{PC}(R) = \emptyset. \qquad (26.55)$$

But if there were some $0 \leq j < u - q$ such that $R^j(V_n(z, 2\tau\Gamma_R)) \cap \operatorname{Crit}(R) \neq \emptyset$, then (26.55) would be contradicted. Hence,

$$R^j(V_n(z, 2\tau\Gamma_R)) \cap \operatorname{Crit}(R) = \emptyset, \quad \forall 0 \leq j < u - q. \qquad (26.56)$$

Having this, (26.51) and (26.54), it follows from Koebe's distortion theorem (Theorem 23.1.9) that

$$\frac{\operatorname{diam}(Q_n^{(1)})}{\operatorname{diam}(Q_n^{(2)})} \geq K^{-2} \frac{\operatorname{diam}(R^{(u+1)-(q+1)}(Q_n^{(1)}))}{\operatorname{diam}(R^{(u+1)-(q+1)}(Q_n^{(2)}))} = K^{-2} \frac{\operatorname{diam}(R^{(u+1)-k}(Q_n^{(1)}))}{\operatorname{diam}(R^{(u+1)-k}(Q_n^{(2)}))}. \qquad (26.57)$$

We deduce from (26.57) and (26.49) that

$$\frac{\operatorname{diam}(Q_n^{(1)})}{\operatorname{diam}(Q_n^{(2)})} \geq (EK^2)^{-2} \frac{\operatorname{diam}(Q^{(1)})}{\operatorname{diam}(Q^{(2)})}$$

and the proof of Lemma 26.3.11 is complete. □

26.4 Exercises

Exercise 26.4.1. Let $P(z) = (3z - z^3)/2$. Show that all the critical points of P are superattracting fixed points, and hence that P is expanding on its Julia set. Show also that all components of its Fatou set are simply connected.

Exercise 26.4.2. Let $Q_c(z) = z^2 + c$. Show that if $|z| \geq |c|$, then $|Q_c(z)| \geq (|c| - 1)|z|$. Deduce that if $|c| > 2$, then $\mathcal{J}(Q_c)$ is a Cantor set and Q_c is expanding and $c \notin \mathcal{M}$, where $\mathcal{M}$ denotes the Mandelbrot set. Conclude that $\mathcal{M} \subseteq \overline{B(0,2)}$.

Exercise 26.4.3. Show that $(1/4, \infty) \subseteq \mathbb{C} \setminus \mathcal{M}$.

Exercise 26.4.4. Let $c_1, c_2 \in \mathbb{C} \setminus \mathcal{M}$. Show that the polynomials Q_{c_1} and Q_{c_2} are topologically conjugate on their Julia sets. Which of them are conjugate on $\widehat{\mathbb{C}}$ by Möbius transformations?

Exercise 26.4.5. Prove that the complement of the immediate basin of attraction of a parabolic point is either connected or else has uncountably many connected components.

Exercise 26.4.6. Show that every quadratic polynomial of the form $z \mapsto \lambda z(1+z)$, where λ is a root of unity, is a parabolic rational function.

Exercise 26.4.7. Note that every quadratic polynomial has at most one rationally indifferent periodic orbit. Show that for every $p \in \mathbb{N}$ there exists a parabolic quadratic polynomial with a rationally indifferent periodic point of prime period p.

Exercise 26.4.8. Find an expansive constant for the parabolic polynomial $Q_{1/4}$.

Exercise 26.4.9 (Cf. [94], Exercise 10e). Show that the only fixed point of the rational function $T(z) = z - 1/z$ is the point at infinity, with local degree 3. Show that the two parabolic basins are the upper and lower half-planes and that $\mathcal{J}(T) = \mathbb{R} \cup \{\infty\}$. Is this rational function parabolic?

Exercise 26.4.10 (Cf. [94], Exercise 10e). Show that ∞ is a rationally indifferent periodic point of the rational function $T(z) = z - 1/z + 1$, with multiplicity 2, and that $\mathcal{J}(T)$ is a Cantor set within $\mathbb{R} \cup \{\infty\}$. Is this rational function parabolic?

Exercise 26.4.11 (Cf. [94], Exercise 10e). Show that the rational function $T(z) = z - 1/z - 2$ has a rationally indifferent periodic point at ∞ with multiplicity 2. Show further that $\mathcal{J}(T) = [0, \infty]$. Is this rational function parabolic?

Exercise 26.4.12 (Cf. [94], Exercise 10e). Show that $T(z) = z + 1/(1 + z^2)$ has a rationally indifferent periodic point at ∞ with multiplicity 4. Show that one of the three immediate basins of attraction of ∞ contains $\mathbb{R}$, and hence nearly disconnects the Riemann sphere. Is this rational function parabolic?

Exercise 26.4.13. Consider the family of polynomials $z \mapsto z(1 - z^p)$, $p \in \mathbb{N}$. Which of them are parabolic?

27 Equilibrium states for rational functions and Hölder continuous potentials with pressure gap

In this chapter, $T : \widehat{\mathbb{C}} \to \widehat{\mathbb{C}}$ is an arbitrary rational function with $\deg(T) \geq 2$ while $\varphi : \mathcal{J}(T) \to \mathbb{R}$ is a Hölder continuous function (a potential) for which $\mathrm{P}(T, \varphi) > \sup \varphi$. Our ultimate goal is to prove the existence and uniqueness of an equilibrium state for each such potential and to explore some of that state's ergodic and dynamical properties. The full statement is given in Theorem 27.8.3, first proved by Denker and Urbański [36]. Another proof was later provided by Przytycki [106]. Our proof basically stems from [36], although we incorporate some reasoning from [106] and [146]. We discuss the proof's structure and main ingredients in greater detail in a number of remarks in the last section. We also discuss in these remarks many matters related to Theorem 27.8.3.

27.1 Bad and good inverse branches

Let $T : \widehat{\mathbb{C}} \to \widehat{\mathbb{C}}$ be a rational function with $d := \deg(T) \geq 2$. For any $z \in \widehat{\mathbb{C}}$, observe that

$$\deg_z(T) \leq \deg(T) =: d \tag{27.1}$$

and

$$\deg_z(T^n) = \prod_{j=0}^{n-1} \deg_{T^j(z)}(T), \quad \forall n \in \mathbb{N}. \tag{27.2}$$

Let $z \in \mathcal{J}(T)$. As T has no superattracting periodic point in $\mathcal{J}(T)$, it holds that

$$\#\{n \geq 0 \,|\, T^n(z) \in \mathrm{Crit}(T)\} \leq \#\mathrm{Crit}(T) \leq 2d - 2. \tag{27.3}$$

It follows from (25.2) and (27.1)–(27.3) that

$$\deg_z(T^n) \leq d^{2d-2}, \quad \forall n \in \mathbb{N},\ \forall z \in \mathcal{J}(T). \tag{27.4}$$

In fact, a much better upper estimate than (27.4) is

$$\deg_z(T^n) \leq \Delta(T) := \prod_{c \in \mathrm{Crit}(T) \cap \mathcal{J}(T)} \deg_c(T), \quad \forall n \in \mathbb{N},\ \forall z \in \mathcal{J}(T). \tag{27.5}$$

Let ρ be the spherical metric on $\widehat{\mathbb{C}}$ normalized so that

$$A_\rho(\widehat{\mathbb{C}}) = 1, \tag{27.6}$$

where A_ρ is the two-dimensional volume (area) on $\widehat{\mathbb{C}}$ induced by ρ.

https://doi.org/10.1515/9783110769876-027

Let $R > 0$ be so small that

$$B_\rho(\mathcal{J}(T), 2R) \cap (\mathrm{Crit}(T) \setminus \mathcal{J}(T)) = \emptyset. \tag{27.7}$$

The starting point of all our considerations in this section and on which virtually all hinges in this chapter, is the following technical fact.

Lemma 27.1.1. *Let $T : \widehat{\mathbb{C}} \to \widehat{\mathbb{C}}$ be a rational function with $d := \deg(T) \geq 2$. Let also $\gamma \in (0,1)$ and $s \in \mathbb{N}$. For every $z \in \mathcal{J}(T)$, there exist $r_z = r_z(\gamma, s) \in (0, R]$ and $C_z = C_z(\gamma, s) \geq 1$ such that for every $n \geq 0$ there is a family $W_n(\gamma, s; z)$ of connected components of $T^{-n}(B(z, r_z))$ with the following properties:*

(a_n) *If $0 \leq n \leq s$, then the family $W_n(\gamma, s; z)$ consists of all the connected components of $T^{-n}(B(z, r_z))$.*

(b_n) $\max\{\mathrm{diam}(V) \mid V \in W_n(\gamma, s; z)\} \leq C_z \gamma^{n/2}$.

(b_n^*) *If $n > s$,*

$$\min\Big\{\inf_{\xi \in V} |(T^{n-s})'(\xi)|_\rho : V \in W_n(\gamma, s; z)\Big\} \geq (C_z \gamma^{n/2})^{-1}.$$

(c_n)

$$W_n(\gamma, s; z) \subseteq Z_n(\gamma, s; z) \tag{27.8}$$

and

$$\#(Z_n(\gamma, s; z) \setminus W_n(\gamma, s; z)) \leq d\gamma^{-n} + d(2d - 2), \tag{27.9}$$

where $Z_n(\gamma, s; z)$ is the family of all connected components of all sets of the form $T^{-1}(V)$ with $V \in W_{n-1}(\gamma, s; z)$.

(d_n) *If $0 \leq n \leq s$, then $Z_n(\gamma, s; z) = W_n(\gamma, s; z)$.*

(e_n) *Each $V \in W_n(\gamma, s; z)$ is a simply connected open subset of $\widehat{\mathbb{C}}$.*

(f_n) *If $n > s$, then*

$$V \cap \mathrm{Crit}(T) = \emptyset, \quad \forall V \in W_n(\gamma, s; z).$$

(g_n) *If $n > s$, then the map $T|_V : V \to T(V)$ is a conformal homeomorphism for every $V \in W_n(\gamma, s; z)$.*

(h_n) *If $V \in W_n(\gamma, s; z)$, then $\deg(T^n|_V : V \to T^n(V)) \leq \Delta(T) := \prod_{c \in \mathrm{Crit}(T) \cap \mathcal{J}(T)} \deg_c(T)$.*

Proof. Let $\delta > 0$ come from the application of Lemma 25.1.5 with $n = s + 1$ and $\varepsilon = \gamma$. Since $\#\mathrm{Crit}(T^s) < \infty$ and $\#(T^{-s}(z)) < \infty$, that lemma implies the existence of $r_z \in (0, \delta/4)$ such that

$$\mathrm{Crit}(T^s) \cap W \subseteq T^{-s}(z) \quad \text{and} \quad \#(W \cap T^{-s}(z)) = 1$$

for all connected components W of $T^{-s}(B(z, 2r_z))$. Then

$$\mathrm{Crit}(T^k) \cap W \subseteq T^{-k}(z) \quad \text{and} \quad \#(W \cap T^{-k}(z)) = 1 \tag{27.10}$$

for every connected component W of $T^{-k}(B(z, 2r_z))$ and every $0 \le k \le s$. In particular, for such k's,

$$\mathrm{Crit}(T^k) \cap (V^+ \setminus V) = \emptyset \tag{27.11}$$

for every connected component V of $T^{-k}(B(z, r_z))$, where V^+ is the unique connected component of $T^{-k}(B(z, 2r_z))$ containing V.

We first construct a family $W_n'(\gamma, s; z)$ of connected components of $T^{-n}(B(z, 2r_z))$ with the following properties:

(a_n') If $0 \le n \le s$, then the family $W_n'(\gamma, s; z)$ consists of all the connected components of $T^{-n}(B(z, 2r_z))$.

(b_n') $\max\{A_\rho(V') \mid V' \in W_n'(\gamma, s; z)\} \le \gamma^n$.

(c_n')

$$W_n'(\gamma, s; z) \subseteq Z_n'(\gamma, s; z) \tag{27.12}$$

and

$$\#(Z_n'(\gamma, s; z) \setminus W_n'(\gamma, s; z)) \le \gamma^{-n} + (2d - 2), \tag{27.13}$$

where $Z_n'(\gamma, s; z)$ is the family of all connected components of all sets of the form $T^{-1}(V')$ with $V' \in W_{n-1}'(\gamma, s; z)$.

(d_n') If $0 \le n \le s$, then $Z_n'(\gamma, s; z) = W_n'(\gamma, s; z)$.

(e_n') Each $V' \in W_n'(\gamma, s; z)$ is a simply connected open subset of $\widehat{\mathbb{C}}$.

(f_n') If $n > s$, then

$$V' \cap \mathrm{Crit}(T) = \emptyset, \quad \forall V' \in W_n'(\gamma, s; z).$$

(g_n') If $n > s$, then the map $T|_{V'} : V' \to T(V')$ is a conformal homeomorphism for every $V' \in W_n'(\gamma, s; z)$.

(h_n') If $V' \in W_n'(\gamma, s; z)$, then $\deg(T^n|_{V'} : V' \to T^n(V')) \le \Delta(T)$.

Starting the construction, for every $0 \le n \le s$ let $W_n'(\gamma, s; z)$ consist of all connected components of $T^{-n}(B(z, 2r_z))$ and set

$$Z_n'(\gamma, s; z) = W_n'(\gamma, s; z), \quad \forall 0 \le n \le s. \tag{27.14}$$

Thus, for all $0 \le n \le s$, properties (a_n'), (c_n') and (d_n') are trivially satisfied; property (b_n') can be fulfilled by reducing r_z if necessary so that by Lemma 25.1.5 all sets in $W_n'(\gamma, s; z)$ have areas not exceeding γ^s; property (e_n') follows from Lemma 25.1.5 as $2r_z < \delta$; and (h_n') is a consequence of (27.10) and (27.5); finally, properties (f_n') and (g_n') are vacuously fulfilled. In summary, properties (a_n')–(h_n') hold for every $0 \le n \le s$.

Proceeding by induction, suppose that the family $W'_n(\gamma, s; z)$ has been constructed for some $n \geq s$. Let $Z'_{n+1}(\gamma, s; z)$ be the family of all connected components of all sets of the form $T^{-1}(V')$ with $V' \in W'_n(\gamma, s; z)$. Define

$$W'_{n+1}(\gamma, s; z) := \{V' \in Z'_{n+1}(\gamma, s; z) : A_\rho(V') \leq \gamma^{n+1} \text{ and } V' \cap \mathrm{Crit}(T) = \emptyset\}. \tag{27.15}$$

We want to show that properties (a'_{n+1})–(h'_{n+1}) hold. Properties (a'_{n+1}) and (d'_{n+1}) are vacuously satisfied. Properties (b'_{n+1}) and (f'_{n+1}) follow directly from the definition of $W'_{n+1}(\gamma, s; z)$. Concerning property (c'_{n+1}), observe that

$$Z'_{n+1}(\gamma, s; z) \setminus W'_{n+1}(\gamma, s; z) = \{V' \in Z'_{n+1}(\gamma, s; z) \mid A_\rho(V') > \gamma^{n+1} \text{ or } V' \cap \mathrm{Crit}(T) \neq \emptyset\}. \tag{27.16}$$

Since all sets in $Z'_{n+1}(\gamma, s; z)$ are mutually disjoint, relation (27.6) imposes that

$$\#\{V' \in Z'_{n+1}(\gamma, s; z) \mid A_\rho(V') > \gamma^{n+1}\} \leq \gamma^{-(n+1)}. \tag{27.17}$$

Moreover,

$$\#\{V' \in Z'_{n+1}(\gamma, s; z) \mid V' \cap \mathrm{Crit}(T) \neq \emptyset\} \leq \#\mathrm{Crit}(T) \leq 2d - 2. \tag{27.18}$$

Therefore,

$$\#(Z'_{n+1}(\gamma, s; z) \setminus W'_{n+1}(\gamma, s; z)) \leq \gamma^{-(n+1)} + (2d - 2), \tag{27.19}$$

which means that (27.13) in (c'_{n+1}) is fulfilled. Of course, by the definitions of $Z'_{n+1}(\gamma, s; z)$ and $W'_{n+1}(\gamma, s; z)$, all of the other requirements of (c'_{n+1}) also hold.

It remains to establish (e'_{n+1}), (g'_{n+1}) and (h'_{n+1}). Both (e'_{n+1}) and (g'_{n+1}) follow from Corollary 22.5.4 by invoking (e'_n) and the definition of W'_{n+1}, while (h'_{n+1}) follows from (h'_n) and (g'_{n+1}). The inductive construction of the family $W'_n(\gamma, s; z)$ is complete.

Now, for every $n \geq 0$ the family $W_n(\gamma, s; z)$ is defined to be comprised of all connected components of the sets of the form $V' \cap T^{-n}(B(z, r_z))$, where $V' \in W'_n(\gamma, s; z)$. Properties ($\mathrm{a}_n$), ($\mathrm{f}_n$) and ($\mathrm{g}_n$) are then automatically satisfied. For each $0 \leq n \leq s$, property (e_n) follows in exactly the same way as (e'_n) did, i. e., from Lemma 25.1.5 and the choice of V', while for $n > s$ it follows from (e_{n-1}) and (g_n). Property (h_n) ensues directly from (h'_n).

Left to show are (b_n)–(b^*_n), (c_n) and (d_n). We shall first demonstrate that (27.8) in (c_n) holds. Let $V \in W_n(\gamma, s; z)$. Then there exists a unique $V' \in W'_n(\gamma, s; z)$ such that V is a connected component of $V' \cap T^{-n}(B(z, r_z))$. Therefore,

$$T(V) \subseteq T(V') \cap T^{-(n-1)}(B(z, r_z)). \tag{27.20}$$

By virtue of (27.12) and the definition of $Z'_n(\gamma, s; z)$, we know that $T(V') \in W'_{n-1}(\gamma, s; z)$. Thus, there exists a unique element $W \in W_{n-1}(\gamma, s; z)$ such that

$$T(V) \subseteq W \subseteq T(V') \cap T^{-(n-1)}(B(z, r_z)).$$

As $W \in W_{n-1}(\gamma, s; z)$, there exists a unique element $W' \in W'_{n-1}(\gamma, s; z)$ such that W is a connected component of $W' \cap T^{-n}(B(z, r_z))$. In particular,

$$W \subseteq W'.$$

Since $V \subseteq T^{-1}(W)$ and V is connected, there exists a unique connected component V^+ of $T^{-1}(W)$ containing V. In particular,

$$V^+ \in Z_n(\gamma, s; z) \tag{27.21}$$

and

$$V^+ \subseteq T^{-1}(W').$$

Because V^+ is connected and $V^+ \subseteq T^{-1}(W')$, there exists a unique connected component V^{++} of $T^{-1}(W')$ containing V^+. In particular,

$$V^{++} \in Z'_n(\gamma, s; z).$$

Given that $V^{++} \supseteq V^+ \supseteq V$ and $V \in W_n(\gamma, s; z)$, we deduce that

$$V^{++} \in W'_n(\gamma, s; z).$$

As, moreover, V^+ is connected and

$$V^+ \subseteq V^{++} \cap T^{-1}(W) \subseteq V^{++} \cap T^{-n}(B(z, r_z)),$$

there is a unique element $\widehat{W}$ of $W_n(\gamma, s; z)$ containing V^+. Hence,

$$V \subseteq V^+ \subseteq \widehat{W}.$$

Given that $V, \widehat{W} \in W_n(\gamma, s; z)$, it follows that $V = \widehat{W}$ and thereby $V = V^+$. By virtue of (27.21), we conclude that $V \in Z_n(\gamma, s; z)$, i. e., (27.8) is established.

We will return to the other parts of (c_n) shortly. For now, we shall prove that (d_n) holds. In view of (27.8), it suffices to show that

$$Z_n(\gamma, s; z) \subseteq W_n(\gamma, s; z), \quad \forall 0 \le n \le s. \tag{27.22}$$

Fix $0 \le n \le s$ and $V \in Z_n(\gamma, s; z)$. Then

$$T(V) \in W_{n-1}(\gamma, s; z). \tag{27.23}$$

So, there exists a unique $H \in W'_{n-1}(\gamma, s; z)$ such that $T(V)$ is a connected component of $H \cap T^{-(n-1)}(B(z, r_z))$. In particular,

$$V \subseteq T^{-1}(H) \cap T^{-n}(B(z, r_z)).$$

As V is connected, there is a unique connected component G of $T^{-1}(H)$ containing V. Hence, $G \in Z'_n(\gamma, s; z)$ and, as $V \in Z_n(\gamma, s; z)$, we have that

$$G \in W'_n(\gamma, s; z).$$

Also, since V is connected and

$$V \subseteq G \cap T^{-n}(B(z, r_z)),$$

there exists a unique connected component V^+ of $G \cap T^{-n}(B(z, r_z))$ containing V. Thus,

$$V^+ \in W_n(\gamma, s; z). \tag{27.24}$$

Since

$$T(V^+) \subseteq T^{-(n-1)}(B(z, r_z)) \cap T(G) = T^{-(n-1)}(B(z, r_z)) \cap H,$$

since $T(V^+)$ is connected and since $H \in W'_{n-1}(\gamma, s; z)$, there is a unique element of $W_{n-1}(\gamma, s; z)$ containing $T(V^+)$. But since also $T(V^+) \cap T(V) \neq \emptyset$ (in fact, $T(V) \subseteq T(V^+)$), it follows from (27.23) that $T(V^+) \subseteq T(V)$. Thus,

$$T(V) = T(V^+).$$

Therefore,

$$V \subseteq V^+ \subseteq T^{-1}(T(V^+)) = T^{-1}(T(V)). \tag{27.25}$$

But as $V \in Z_n(\gamma, s; z)$, it ensues from (27.23) and the definition of $Z_n(\gamma, s; z)$ that V is a connected component of $T^{-1}(T(V))$. Since V^+ is connected, it follows from (27.25) that $V = V^+$. This and (27.24) yield that $V \in W_n(\gamma, s; z)$. This means that (27.22) is proved, and so is (d_n) simultaneously.

Now, aiming to prove (27.9) in (c_n), we shall first show by induction that

$$\#W_n(\gamma, s; z) = \#W'_n(\gamma, s; z), \quad \forall n \geq 0. \tag{27.26}$$

For $n = 0$, this is true since $W_0(\gamma, s; z) = \{B(z, r_z)\}$ and $W'_0(\gamma, s; z) = \{B(z, 2r_z)\}$. So, suppose that (27.26) holds for some $n \geq 0$. Since each element of $W'_{n+1}(\gamma, s; z)$ contains at least one element of $W_{n+1}(\gamma, s; z)$ and since each element of $W_{n+1}(\gamma, s; z)$ is contained in exactly one element of $W'_{n+1}(\gamma, s; z)$, it suffices to show that each element V' of $W'_{n+1}(\gamma, s; z)$ contains exactly one element of $W_{n+1}(\gamma, s; z)$. For this purpose, let

$$W_{n+1}(V') := \{U \in W_{n+1}(\gamma, s; z) \mid U \subseteq V'\}.$$

Then $T(U) \in W_n(\gamma, s; z)$ and $T(U) \subseteq T(V') \in W'_n(\gamma, s; z)$. By the inductive hypothesis, all the images $T(U)$, $U \in W_{n+1}(V')$, are then equal. Denote them by V. Then the degrees $\deg(T|_{V'} : V' \to T(V'))$ and $\deg(T|_U : U \to V)$, $U \in W_{n+1}(V')$, are well-defined by virtue of Theorem 22.1.3. Thus,

$$\sum_{U \in W_{n+1}(V')} \deg(T|_U) = \deg(T|_{V'}). \tag{27.27}$$

If $n \geq s$, then we deduce from (27.27) and (f'_{n+1}) (or (g'_{n+1})) that $\#W_{n+1}(V') = 1$. So, let's consider the case $0 \leq n < s$. By (e'_{n+1}), (e'_n), (e_{n+1}) and (e_n), respectively, the sets V', $T(V')$, U and V are simply connected for every $U \in W_{n+1}(V')$. It follows from Corollary 22.5.3 that

$$\sum_{c \in V' \cap \mathrm{Crit}(T)} [\deg_c(T) - 1] = \deg(T|_{V'}) - 1. \tag{27.28}$$

Similarly, it ensues from Corollary 22.5.3 and (27.27) that

$$\sum_{U \in W_{n+1}(V')} \sum_{c \in U \cap \mathrm{Crit}(T)} [\deg_c(T) - 1] = \sum_{U \in W_{n+1}(V')} [\deg(T|_U) - 1] = \deg(T|_{V'}) - \#W_{n+1}(V'). \tag{27.29}$$

But, because of (27.11), we have the disjoint union $\bigcup_{U \in W_{n+1}(V')} U \cap \mathrm{Crit}(T) = V' \cap \mathrm{Crit}(T)$ and thus

$$\sum_{U \in W_{n+1}(V')} \sum_{c \in U \cap \mathrm{Crit}(T)} [\deg_c(T) - 1] = \sum_{c \in V' \cap \mathrm{Crit}(T)} [\deg_c(T) - 1]. \tag{27.30}$$

We infer from (27.28)–(27.30) that

$$\#W_{n+1}(V') = 1.$$

The proof of (27.26) is complete. By virtue of (27.26), we get the following.

Claim 1. *For every $n \geq 0$, each element of $W_n(\gamma, s; z)$ is contained in exactly one element of $W'_n(\gamma, s; z)$ and each element of $W'_n(\gamma, s; z)$ contains exactly one element of $W_n(\gamma, s; z)$.*

We are now ready to prove (27.9). As each element of $Z'_n(\gamma, s; z)$ contains at most d elements of $Z_n(\gamma, s; z)$, we conclude by means of Claim 1 and (27.13) that

$$\#(Z_n(\gamma, s; z) \setminus W_n(\gamma, s; z)) \leq d\,\#(Z'_n(\gamma, s; z) \setminus W'_n(\gamma, s; z)) \leq d\gamma^{-n} + d(2d - 2),$$

meaning that (27.9) holds. Since (27.8) was already established, we know that (c_n) is satisfied.

We are thus left to show that (b_n)–(b_n^*) are valid. For every $n \geq 0$ and $V \in W_n(\gamma, s; z)$, denote by V^+ the unique element of $W_n'(\gamma, s; z)$ such that V is a connected component of $V^+ \cap T^{-n}(B(z, r_z))$. So V^+ is a connected component of $T^{-n}(B(z, 2r_z))$. Since $r_z < 2r_z$, we have

$$\overline{V} \subseteq V^+. \tag{27.31}$$

Recall that $\#W_s(\gamma, s; z) < \infty$. Using (27.31) and (e_s'), we gather from Koebe's distortion theorem (Theorem 23.1.9) that there exists a constant $K \geq 1$ such that

$$K^{-1} \leq \frac{|H'(x)|}{|H'(y)|} \leq K \tag{27.32}$$

for all $W \in W_s(\gamma, s; z)$, all points $x, y \in \overline{W}$ and all univalent maps $H : W^+ \to \widehat{\mathbb{C}}$. For any connected open set $G \subseteq \widehat{\mathbb{C}}$ and any points $a, b \in G$, denote by $\ell_G(a, b)$ the infimum of the lengths of all rectifiable curves joining a and b in G. Denote also

$$\ell(G) := \sup\{\ell_G(a, b) \mid a, b \in G\} \in [0, \infty].$$

As the boundary of a set $W \in W_s(\gamma, s; z)$ is a real-analytic Jordan curve, we know that $\ell(W) < \infty$. Then

$$D := \max\{\ell(W) \mid W \in W_s(\gamma, s; z)\} < \infty. \tag{27.33}$$

Continuing the proof of (b_n), we next note that we can make this property hold for all $0 \leq n \leq s$ by taking $C_z \geq 1$ large enough. So, suppose that $n > s$ and $V \in W_n(\gamma, s; z)$. As the analytic map $T^{n-s} : V^+ \to T^{n-s}(V^+) \in W_s'(\gamma, s; z)$ is injective by (g_k), $k = n, n-1, \ldots, s+1$, we get from (b_n') that

$$\gamma^n \geq A_\rho(V^+) \geq A_\rho(V) \geq \|(T^{n-s}|_V)'\|_\infty^{-2} A_\rho(T^{n-s}(V)) \geq A\|(T^{n-s}|_V)'\|_\infty^{-2},$$

where

$$A := \min\{A_\rho(W) \mid W \in W_s(\gamma, s; z)\} > 0.$$

Equivalently,

$$\|(T^{n-s}|_V)'\|_\infty^{-1} \leq A^{-1/2}\gamma^{n/2}. \tag{27.34}$$

Using (27.33), and (27.32) applied to the univalent map $T^{n-s} : V^+ \to T(V^+) \in W_s'(\gamma, s; z)$, we obtain that

$$\operatorname{diam}(V) \leq \ell(V) \leq K\|(T^{n-s}|_V)'\|_\infty^{-1}\ell(T^{n-s}(V)) \leq KD\|(T^{n-s}|_V)'\|_\infty^{-1} \leq KDA^{-1/2}\gamma^{n/2}.$$

This proves item (b_n), while combining (27.34) with (27.32) proves (b_n^*). The proof of Lemma 27.1.1 is complete. □

Since all $V \in Z_n(\gamma, s; z)$, $n \geq 0$, have real-analytic Jordan curves for boundary and since the union $\bigcup_{n=0}^{s} Z_n(\gamma, s; z)$ is a finite set, there exists a constant $L(z) = L(\gamma, s; z) \geq 1$ such that

$$\ell_V(x,y) \leq L(z)|x-y| \tag{27.35}$$

for all $0 \leq n \leq s$, all $V \in Z_n(\gamma, s; z)$ and all $x, y \in V$, where $\ell_V(x,y)$ is the infimum of the lengths of all rectifiable curves joining x to y within V.

Recall that $(H_\alpha(\mathcal{J}(T)), \|\cdot\|_\alpha)$ denotes the Banach space of all Hölder continuous functions with exponent α defined on $\mathcal{J}(T)$ and equipped with the norm $\|\varphi\|_\alpha := v_\alpha(\varphi) + \|\varphi\|_\infty$, where $v_\alpha(\varphi)$ stands for the least constant for which φ is Hölder with exponent α while $(C(\mathcal{J}(T)), \|\cdot\|_\infty)$ is the Banach space of all continuous functions on $\mathcal{J}(T)$ endowed with the usual supremum norm (see Section 13.1 and Exercise 13.11.2).

Lemma 27.1.1 will now start bearing fruits. The first one is the following technical fact about the ergodic sums of Hölder continuous potentials, which will play a crucial role in subsequent considerations.

Lemma 27.1.2. *In addition to Lemma 27.1.1's hypotheses, let $\varphi : \mathcal{J}(T) \to \mathbb{R}$ be a Hölder continuous potential. For every $z \in \mathcal{J}(T)$ there is a constant $\Gamma_z(\varphi) \geq 1$ such that*

$$|S_n\varphi(y) - S_n\varphi(x)| \leq \Gamma_z(\varphi) \tag{27.36}$$

for every $n \in \mathbb{N}$, every $V \in W_n(\gamma, s; z)$ and all $x, y \in V$.

Furthermore, for every $\varepsilon > 0$ there exists $\delta_z(\varphi) > 0$ such that

$$|S_n\varphi(y) - S_n\varphi(x)| < \varepsilon, \tag{27.37}$$

for every $n \in \mathbb{N}$, every $V \in W_n(\gamma, s; z)$ and all $x, y \in V$ for which $|T^n(y) - T^n(x)| \leq \delta_z(\varphi)$.

Proof. Fix $n \in \mathbb{N}$, $V \in W_n(\gamma, s; z)$ and $x, y \in V$. Recall that $V \in W_n(\gamma, s; z)$ implies that $T^j(V)$ is contained in an element of $W_{n-j}(\gamma, s; z)$, for every $0 \leq j \leq n$. We first prove (27.36). Use of Lemma 27.1.1(b_n) and of the Hölder continuity of φ (say, with exponent α and Hölder constant $v_\alpha(\varphi)$) yields that

$$\begin{aligned}
|S_n\varphi(y) - S_n\varphi(x)| &= \left|\sum_{j=0}^{n-1}(\varphi \circ T^j(y) - \varphi \circ T^j(x))\right| \leq \sum_{j=0}^{n-1}|\varphi(T^j(y)) - \varphi(T^j(x))| \\
&\leq v_\alpha(\varphi)\sum_{j=0}^{n-1}|T^j(y) - T^j(x)|^\alpha \leq v_\alpha(\varphi)\sum_{j=0}^{n-1}\operatorname{diam}^\alpha(T^j(V)) \\
&\leq v_\alpha(\varphi)C_z^\alpha\sum_{j=0}^{n-1}\gamma^{(n-j)\alpha/2} \leq \Gamma_z(\varphi),
\end{aligned}$$

where

$$\Gamma_z(\varphi) = v_\alpha(\varphi)C_z^\alpha(1-\gamma^{\alpha/2})^{-1}. \tag{27.38}$$

This establishes (27.36).

The proof of (27.37) is somewhat similar but subtler. Set

$$\Sigma_T^\alpha := v_\alpha(\varphi) \sum_{j=\max\{0,n-s\}}^{n-1} |T^j(y) - T^j(x)|^\alpha.$$

Using Lemma 27.1.1(b_n^*), the Hölder continuity of φ and (27.35), we get (with the convention that if $n-s-1<0$, then any summation from 0 to $n-s-1$ is null):

$$\begin{aligned}
|S_n\varphi(y) - S_n\varphi(x)| &\le v_\alpha(\varphi)\sum_{j=0}^{n-s-1} |T^j(y)-T^j(x)|^\alpha + \Sigma_T^\alpha \\
&\le v_\alpha(\varphi)\sum_{j=0}^{n-s-1} \ell^\alpha_{T^j(V)}(T^j(y),T^j(x)) + \Sigma_T^\alpha \\
&\le v_\alpha(\varphi)\sum_{j=0}^{n-s-1} \Big[\inf_{\xi\in T^j(V)} |(T^{n-s-j})'(\xi)|\Big]^{-\alpha} \ell^\alpha_{T^{n-s}(V)}(T^{n-s}(y),T^{n-s}(x)) + \Sigma_T^\alpha \\
&\le v_\alpha(\varphi)\ell^\alpha_{T^{n-s}(V)}(T^{n-s}(y),T^{n-s}(x)) \sum_{j=0}^{n-s-1} C_z^{\alpha/2}\gamma^{(n-j)\alpha/2} + \Sigma_T^\alpha \\
&\le v_\alpha(\varphi)C_z^{\alpha/2}(1-\gamma^{\alpha/2})^{-1}\ell^\alpha_{T^{n-s}(V)}(T^{n-s}(y),T^{n-s}(x)) + \Sigma_T^\alpha \\
&\le v_\alpha(\varphi)C_z^{\alpha/2}(1-\gamma^{\alpha/2})^{-1}(L(z))^\alpha|T^{n-s}(y)-T^{n-s}(x)|^\alpha + \Sigma_T^\alpha.
\end{aligned}$$

But Lemma 25.1.7 affirms that

$$\lim_{r\to 0}\max\{\operatorname{diam}(V) : V \text{ connected component of } T^{-k}(B(z,r)) \text{ for some } 0\le k\le s\} = 0.$$

So, recalling that $V \in W_n(\gamma, s; z)$ implies that $T^j(V)$ is an element of $W_{n-j}(\gamma, s; z)$, for any $0 \le j \le n$, there exists $\delta_z(\varphi) > 0$ such that

$$v_\alpha(\varphi)C_z^{\alpha/2}(1-\gamma^{\alpha/2})^{-1}(L(z))^\alpha|T^{n-s}(y)-T^{n-s}(x)|^\alpha < \varepsilon/2$$

and

$$\Sigma_T^\alpha := v_\alpha(\varphi) \sum_{j=\max\{0,n-s\}}^{n-1} |T^j(y) - T^j(x)|^\alpha < \varepsilon/2$$

whenever $|T^n(y) - T^n(x)| \le \delta_z(\varphi)$. Then we conclude that

$$|S_n\varphi(y) - S_n\varphi(x)| < \varepsilon/2 + \varepsilon/2 = \varepsilon,$$

and (27.37) is proved. □

As an immediate consequence of this lemma, we get the following.

Corollary 27.1.3. *Under Lemma 27.1.2's hypotheses and notation, we have that*

$$e^{-\Gamma_z(\varphi)} \le \frac{\exp(S_n\varphi(y))}{\exp(S_n\varphi(x))} \le e^{\Gamma_z(\varphi)} \tag{27.39}$$

for every $n \in \mathbb{N}$, *every* $V \in W_n(\gamma, s; z)$ *and all* $x, y \in V$.

In Section 27.6, we will need a generalization of the previous two results. Let $\varphi : \mathcal{J}(T) \to \mathbb{R}$ be a Hölder continuous potential. Given two numbers $\eta, H \ge 0$, write

$$D_{\eta,H}(\varphi) := \{\psi \in H_\alpha(\mathcal{J}(T)) : \|\psi - \varphi\|_\infty \le \eta \text{ and } v_\alpha(\psi) \le H\}. \tag{27.40}$$

Observe that $D_{\eta,H}(\varphi)$ is a compact subset of the Banach space $(C(\mathcal{J}(T)), \|\cdot\|_\infty)$. Unless otherwise stated, we consider $D_{\eta,H}(\varphi)$ to be endowed with the topology inherited from $(C(\mathcal{J}(T)), \|\cdot\|_\infty)$.

Lemma 27.1.4. *In addition to Lemma 27.1.1's hypotheses, let* $\varphi : \mathcal{J}(T) \to \mathbb{R}$ *be a Hölder continuous potential and* $\eta, H \ge 0$. *Then for every* $z \in \mathcal{J}(T)$ *there exists a constant* $\Gamma_z(\varphi, \eta, H) \ge 1$ *such that*

$$|S_n\psi(y) - S_n\psi(x)| \le \Gamma_z(\varphi, \eta, H) \tag{27.41}$$

for all $\psi \in D_{\eta,H}(\varphi)$, *every* $n \in \mathbb{N}$, *every* $V \in W_n(\gamma, s; z)$ *and all* $x, y \in V$.

Furthermore, for every $\varepsilon > 0$ *there exists* $\delta_z(\varphi, \eta, H) > 0$ *such that*

$$|S_n\psi(y) - S_n\psi(x)| < \varepsilon \tag{27.42}$$

for all $\psi \in D_{\eta,H}(\varphi)$, *every* $n \in \mathbb{N}$, *every* $V \in W_n(\gamma, s; z)$ *and all* $x, y \in V$ *for which* $|T^n(y) - T^n(x)| \le \delta_z(\varphi, \eta, H)$.

Proof. The key is to observe in (27.38) that the sole dependence on φ of the constant $\Gamma_z(\varphi)$ is the term $v_\alpha(\varphi)$. But like the norms $\|\varphi\|_\alpha$ and $\|\varphi\|_\infty$, the term $v_\alpha(\varphi)$ depends continuously on φ. Given the compactness of $D_{\eta,H}(\varphi)$, this implies that $\Gamma_z(\varphi, \eta, H) := \max_{\psi \in D_{\eta,H}(\varphi)} \Gamma_z(\psi) < \infty$. A similar argument holds to secure $\delta_z(\varphi, \eta, H) > 0$. □

Corollary 27.1.5. *Under Lemma 27.1.4's hypotheses and notation, we have that*

$$e^{-\Gamma_z(\varphi,\eta,H)} \le \frac{\exp(S_n\varphi(y))}{\exp(S_n\varphi(x))} \le e^{\Gamma_z(\varphi,\eta,H)} \tag{27.43}$$

for all $\psi \in D_{\eta,H}(\varphi)$, *every* $n \in \mathbb{N}$, *every* $V \in W_n(\gamma, s; z)$ *and all* $x, y \in V$.

27.2 The transfer operator $\mathcal{L}_\varphi : C(\mathcal{J}(T)) \longrightarrow C(\mathcal{J}(T))$: its lower and upper bounds

Let $T : \widehat{\mathbb{C}} \to \widehat{\mathbb{C}}$ be a rational function with $\deg(T) \geq 2$ and $\varphi : \mathcal{J}(T) \to \mathbb{R}$ be a continuous potential. As in Chapter 13, we define a transfer operator $\mathcal{L}_\varphi : C(\mathcal{J}(T)) \to C(\mathcal{J}(T))$ associated to the potential φ by the formula

$$\mathcal{L}_\varphi(g)(z) := \sum_{w \in T^{-1}(z)} g(w) e^{\varphi(w)} \deg_w(T), \quad \forall g \in C(\mathcal{J}(T)), \forall z \in \mathcal{J}(T). \tag{27.44}$$

It is immediate that $\mathcal{L}_\varphi$ is a positive bounded linear operator on $C(\mathcal{J}(T))$. Iterating (27.44) yields for every $n \in \mathbb{N}$ that

$$\begin{aligned} \mathcal{L}_\varphi^n(g)(z) &= \sum_{w \in T^{-n}(z)} g(w) \exp(S_n\varphi(w)) \prod_{j=0}^{n-1} \deg_{T^j(w)}(T) \\ &= \sum_{w \in T^{-n}(z)} g(w) \exp(S_n\varphi(w)) \deg_w(T^n). \end{aligned} \tag{27.45}$$

Furthermore, observe that

$$\|\mathcal{L}_\varphi^n\|_\infty = \|\mathcal{L}_\varphi^n(\mathbb{1})\|_\infty < \infty, \quad \forall n \in \mathbb{N}. \tag{27.46}$$

In this section, we ultimately aim at finding lower and upper bounds on the iterates $\mathcal{L}_\varphi^n(\mathbb{1})$, $n \in \mathbb{N}$. These bounds will depend on n exponentially. This will be accomplished by introducing the normalization $\widehat{\mathcal{L}}_\varphi$ of $\mathcal{L}_\varphi$ and by bounding the iterates $\widehat{\mathcal{L}}_\varphi^n(\mathbb{1})$ by two positive constants. But this requires several preliminary results.

27.2.1 First lower bounds on $\mathcal{L}_\varphi$

Lemma 27.2.1. *If $T : \widehat{\mathbb{C}} \to \widehat{\mathbb{C}}$ is a rational function with $d := \deg(T) \geq 2$, then for all $\delta > 0$, all $\varepsilon > 0$, and all $n = n(\delta, \varepsilon) \in \mathbb{N}$ large enough, there exists a finite open cover $\mathcal{V}_n(\delta, \varepsilon)$ of $\mathcal{J}(T)$ consisting of open balls such that:*
(a) *$\#\mathcal{V}_n(\delta, \varepsilon) \leq e^{\delta n}$, and*
(b) *$\operatorname{diam}(T^i(U)) < \varepsilon$ for every $V \in \mathcal{V}_n(\delta, \varepsilon)$, every connected component U of $T^{-n}(V)$, and all $0 \leq i \leq n$.*

Proof. Recall that $\operatorname{Critv}(T^n) := T^n(\operatorname{Crit}(T^n)) = \bigcup_{k=1}^n T^k(\operatorname{Crit}(T))$ is the set of all critical values of T^n and satisfies

$$\#\operatorname{Critv}(T^n) \leq \#\operatorname{Crit}(T) \cdot n \leq (2d-2)n. \tag{27.47}$$

Set $\delta \in (0, 1)$ and

$$\chi_n := \exp(-e^{n\delta/4}), \quad \forall n \in \mathbb{N}.$$

Let $C \geq 3$ be a constant whose value will be determined later in the course of the proof. In particular, it will depend on ε.

Claim 1. *There exists a finite set $Q \subseteq \mathcal{J}(T)$ such that for every $n \in \mathbb{N}$ large enough there are a finite set $F_n \subseteq \widehat{\mathbb{C}}$ and radii $s_z > 0$, $z \in F_n$, such that*

$$\#F_n \leq e^{n\delta/2}, \tag{27.48}$$

$$\bigcup_{z \in F_n} B(z, s_z) \supseteq \mathcal{J}(T) \setminus B(\mathrm{Critv}(T^n) \cup Q, \chi_n), \tag{27.49}$$

$$\mathrm{Critv}(T^n) \cap \bigcup_{z \in F_n} B(z, Cs_z) = \emptyset, \tag{27.50}$$

and

$$\mathcal{J}(T) \cap B(z, s_z) \neq \emptyset, \quad \forall z \in F_n. \tag{27.51}$$

Proof of Claim 1. Let $R' = \frac{1}{4}\mathrm{diam}(\widehat{\mathbb{C}})$ and $Q \subseteq \mathcal{J}(T)$ be a finite set such that

$$B(Q, R') \supseteq \mathcal{J}(T)$$

and $0 < \delta := \sup\{\mathrm{dist}(w, Q) : w \in \widehat{\mathbb{C}}\} < \delta_T$ so small that (thanks to Lemma 24.1.13)

$$\sup_{k \in \mathbb{N}} \sup\{\mathrm{diam}_\rho(T_w^{-k}(B_\rho(z, \delta))) : z \in F_k(\delta),\ w \in T_*^{-k}(z)\} \leq \pi/2.$$

Fix $n \in \mathbb{N}$. Apply Lemma 25.4.1 with $\varepsilon = \chi_n$, $\beta = C^{-1}$, and

$$Z = Z_n := \mathrm{Critv}(T^n) \cup Q. \tag{27.52}$$

This produces a finite set $F_n' \subseteq B(Z_n, R')$ such that

$$\bigcup_{z \in F_n'} B(z, C^{-1}\mathrm{dist}(z, Z_n)) \supseteq B(Z_n, R') \setminus B(Z_n, \chi_n) \supseteq B(Q, R') \setminus B(Z_n, \chi_n) \supseteq \mathcal{J}(T) \setminus B(Z_n, \chi_n) \tag{27.53}$$

and

$$\#F_n' \leq A(C^{-1}) \cdot (\#Z_n)^2 \log(R'/\chi_n) = A(C^{-1}) \cdot (\#Z_n)^2 (\log R' + e^{n\delta/4}) \leq e^{n\delta/2},$$

where, thanks to (27.47), the last inequality holds for all $n \in \mathbb{N}$ large enough. Let

$$F_n := \{z \in F_n' \mid \mathcal{J}(T) \cap B(z, C^{-1}\mathrm{dist}(z, Z_n)) \neq \emptyset\}.$$

By virtue of (27.52)–(27.53), we have that

$$\bigcup_{z \in F_n} B(z, C^{-1}\mathrm{dist}(z, Z_n)) \supseteq \mathcal{J}(T) \setminus B(Z_n, \chi_n).$$

Setting $s_z := C^{-1}\mathrm{dist}(z, Z_n)$ for every $z \in F_n$, Claim 1 ensues. ∎

Now define

$$\mathcal{V}_n(\delta,\varepsilon) := \{B(w,\chi_n) \mid w \in \mathrm{Critv}(T^n) \cup Q\} \cup \{B(z,s_z) \mid z \in F_n\}. \tag{27.54}$$

By (27.49), observe that $\mathcal{V}_n(\delta,\varepsilon)$ is a finite open cover of $\mathcal{J}(T)$. By (27.51) and the definition of Q, we also know that all atoms of this cover intersect the Julia set.

Let us show that item (a) holds. It follows from (27.47)–(27.48) that

$$\#\mathcal{V}_n(\delta,\varepsilon) \le (2d-2)n + \#Q + e^{n\delta/2} \le e^{\delta n}$$

for all $n \in \mathbb{N}$ large enough, which confirms (a)'s validity.

Moving on to (b), let $n \in \mathbb{N}$ be so large as required thus far in this proof. First, we will prove (b) for those $V \in \mathcal{V}_n(\delta,\varepsilon)$ of the form $V = B(w,\chi_n)$, where $w \in \mathrm{Critv}(T^n) \cup Q$. Take any connected component U of $T^{-n}(B(w,\chi_n))$. Pick $0 < \kappa < \delta/[16\log(2d_T)]$, where d_T originates from Lemma 25.1.7. We show by induction that

$$\mathrm{diam}(T^{n-i}(U)) \le (2\chi_n)^{(2d_T)^{-\kappa i}} < \frac{1}{2}\min\{r,\varepsilon\} \tag{27.55}$$

for every $i = 0,1,\ldots,n$, where $r \in (0,1/2]$ emerges from Lemma 25.1.7.

Indeed, if $i = 0$ then

$$\mathrm{diam}(T^{n-i}(U)) = \mathrm{diam}(B(w,\chi_n)) = 2\chi_n = (2\chi_n)^{(2d_T)^{-\kappa 0}} < \frac{1}{2}\min\{r,\varepsilon\}$$

provided that $n \in \mathbb{N}$ is large enough. For the inductive step, suppose that (27.55) holds for all $0 \le i \le j$ for some $j \le n-1$. Then

$$T^{n-j}(U) = T(T^{n-(j+1)}(U)).$$

Choose $n \ge \kappa^{-1}$. It follows from Lemma 25.1.7 and the inductive hypothesis that

$$\begin{aligned}
\mathrm{diam}(T^{n-(j+1)}(U)) &\le [\mathrm{diam}(T^{n-j}(U))]^{(2d_T)^{-1}} \\
&\le [(2\chi_n)^{(2d_T)^{-\kappa j}}]^{(2d_T)^{-1}} \\
&= (2\chi_n)^{(2d_T)^{-(\kappa j+1)}} \\
&= [2\exp(-e^{n\delta/4})]^{(2d_T)^{-(\kappa j+1)}} \\
&= 2^{(2d_T)^{-(\kappa j+1)}} \exp(-e^{n\delta/4}\cdot(2d_T)^{-(\kappa j+1)}) \\
&\le 2\exp\Big(-\exp(n\delta/4 - (\kappa j+1)\log(2d_T))\Big) \\
&\le 2\exp\Big(-\exp(n[\delta/4 - 2\kappa\log(2d_T)])\Big) \\
&\le 2\exp(-\exp(n\delta/8)) \\
&< \frac{1}{2}\min\{r,\varepsilon\},
\end{aligned}$$

where the last inequality holds as long as $n \in \mathbb{N}$ is large enough depending on δ, r and ε. The inductive proof of (27.55) is complete and item (b) holds for all $V = B(w, \chi_n)$.

Now, let's prove (b) for those $V \in \mathcal{V}_n(\delta, \varepsilon)$ of the form $V = B(z, s_z)$, where $z \in F_n$. Take any connected component U of $T^{-n}(B(z, s_z))$. Let $\widehat{U}$ be the unique connected component of $T^{-n}(B(z, Cs_z))$ containing U. By the definition of s_z (more precisely, (27.50)) and by Corollary 22.5.4, the map

$$T^n|_{\widehat{U}} : \widehat{U} \to B(z, Cs_z)$$

is univalent. So, for any $0 \le i \le n$ the map

$$T^{n-i} : T^i(\widehat{U}) \to B(z, Cs_z)$$

is also univalent. Denote its inverse by $T_{\widehat{U}}^{-(n-i)} : B(z, Cs_z) \to T^i(\widehat{U})$. This map is univalent, too. Since $C \ge 3$, we have

$$B(z, s_z) \subseteq B(z, Cs_z/2) \subseteq B(z, Cs_z).$$

Hence, by virtue of Koebe's distortion theorem (Theorem 23.1.9),

$$\frac{|(T_{\widehat{U}}^{-(n-i)})'(y)|}{|(T_{\widehat{U}}^{-(n-i)})'(x)|} \le K \tag{27.56}$$

for all $x, y \in B(z, Cs_z/2)$, where $K = k_{\delta,\pi/2}(1/2) \ge 1$ is the Koebe constant (alternatively, one can take $K = k_{Cs_z,\pi/2}(1/2)$). Thus,

$$\begin{aligned} \operatorname{diam}(T_{\widehat{U}}^{-(n-i)}(B(z, s_z))) &\le \operatorname{diam}\Big(B(T_{\widehat{U}}^{-(n-i)}(z), K|(T_{\widehat{U}}^{-(n-i)})'(z)|s_z)\Big) \\ &= 2K|(T_{\widehat{U}}^{-(n-i)})'(z)|s_z. \end{aligned} \tag{27.57}$$

On the other hand, by Koebe's $\frac{1}{4}$-theorem (Theorem 23.1.3),

$$T_{\widehat{U}}^{-(n-i)}(B(z, Cs_z/2)) \supseteq B(T_{\widehat{U}}^{-(n-i)}(z), C|(T_{\widehat{U}}^{-(n-i)})'(z)|s_z/8).$$

But since $B(T_{\widehat{U}}^{-(n-i)}(z), C|(T_{\widehat{U}}^{-(n-i)})'(z)|s_z/8) \subseteq \widehat{\mathbb{C}}$, we have that

$$\frac{1}{4}C|(T_{\widehat{U}}^{-(n-i)})'(z)|s_z \le \operatorname{diam}(\widehat{\mathbb{C}}). \tag{27.58}$$

By (27.57)–(27.58),

$$\operatorname{diam}(T^i(U)) = \operatorname{diam}(T_{\widehat{U}}^{-(n-i)}(B(z, s_z))) \le 8KC^{-1}\operatorname{diam}(\widehat{\mathbb{C}}) < \varepsilon,$$

provided that $C > 8K\varepsilon^{-1}\operatorname{diam}(\widehat{\mathbb{C}})$. This proves that (b) holds for any $V = B(z, s_z)$. In conjunction with (27.55), we have demonstrated that (b) is true for all $V \in V_n(\delta, \varepsilon)$ (cf. (27.54)) provided that n is large enough. The proof of Lemma 27.2.1 is complete. □

We can now provide a lower bound for the values of $\mathcal{L}_\varphi^n(\mathbb{1})$ at a sequence of points in the Julia set.

Lemma 27.2.2. *Let $T : \widehat{\mathbb{C}} \to \widehat{\mathbb{C}}$ be a rational function with $\deg(T) \geq 2$ and $\varphi : \mathcal{J}(T) \to \mathbb{R}$ a continuous potential. For every $\delta > 0$, there exists for each $n \in \mathbb{N}$ large enough a point $x_n \in \mathcal{J}(T)$ for which*

$$\mathcal{L}_\varphi^n(\mathbb{1})(x_n) \geq \exp(n(\mathrm{P}(\varphi) - \delta)).$$

Proof. By Theorem 11.2.1 (Bowen's pressure) and the uniform continuity of φ, there exists $\varepsilon > 0$ such that for every $n \in \mathbb{N}$ large enough there is a maximal $(n, 2\varepsilon)$-separated set $E_n(2\varepsilon) \subseteq \mathcal{J}(T)$ for which

$$\sum_{y \in E_n(2\varepsilon)} \exp(S_n\varphi(y)) \geq \exp(n(\mathrm{P}(\varphi) - \delta/3)) \tag{27.59}$$

and

$$w, z \in \mathcal{J}(T),\ |w - z| \leq \varepsilon \quad \Longrightarrow \quad |\varphi(w) - \varphi(z)| < \delta/3. \tag{27.60}$$

For every $n \in \mathbb{N}$ so large as required in Lemma 27.2.1, let $\mathcal{V}_n(\delta/3, \varepsilon)$ be the finite open cover of $\mathcal{J}(T)$ constructed in that lemma. By (27.59) and Lemma 27.2.1(a), there exists $V \in \mathcal{V}_n(\delta/3, \varepsilon)$ such that

$$\begin{aligned}\sum_{y \in E_n(2\varepsilon) \cap T^{-n}(V)} \exp(S_n\varphi(y)) &\geq \frac{1}{\#\mathcal{V}_n(\delta/3, \varepsilon)} \sum_{y \in E_n(2\varepsilon)} \exp(S_n\varphi(y)) \\ &\geq \exp(n(\mathrm{P}(\varphi) - 2\delta/3)).\end{aligned} \tag{27.61}$$

Let x_n be an arbitrary point in $V \cap \mathcal{J}(T)$. For every $y \in E_n(2\varepsilon) \cap T^{-n}(V)$, denote by $\widehat{y}$ an arbitrary point of the set $T^{-n}(x_n)$ belonging to the same connected component of $T^{-n}(V)$ as y. By Lemma 27.2.1(b), we have

$$|T^i(y) - T^i(\widehat{y})| \leq \varepsilon, \quad \forall i = 0, \ldots, n.$$

Hence, the function

$$E_n(2\varepsilon) \cap T^{-n}(V) \ni y \longmapsto \widehat{y} \in T^{-n}(x_n) \tag{27.62}$$

is 1-to-1. With the use of (27.60), we also know that

$$S_n\varphi(\widehat{y}) - S_n\varphi(y) > -n\delta/3.$$

Using this, (27.62) and (27.61), we get

$$\begin{aligned}\mathcal{L}_\varphi^n(\mathbb{1})(x_n) &\geq \sum_{y\in E_n(2\varepsilon)\cap T^{-n}(V)} \exp(S_n\varphi(\hat{y})) \\ &\geq \sum_{y\in E_n(2\varepsilon)\cap T^{-n}(V)} \exp(S_n\varphi(y))\exp(-n\delta/3) \\ &\geq \exp(n(\mathrm{P}(\varphi)-\delta)).\end{aligned}$$

□

27.2.2 Auxiliary operators L_β

Throughout this section, assume that

$$\beta > \sup\varphi. \tag{27.63}$$

Take any

$$\gamma := \gamma_\beta \in (e^{\sup\varphi-\beta}, 1). \tag{27.64}$$

Let d and $\Delta(T)$ be as defined in (27.1) and (27.5). Pick $s \in \mathbb{N}$ so large that

$$\gamma^{-s} \geq d(2d-2)\Delta(T) \tag{27.65}$$

and

$$2d(2d-2)\Delta(T)(1-e^{-\eta})^{-1}e^{-\eta(s+1)} < 1/2, \tag{27.66}$$

where

$$\eta := \log\gamma - \sup\varphi + \beta > 0. \tag{27.67}$$

For every $n \in \mathbb{N}$ and every $z \in \mathcal{J}(T)$, define the function $B_n(\beta, z, \cdot) : B(z, r_z) \to \mathbb{R}$ by

$$B_n(\beta, z, \xi) := e^{-\beta n} \sum_{V\in Z_n(\gamma,s;z)\setminus W_n(\gamma,s;z)} \sum_{w\in T^{-n}(\xi)\cap V} \exp(S_n\varphi(w)), \tag{27.68}$$

where $W_n(\gamma, s; z)$, $Z_n(\gamma, s; z)$ and r_z originate from Lemma 27.1.1. The symbol B in this definition alludes to the fact that the above sum is taken over "bad" inverse branches/components of iterates of T in the sense that there is no distortion control over them. Using (27.67) and items (c_n) and (h_n) of Lemma 27.1.1, we get for every $\xi \in B(z, r_z)$ that

$$\begin{aligned}B_n(\beta, z, \xi) &\leq e^{-\beta n} \cdot \exp(n\sup\varphi) \cdot \#(Z_n(\gamma, s; z) \setminus W_n(\gamma, s; z)) \cdot \Delta(T) \\ &\leq \exp(n(\sup\varphi - \beta)) \cdot 2d(2d-2)\gamma^{-n} \cdot \Delta(T) \\ &= \exp(n(\sup\varphi - \beta - \log\gamma)) \cdot 2d(2d-2)\Delta(T) \\ &= 2d(2d-2)\Delta(T)e^{-\eta n}.\end{aligned} \tag{27.69}$$

Similarly, define the function $G_n(\beta, z, \cdot) : B(z, r_z) \to \mathbb{R}$ by

$$G_n(\beta, z, \xi) := e^{-\beta n} \sum_{V \in W_n(\gamma, s; z)} \sum_{w \in T^{-n}(\xi) \cap V} \exp(S_n\varphi(w)). \tag{27.70}$$

The symbol G alludes to the fact that the above sum is taken over "good" inverse branches/components of iterates of T in the sense that there is good distortion control over them.

We now define a multiple of the transfer operator $\mathcal{L}_\varphi$ by

$$L_\beta := e^{-\beta}\mathcal{L}_\varphi : C(\mathcal{J}(T)) \to C(\mathcal{J}(T)).$$

For every $n > s$ and $\xi \in B(z, r_z)$, observe that

$$\begin{aligned} L_\beta^n(\mathbb{1})(\xi) = & \sum_{k=s+1}^{n} \sum_{V \in Z_k(\gamma, s; z) \setminus W_k(\gamma, s; z)} e^{-\beta k} \sum_{w \in T^{-k}(\xi) \cap V} \exp(S_k\varphi(w)) L_\beta^{n-k}(\mathbb{1})(w) \\ & + G_n(\beta, z, \xi). \end{aligned} \tag{27.71}$$

Indeed, recall that $L_\beta^n(\mathbb{1})(\xi)$ consists of a sum over all the preimages of ξ under T^n. These preimages can be split according to the first time k at which they belong to a "bad" branch (from time 0 at ξ and going backwards in time). The sum $G_n(\beta, z, \xi)$ accounts for all the preimages that belong only to "good" branches (i. e., that belong to a component in $W_n(\gamma, s; \xi)$ by Lemma 27.1.1(g_n)). The rest is the sum over all preimages that belong to "bad" branches. Each such preimage w belongs to a "good" branch until some maximal time $s \le k - 1 < n$ and belongs to a "bad" branch for the first time at time k (so belongs to $Z_k(\gamma, s; z) \setminus W_k(\gamma, s; z)$ by Lemma 27.1.1(c_k)); from that time on, i. e., for the remaining time $n - k$, one takes $L_\beta^{n-k}(\mathbb{1})(w)$ and one multiplies it by $e^{-\beta k} \exp(S_k\varphi(w))$ to reckon for the trajectory along "good" branches from time 0 until time $k - 1$.

Since $\mathcal{J}(T)$ is compact, there exists a finite set $Y_{\gamma,s} \subseteq \mathcal{J}(T)$ such that

$$\bigcup_{z \in Y_{\gamma,s}} B(z, r_z) \supseteq \mathcal{J}(T). \tag{27.72}$$

Under a uniform boundedness condition on the family of functions $\{G_n(\beta, z, \cdot)\}_{n=1}^{\infty}$, we now derive a uniform upper bound on the supremum norms of the $\{L_\beta^n(\mathbb{1})\}_{n=1}^{\infty}$.

Lemma 27.2.3. *Let $T : \widehat{\mathbb{C}} \to \widehat{\mathbb{C}}$ be a rational function with $d := \deg(T) \ge 2$ and let $\varphi : \mathcal{J}(T) \to \mathbb{R}$ be a Hölder continuous potential. If*

$$K_\beta := \sup\Big\{ \sup_{\xi \in B(z, r_z)} G_n(\beta, z, \xi) \,\Big|\, n \in \mathbb{N},\ z \in Y_{\gamma,s} \Big\} < \infty \tag{27.73}$$

and

$$M_\beta(\varphi) := \max\Big\{K_\beta, \max_{0\le j\le s}\|L_\beta^j(\mathbb{1})\|_\infty\Big\}, \tag{27.74}$$

then

$$\sup_{n\ge 0}\|L_\beta^n(\mathbb{1})\|_\infty \le N_\beta(\varphi) := M_\beta(\varphi)\big[1 - 2d(2d-2)\Delta(T)(1-e^{-\eta})^{-1}e^{-\eta(s+1)}\big]^{-1}. \tag{27.75}$$

Proof. The first crucial observation is that for every $n > s$, every $z \in Y_{\gamma,s}$ and every $\xi \in B(z, r_z)$, we obtain by means of (27.68)–(27.69) and (27.71) that

$$\begin{aligned} L_\beta^n(\mathbb{1})(\xi) &= \sum_{k=s+1}^{n} \sum_{V\in Z_k(\gamma,s;z)\setminus W_k(\gamma,s;z)} e^{-\beta k} \sum_{w\in T^{-k}(\xi)\cap V} \exp(S_k\varphi(w))L_\beta^{n-k}(\mathbb{1})(w) \\ &\quad + G_n(\beta, z, \xi) \\ &\le \sum_{k=s+1}^{n} \sum_{V\in Z_k(\gamma,s;z)\setminus W_k(\gamma,s;z)} e^{-\beta k} \sum_{w\in T^{-k}(\xi)\cap V} \exp(S_k\varphi(w))\|L_\beta^{n-k}(\mathbb{1})\|_\infty \\ &\quad + G_n(\beta, z, \xi) \\ &\le \sum_{k=s+1}^{n} B_k(\beta, z, \xi)\|L_\beta^{n-k}(\mathbb{1})\|_\infty + G_n(\beta, z, \xi) \\ &\le 2d(2d-2)\Delta(T)\sum_{k=s+1}^{n} e^{-\eta k}\|L_\beta^{n-k}(\mathbb{1})\|_\infty + G_n(\beta, z, \xi). \end{aligned} \tag{27.76}$$

Proceeding by induction, note that (27.75) holds for all $0 \le n \le s$ by the very definition of $N_\beta(\varphi)$. So, suppose that $n > s$ and that (27.75) holds for all integers from 0 to $n-1$. We can then continue (27.76) as follows:

$$\begin{aligned} L_\beta^n(\mathbb{1})(\xi) &\le 2d(2d-2)\Delta(T)N_\beta(\varphi)\sum_{k=s+1}^{n} e^{-\eta k} + M_\beta(\varphi) \\ &\le 2d(2d-2)\Delta(T)N_\beta(\varphi)(1-e^{-\eta})^{-1}e^{-\eta(s+1)} + M_\beta(\varphi) \\ &= [N_\beta(\varphi) - M_\beta(\varphi)] + M_\beta(\varphi) = N_\beta(\varphi). \end{aligned}$$

Invoking (27.72), we conclude that $\|L_\beta^n(\mathbb{1})\|_\infty \le N_\beta(\varphi)$. □

We now demonstrate that the uniform boundedness condition (27.73) on the family of functions $\{G_n(\beta, z, \cdot)\}_{n=1}^\infty$ is not fulfilled when $\sup\varphi < \beta < \mathrm{P}(\varphi)$. This happens whenever the potential φ exhibits a pressure gap (cf. Definition 13.7.18 and Remarks 27.9.1–27.9.3). According to Theorem 13.7.30, every Hölder continuous potential generates a pressure gap on a transitive, open, distance expanding system (on an infinite space). So, this property holds for all expanding rational functions.

Lemma 27.2.4. *Let $T : \widehat{\mathbb{C}} \to \widehat{\mathbb{C}}$ be a rational function with $\deg(T) \ge 2$ and $\varphi : \mathcal{J}(T) \to \mathbb{R}$ a Hölder continuous potential such that $\mathrm{P}(\varphi) > \sup\varphi$. If $\sup\varphi < \beta < \mathrm{P}(\varphi)$, then*

$$K_\beta := \sup\Big\{\sup_{\xi\in B(z,r_z)} G_n(\beta,z,\xi)\,\Big|\, n\in\mathbb{N},\ z\in Y_{\gamma,s}\Big\} = \infty. \tag{27.77}$$

Proof. Seeking a contradiction, suppose that $K_\beta < \infty$. It follows from Lemma 27.2.3 that $\|L_\beta^n(\mathbb{1})\|_\infty \le N_\beta(\varphi)$ for every $n \ge 0$. Equivalently,

$$\mathcal{L}_\varphi^n(\mathbb{1})(\xi) \le N_\beta(\varphi)e^{\beta n}, \quad \forall n\ge 0,\ \forall \xi\in\mathcal{J}(T). \tag{27.78}$$

Take $\delta > 0$ so small that $\mathrm{P}(\varphi) - \delta > \beta$. According to Lemma 27.2.2, there exists a constant $C > 0$ such that for every $n \in \mathbb{N}$ there is $x_n \in \mathcal{J}(T)$ for which

$$\mathcal{L}_\varphi^n(\mathbb{1})(x_n) \ge C\exp(n(\mathrm{P}(\varphi)-\delta)).$$

Combining this with (27.78), we deduce that

$$C\exp(n(\mathrm{P}(\varphi)-\delta)) \le N_\beta(\varphi)e^{\beta n}, \quad \forall n\in\mathbb{N}.$$

Equivalently,

$$\exp(n(\mathrm{P}(\varphi)-\delta-\beta)) \le C^{-1}N_\beta(\varphi), \quad \forall n\in\mathbb{N}.$$

However, this is impossible since $\mathrm{P}(\varphi) - \delta - \beta > 0$. □

27.2.3 Final bounds on $\mathcal{L}_\varphi$ and its normalization $\widehat{\mathcal{L}}_\varphi$; the eigenmeasure m_φ

Per the notation used throughout this book, let $M(\mathcal{J}(T))$ denote the topological space of all Borel probability measures on $\mathcal{J}(T)$ endowed with the weak* topology inherited from the topological vector space $C(\mathcal{J}(T),\mathbb{R})^*$ of all real-valued bounded linear functionals on $\mathcal{J}(T)$. We know that $C(\mathcal{J}(T),\mathbb{R})^*$ is locally convex with respect to the weak* topology and $M(\mathcal{J}(T))$ is a convex compact subset of $C(\mathcal{J}(T),\mathbb{R})^*$ (see Theorem A.1.24). Let $\mathcal{L}_\varphi^* : C(\mathcal{J}(T),\mathbb{R})^* \to C(\mathcal{J}(T),\mathbb{R})^*$ be the dual operator of $\mathcal{L}_\varphi$; it is defined by

$$\mathcal{L}_\varphi^*(m)(g) := m(\mathcal{L}_\varphi(g)), \quad \forall m\in C(\mathcal{J}(T),\mathbb{R})^*,\ \forall g\in C(\mathcal{J}(T),\mathbb{R}).$$

Since $\mathcal{L}_\varphi(\mathbb{1})$ is a positive continuous function on $\mathcal{J}(T)$, the operator

$$M(\mathcal{J}(T)) \ni m \longmapsto \frac{\mathcal{L}_\varphi^*(m)}{\mathcal{L}_\varphi^*(m)(\mathbb{1})} \in M(\mathcal{J}(T))$$

is positive, linear and continuous. This was already checked in another context (cf. Theorem 13.6.2). Therefore, the Schauder–Tychonov fixed-point theorem asserts that there is $m_\varphi \in M(\mathcal{J}(T))$ such that

$$\frac{\mathcal{L}_\varphi^*(m_\varphi)}{\mathcal{L}_\varphi^*(m_\varphi)(\mathbb{1})} = m_\varphi.$$

Writing

$$\lambda_\varphi := \mathcal{L}_\varphi^*(m_\varphi)(\mathbb{1}) > 0,$$

this means that

$$\mathcal{L}_\varphi^*(m_\varphi) = \lambda_\varphi m_\varphi. \tag{27.79}$$

The corresponding normalized operator

$$\widehat{\mathcal{L}}_\varphi := \lambda_\varphi^{-1}\mathcal{L}_\varphi : C(\mathcal{J}(T)) \to C(\mathcal{J}(T)) \tag{27.80}$$

has a dual $\widehat{\mathcal{L}}_\varphi^*$ which fixes m_φ:

$$\widehat{\mathcal{L}}_\varphi^*(m_\varphi) = m_\varphi. \tag{27.81}$$

Equivalently,

$$m_\varphi(\widehat{\mathcal{L}}_\varphi(g)) = m_\varphi(g), \quad \forall g \in C(\mathcal{J}(T)). \tag{27.82}$$

That is,

$$\int_{\mathcal{J}(T)} \widehat{\mathcal{L}}_\varphi(g)\, dm_\varphi = \int_{\mathcal{J}(T)} g\, dm_\varphi, \quad \forall g \in C(\mathcal{J}(T)).$$

Observe that $\widehat{\mathcal{L}}_\varphi = L_\beta$ when $\beta = \log \lambda_\varphi$.

Iterating (27.81)–(27.82) yields

$$\widehat{\mathcal{L}}_\varphi^{*n}(m_\varphi) = m_\varphi \quad \text{and} \quad m_\varphi(\widehat{\mathcal{L}}_\varphi^n(g)) = m_\varphi(g), \quad \forall n \in \mathbb{N},\ \forall g \in C(\mathcal{J}(T)). \tag{27.83}$$

So, $\widehat{\mathcal{L}}_\varphi^{*n} = (\widehat{\mathcal{L}}_\varphi^n)^*$.

In Subsection 13.6.2, we have observed that the eigenmeasures of the normalized transfer operator's dual are conformal measures. This is the case here too. For the sake of completeness, let us repeat this classical demonstration (cf. last part of the proof of Lemma 13.6.13.) Let $n \in \mathbb{N}$ and B be a Borel set such that $T^n|_B$ is injective. Because $\widehat{\mathcal{L}}_\varphi^*$ fixes m_φ, we obtain that (using (27.83) and (27.45))

$$\int_B g\, dm_\varphi = \int_B g\, d\widehat{\mathcal{L}}_\varphi^{n*}(m_\varphi) = \lambda_\varphi^{-n} \int_{\mathcal{J}(T)} \mathcal{L}_\varphi^n(\mathbb{1}_B \cdot g)\, dm_\varphi = \lambda_\varphi^{-n} \int_{\mathcal{J}(T)} \sum_{y \in T^{-n}(x)} \mathbb{1}_B(y) g(y) e^{S_n\varphi(y)}\, dm_\varphi(x).$$

Letting $g = \lambda_\varphi^n e^{-S_n\varphi}$ results in

$$\int_B \lambda_\varphi^n e^{-S_n\varphi}\, dm_\varphi = \int_{\mathcal{J}(T)} \sum_{y\in T^{-n}(x)} \mathbb{1}_B(y)\, dm_\varphi(x) = m_\varphi(T^n(B)). \tag{27.84}$$

That is, m_φ is $(\lambda_\varphi e^{-\varphi})$-conformal in the sense of Chapter 13. On the other hand, for any set $A \subseteq T^n(B)$ we have

$$m_\varphi(T^n|_B^{-1}(A)) = \lambda_\varphi^{-n} \int_A e^{(S_n\varphi)\circ T^n|_B^{-1}}\, dm_\varphi. \tag{27.85}$$

In Chapters 13, 16, 19 and 21, we witnessed the crucial role played by conformal measures. We will see their power at work later in this chapter (in the proof of Lemma 27.5.2) and much more substantially in Chapters 29–32.

Our next goal is to show that $\lambda_\varphi \geq e^{\mathrm{P}(\varphi)}$. We proceed as follows. Fix $z \in \mathcal{J}(T)$ and $\xi_1, \xi_2 \in \mathcal{J}(T) \cap B(z, r_z)$. For every $n \geq 0$ and every $V \in W_n(\gamma, s; z)$, pick $z_n(V) \in V$. By virtue of (27.36) in Lemma 27.1.2 and Lemma 27.1.1(h_n), we get for every $n \geq 0$ that

$$\begin{aligned} G_n(0, z, \xi_2) &\leq e^{\Gamma_z(\varphi)} \sum_{V\in W_n(\gamma,s;z)} \sum_{w\in T^{-n}(\xi_2)\cap V} \exp(S_n\varphi(z_n(V))) \\ &\leq \Delta(T) e^{\Gamma_z(\varphi)} \sum_{V\in W_n(\gamma,s;z)} \exp(S_n\varphi(z_n(V))) \\ &\leq \Delta(T) e^{2\Gamma_z(\varphi)} \sum_{V\in W_n(\gamma,s;z)} \sum_{w\in T^{-n}(\xi_1)\cap V} \exp(S_n\varphi(\xi_1)) \\ &\leq \Delta(T) e^{2\Gamma_z(\varphi)} G_n(0, z, \xi_1). \end{aligned}$$

Therefore, for every $\beta \in \mathbb{R}$, every $n \geq 0$, every $z \in \mathcal{J}(T)$, and all $\xi_1, \xi_2 \in \mathcal{J}(T) \cap B(z, r_z)$, we obtain that

$$\left(\Delta(T) e^{2\Gamma_z(\varphi)}\right)^{-1} \leq \frac{G_n(\beta, z, \xi_2)}{G_n(\beta, z, \xi_1)} \leq \Delta(T) e^{2\Gamma_z(\varphi)}.$$

In particular,

$$\left(C_{\gamma,s}(\varphi)\right)^{-1} \leq \frac{G_n(\beta, z, \xi_2)}{G_n(\beta, z, \xi_1)} \leq C_{\gamma,s}(\varphi) \tag{27.86}$$

for every $\beta \in \mathbb{R}$, every $n \geq 0$, every $z \in Y_{\gamma,s}$, and all $\xi_1, \xi_2 \in \mathcal{J}(T) \cap B(z, r_z)$, where

$$C_{\gamma,s}(\varphi) := \Delta(T) \max\{e^{2\Gamma_w(\varphi)} \mid w \in Y_{\gamma,s}\} \in [1, \infty). \tag{27.87}$$

Denote

$$E_\varphi := \min\{m_\varphi(B(w, r_w)) \mid w \in Y_{\gamma,s}\} > 0. \tag{27.88}$$

The positivity of this quantity is a consequence of the fact that $\mathrm{supp}(m_\varphi) = \mathcal{J}(T)$. (Note: While $T|_{\mathcal{J}(T)}$ is not necessarily expanding, it is strongly transitive (in fact, topologically exact), open, and admits finitely many inverses branches of any given order defined on sufficiently small balls that form a cover of $\mathcal{J}(T)$. These are the fundamental properties that ensure that m_φ has full support. For more information, see Proposition 13.6.14, Corollary 13.6.5 and Proposition 13.5.2.)

Lemma 27.2.5. *Let $T : \widehat{\mathbb{C}} \to \widehat{\mathbb{C}}$ be a rational function with $\deg(T) \geq 2$ and $\varphi : \mathcal{J}(T) \to \mathbb{R}$ a Hölder continuous potential such that $\mathrm{P}(\varphi) > \sup \varphi$. Then*

$$K_{\log \lambda_\varphi} := \sup\Big\{ \sup_{\xi \in B(z,r_z)} G_n(\log \lambda_\varphi, z, \xi) \,\Big|\, n \in \mathbb{N},\ z \in Y_{\gamma,s} \Big\} \leq C_{\gamma,s}(\varphi) E_\varphi^{-1}.$$

Proof. Applying (27.83) with $g = \mathbb{1}$ and using (27.71) and (27.86), we get for every $n \geq 0$ and every $z \in Y_{\gamma,s}$ that

$$\begin{aligned} 1 = m_\varphi(\mathbb{1}) = m_\varphi(\widehat{\mathcal{L}}_\varphi^n(\mathbb{1})) &= \int_{\mathcal{J}(T)} \widehat{\mathcal{L}}_\varphi^n(\mathbb{1})\, dm_\varphi \geq \int_{B(z,r_z)} \widehat{\mathcal{L}}_\varphi^n(\mathbb{1})\, dm_\varphi \\ &\geq \int_{B(z,r_z)} G_n(\log \lambda_\varphi, z, \xi)\, dm_\varphi(\xi) \\ &\geq \Big[\inf_{\xi \in B(z,r_z)} G_n(\log \lambda_\varphi, z, \xi) \Big] \cdot m_\varphi(B(z, r_z)) \\ &\geq (C_{\gamma,s}(\varphi))^{-1} \Big[\sup_{\xi \in B(z,r_z)} G_n(\log \lambda_\varphi, z, \xi) \Big] \cdot E_\varphi. \end{aligned}$$

Hence,

$$\sup_{\xi \in B(z,r_z)} G_n(\log \lambda_\varphi, z, \xi) \leq C_{\gamma,s}(\varphi) E_\varphi^{-1}.$$ □

We draw two repercussions of the uniform boundedness of the functions $\{G_n(\log \lambda_\varphi, z, \cdot)\}_{n=1}^\infty$. First, the eigenvalue λ_φ is larger than or equal to the exponential of the pressure of the system under the potential φ. Second, the supremum norms of the iterates of the normalized operator $\{\widehat{\mathcal{L}}_\varphi^n(\mathbb{1})\}_{n=1}^\infty$ are uniformly bounded.

Lemma 27.2.6. *Let $T : \widehat{\mathbb{C}} \to \widehat{\mathbb{C}}$ be a rational function with $\deg(T) \geq 2$ and $\varphi : \mathcal{J}(T) \to \mathbb{R}$ a Hölder continuous potential such that $\mathrm{P}(\varphi) > \sup \varphi$. Then*

$$\log \lambda_\varphi \geq \mathrm{P}(\varphi). \tag{27.89}$$

Proof. By way of contradiction, suppose that $\log \lambda_\varphi < \mathrm{P}(\varphi)$. It follows from Lemma 27.2.4 that

$$\sup\Big\{ \sup_{\xi \in B(z,r_z)} G_n(\log \lambda_\varphi, z, \xi) \,\Big|\, n \in \mathbb{N},\ z \in Y_{\gamma,s} \Big\} = \infty.$$

This contradicts Lemma 27.2.5. □

Lemma 27.2.7. *Let $T : \widehat{\mathbb{C}} \to \widehat{\mathbb{C}}$ be a rational function with* $\deg(T) \geq 2$ *and* $\varphi : \mathcal{J}(T) \to \mathbb{R}$ *a Hölder continuous potential such that* $\mathrm{P}(\varphi) > \sup \varphi$. *Then*

$$u_\varphi := \sup_{n\geq 0} \|\widehat{\mathcal{L}}_\varphi^n(\mathbb{1})\|_\infty < \infty. \tag{27.90}$$

Proof. Lemma 27.2.5 ensures that Lemma 27.2.3 applies with $\beta = \log \lambda_\varphi$ and yields

$$\sup_{n\geq 0} \|\widehat{\mathcal{L}}_\varphi^n(\mathbb{1})\|_\infty \leq N_{\log \lambda_\varphi}(\varphi) < \infty.$$

□

An immediate consequence of this lemma is the following.

Corollary 27.2.8. *Let $T : \widehat{\mathbb{C}} \to \widehat{\mathbb{C}}$ be a rational function with* $\deg(T) \geq 2$ *and* $\varphi : \mathcal{J}(T) \to \mathbb{R}$ *a Hölder continuous potential such that* $\mathrm{P}(\varphi) > \sup \varphi$. *Then*

$$\sup_{n\geq 0} \|\widehat{\mathcal{L}}_\varphi^n(g)\|_\infty \leq u_\varphi \|g\|_\infty < \infty, \quad \forall g \in C(\mathcal{J}(T)).$$

We now prove the lower bound counterpart of Lemma 27.2.7.

Lemma 27.2.9. *Let $T : \widehat{\mathbb{C}} \to \widehat{\mathbb{C}}$ be a rational function with* $\deg(T) \geq 2$ *and* $\varphi : \mathcal{J}(T) \to \mathbb{R}$ *a Hölder continuous potential such that* $\mathrm{P}(\varphi) > \sup \varphi$. *Then*

$$\ell_\varphi := \inf_{n\geq 0} \inf_{\xi \in \mathcal{J}(T)} \widehat{\mathcal{L}}_\varphi^n(\mathbb{1})(\xi) > 0. \tag{27.91}$$

Proof. Set $\beta = \log \lambda_\varphi$ and γ according to (27.64). Regarding s, we require that it satisfy (27.65) as well as $2d(2d-2)\Delta(T)(1-e^{-\eta})^{-1}e^{-\eta(s+1)} \cdot u_\varphi < 1/2$ rather than (27.66). As before, η is defined per (27.67). From (27.76) and Lemma 27.2.7, we get for every $n > s$, every $z \in Y_{\gamma,s}$ and every $\xi \in B(z, r_z)$, that

$$\begin{aligned}
\widehat{\mathcal{L}}_\varphi^n(\mathbb{1})(\xi) &\leq 2d(2d-2)\Delta(T) \sum_{k=s+1}^{n} e^{-\eta k} \|\widehat{\mathcal{L}}_\varphi^{n-k}(\mathbb{1})\|_\infty + G_n(\log \lambda_\varphi, z, \xi) \\
&\leq 2d(2d-2)\Delta(T) u_\varphi \sum_{k=s+1}^{n} e^{-\eta k} + G_n(\log \lambda_\varphi, z, \xi) \\
&\leq 2d(2d-2)\Delta(T) u_\varphi (1-e^{-\eta})^{-1} e^{-\eta(s+1)} + G_n(\log \lambda_\varphi, z, \xi) \\
&\leq \frac{1}{2} + G_n(\log \lambda_\varphi, z, \xi).
\end{aligned} \tag{27.92}$$

Since $m_\varphi(\widehat{\mathcal{L}}_\varphi^n(\mathbb{1})) = 1$, there exists $\xi_n \in \mathcal{J}(T)$ such that

$$\widehat{\mathcal{L}}_\varphi^n(\mathbb{1})(\xi_n) \geq 1. \tag{27.93}$$

By (27.72), there is $z_n \in Y_{\gamma,s}$ such that $\xi_n \in B(z_n, r_{z_n})$. Using this, (27.92) and (27.93), we deduce that

$$G_n(\log\lambda_\varphi, z_n, \xi_n) \ge 1/2, \quad \forall n > s. \tag{27.94}$$

As the map $T : \mathcal{J}(T) \to \mathcal{J}(T)$ is topologically exact (per Corollary 24.3.5), for every $z \in Y_{\gamma,s}$ there exists $q_z > s$ such that

$$T^{q_z}(B(z, r_z)) \supseteq \mathcal{J}(T).$$

Letting $q = \max\{q_z : z \in Y_{\gamma,s}\} > s$, we infer that

$$T^q(B(z, r_z)) \supseteq \mathcal{J}(T), \quad \forall z \in Y_{\gamma,s}.$$

Consequently, for every $w \in \mathcal{J}(T)$ and every $n \ge 0$ there exists

$$w_n \in B(z_n, r_{z_n}) \cap T^{-q}(w). \tag{27.95}$$

Using this, (27.71) with $\beta = \log\lambda_\varphi$, (27.86) and (27.94), we obtain that

$$\begin{aligned}\widehat{\mathcal{L}}_\varphi^{n+q}(\mathbb{1})(w) &\ge \lambda_\varphi^{-q}\exp(S_q\varphi(w_n))\widehat{\mathcal{L}}_\varphi^n(\mathbb{1})(w_n) \ge \lambda_\varphi^{-q}\exp(q\inf\varphi)G_n(\log\lambda_\varphi, z_n, w_n)\\ &\ge \lambda_\varphi^{-q}\exp(q\inf\varphi)(C_{\gamma,s}(\varphi))^{-1}G_n(\log\lambda_\varphi, z_n, \xi_n)\\ &\ge \frac{1}{2}(C_{\gamma,s}(\varphi))^{-1}\lambda_\varphi^{-q}\exp(q\inf\varphi).\end{aligned}$$

Hence,

$$\ell_\varphi \ge \min\left\{\frac{1}{2}(C_{\gamma,s}(\varphi))^{-1}\lambda_\varphi^{-q}\exp(q\inf\varphi), \min_{0\le k\le q}\inf_{\xi\in\mathcal{J}(T)}\widehat{\mathcal{L}}_\varphi^k(\mathbb{1})(\xi)\right\} > 0. \qquad \square$$

We end this section with a straightforward but important ramification of Lemmas 27.2.7 and 27.2.9.

Lemma 27.2.10. *Let $T : \widehat{\mathbb{C}} \to \widehat{\mathbb{C}}$ be a rational function with* $\deg(T) \ge 2$ *and* $\varphi : \mathcal{J}(T) \to \mathbb{R}$ *a Hölder continuous potential such that* $\mathrm{P}(\varphi) > \sup\varphi$. *Then*

$$\lim_{n\to\infty}\frac{1}{n}\log\mathcal{L}_\varphi^n(\mathbb{1})(z) = \log\lambda_\varphi, \quad \forall z \in \mathcal{J}(T). \tag{27.96}$$

27.3 Equicontinuity of iterates of $\widehat{\mathcal{L}}_\varphi$; “Gibbs states” m_φ and μ_φ

We now look at the equicontinuity of families of functions generated by the iterates of the normalized operator, like we did for distance expanding maps (cf. Theorem 13.7.4).

Lemma 27.3.1. *Let $T : \widehat{\mathbb{C}} \to \widehat{\mathbb{C}}$ be a rational function with* $\deg(T) \ge 2$ *and* $\varphi : \mathcal{J}(T) \to \mathbb{R}$ *a Hölder continuous potential such that* $\mathrm{P}(\varphi) > \sup\varphi$. *Then for every function $g \in C(\mathcal{J}(T))$, the family $\{\widehat{\mathcal{L}}_\varphi^n(g)\}_{n=0}^\infty$ is equicontinuous and pointwise bounded in $C(\mathcal{J}(T))$. That is, it is* *relatively compact in $C(\mathcal{J}(T))$. In other terms, the operator $\widehat{\mathcal{L}}_\varphi : C(\mathcal{J}(T)) \to C(\mathcal{J}(T))$ is almost periodic.*

Proof. The pointwise boundedness was the object of Corollary 27.2.8. Once equicontinuity is established, the rest will follow from the Arzelà–Ascoli theorem (Theorem A.2.3) and the definition of almost periodicity of an operator.

As regards equicontinuity, fix $\varepsilon > 0$. Set $\beta = \log \lambda_\varphi$ and γ according to (27.64). Concerning s, we require that it satisfy (27.65) as well as

$$2d(2d-2)\Delta(T)(1-e^{-\eta})^{-1}e^{-\eta(s+1)} \cdot u_\varphi \|g\|_\infty < \varepsilon/4, \tag{27.97}$$

where u_φ comes from (27.90) in Lemma 27.2.7 while η is still defined per (27.67). Generalizing naturally (27.70), define for every $n \in \mathbb{N}$, every $z \in \mathcal{J}(T)$ and every $\xi \in B(z, r_z)$ the function

$$G_n(\log \lambda_\varphi, z, \xi; g) := \lambda_\varphi^{-n} \sum_{V \in W_n(\gamma,s;z)} \sum_{w \in T^{-n}(\xi)\cap V} \exp(S_n\varphi(w))g(w). \tag{27.98}$$

For every $n > s$, every $z \in Y_{\gamma,s}$ and every $\xi \in B(z, r_z)$, relation (27.71) generalizes to

$$\begin{aligned}\widehat{\mathcal{L}}_\varphi^n(g)(\xi) = \sum_{k=s+1}^{n} \sum_{V \in Z_k(\gamma,s;z)\setminus W_k(\gamma,s;z)} \lambda_\varphi^{-k} \sum_{w \in T^{-k}(\xi)\cap V} \exp(S_k\varphi(w))\widehat{\mathcal{L}}_\varphi^{n-k}(g)(w)\\ + G_n(\log \lambda_\varphi, z, \xi; g).\end{aligned} \tag{27.99}$$

Denote the first summand by $H_n(\log \lambda_\varphi, z, \xi; g)$. Using Corollary 27.2.8, (27.69) and (27.97), we infer that

$$\begin{aligned} H_n(\log \lambda_\varphi, z, \xi; g) &\le u_\varphi \|g\|_\infty H_n(\log \lambda_\varphi, z, \xi; \mathbb{1}) \\ &= u_\varphi \|g\|_\infty \sum_{k=s+1}^{n} B_k(\log \lambda_\varphi, z, \xi) \\ &\le u_\varphi \|g\|_\infty \cdot 2d(2d-2)\Delta(T) \sum_{k=s+1}^{n} e^{-\eta k} \\ &\le u_\varphi \|g\|_\infty \cdot 2d(2d-2)\Delta(T)(1-e^{-\eta})^{-1}e^{-\eta(s+1)} \\ &< \varepsilon/4. \end{aligned} \tag{27.100}$$

We now deal with $G_n(\log \lambda_\varphi, z, \xi; g)$, the second summand in (27.99). Keep $n > s$ and $z \in Y_{\gamma,s}$. Fix two points $x, y \in B(z, r_z)$. Consider $V \in W_n(\gamma, s; z)$. As the degree of the map $T^n|_V : V \to T^n(V)$ is well-defined, there exists a bijection

$$b_{V,x,y} : V \cap T^{-n}(x) \to V \cap T^{-n}(y),$$

where each element in $V \cap T^{-n}(x)$ and $V \cap T^{-n}(y)$ is counted according to its degree with respect to the map T^n. Starting with (27.98) and using the mean value theorem and Corollary 27.1.3, we can write:

$$
\begin{aligned}
&\big|G_n(\log\lambda_\varphi, z, y; g) - G_n(\log\lambda_\varphi, z, x; g)\big| \\
&= \Bigg|\lambda_\varphi^{-n} \sum_{V\in W_n(\gamma,s;z)} \sum_{w\in T^{-n}(x)\cap V} \big[e^{S_n\varphi(w)}g(w) - e^{S_n\varphi(b_{V,x,y}(w))}g(b_{V,x,y}(w))\big]\Bigg| \\
&= \Bigg|\lambda_\varphi^{-n} \sum_{V\in W_n(\gamma,s;z)} \sum_{w\in T^{-n}(x)\cap V} \left[\begin{array}{l} e^{S_n\varphi(w)}\cdot[g(w) - g(b_{V,x,y}(w))] \\ \quad + g(b_{V,x,y}(w))\cdot[e^{S_n\varphi(w)} - e^{S_n\varphi(b_{V,x,y}(w))}] \end{array}\right]\Bigg| \\
&\le \lambda_\varphi^{-n} \sum_{V\in W_n(\gamma,s;z)} \sum_{w\in T^{-n}(x)\cap V} \left[\begin{array}{l} e^{S_n\varphi(w)}|g(w) - g(b_{V,x,y}(w))| \\ \quad + \|g\|_\infty|e^{S_n\varphi(w)} - e^{S_n\varphi(b_{V,x,y}(w))}| \end{array}\right] \\
&\le \lambda_\varphi^{-n} \sum_{V\in W_n(\gamma,s;z)} \sum_{w\in T^{-n}(x)\cap V} \left[\begin{array}{l} e^{S_n\varphi(w)}|g(w) - g(b_{V,x,y}(w))| \\ \quad + \left[\begin{array}{l} \|g\|_\infty \max\{e^{S_n\varphi(w)}, e^{S_n\varphi(b_{V,x,y}(w))}\} \\ \quad \cdot |S_n\varphi(w) - S_n\varphi(b_{V,x,y}(w))| \end{array}\right] \end{array}\right] \\
&\le \lambda_\varphi^{-n} \sum_{V\in W_n(\gamma,s;z)} \sum_{w\in T^{-n}(x)\cap V} e^{S_n\varphi(w)} \left[\begin{array}{l} |g(w) - g(b_{V,x,y}(w))| \\ \quad + e^{\Gamma(\varphi)}\|g\|_\infty|S_n\varphi(w) - S_n\varphi(b_{V,x,y}(w))| \end{array}\right],
\end{aligned}
\tag{27.101}
$$

where $\Gamma(\varphi) = \max\{\Gamma_z(\varphi) : z \in Y_{\gamma,s}\}$ and $\Gamma_z(\varphi), z \in Y_{\gamma,s}$, come from Lemma 27.1.2. Because the function $g : \mathcal{J}(T) \to \mathbb{R}$ is uniformly continuous, because of (27.37) in Lemma 27.1.2 and because $\#Y_{\gamma,s} < \infty$, there exists $\delta > 0$ such that

$$|a - b| < \delta \Longrightarrow u_\varphi|g(a) - g(b)| < \varepsilon/4$$

and

$$e^{\Gamma(\varphi)}\|g\|_\infty u_\varphi|S_n\varphi(a) - S_n\varphi(b)| < \varepsilon/4,$$

if, in addition, $a, b \in V$ for the same $z \in Y_{\gamma,s}$ and the same $V \in W_n(\gamma, s; z)$. Using this and (27.90) in Lemma 27.2.7, we can continue (27.101) as follows:

$$
\begin{aligned}
&\big|G_n(\log\lambda_\varphi, z, y; g) - G_n(\log\lambda_\varphi, z, x; g)\big| \\
&\quad \le G_n(\log\lambda_\varphi, z, x)\left[\frac{\varepsilon}{4}u_\varphi^{-1} + e^{\Gamma(\varphi)}\|g\|_\infty\frac{\varepsilon}{4}e^{-\Gamma(\varphi)}\|g\|_\infty^{-1}u_\varphi^{-1}\right] \le u_\varphi\left[\frac{\varepsilon}{2}u_\varphi^{-1}\right] = \frac{\varepsilon}{2}.
\end{aligned}
$$

Combining this estimate with (27.100) and (27.99) yields

$$
\begin{aligned}
|\widehat{\mathcal{L}}_\varphi^n(g)(y) - \widehat{\mathcal{L}}_\varphi^n(g)(x)| &\le |H_n(\log\lambda_\varphi, z, y; g)| + |H_n(\log\lambda_\varphi, z, x; g)| \\
&\quad + |G_n(\log\lambda_\varphi, z, y; g) - G_n(\log\lambda_\varphi, z, x; g)| \\
&< \varepsilon/4 + \varepsilon/4 + \varepsilon/2 = \varepsilon
\end{aligned}
$$

for all $n > s$, $z \in Y_{\gamma,s}$, and $x, y \in B(z, r_z)$. The family $\{\widehat{\mathcal{L}}_\varphi^n(g)\}_{n=0}^\infty$ is thereby uniformly equicontinuous. □

We now identify a fixed point for the normalized operator. This standard argument, already given in the context of distance expanding maps (cf. Lemma 13.7.5), is included here for the sake of completeness and independence of this section.

Proposition 27.3.2. *Let $T : \widehat{\mathbb{C}} \to \widehat{\mathbb{C}}$ be a rational function of $\deg(T) \geq 2$ and $\varphi : \mathcal{J}(T) \to \mathbb{R}$ a Hölder continuous potential such that $\mathrm{P}(\varphi) > \sup \varphi$. Then there is $\rho_\varphi \in C(\mathcal{J}(T))$ such that*

$$\widehat{\mathcal{L}}_\varphi(\rho_\varphi) = \rho_\varphi, \quad 0 < \ell_\varphi \leq \rho_\varphi \leq u_\varphi < \infty, \quad \textit{and} \quad m_\varphi(\rho_\varphi) := \int_{\mathcal{J}(T)} \rho_\varphi \, dm_\varphi = 1.$$

Proof. It ensues from Lemmas 27.2.7 and 27.2.9 that

$$\ell_\varphi \leq \frac{1}{n} \sum_{j=0}^{n-1} \widehat{\mathcal{L}}_\varphi^j(\mathbb{1})(x) \leq u_\varphi, \quad \forall n \in \mathbb{N},\ \forall x \in \mathcal{J}(T). \tag{27.102}$$

Hence, the family $\{\frac{1}{n}\sum_{j=0}^{n-1} \widehat{\mathcal{L}}_\varphi^j(\mathbb{1})\}_{n=1}^\infty$ is uniformly bounded and thereby pointwise bounded. Moreover, it follows from Lemma 27.3.1 that the family $\{\frac{1}{n}\sum_{j=0}^{n-1} \widehat{\mathcal{L}}_\varphi^j(\mathbb{1})\}_{n=1}^\infty$ is equicontinuous. Thus, all the hypotheses of the Arzelà–Ascoli theorem (Theorem A.2.3) are satisfied. According to that theorem, there exists a strictly increasing sequence $(n_k)_{k=1}^\infty$ such that the sequence $\{\frac{1}{n_k}\sum_{j=0}^{n_k-1} \widehat{\mathcal{L}}_\varphi^j(\mathbb{1})\}_{k=1}^\infty$ converges uniformly. Denote its limit by $\rho_\varphi : \mathcal{J}(T) \to [0, \infty)$. As a uniform limit of continuous functions on a compact metric space, the function ρ_φ is also continuous, i. e., $\rho_\varphi \in C(\mathcal{J}(T))$. It immediately follows from (27.102) that

$$\ell_\varphi \leq \rho_\varphi \leq u_\varphi.$$

Furthermore, using (27.83) with $g = \mathbb{1}$ and Lebesgue's dominated convergence theorem, we have that

$$m_\varphi(\rho_\varphi) = \lim_{k\to\infty} m_\varphi\left(\frac{1}{n_k}\sum_{j=0}^{n_k-1} \widehat{\mathcal{L}}_\varphi^j(\mathbb{1})\right) = \lim_{k\to\infty} \frac{1}{n_k}\sum_{j=0}^{n_k-1} m_\varphi(\widehat{\mathcal{L}}_\varphi^j(\mathbb{1})) = \lim_{k\to\infty} \frac{1}{n_k}\sum_{j=0}^{n_k-1} m_\varphi(\mathbb{1}) = 1.$$

Finally, using Lemma 27.2.7 and the continuity and linearity of $\widehat{\mathcal{L}}_\varphi$, we obtain that

$$\begin{aligned}
\widehat{\mathcal{L}}_\varphi(\rho_\varphi) &= \widehat{\mathcal{L}}_\varphi\left(\lim_{k\to\infty} \frac{1}{n_k}\sum_{j=0}^{n_k-1} \widehat{\mathcal{L}}_\varphi^j(\mathbb{1})\right) = \lim_{k\to\infty} \frac{1}{n_k}\sum_{j=0}^{n_k-1} \widehat{\mathcal{L}}_\varphi^{j+1}(\mathbb{1}) \\
&= \lim_{k\to\infty} \frac{1}{n_k}\left[\widehat{\mathcal{L}}_\varphi^{n_k}(\mathbb{1}) + \sum_{j=0}^{n_k-1} \widehat{\mathcal{L}}_\varphi^j(\mathbb{1}) - \mathbb{1}\right] = \lim_{k\to\infty} \frac{1}{n_k}\sum_{j=0}^{n_k-1} \widehat{\mathcal{L}}_\varphi^j(\mathbb{1}) = \rho_\varphi.
\end{aligned}$$

□

Thanks to this proposition, the usual reasoning, already outlined for distance expanding maps (cf. Theorem 13.7.7), yields an invariant measure.

Theorem 27.3.3. *Let $T : \widehat{\mathbb{C}} \to \widehat{\mathbb{C}}$ be a rational function with $\deg(T) \geq 2$ and $\varphi : \mathcal{J}(T) \to \mathbb{R}$ a Hölder continuous potential such that $\mathrm{P}(\varphi) > \sup \varphi$. Then $\mu_\varphi := \rho_\varphi m_\varphi$ is a T-invariant Borel probability measure on $\mathcal{J}(T)$ which is equivalent to m_φ.*

Proof. It follows from the second and third relations in Proposition 27.3.2 that μ_φ is equivalent to m_φ and that $\mu_\varphi \in M(\mathcal{J}(T))$. Using (27.82) and the first relation in Proposition 27.3.2, we deduce for every $g \in C(\mathcal{J}(T))$ that

$$\begin{aligned}(\mu_\varphi \circ T^{-1})(g) &= \mu_\varphi(g \circ T) = m_\varphi(\rho_\varphi \cdot g \circ T) = m_\varphi(\widehat{\mathcal{L}}_\varphi(\rho_\varphi \cdot g \circ T)) \\ &= m_\varphi(g \cdot \widehat{\mathcal{L}}_\varphi(\rho_\varphi)) = m_\varphi(\rho_\varphi g) \\ &= \mu_\varphi(g).\end{aligned}$$

Therefore, $\mu_\varphi \circ T^{-1} = \mu_\varphi$. □

We sometimes refer to the measures m_φ and μ_φ as “Gibbs states” for the potential $\varphi : \mathcal{J}(T) \to \mathbb{R}$ because they resemble, to some extent, the Gibbs states exposed for distance expanding maps in Chapter 13. They behave exactly as ordinary Gibbs states with respect to good inverse branches G_n and are genuine Gibbs states whenever the rational function T is expanding. Indeed, in Section 27.5 we will prove that μ_φ is an equilibrium state for φ, and if T is expanding then μ_φ is a Gibbs state according to Chapter 13.

27.4 Spectral properties of the transfer operator $\widehat{\mathcal{L}}_\varphi$ and their dynamical consequences

We can now quite easily prove the following about the eigenvalue 1 and the eigenvector ρ_φ of $\widehat{\mathcal{L}}_\varphi$ (cf. Theorem 13.8.18).

Lemma 27.4.1. *Let $T : \widehat{\mathbb{C}} \to \widehat{\mathbb{C}}$ be a rational function with $\deg(T) \geq 2$ and $\varphi : \mathcal{J}(T) \to \mathbb{R}$ a Hölder continuous potential such that $\mathrm{P}(\varphi) > \sup \varphi$. Then the geometric multiplicity of the eigenvalue 1 of the operator $\widehat{\mathcal{L}}_\varphi : C(\mathcal{J}(T)) \to C(\mathcal{J}(T))$ is equal to 1. More precisely,*

$$\mathrm{Ker}(\widehat{\mathcal{L}}_\varphi - \mathrm{Id}) = \mathbb{C}\rho_\varphi.$$

Proof. We first establish the following fact.

Claim 1*. *If $h : \mathcal{J}(T) \to [0, \infty)$ is a continuous function such that $\widehat{\mathcal{L}}_\varphi(h) = h$ and $h(w) = 0$ for some $w \in \mathcal{J}(T)$, then $h \equiv 0$.*

Proof of Claim 1.* Using (27.45), we have for every $n \in \mathbb{N}$ that

$$0 = h(w) = \widehat{\mathcal{L}}_\varphi^n(h)(w) = \lambda_\varphi^{-n} \sum_{z \in T^{-n}(w)} h(z) \exp(S_n\varphi(z)) \deg_z(T^n).$$

Therefore, $h(z) = 0$ for every $z \in \bigcup_{n=0}^{\infty} T^{-n}(w)$. Since h is continuous, it ensues that $h(z) = 0$ for every $z \in \overline{\bigcup_{n=0}^{\infty} T^{-n}(w)}$. As $T|_{\mathcal{J}(T)}$ is topologically exact, it is strongly transitive, i. e., the backward orbit of any Julia point is dense in the Julia set (see Definition 1.5.14). So, the claim follows. ■

Moving on to the proof of the lemma itself, assume that $g \in C(\mathcal{J}(T))$ is an eigenfunction associated with the eigenvalue 1 of the operator $\widehat{\mathcal{L}}_\varphi : C(\mathcal{J}(T)) \to C(\mathcal{J}(T))$. Given that $\widehat{\mathcal{L}}_\varphi$ preserves the vector subspace $C(\mathcal{J}(T), \mathbb{R})$ consisting of all real-valued functions, both the real and imaginary parts of g are also eigenfunctions associated with the eigenvalue 1. So, we may assume without loss of generality that g is real-valued. Because the function $g/\rho_\varphi : \mathcal{J}(T) \to \mathbb{R}$ is continuous and $\mathcal{J}(T)$ is compact, this function attains its minimum value; denote this latter by s. Then $(g - s\rho_\varphi)(z) \geq 0$ for all $z \in \mathcal{J}(T)$ and there is $\xi \in \mathcal{J}(T)$ such that $(g - s\rho_\varphi)(\xi) = 0$. Since $g - s\rho_\varphi$ is an eigenfunction associated with the eigenvalue 1, too, it follows from Claim 1* that $g - s\rho_\varphi \equiv 0$; equivalently, $g = s\rho_\varphi$. □

We now introduce a fully normalized version of the operator $\widehat{\mathcal{L}}_\varphi$, namely

$$\Lambda_\varphi(g) := \frac{1}{\rho_\varphi} \widehat{\mathcal{L}}_\varphi(\rho_\varphi g). \tag{27.103}$$

This operator was denoted by $\widehat{\mathcal{L}}_{\hat{\varphi}}$ in Section 13.8 but we lighten notation here. Like $\widehat{\mathcal{L}}_\varphi$, this positive bounded linear operator acts on the Banach space $C(\mathcal{J}(T))$. It is conjugate with $\widehat{\mathcal{L}}_\varphi$ via the Banach isomorphism $g \mapsto \rho_\varphi g$. By induction on $n \in \mathbb{N}$,

$$\Lambda_\varphi^n(g) = \frac{1}{\rho_\varphi} \widehat{\mathcal{L}}_\varphi^n(\rho_\varphi g). \tag{27.104}$$

Using this and (27.83) (with g replaced by $\rho_\varphi g$), notice that

$$\mu_\varphi(\Lambda_\varphi^n(g)) = \int_{\mathcal{J}(T)} \rho_\varphi \Lambda_\varphi^n(g)\, dm_\varphi = \int_{\mathcal{J}(T)} \widehat{\mathcal{L}}_\varphi^n(\rho_\varphi g)\, dm_\varphi = \int_{\mathcal{J}(T)} \rho_\varphi g\, dm_\varphi = \mu_\varphi(g). \tag{27.105}$$

So, μ_φ takes the role for Λ_φ that m_φ played for $\widehat{\mathcal{L}}_\varphi$.

Substituting the function $\mathbb{1}$ in (27.103) and using Proposition 27.3.2 results in

$$\Lambda_\varphi(\mathbb{1}) = \frac{1}{\rho_\varphi} \widehat{\mathcal{L}}_\varphi(\rho_\varphi) = \frac{\rho_\varphi}{\rho_\varphi} = \mathbb{1}, \tag{27.106}$$

i. e., $\mathbb{1}$ is an eigenfunction corresponding to the eigenvalue 1 of Λ_φ. From this, we obtain for every $g \in C(\mathcal{J}(T))$ and $n \in \mathbb{N}$ that

$$\|\Lambda_\varphi^n(g)\|_\infty \leq \|\Lambda_\varphi^n(\|g\|_\infty \mathbb{1})\|_\infty = \|\Lambda_\varphi^n(\mathbb{1})\|_\infty \|g\|_\infty = \|g\|_\infty. \tag{27.107}$$

In Theorem 13.6.3, we established that the linear operator $\mathcal{L}_\varphi$, and hence Λ_φ, can be extended to $L^1(\mu_\varphi)$. From (27.105), we have for any $g \in L^1(\mu_\varphi)$ that

$$\|\Lambda_\varphi^n(g)\|_{L^1(\mu_\varphi)} := \mu_\varphi(|\Lambda_\varphi^n(g)|) \le \mu_\varphi(\Lambda_\varphi^n(|g|)) = \mu_\varphi(|g|) =: \|g\|_{L^1(\mu_\varphi)}. \tag{27.108}$$

The reader may want to show that Lemmas 27.2.7–27.3.1, as well as 27.4.1, have analogs for the fully normalized operator Λ_φ. Among those, the following is an immediate consequence of Lemma 27.3.1, relation (27.104) and the continuity of ρ_φ from Proposition 27.3.2.

Lemma 27.4.2. *Let $T : \widehat{\mathbb{C}} \to \widehat{\mathbb{C}}$ be a rational function with $\deg(T) \ge 2$ and $\varphi : \mathcal{J}(T) \to \mathbb{R}$ a Hölder continuous potential such that $\mathrm{P}(\varphi) > \sup\varphi$. Then for every function $g \in C(\mathcal{J}(T))$, the family of iterates $\{\Lambda_\varphi^n(g)\}_{n=0}^\infty$ is equicontinuous and pointwise bounded in $C(\mathcal{J}(T))$. That is, it is relatively compact in $C(\mathcal{J}(T))$. In other terms, the operator $\Lambda_\varphi : C(\mathcal{J}(T)) \to C(\mathcal{J}(T))$ is almost periodic.*

It turns out that the sequence $(\Lambda_\varphi^n(g))_{n=0}^\infty$ converges uniformly and we now identify its limit (cf. Lemma 13.8.7).

Lemma 27.4.3. *Let $T : \widehat{\mathbb{C}} \to \widehat{\mathbb{C}}$ be a rational function with $\deg(T) \ge 2$ and $\varphi : \mathcal{J}(T) \to \mathbb{R}$ a Hölder continuous potential such that $\mathrm{P}(\varphi) > \sup\varphi$. Then for every function $g \in C(\mathcal{J}(T))$, the sequence $(\Lambda_\varphi^n(g))_{n=0}^\infty$ converges uniformly in the Banach space $C(\mathcal{J}(T))$ and*

$$\lim_{n\to\infty} \Lambda_\varphi^n(g) = \mu_\varphi(g)\mathbb{1}.$$

Proof. Splitting g into its real and imaginary parts, we may assume without loss of generality that $g \in C(\mathcal{J}(T),\mathbb{R})$. Fix a strictly increasing sequence $(n_k)_{k=1}^\infty$ of natural numbers. By Lemma 27.4.2, there exists a strictly increasing subsequence $(n_{k_j})_{j=1}^\infty$ such that the sequence $(\Lambda_\varphi^{n_{k_j}}(g))_{j=1}^\infty$ converges uniformly in $C(\mathcal{J}(T),\mathbb{R})$. Denote its limit by $g^* \in C(\mathcal{J}(T),\mathbb{R})$.

For every $f \in C(\mathcal{J}(T),\mathbb{R})$ and every $z \in \mathcal{J}(T)$, we have that

$$\Lambda_\varphi(f)(z) \le \Lambda_\varphi(\sup f\cdot\mathbb{1})(z) = \sup f\cdot\Lambda_\varphi(\mathbb{1})(z) = \sup f\cdot\mathbb{1}(z) = \sup f.$$

Thus,

$$\sup\Lambda_\varphi(f) \le \sup f.$$

Consequently, the sequence $(\sup\Lambda_\varphi^n(f))_{n=0}^\infty$ is monotone decreasing. In particular, the sequence $(\sup\Lambda_\varphi^n(g^*))_{n=0}^\infty$ is monotone decreasing and

$$\sup\Lambda_\varphi^n(g^*) \le \sup g^*, \quad \forall n \ge 0. \tag{27.109}$$

Furthermore, by the uniform convergence, $\sup g^* = \lim_{j\to\infty}\sup\Lambda_\varphi^{n_{k_j}}(g)$. It follows that

$$\lim_{n\to\infty}\sup\Lambda_\varphi^n(g) = \inf_{n\ge0}\sup\Lambda_\varphi^n(g) = \sup g^*.$$

Hence,

$$\sup \Lambda_\varphi^n(g^*) \geq \sup g^*, \quad \forall n \geq 0. \tag{27.110}$$

Inequalities (27.109)–(27.110) yield

$$\sup \Lambda_\varphi^n(g^*) = \sup g^*, \quad \forall n \geq 0.$$

Since $\mathcal{J}(T)$ is compact, for every $n \geq 0$ there exists a point $\xi_n \in \mathcal{J}(T)$ such that

$$\sup \Lambda_\varphi^n(g^*) = \Lambda_\varphi^n(g^*)(\xi_n).$$

Therefore,

$$\sup g^* = \Lambda_\varphi^n(g^*)(\xi_n) = \sum_{z \in T^{-n}(\xi_n)} \frac{\lambda_\varphi^{-n}}{\rho_\varphi(\xi_n)} \rho_\varphi(z) \exp(S_n\varphi(z)) \deg_z(T^n) g^*(z).$$

By (27.106), the right-hand side of the above equation is a convex combination of numbers smaller than or equal to $\sup g^*$. Consequently,

$$g^*(z) = \sup g^*, \quad \forall z \in T^{-n}(\xi_n),\ \forall n \geq 0.$$

As the function $g^* : \mathcal{J}(T) \to \mathbb{R}$ is continuous, it follows from Proposition 25.6.3 that g^* is constant. From (27.105) and Lebesgue's dominated convergence theorem, we then get that

$$g^* = \mu_\varphi(g^*) = \mu_\varphi\Big(\lim_{j\to\infty} \Lambda_\varphi^{n_{k_j}}(g)\Big) = \lim_{j\to\infty} \mu_\varphi(\Lambda_\varphi^{n_{k_j}}(g)) = \lim_{j\to\infty} \mu_\varphi(g) = \mu_\varphi(g).$$

So, we have proved that every subsequence $(n_k)_{k=1}^\infty$ contains a subsubsequence $(n_{k_j})_{j=1}^\infty$ such that $(\Lambda_\varphi^{n_{k_j}}(g))_{j=1}^\infty$ converges to $\mu_\varphi(g)\mathbb{1}$ in $C(\mathcal{J}(T))$. This means that the sequence $(\Lambda_\varphi^n(g))_{n=0}^\infty$ converges uniformly. □

Let

$$\mathbb{C}^\perp := \{g \in C(\mathcal{J}(T)) \,|\, \mu_\varphi(g) = 0\}.$$

As a straightforward repercussion of Lemma 27.4.3 and relations (27.105)–(27.106), the space $C(\mathcal{J}(T))$ is the direct sum of two simply expressed, Λ_φ-invariant subspaces and the Λ_φ-orbits of its points converge as follows (cf. Theorem 13.8.8).

Theorem 27.4.4. *Let $T : \widehat{\mathbb{C}} \to \widehat{\mathbb{C}}$ be a rational function with $\deg(T) \geq 2$ and $\varphi : \mathcal{J}(T) \to \mathbb{R}$ a Hölder continuous potential such that $\mathrm{P}(\varphi) > \sup \varphi$. Then:*

(a) *$\mathbb{C}\mathbb{1}$ and $\mathbb{C}^\perp$ are both closed vector subspaces (so are Banach subspaces) of $C(\mathcal{J}(T))$.*

(b) *$\Lambda_\varphi(\mathbb{C}\mathbb{1}) = \mathbb{C}\mathbb{1}$ while $\Lambda_\varphi(\mathbb{C}^\perp) \subseteq \mathbb{C}^\perp$.*

(c) $C(\mathcal{J}(T)) = \mathbb{C}\mathbb{1} \bigoplus \mathbb{C}^\perp$.
(d) *If* $g \in C(\mathcal{J}(T))$, *then the sequence* $(\Lambda_\varphi^n(g))_{n=0}^\infty$ *converges uniformly, i. e., in the Banach space* $C(\mathcal{J}(T))$, *and*

$$\lim_{n\to\infty} \Lambda_\varphi^n(g) = \mu_\varphi(g)\mathbb{1}.$$

(e) *In particular, if* $g \in \mathbb{C}^\perp$ *then*

$$\lim_{n\to\infty} \Lambda_\varphi^n(g) = 0.$$

Let

$$(\mathbb{C}\rho_\varphi)^\perp := \{g \in C(\mathcal{J}(T)) \,|\, m_\varphi(g) = 0\}.$$

As a direct ramification of Theorem 27.4.4 and (27.104), the space $C(\mathcal{J}(T))$ is the direct sum of two simply expressed, $\widehat{\mathcal{L}}_\varphi$-invariant subspaces and the $\widehat{\mathcal{L}}_\varphi$-orbits of its points converge as follows (cf. Theorem 13.8.9).

Theorem 27.4.5. *Let* $T : \widehat{\mathbb{C}} \to \widehat{\mathbb{C}}$ *be a rational function with* $\deg(T) \geq 2$ *and* $\varphi : \mathcal{J}(T) \to \mathbb{R}$ *a Hölder continuous potential such that* $\mathrm{P}(\varphi) > \sup \varphi$. *Then:*
(a) $\mathbb{C}\rho_\varphi$ *and* $(\mathbb{C}\rho_\varphi)^\perp$ *are closed vector subspaces (so are Banach subspaces) of* $C(\mathcal{J}(T))$.
(b) $\widehat{\mathcal{L}}_\varphi(\mathbb{C}\rho_\varphi) = \mathbb{C}\rho_\varphi$ *while* $\widehat{\mathcal{L}}_\varphi((\mathbb{C}\rho_\varphi)^\perp) \subseteq (\mathbb{C}\rho_\varphi)^\perp$.
(c) $C(\mathcal{J}(T)) = (\mathbb{C}\rho_\varphi) \bigoplus (\mathbb{C}\rho_\varphi)^\perp$.
(d) *If* $g \in C(\mathcal{J}(T))$ *then the sequence* $(\widehat{\mathcal{L}}_\varphi^n(g))_{n=0}^\infty$ *converges uniformly and*

$$\lim_{n\to\infty} \widehat{\mathcal{L}}_\varphi^n(g) = m_\varphi(g)\rho_\varphi.$$

(e) *In particular, if* $g \in (\mathbb{C}\rho_\varphi)^\perp$ *then*

$$\lim_{n\to\infty} \widehat{\mathcal{L}}_\varphi^n(g) = 0.$$

In Theorem 27.4.4, we described the uniform convergence of the iterates of a function $g \in C(\mathcal{J}(T))$ under the operator Λ_φ. We now establish the $L^1(\mu_\varphi)$-convergence of the iterates of a function $g \in L^1(\mu_\varphi)$ (cf. Theorem 13.8.12).

Theorem 27.4.6. *Let* $T : \widehat{\mathbb{C}} \to \widehat{\mathbb{C}}$ *be a rational function with* $\deg(T) \geq 2$ *and* $\varphi : \mathcal{J}(T) \to \mathbb{R}$ *a Hölder continuous potential such that* $\mathrm{P}(\varphi) > \sup \varphi$. *For every* $g \in L^1(\mu_\varphi)$,

$$\lim_{n\to\infty} \|\Lambda_\varphi^n(g) - \mu_\varphi(g)\mathbb{1}\|_{L^1(\mu_\varphi)} = 0.$$

Proof. Fix $g \in L^1(\mu_\varphi)$ and $\varepsilon > 0$. Since $C(\mathcal{J}(T))$ is a dense subset of the Banach space $L^1(\mu_\varphi)$, there exists $f \in C(\mathcal{J}(T))$ such that

$$\|g - f\|_{L^1(\mu_\varphi)} < \varepsilon/2.$$

Then

$$|\mu_\varphi(g) - \mu_\varphi(f)| < \varepsilon/2.$$

Applying Lemma 27.4.3 and relation (27.108), we obtain that

$$\begin{aligned}
&\limsup_{n\to\infty}\|\Lambda_\varphi^n(g) - \mu_\varphi(g)\mathbb{1}\|_{L^1(\mu_\varphi)} \\
&\quad = \limsup_{n\to\infty}\|\Lambda_\varphi^n(g-f) + (\Lambda_\varphi^n(f) - \mu_\varphi(f)\mathbb{1}) + (\mu_\varphi(f) - \mu_\varphi(g))\mathbb{1}\|_{L^1(\mu_\varphi)} \\
&\quad \le \limsup_{n\to\infty}\|\Lambda_\varphi^n(g-f)\|_{L^1(\mu_\varphi)} + \limsup_{n\to\infty}\|\Lambda_\varphi^n(f) - \mu_\varphi(f)\mathbb{1}\|_{L^1(\mu_\varphi)} \\
&\qquad + \|(\mu_\varphi(f) - \mu_\varphi(g))\mathbb{1}\|_{L^1(\mu_\varphi)} \\
&\quad \le \|g-f\|_{L^1(\mu_\varphi)} + \limsup_{n\to\infty}\|\Lambda_\varphi^n(f) - \mu_\varphi(f)\mathbb{1}\|_\infty + |\mu_\varphi(f) - \mu_\varphi(g)| \\
&\quad < \varepsilon/2 + 0 + \varepsilon/2 = \varepsilon.
\end{aligned}$$

The arbitrariness of $\varepsilon > 0$ leads to the conclusion. □

As a fairly immediate consequence of this theorem, we obtain the following remarkable result (cf. Theorem 13.8.13).

Theorem 27.4.7. *Let $T : \widehat{\mathbb{C}} \to \widehat{\mathbb{C}}$ be a rational function of $\deg(T) \ge 2$ and $\varphi : \mathcal{J}(T) \to \mathbb{R}$ a Hölder continuous potential such that $\mathrm{P}(\varphi) > \sup\varphi$. Then the measure-preserving dynamical system (T, μ_φ) is totally ergodic. That is, all iterates $T^q : \mathcal{J}(T) \to \mathcal{J}(T)$, $q \in \mathbb{N}$, are ergodic with respect to the measure μ_φ, including T itself. Similarly, the measure m_φ is totally ergodic.*

Proof. Fix $q \in \mathbb{N}$. Suppose that $A \subseteq \mathcal{J}(T)$ is a completely T^q-invariant Borel set, i. e., $T^{-q}(A) = A$. Then $T^{-qn}(A) = A$ for all $n \ge 0$. Equivalently, $\mathbb{1}_A \circ T^{qn} = \mathbb{1}_A$ for all $n \ge 0$. Therefore,

$$\Lambda_\varphi^{qn}(\mathbb{1}_A) = \Lambda_\varphi^{qn}(\mathbb{1}_A \circ T^{qn}) = \mathbb{1}_A \cdot \Lambda_\varphi^{qn}(\mathbb{1}) = \mathbb{1}_A.$$

It follows from Theorem 27.4.6 that $\mathbb{1}_A = \mu_\varphi(A)\mathbb{1}$ μ_φ-almost everywhere. Thus, $\mu_\varphi(A) \in \{0, 1\}$, yielding the ergodicity of T^q. The total ergodicity of m_φ ensues from that of μ_φ and Theorem 27.3.3. □

In fact, we can easily prove more (cf. Theorem 13.8.14).

Theorem 27.4.8. *Let $T : \widehat{\mathbb{C}} \to \widehat{\mathbb{C}}$ be a rational function with $\deg(T) \ge 2$ and $\varphi : \mathcal{J}(T) \to \mathbb{R}$ a Hölder continuous potential such that $\mathrm{P}(\varphi) > \sup\varphi$. Then the measure-preserving dynamical system $(\mathcal{J}(T), T, \mu_\varphi)$ is metrically exact, and hence its Rokhlin natural extension is a K-system.*

Proof. Let $\mathcal{B} = \mathcal{B}(\mathcal{J}(T))$ be the Borel σ-algebra of $\mathcal{J}(T)$. By Definition 8.4.4, the measure-preserving dynamical system $(\mathcal{J}(T), T, \mu_\varphi)$ is metrically exact if $\lim_{n\to\infty} \mu_\varphi(T^n(B)) = 1$ for all $B \in \mathcal{B}$. Per Proposition 8.4.5, this holds if and only if the tail σ-algebra

$$\mathrm{Tail}_T(\mathcal{B}) := \bigcap_{n=0}^{\infty} T^{-n}(\mathcal{B})$$

consists only of sets of μ_φ-measure 0 or 1. So, let $A \in \mathrm{Tail}_T(\mathcal{B})$. For every $n \geq 0$, there exists a set $A_n \in \mathcal{B}$ such that $A = T^{-n}(A_n)$. Then

$$\Lambda_\varphi^n(\mathbb{1}_A) = \Lambda_\varphi^n(\mathbb{1}_{A_n} \circ T^n) = \mathbb{1}_{A_n} \cdot \Lambda_\varphi^n(\mathbb{1}) = \mathbb{1}_{A_n}.$$

It ensues from Theorem 27.4.6 that $\mathbb{1}_{A_n}$ converges in $L^1(\mu_\varphi)$ to the constant function $\mu_\varphi(A)\mathbb{1}$. But since $\mathbb{1}_{A_n}$ has only two values, namely 0 and 1, we conclude that

$$\mu_\varphi(A) \in \{0, 1\}.$$

Thus, the metric exactness of the measure-preserving system $(\mathcal{J}(T), T, \mu_\varphi)$ is ascertained. That its Rokhlin natural extension is K-mixing is the object of Theorem 8.4.6. □

27.5 Equilibrium states for Hölder potentials $\varphi : \mathcal{J}(T) \to \mathbb{R}$

27.5.1 The "Gibbs state" μ_φ is an equilibrium state for φ

As the only invariant Borel probability measure absolutely continuous with respect to an ergodic invariant Borel probability measure, is this ergodic measure itself, we establish from Theorems 27.4.7 and 27.3.3 the following fact.

Theorem 27.5.1. *Let $T : \widehat{\mathbb{C}} \to \widehat{\mathbb{C}}$ be a rational function with* $\deg(T) \geq 2$ *and $\varphi : \mathcal{J}(T) \to \mathbb{R}$ a Hölder continuous potential such that* $\mathrm{P}(\varphi) > \sup \varphi$. *Then μ_φ is the only T-invariant Borel probability measure on $\mathcal{J}(T)$ which is absolutely continuous with respect to m_φ.*

Our next objective, the main one in this subsection and one of the chief results in this chapter, is to show that μ_φ is an equilibrium state for the potential φ. The following lemma provides the desired lower bound on the free energy of T with respect to μ_φ.

Lemma 27.5.2. *Let $T : \widehat{\mathbb{C}} \to \widehat{\mathbb{C}}$ be a rational function with* $\deg(T) \geq 2$ *and $\varphi : \mathcal{J}(T) \to \mathbb{R}$ a Hölder continuous potential such that* $\mathrm{P}(\varphi) > \sup \varphi$. *Then*

$$\mathrm{h}_{\mu_\varphi}(T) + \int_{\mathcal{J}(T)} \varphi \, d\mu_\varphi \geq \log \lambda_\varphi. \tag{27.111}$$

Proof. Fix $\varepsilon > 0$. As the functions $\varphi : \mathcal{J}(T) \to \mathbb{R}$, $\rho_\varphi : \mathcal{J}(T) \to [\ell_\varphi, u_\varphi] \subseteq (0,\infty)$ and $1/\rho_\varphi : \mathcal{J}(T) \to [1/u_\varphi, 1/\ell_\varphi]$ are uniformly continuous on the compact set $\mathcal{J}(T)$, there exists $\delta > 0$ such that

$$|e^{\varphi(y)} - e^{\varphi(x)}| < \varepsilon \inf e^{\varphi} \quad \text{and} \quad |\rho_\varphi(y) - \rho_\varphi(x)| < \varepsilon \inf \rho_\varphi \tag{27.112}$$

whenever $x, y \in \mathcal{J}(T)$ and $|x - y| < \delta$. Let $\alpha = \alpha(\delta)$ be the Borel partition of $\mathcal{J}(T)$ constructed per Lemma 25.6.4. So, $\operatorname{diam}(\alpha) < \delta$ and $T|_A : A \to \widehat{\mathbb{C}}$ is injective for every $A \in \alpha$. Fix $z \in \mathcal{J}(T)$. Recall that $\alpha(z)$ is the atom in α where z lies. Using Chapter 9's notation, for every $n \in \mathbb{N}$ we then have

$$\alpha_1^n(z) := \bigcap_{j=1}^{n-1} T^{-j}(\alpha(T^j(z))) = T^{-1}\Big(\bigcap_{j=0}^{n-2} T^{-j}(\alpha(T^{j+1}(z)))\Big)$$

and

$$\alpha^n(z) := \alpha_0^n(z) = \alpha(z) \cap \alpha_1^n(z).$$

Using (27.85) and the left relation in (27.112), we obtain that

$$\begin{aligned}
m_\varphi(\alpha^n(z)) &= m_\varphi\Big(\alpha(z) \cap T^{-1}\Big(\bigcap_{j=0}^{n-2} T^{-j}(\alpha(T^{j+1}(z)))\Big)\Big) \\
&= m_\varphi\Big((T|_{\alpha(z)})^{-1}\Big(T(\alpha(z)) \cap \bigcap_{j=0}^{n-2} T^{-j}(\alpha(T^{j+1}(z)))\Big)\Big) \\
&= \int\limits_{T(\alpha(z)) \cap \bigcap_{j=0}^{n-2} T^{-j}(\alpha(T^{j+1}(z)))} \lambda_\varphi^{-1} \exp(\varphi \circ (T|_{\alpha(z)})^{-1})\, dm_\varphi \\
&\le \lambda_\varphi^{-1}(1+\varepsilon) e^{\varphi(z)} m_\varphi\Big(T(\alpha(z)) \cap \bigcap_{j=0}^{n-2} T^{-j}(\alpha(T^{j+1}(z)))\Big) \\
&\le (1+\varepsilon)\lambda_\varphi^{-1} e^{\varphi(z)} m_\varphi\Big(\bigcap_{j=0}^{n-2} T^{-j}(\alpha(T^{j+1}(z)))\Big).
\end{aligned}$$

Using the right relation in (27.112) and the T-invariance of μ_φ, we get that

$$\begin{aligned}
\mu_\varphi(\alpha^n(z)) &= \int\limits_{\alpha^n(z)} \rho_\varphi \, dm_\varphi \le (1+\varepsilon)\rho_\varphi(z) m_\varphi(\alpha^n(z)) \\
&\le (1+\varepsilon)^2 \lambda_\varphi^{-1} e^{\varphi(z)} \rho_\varphi(z) m_\varphi\Big(\bigcap_{j=0}^{n-2} T^{-j}(\alpha(T^{j+1}(z)))\Big) \\
&= (1+\varepsilon)^2 \lambda_\varphi^{-1} e^{\varphi(z)} \int\limits_{\bigcap_{j=0}^{n-2} T^{-j}(\alpha(T^{j+1}(z)))} \rho_\varphi(z)\, dm_\varphi(w)
\end{aligned}$$

$$
\begin{aligned}
&\le (1+\varepsilon)^2 \lambda_\varphi^{-1} e^{\varphi(z)} \int_{\bigcap_{j=0}^{n-2} T^{-j}(\alpha(T^{j+1}(z)))} (1+\varepsilon)\rho_\varphi(w)\, dm_\varphi(w) \\
&\le (1+\varepsilon)^3 \lambda_\varphi^{-1} e^{\varphi(z)} \int_{\bigcap_{j=0}^{n-2} T^{-j}(\alpha(T^{j+1}(z)))} d\mu_\varphi(w) \\
&= (1+\varepsilon)^3 \lambda_\varphi^{-1} e^{\varphi(z)} \mu_\varphi\Big(\bigcap_{j=0}^{n-2} T^{-j}(\alpha(T^{j+1}(z)))\Big) \\
&= (1+\varepsilon)^3 \lambda_\varphi^{-1} e^{\varphi(z)} \mu_\varphi\Big(\bigcap_{j=0}^{n-2} T^{-(j+1)}(\alpha(T^{j+1}(z)))\Big) \\
&= (1+\varepsilon)^3 \lambda_\varphi^{-1} e^{\varphi(z)} \mu_\varphi(\alpha_1^n(z)).
\end{aligned}
$$

Thus,

$$
(\mu_\varphi)_{\alpha_1^n(z)}(\alpha^n(z)) := \frac{\mu_\varphi(\alpha^n(z) \cap \alpha_1^n(z))}{\mu_\varphi(\alpha_1^n(z))} = \frac{\mu_\varphi(\alpha^n(z))}{\mu_\varphi(\alpha_1^n(z))} \le (1+\varepsilon)^3 \lambda_\varphi^{-1} e^{\varphi(z)}.
$$

Hence,

$$
I_{\mu_\varphi}(\alpha^n | \alpha_1^n)(z) := -\log((\mu_\varphi)_{\alpha_1^n(z)}(\alpha^n(z))) \ge -3\log(1+\varepsilon) + \log\lambda_\varphi - \varphi(z).
$$

So,

$$
\mathrm{H}_{\mu_\varphi}(\alpha^n | \alpha_1^n) := \int_{\mathcal{J}(T)} I_{\mu_\varphi}(\alpha^n | \alpha_1^n)\, d\mu_\varphi \ge -3\log(1+\varepsilon) + \log\lambda_\varphi - \int_{\mathcal{J}(T)} \varphi\, d\mu_\varphi.
$$

Therefore (cf. Definition 9.4.10),

$$
\mathrm{h}_{\mu_\varphi}(T) \ge \mathrm{h}_{\mu_\varphi}(T, \alpha) := \lim_{n\to\infty} \mathrm{H}_{\mu_\varphi}(\alpha^n | \alpha_1^n) \ge -3\log(1+\varepsilon) + \log\lambda_\varphi - \int_{\mathcal{J}(T)} \varphi\, d\mu_\varphi.
$$

Letting $\varepsilon \searrow 0$ leads to the result. □

As an immediate consequence of this lemma, Lemma 27.2.6 and the variational principle (Theorem 12.1.1), we reach our objective (cf. Proposition 13.2.9).

Theorem 27.5.3. *Let $T : \widehat{\mathbb{C}} \to \widehat{\mathbb{C}}$ be a rational function with $\deg(T) \ge 2$ and $\varphi : \mathcal{J}(T) \to \mathbb{R}$ a Hölder continuous potential such that $\mathrm{P}(\varphi) > \sup\varphi$. Then $\mathrm{P}(\varphi) = \log\lambda_\varphi$ and μ_φ is an equilibrium state for the potential φ.*

Finally, we confirm the uniqueness of m_φ, μ_φ and ρ_φ (cf. Proposition 13.8.10).

Proposition 27.5.4. *Let $T : \widehat{\mathbb{C}} \to \widehat{\mathbb{C}}$ be a rational function of $\deg(T) \ge 2$ and $\varphi : \mathcal{J}(T) \to \mathbb{R}$ a Hölder continuous potential such that $\mathrm{P}(\varphi) > \sup\varphi$. Then:*

(a) *There exists a unique pair* (λ, m), *with* $\lambda \in [0,\infty)$ *and* $m \in M(\mathcal{J}(T))$ *such that* $\mathcal{L}_\varphi^*(m) = \lambda m$; *namely,* $\lambda = e^{\mathrm{P}(\varphi)}$ *and* $m = m_\varphi$. *In particular,* $\lambda > 0$.
(b) *There exists a unique* $\rho \in C(\mathcal{J}(T))$ *such that* $\widehat{\mathcal{L}}_\varphi(\rho) = \rho$, $\rho \geq 0$, *and* $\int_{\mathcal{J}(T)} \rho\, dm_\varphi = 1$; *namely* $\rho = \rho_\varphi$.

Proof. From Subsection 27.2.3 on, all that was assumed about m_φ and λ_φ was that m_φ is an eigenmeasure of the dual operator $\mathcal{L}_\varphi^* : C(\mathcal{J}(T))^* \to C(\mathcal{J}(T))^*$ with eigenvalue $\lambda_\varphi > 0$, that is, $\mathcal{L}_\varphi^*(m_\varphi) = \lambda_\varphi m_\varphi$ (cf. (27.79)). Thus, all the results from that point on apply to any pair (λ, m), with $\lambda \in (0,\infty)$ and $m \in M(\mathcal{J}(T))$ such that $\mathcal{L}_\varphi^*(m) = \lambda m$.

The fact that $\lambda = e^{\mathrm{P}(\varphi)}$, and hence is unique, follows from the observation that, when proving that $\lambda_\varphi = e^{\mathrm{P}(\varphi)}$, all that we assumed about λ_φ was that $\mathcal{L}_\varphi^*(m_\varphi) = \lambda_\varphi m_\varphi$ for some $m_\varphi \in M(\mathcal{J}(T))$.

Likewise, assuming only the equality $\mathcal{L}_\varphi^*(m_\varphi) = \lambda_\varphi m_\varphi$ for some $m_\varphi \in M(\mathcal{J}(T))$, we derived Theorem 27.4.5. From part (d) of that result, we deduce that

$$\rho_\varphi = \lim_{n\to\infty} \widehat{\mathcal{L}}_\varphi^n(\mathbb{1}),$$

which is completely independent of m_φ (given that λ_φ has already been shown to be unique). So ρ_φ is unique, i.e. $\rho = \rho_\varphi$, and the above is a formula for it. This yields part (b).

From this and Theorem 27.4.5(d), we infer that

$$m_\varphi(g) = \lim_{n\to\infty} \frac{\widehat{\mathcal{L}}_\varphi^n(g)(z)}{\widehat{\mathcal{L}}_\varphi^n(\mathbb{1})(z)}, \qquad \forall z \in \mathcal{J}(T).$$

This establishes the uniqueness of m_φ. □

27.6 Continuous dependence on φ of the "Gibbs states" m_φ, μ_φ and of the density ρ_φ

Our ultimate goal now is to show that μ_φ is the unique equilibrium state for the potential φ. We will achieve this by proving in the next section that the pressure function is differentiable. It is a fairly general fact that such differentiability gives rise to uniqueness of equilibria. As a preparation, we establish in this section some interesting results leading to the continuous dependence on the potential φ of the "Gibbs states" m_φ, μ_φ, and of the Radon–Nikodym derivative ρ_φ.

Let $T : \widehat{\mathbb{C}} \to \widehat{\mathbb{C}}$ be a rational function with $\deg(T) \geq 2$ and $\varphi : \mathcal{J}(T) \to \mathbb{R}$ a Hölder continuous potential such that $\mathrm{P}(\varphi) > \sup \varphi$. As the function $C(\mathcal{J}(T)) \ni \psi \mapsto \mathrm{P}(\psi) \in \mathbb{R}$ is continuous (even Lipschitz continuous with Lipschitz constant 1; see Exercise 11.5.7 or Corollary 12.1.11), there exists $\eta_\varphi > 0$ such that

$$\mathrm{P}(\psi) > \sup \psi \quad \text{for every} \quad \psi \in \overline{B_\infty(\varphi, \eta_\varphi)} = \{\psi \in C(\mathcal{J}(T)) : \|\psi - \varphi\|_\infty \leq \eta_\varphi\}.$$

Prior to stating the next result, recall the definition of $D_{\eta,H}(\varphi)$ in (27.40). When $\eta = \eta_\varphi$, we shorten the notation $D_{\eta_\varphi,H}(\varphi)$ to $D_H(\varphi)$.

Lemma 27.6.1. *Let $T : \widehat{\mathbb{C}} \to \widehat{\mathbb{C}}$ be a rational function with $\deg(T) \geq 2$ and $\varphi : \mathcal{J}(T) \to \mathbb{R}$ a Hölder continuous potential such that $\mathrm{P}(\varphi) > \sup\varphi$. For any $H \geq 0$, the function*

$$D_H(\varphi) \ni \psi \longmapsto m_\psi \in M(\mathcal{J}(T))$$

is continuous when $M(\mathcal{J}(T))$ is equipped with the weak* topology.

Proof. Let $(\psi_n)_{n=1}^\infty$ be a sequence of functions in $D_H(\varphi)$ converging uniformly to some function $\psi : \mathcal{J}(T) \to \mathbb{R}$. Let $(m_{\psi_{n_k}})_{k=1}^\infty$ be any convergent subsequence of the sequence $(m_{\psi_n})_{n=1}^\infty$. Denote its limit by m. The function

$$C(\mathcal{J}(T)) \ni \chi \longmapsto \mathcal{L}_\chi \in L(C(\mathcal{J}(T)))$$

is obviously continuous. Recalling that $C(\mathcal{J}(T)) \ni \chi \mapsto \mathrm{P}(\chi) \in \mathbb{R}$ is continuous too, we deduce that

$$m = \lim_{k\to\infty} m_{\psi_{n_k}} = \lim_{k\to\infty} e^{-\mathrm{P}(\psi_{n_k})}\mathcal{L}^*_{\psi_{n_k}}(m_{\psi_{n_k}}) = e^{-\mathrm{P}(\psi)}\mathcal{L}^*_\psi(m).$$

So, $\mathcal{L}^*_\psi(m) = e^{\mathrm{P}(\psi)}m$, whence $m = m_\psi$ by Proposition 27.5.4(a). □

In Lemma 27.2.7, we established that the supremum norms of the iterates of the normalized operator are uniformly bounded from above and called that bound u_φ. We now show that there is a bound that works for all $\psi \in D_H(\varphi)$ and call it $u_{\varphi,H}$.

Lemma 27.6.2. *Let $T : \widehat{\mathbb{C}} \to \widehat{\mathbb{C}}$ be a rational function with $\deg(T) \geq 2$ and $\varphi : \mathcal{J}(T) \to \mathbb{R}$ a Hölder continuous potential such that $\mathrm{P}(\varphi) > \sup\varphi$. For any $H \geq 0$,*

$$u_{\varphi,H} := \sup\{\|\widehat{\mathcal{L}}^n_\psi(\mathbb{1})\|_\infty : \psi \in D_H(\varphi),\ n \geq 0\} < \infty.$$

Proof. It follows from Lemma 27.6.1 and the compactness of $M(\mathcal{J}(T))$ that

$$E_{\varphi,H} := \inf\{E_\psi \mid \psi \in D_H(\varphi)\} > 0,$$

where the quantities E_ψ are defined by (27.88).

Moreover, by definition of $D_H(\varphi)$, the function $D_H(\varphi) \ni \psi \mapsto \mathrm{P}(\psi) - \sup\psi$ is strictly positive at each point. The continuity of that function and the compactness of $D_H(\varphi)$ imply that this function attains a strictly positive minimum. Call this minimum β'. Take $\gamma \in (e^{-\beta'}, 1)$ in lieu of (27.64). Pick s according to (27.65)–(27.66). Observe that the results brought about in Subsection 27.2.2 apply with $\beta = \log\lambda_\psi = \mathrm{P}(\psi)$ (and corresponding η per (27.67)) for every $\psi \in D_H(\varphi)$. Additionally, thanks to Lemma 27.1.4 we can replace $C_{\gamma,s}(\varphi)$ in (27.87) by the constant

$$C(\varphi,H)=C_{\gamma,s}(\varphi,H):=\Delta(T)\max\{e^{2\Gamma_w(\varphi,\eta_\varphi,H)}\mid w\in Y_{\gamma,s}\}\in[1,\infty).$$

It ensues from Lemma 27.2.5 that

$$G_{\varphi,H}:=\sup\{G_n(\log\lambda_\psi,z,\xi)\mid n\in\mathbb{N},\ \psi\in D_H(\varphi),\ z\in Y_{\gamma,s},\ \xi\in B(z,r_z)\}\le E_{\varphi,H}^{-1}C(\varphi,H).$$

We can further replace $M_\beta(\varphi)$ and $N_\beta(\varphi)$ from (27.74)–(27.75) by

$$M_{\varphi,H}:=\sup\{M_{\log\lambda_\psi}(\psi)\mid\psi\in D_H(\varphi)\}\le\max\Big\{G_{\varphi,H},\ \sup_{\psi\in D_H(\varphi)}\max_{0\le j\le s}\|\widehat{\mathcal{L}}_\psi^j(\mathbb{1})\|_\infty\Big\}<\infty$$

and

$$N_{\varphi,H}:=M_{\varphi,H}\big[1-2d(2d-2)\Delta(T)(1-e^{-\eta})^{-1}e^{-\eta(s+1)}\big]^{-1}<\infty.$$

Lemma 27.2.3 implies that $u_{\varphi,H}\le N_{\varphi,H}<\infty$. □

Aiming at generalizing Lemma 27.3.1, we now turn our attention to the relative compactness of sets of functions $\widehat{\mathcal{L}}_\psi^n(g)$.

Lemma 27.6.3. *Let $T:\widehat{\mathbb{C}}\to\widehat{\mathbb{C}}$ be a rational function with* $\deg(T)\ge 2$ *and $\varphi:\mathcal{J}(T)\to\mathbb{R}$ a Hölder continuous potential such that* $\mathrm{P}(\varphi)>\sup\varphi$. *Then for every $H\ge 0$ and every relatively compact set $K\subseteq C(\mathcal{J}(T))$, the set*

$$\{\widehat{\mathcal{L}}_\psi^n(g):\psi\in D_H(\varphi),\ g\in K,\ n\ge 0\}\subseteq C(\mathcal{J}(T))$$

is relatively compact.

Proof. By Lemma 27.6.2, the said set is uniformly bounded and thereby pointwise bounded.

Suppose first that K is a singleton, i. e., $K=\{g\}$ for some $g\in C(\mathcal{J}(T))$. With slight appropriate modifications which we indicate below, the proof of Lemma 27.3.1 establishes that the set

$$\{\widehat{\mathcal{L}}_\psi^n(g):\psi\in D_H(\varphi),\ n\ge 0\}\subseteq C(\mathcal{J}(T))$$

is equicontinuous, and hence relatively compact according to the Arzelà–Ascoli theorem (Theorem A.2.3). Indeed, as in the proof of Lemma 27.6.2, take $\gamma\in(e^{-\beta'},1)$. However, pick s so that it satisfy (27.65) as well as (27.97) with u_φ replaced by $u_{\varphi,H}$ from Lemma 27.6.2. Observe once again that the results brought about in Subsection 27.2.2 apply with $\beta=\log\lambda_\psi=\mathrm{P}(\psi)$ for every $\psi\in D_H(\varphi)$. Accordingly, replace φ by $\psi\in D_H(\varphi)$ in the rest of the proof of Lemma 27.3.1. Next, replace all occurrences of u_φ by $u_{\varphi,H}$. Replace also $\Gamma(\varphi)$ by $\Gamma(\varphi,H):=\max\{\Gamma_z(\varphi,\eta_\varphi,H):z\in Y_{\gamma,s}\}$ and correspondingly use Lemma 27.1.4 rather than Lemma 27.1.2.

Now consider the general case. Choose an arbitrary sequence

$$(\widehat{\mathcal{L}}_{\psi_j}^{n_j}(g_j))_{j=1}^{\infty},$$

where $(n_j)_{j=1}^{\infty}$ is a sequence of nonnegative integers, $(\psi_j)_{j=1}^{\infty}$ a sequence in $D_H(\varphi)$ and $(g_j)_{j=1}^{\infty}$ a sequence in K. As the set K is relatively compact and $D_H(\varphi)$ is compact, by passing to a subsequence, we may assume without loss of generality that $(g_j)_{j=1}^{\infty}$ converges uniformly to some $g \in C(\mathcal{J}(T))$ and $(\psi_j)_{j=1}^{\infty}$ converges uniformly to some $\psi \in D_H(\varphi)$. Using Lemma 27.6.2, we then have that

$$\|\widehat{\mathcal{L}}_{\psi_j}^{n_j}(g_j) - \widehat{\mathcal{L}}_{\psi_j}^{n_j}(g)\|_\infty = \|\widehat{\mathcal{L}}_{\psi_j}^{n_j}(g_j - g)\|_\infty \le u_{\varphi,H}\|g_j - g\|_\infty.$$

Since $\lim_{j\to\infty} u_{\varphi,H}\|g_j - g\|_\infty = 0$ and since, in view of the already proved singleton part of our lemma, the sequence $(\widehat{\mathcal{L}}_{\psi_j}^{n_j}(g))_{j=1}^{\infty}$ is relatively compact, we conclude that $(\widehat{\mathcal{L}}_{\psi_j}^{n_j}(g_j))_{j=1}^{\infty}$ contains a convergent subsequence. □

Lemma 27.6.3 (applied with $K = \{\mathbb{1}\}$) combined with Proposition 27.4.5(e) (applied with $g := \mathbb{1}$) allow us to attest the relative compactness of the set of Radon–Nikodym derivatives $\rho_\psi = d\mu_\psi/dm_\psi$.

Lemma 27.6.4. *Let $T : \widehat{\mathbb{C}} \to \widehat{\mathbb{C}}$ be a rational function with $\deg(T) \ge 2$ and $\varphi : \mathcal{J}(T) \to \mathbb{R}$ a Hölder continuous potential such that $\mathrm{P}(\varphi) > \sup\varphi$. Then for every $H \ge 0$ the set*

$$\{\rho_\psi : \psi \in D_H(\varphi)\} \subseteq C(\mathcal{J}(T))$$

is relatively compact.

We can now assert the continuity of the Radon–Nikodym derivatives ρ_ψ as well as the continuity of the "Gibbs states" m_ψ and μ_ψ (cf. Theorem 13.8.21).

Proposition 27.6.5. *Let $T : \widehat{\mathbb{C}} \to \widehat{\mathbb{C}}$ be a rational function of $\deg(T) \ge 2$ and $\varphi : \mathcal{J}(T) \to \mathbb{R}$ a Hölder continuous potential such that $\mathrm{P}(\varphi) > \sup\varphi$. Then for every $H \ge 0$ the function*

$$D_H(\varphi) \ni \psi \longmapsto \rho_\psi \in C(\mathcal{J}(T))$$

is continuous with respect to the topology of uniform convergence on $C(\mathcal{J}(T))$ while the functions

$$D_H(\varphi) \ni \psi \longmapsto m_\psi \in M(\mathcal{J}(T)) \quad \text{and} \quad D_H(\varphi) \ni \psi \longmapsto \mu_\psi \in M(\mathcal{J}(T))$$

are continuous with respect to the weak* *topology on the set $M(\mathcal{J}(T))$ of Borel probability measures on $\mathcal{J}(T)$.*

Proof. Let $(\psi_n)_{n=1}^{\infty}$ be a sequence of functions in $D_H(\varphi)$ converging uniformly to some function $\psi : \mathcal{J}(T) \to \mathbb{R}$, which belongs to $D_H(\varphi)$ as this set is compact.

The assertion about $\chi \mapsto m_\chi$ has already been proved in Lemma 27.6.1. The statement on $\chi \mapsto \mu_\chi$ will follow once we establish the continuity of $\chi \mapsto \rho_\chi$.

To this end, suppose for a contradiction that this latter fails at a function $\psi \in D_H(\varphi)$. This means that there exist $\delta > 0$ and a sequence $(\psi_k)_{k=1}^\infty$ of functions in $D_H(\varphi)$ that converges uniformly to the function ψ and such that

$$\|\rho_{\psi_k} - \rho_\psi\|_\infty > \delta, \quad \forall k \in \mathbb{N}. \tag{27.113}$$

By virtue of Lemma 27.6.4 and by passing to a subsequence, we may assume that the sequence $(\rho_{\psi_k})_{k=1}^\infty$ converges uniformly to some function $\rho \in C(\mathcal{J}(T))$. By (27.113),

$$\rho \neq \rho_\psi. \tag{27.114}$$

On the other hand, the continuity of the functions $\chi \mapsto \mathcal{L}_\chi$ and $\chi \mapsto \mathrm{P}(\chi)$ entails that

$$\mathcal{L}_\psi(\rho) = e^{\mathrm{P}(\psi)}\rho.$$

Moreover, using the continuity of $\chi \mapsto m_\chi$, we conclude that

$$\int_{\mathcal{J}(T)} \rho \, dm_\psi = \lim_{k\to\infty} \int_{\mathcal{J}(T)} \rho_{\psi_k} \, dm_{\psi_k} = \lim_{k\to\infty} 1 = 1.$$

So, $\rho = \rho_\psi$ according to Proposition 27.5.4(b), contrary to (27.114).

The continuity of $\chi \mapsto \mu_\chi$ is a direct consequence of the continuity of the eigenmeasures $\chi \mapsto m_\chi$ and of the eigenfunctions $\chi \mapsto \rho_\chi$. Indeed, if $(\psi_n)_{n=1}^\infty$ is a sequence in $D_H(\varphi)$ converging uniformly to $\psi \in D_H(\varphi)$, then for every function $g \in C(\mathcal{J}(T))$ we get

$$\lim_{n\to\infty} \int_{\mathcal{J}(T)} g \, d\mu_{\psi_n} = \lim_{n\to\infty} \int_{\mathcal{J}(T)} g\rho_{\psi_n} \, dm_{\psi_n} = \int_{\mathcal{J}(T)} g\rho_\psi \, dm_\psi = \int_{\mathcal{J}(T)} g \, d\mu_\psi.$$

So, the sequence $(\mu_{\psi_n})_{n=1}^\infty$ converges to μ_ψ in the weak* topology on $M(C(\mathcal{J}(T)))$. □

27.7 Differentiability of topological pressure

In this section, we demonstrate that the pressure function is differentiable. This will be the key ingredient to prove the uniqueness of equilibrium states.

Lemma 27.7.1. *Let $T : \widehat{\mathbb{C}} \to \widehat{\mathbb{C}}$ be a rational function with $\deg(T) \geq 2$ and $\varphi : \mathcal{J}(T) \to \mathbb{R}$ a Hölder continuous potential such that $\mathrm{P}(\varphi) > \sup \varphi$. Fix $H \geq 0$. If $\psi \in D_H(\varphi)$ and $g \in C(\mathcal{J}(T))$, then*

$$\lim_{n\to\infty} \left\| \frac{\frac{1}{n}\mathcal{L}_\psi^n(S_n g)}{\mathcal{L}_\psi^n(\mathbb{1})} - \left(\int_{\mathcal{J}(T)} g \, d\mu_\psi\right)\mathbb{1} \right\|_\infty = 0.$$

Proof. Observe that

$$\frac{\frac{1}{n}\mathcal{L}_\psi^n(S_n g)}{\mathcal{L}_\psi^n(\mathbb{1})} = \frac{\frac{1}{n}\sum_{j=0}^{n-1}\widehat{\mathcal{L}}_\psi^n(g\circ T^j)}{\widehat{\mathcal{L}}_\psi^n(\mathbb{1})} = \frac{\frac{1}{n}\sum_{j=0}^{n-1}\widehat{\mathcal{L}}_\psi^{n-j}(g\widehat{\mathcal{L}}_\psi^j(\mathbb{1}))}{\widehat{\mathcal{L}}_\psi^n(\mathbb{1})}. \tag{27.115}$$

It follows from Lemma 27.6.3 that

$$\text{the set } \{g\widehat{\mathcal{L}}_\psi^j(\mathbb{1}) : j \geq 0\} \subseteq C(\mathcal{J}(T)) \text{ is relatively compact.}$$

Applying Lemma 27.6.3 again, we deduce that

$$\text{the set } \{\widehat{\mathcal{L}}_\psi^{n-j}(g\widehat{\mathcal{L}}_\psi^j(\mathbb{1})) : n \in \mathbb{N},\ 0 \leq j < n\} \text{ is relatively compact.}$$

Therefore,

$$\text{the set } \left\{\frac{1}{n}\sum_{j=0}^{n-1}\widehat{\mathcal{L}}_\psi^{n-j}(g\widehat{\mathcal{L}}_\psi^j(\mathbb{1})) : n \in \mathbb{N}\right\} \text{ is relatively compact.} \tag{27.116}$$

Using Lemma 27.6.2, we obtain that

$$\begin{aligned}
&\left\|\widehat{\mathcal{L}}_\psi^{n-j}(g\widehat{\mathcal{L}}_\psi^j(\mathbb{1})) - \left(\int g\,d\mu_\psi\right)\rho_\psi\right\|_\infty \\
&\quad = \left\|\widehat{\mathcal{L}}_\psi^{n-j}(g\widehat{\mathcal{L}}_\psi^j(\mathbb{1}) - g\rho_\psi) + \widehat{\mathcal{L}}_\psi^{n-j}(g\rho_\psi) - \left(\int g\,d\mu_\psi\right)\rho_\psi\right\|_\infty \\
&\quad \leq \|\widehat{\mathcal{L}}_\psi^{n-j}\|_\infty \|g(\widehat{\mathcal{L}}_\psi^j(\mathbb{1}) - \rho_\psi)\|_\infty + \left\|\widehat{\mathcal{L}}_\psi^{n-j}(g\rho_\psi) - \left(\int g\,d\mu_\psi\right)\rho_\psi\right\|_\infty \\
&\quad \leq u_{\varphi,H}\|g\|_\infty \|\widehat{\mathcal{L}}_\psi^j(\mathbb{1}) - \rho_\psi\|_\infty + \left\|\widehat{\mathcal{L}}_\psi^{n-j}(g\rho_\psi) - \left(\int g\,d\mu_\psi\right)\rho_\psi\right\|_\infty.
\end{aligned}$$

By virtue of Theorem 27.4.5(d),

$$\lim_{n\to\infty}\|\widehat{\mathcal{L}}_\psi^n(\mathbb{1}) - \rho_\psi\|_\infty = 0 \quad \text{and} \quad \lim_{n-j\to\infty}\left\|\widehat{\mathcal{L}}_\psi^{n-j}(g\rho_\psi) - \left(\int g\,d\mu_\psi\right)\rho_\psi\right\|_\infty = 0.$$

Therefore,

$$\lim_{\substack{j\to\infty \\ n-j\to\infty}}\left\|\widehat{\mathcal{L}}_\psi^{n-j}(g\widehat{\mathcal{L}}_\psi^j(\mathbb{1})) - \left(\int g\,d\mu_\psi\right)\rho_\psi\right\|_\infty = 0.$$

Consequently,

$$\lim_{n\to\infty}\left\|\frac{1}{n}\sum_{j=0}^{n-1}\widehat{\mathcal{L}}_\psi^{n-j}(g\widehat{\mathcal{L}}_\psi^j(\mathbb{1})) - \left(\int g\,d\mu_\psi\right)\rho_\psi\right\|_\infty = 0.$$

It follows from this and Theorem 27.4.5(d) (applied with $g = \mathbb{1}$) that

$$\lim_{n\to\infty}\left\| \frac{\frac{1}{n}\sum_{j=0}^{n-1}\widehat{\mathcal{L}}_{\psi}^{n-j}(g\widehat{\mathcal{L}}_{\psi}^{j}(\mathbb{1}))}{\widehat{\mathcal{L}}_{\psi}^{n}(\mathbb{1})} - \left(\int g\,d\mu_{\psi}\right)\mathbb{1}\right\|_{\infty} = 0.$$

We are done by invoking (27.115). □

Finally, we assert the differentiability of the pressure function (cf. Theorem 13.10.6).

Proposition 27.7.2. *Let $T : \widehat{\mathbb{C}} \to \widehat{\mathbb{C}}$ be a rational function of $\deg(T) \geq 2$ and $\varphi : \mathcal{J}(T) \to \mathbb{R}$ a Hölder continuous potential such that $\mathrm{P}(\varphi) > \sup\varphi$. If $g : \mathcal{J}(T) \to \mathbb{R}$ is a Hölder continuous function, then the function*

$$\mathbb{R} \ni t \longmapsto \mathrm{P}(\varphi + tg) \in \mathbb{R}$$

is differentiable on some sufficiently small open neighborhood of 0, and

$$\frac{d}{dt}\mathrm{P}(\varphi + tg) = \int_{\mathcal{J}(T)} g\,d\mu_{\varphi+tg}.$$

Proof. Set

$$\varphi_t := \varphi + tg.$$

Obviously, there are $H > 0$ and $\delta > 0$ such that $\varphi_t \in D_H(\varphi)$ for all $t \in [-\delta, \delta]$. It then ensues from Lemma 27.2.10 and Theorem 27.5.3 that for every $z \in \mathcal{J}(T)$,

$$\begin{aligned}\mathrm{P}(\varphi_t) &= \lim_{n\to\infty}\frac{1}{n}\log\mathcal{L}_{\varphi_t}^{n}(\mathbb{1})(z)\\ &= \lim_{n\to\infty}\frac{1}{n}\log\sum_{w\in T^{-n}(z)}\exp(S_n\varphi(w) + tS_ng(w))\deg_w(T^n).\end{aligned} \tag{27.117}$$

Fix an arbitrary $z \in \mathcal{J}(T)$ and for every $n \in \mathbb{N}$ set

$$F_n(t) = \frac{1}{n}\log\mathcal{L}_{\varphi_t}^{n}(\mathbb{1})(z),$$

so that

$$\mathrm{P}(\varphi_t) = \lim_{n\to\infty}F_n(t). \tag{27.118}$$

The following claim follows from Hölder's inequality.

Claim 1. *For every $n \in \mathbb{N}$, the function $[-\delta, \delta] \ni t \mapsto F_n(t) \in \mathbb{R}$ is convex.*

The reader can further convince themselves of the following fact.

Claim 2. *For every $n \in \mathbb{N}$,*

$$F_n'(t) = \frac{\frac{1}{n}\mathcal{L}_{\varphi_t}^n(S_n g)(z)}{\mathcal{L}_{\varphi_t}^n(\mathbb{1})(z)}, \quad \forall t \in (-\delta, \delta).$$

The functions

$$\big((-\delta, \delta) \ni t \longmapsto F_n'(t) \in \mathbb{R} \big)_{n=1}^{\infty}$$

are each monotone by Claim 1. By Claim 2 and Lemma 27.7.1, they converge pointwise to the function

$$(-\delta, \delta) \ni t \longmapsto \int_{\mathcal{J}(T)} g \, d\mu_{\varphi_t},$$

which is continuous in view of Proposition 27.6.5. Hence, by virtue of Dini's theorem (Theorem A.1.10), this convergence is uniform on any compact subset of $(-\delta, \delta)$. Combining this with (27.118), we conclude that the function

$$(-\delta, \delta) \ni t \longmapsto \mathrm{P}(\varphi + tg) \in \mathbb{R}$$

is differentiable and its derivative $\frac{d}{dt}\mathrm{P}(\varphi+tg)$ is equal to $\int g \, d\mu_{\varphi+tg}$, the limit of the derivatives $F_n'(t)$. □

27.8 Uniqueness of equilibrium states for Hölder continuous potentials

In order to make use of Proposition 27.7.2 to conclude the proof of the uniqueness of equilibrium states, we need one general fact. We formulate and prove it now.

Suppose that $(V, \|\cdot\|)$ is a real normed space. Fix a point $x \in V$ and let U be an open neighborhood of x in V. Let

$$P : U \to \mathbb{R}$$

be a function. We call a continuous linear functional

$$F : V \to \mathbb{R}$$

tangent to P at a point $x \in V$ if there exists a number $\eta = \eta(F) > 0$ such that

$$x + B_V(0, \eta) \subseteq U \quad \text{and} \quad F(y) \le P(x+y) - P(x), \quad \forall y \in B_V(0, \eta).$$

We denote the set of all such functionals by $V^*(P; x)$. Sometimes the term *supporting functional* instead of *tangent* is used in the literature.

The following fact is well known. For instance, its short proof is provided as the proof of the second part (implication) of Theorem 3.6.5 in [110]. We include it here for the sake of completeness and convenience of the reader.

Lemma 27.8.1. *Let $(V, \|\cdot\|)$ be a real normed space, let $x \in V$ and let U be an open neighborhood of x in V. Let $P : U \to \mathbb{R}$ be a function. If the function P is differentiable at x in every direction in V, then the set $V^*(P; x)$ contains at most one element.*

Proof. Suppose for a contradiction that $V^*(P; x)$ contains at least two distinct bounded linear functionals, say F and $\widehat{F}$. Then $F(y) - \widehat{F}(y) > 0$ for some $y \in V$. Since $F, \widehat{F}$ are linear, we can find such a point y with $\|y\| < \min\{\eta(F), \eta(\widehat{F})\}$. By hypothesis, P is differentiable at x in the direction given by y. Hence,

$$\lim_{t\to 0} \frac{P(x+ty) - P(x)}{t} = \lim_{t\to 0} \frac{P(x-ty) - P(x)}{-t}.$$

Consequently,

$$\lim_{t\to 0} \frac{P(x+ty) + P(x-ty) - 2P(x)}{t} = 0.$$

On the other hand, as F and $\widehat{F}$ are supporting functionals of P at x, we have for every $0 < t \le 1$ that

$$P(x+ty) - P(x) \ge F(ty) = tF(y) \quad \text{and} \quad P(x-ty) - P(x) \ge \widehat{F}(-ty) = -t\widehat{F}(y).$$

Thus,

$$\limsup_{t\to 0} \frac{P(x+ty) + P(x-ty) - 2P(x)}{t} \ge F(y) - \widehat{F}(y) > 0.$$

This contradiction finishes the proof. □

Finally, we can confirm the uniqueness of equilibrium states. It is worth recalling that the uniqueness of equilibrium states was established in Chapter 13 for transitive, open, distance expanding systems. But not all rational functions are expanding. The proof in the current setting is quite different. It relies on the differentiability of the pressure function. In Chapter 13, we opted for a more straightforward approach. We could have proved uniqueness of equilibrium states through the pressure's differentiability as well but that property (in fact, the pressure's real-analyticity) was derived at a later stage.

Theorem 27.8.2. *If $T : \widehat{\mathbb{C}} \to \widehat{\mathbb{C}}$ is a rational function with $\deg(T) \ge 2$ and $\varphi : \mathcal{J}(T) \to \mathbb{R}$ is a Hölder continuous potential such that $\mathrm{P}(\varphi) > \sup \varphi$, then μ_φ is the unique equilibrium state for φ.*

Proof. Consider the topological pressure function

$$\mathrm{P} : H_\alpha(\mathcal{J}(T)) \to \mathbb{R}.$$

Let ν be an equilibrium state for the potential φ. Restricting ν to the Banach space $H_\alpha(\mathcal{J}(T))$, we know that ν is a bounded linear functional. Since $\nu(\varphi + \psi) = \nu(\varphi) + \nu(\psi)$, by subtracting the equality $\mathrm{h}_\nu(T) + \nu(\varphi) = \mathrm{P}(\varphi)$ from both sides of the inequality $\mathrm{h}_\nu(T) + \nu(\varphi + \psi) \le \mathrm{P}(\varphi + \psi)$, we get

$$\nu(\psi) \le \mathrm{P}(\varphi + \psi) - \mathrm{P}(\varphi).$$

So, since, by Proposition 27.7.2, the pressure function $\mathrm{P} : H_\alpha(\mathcal{J}(T)) \to \mathbb{R}$ is differentiable at φ in every direction in $H_\alpha(\mathcal{J}(T))$, we deduce from Lemma 27.8.1 that there exists at most one equilibrium state for the potential φ. As μ_φ is an equilibrium state, the proof of Theorem 27.8.2 is complete. □

Invoking Theorem 12.2.2 according to which $\mathrm{h}_\mu(T) > 0$ for every equilibrium state μ of φ whenever $\mathrm{P}(\varphi) > \sup\varphi$, we can collect the main results of this chapter in the following statements.

Theorem 27.8.3. *If $T : \widehat{\mathbb{C}} \to \widehat{\mathbb{C}}$ is a rational function with $\deg(T) \ge 2$ and $\varphi : \mathcal{J}(T) \to \mathbb{R}$ is a Hölder continuous potential such that $\mathrm{P}(\varphi) > \sup\varphi$, then:*
(a) *There is a unique equilibrium state for the potential φ. It is μ_φ.*
(b) *There is a unique pair (λ, m), with $\lambda \in [0,\infty)$, $m \in M(\mathcal{J}(T))$ such that $\mathcal{L}_\varphi^*(m) = \lambda m$. Namely, $\lambda = e^{\mathrm{P}(\varphi)}$ and $m = m_\varphi$.*
(c) *There is a unique $\rho \in C(\mathcal{J}(T))$ such that $\mathcal{L}_\varphi(\rho) = e^{\mathrm{P}(\varphi)}\rho$, $\rho \ge 0$, and $\int_{\mathcal{J}(T)} \rho\, dm_\varphi = 1$. Namely, $\rho = \rho_\varphi = d\mu_\varphi/dm_\varphi > 0$.*
(d) *The measure-preserving dynamical system $(\mathcal{J}(T), T, \mu_\varphi)$ is metrically exact and its Rokhlin natural extension is a K-system. In particular, this system is totally ergodic, so ergodic. This implies that m_φ is metrically exact. Note: Metric exactness was introduced in Definition 8.4.4 for invariant measures, but this notion makes sense for non-invariant measures equally.*
(e) *$e^{\mathrm{P}(\varphi)}$ is a simple eigenvalue of the operator $\mathcal{L}_\varphi : C(\mathcal{J}(T)) \to C(\mathcal{J}(T))$ and it is the only eigenvalue of $\mathcal{L}_\varphi$ with modulus equal to $e^{\mathrm{P}(\varphi)}$.*
(f) *If $g \in C(\mathcal{J}(T))$, then the sequence $((e^{-\mathrm{P}(\varphi)}\mathcal{L}_\varphi)^n(g))_{n=0}^\infty$ converges uniformly and*

$$\lim_{n\to\infty} e^{-n\mathrm{P}(\varphi)}\mathcal{L}_\varphi^n(g) = m_\varphi(g)\rho_\varphi.$$

In particular,

$$\lim_{n\to\infty} e^{-n\mathrm{P}(\varphi)}\mathcal{L}_\varphi^n(\mathbb{1}) = \rho_\varphi.$$

(g) $\mathrm{h}_{\mu_\varphi}(T) > 0$.

We now derive a consequence of Theorem 27.8.3. That consequence is very important on its own but will also be used a few times in the sequel.

Theorem 27.8.4. *If $T : \widehat{\mathbb{C}} \to \widehat{\mathbb{C}}$ is a rational function with* $\deg(T) \geq 2$, *then*

$$\mathrm{h}_{\mathrm{top}}(T) = \log(\deg(T)).$$

Proof. The inequality $\mathrm{h}_{\mathrm{top}}(T) \geq \log(\deg(T))$ directly follows from the Misiurewicz–Przytycki theorem (Theorem 7.5.1) while the inequality $\mathrm{h}_{\mathrm{top}}(T) \leq \log(\deg(T))$ directly follows from Lemma 27.2.2 and (27.45) applied to the potential $\varphi \equiv 0$ since then

$$\mathcal{L}_0^n(\mathbb{1})(z) = (\deg(T))^n$$

for every $n \geq 0$ and every $z \in \mathcal{J}(T)$. □

According to Theorem 27.8.4,

$$\mathrm{P}(0) = \mathrm{h}_{\mathrm{top}}(T) = \log(\deg(T)) \geq \log 2 > 0 = \sup 0. \tag{27.119}$$

Therefore, Theorem 27.8.3 applies to $\varphi \equiv 0$. This special case yields the following statement.

Theorem 27.8.5. *If $T : \widehat{\mathbb{C}} \to \widehat{\mathbb{C}}$ is a rational function with* $\deg(T) \geq 2$, *then:*
(a) *There is a unique measure of maximal entropy for T. It is μ_0.*
(b) *There is a unique pair (λ, m), with $\lambda \in [0, \infty)$, $m \in M(\mathcal{J}(T))$ such that $\mathcal{L}_0^*(m) = \lambda m$. Namely, $\lambda = \deg(T)$ and $m = \mu_0$.*
(c) *There is a unique $\rho \in C(\mathcal{J}(T))$ such that $\mathcal{L}_0(\rho) = \deg(T)\rho$, $\rho \geq 0$, and $\int_{\mathcal{J}(T)} \rho \, d\mu_0 = 1$. Namely, $\rho = \mathbb{1}$. Explicitly,*

$$\mathcal{L}_0(\mathbb{1}) = \deg(T)\mathbb{1}.$$

(d) *The measure-preserving dynamical system $(\mathcal{J}(T), T, \mu_0)$ is metrically exact and its Rokhlin natural extension is a K–system. In particular, this system is totally ergodic, so ergodic.*
(e) *$\deg(T)$ is a simple eigenvalue of the operator $\mathcal{L}_0 : C(\mathcal{J}(T)) \to C(\mathcal{J}(T))$ and it is the only eigenvalue of $\mathcal{L}_0$ with modulus equal to $\deg(T)$.*
(f) *If $g \in C(\mathcal{J}(T))$, then the sequence $((\frac{1}{\deg(T)}\mathcal{L}_0)^n(g))_{n=0}^{\infty}$ converges uniformly and*

$$\lim_{n\to\infty} \frac{1}{(\deg(T))^n}\mathcal{L}_0^n(g) = \mu_0(g)\mathbb{1}.$$

(g) $\mathrm{h}_{\mu_0}(T) = \mathrm{h}_{\mathrm{top}}(T) = \mathrm{P}(0) = \log(\deg(T)) \geq \log 2 > 0.$

Proof. This theorem is an immediate consequence of Theorem 27.8.3, Theorem 27.8.4, formula (27.119) and the obvious equality $\mathcal{L}_0(\mathbb{1}) = \deg(T)\mathbb{1}$. □

27.9 Assorted remarks

Remark 27.9.1. The hypothesis $\mathrm{P}(\varphi) > \sup\varphi$ can be replaced by a weaker one, namely

$$\mathrm{P}(S_k\varphi) > \sup(S_k\varphi) \tag{27.120}$$

for some $k \in \mathbb{N}$. Only a few proofs in this chapter would require minor modifications.

Remark 27.9.2. The hypothesis $\mathrm{P}(\varphi) > \sup\varphi$ is not so restrictive. For instance, it is entailed by the hypothesis

$$\sup\varphi - \inf\varphi < \log(\deg(T)).$$

Indeed, in light of Theorem 27.8.4, this inequality means that

$$\sup\varphi - \inf\varphi < \mathrm{h_{top}}(T). \tag{27.121}$$

As $\varphi \geq \psi \Rightarrow \mathrm{P}(\varphi) \geq \mathrm{P}(\psi)$, as $\mathrm{P}(\varphi + c) = \mathrm{P}(\varphi) + c$ for any $c \in \mathbb{R}$ and as $\mathrm{P}(0) = \mathrm{h_{top}}(T)$, we deduce that

$$\mathrm{P}(\varphi) \geq \mathrm{h_{top}}(T) + \inf\varphi > \sup\varphi.$$

Remark 27.9.3. The hypothesis $\mathrm{P}(\varphi) > \sup\varphi$ is necessary in general. Indeed, consider the quadratic polynomial

$$Q_{1/4}(z) = z^2 + 1/4$$

already encountered in the context of the definition of the Mandelbrot set (see Subsection 26.1.1) and which we will come across again, for instance in Example 31.4.6. Let $h = \mathrm{HD}(\mathcal{J}(Q_{1/4}))$ and introduce the Hölder potential $\varphi_h : \mathcal{J}(Q_{1/4}) \to \mathbb{R}$ defined by

$$\varphi_h(z) = -h\log|Q'_{1/4}(z)| = -h\log(2|z|).$$

By the forthcoming Bowen's formula (Theorem 31.2.1(i)),

$$\mathrm{P}(\varphi_h) = 0.$$

Thus,

$$\sup\varphi_h \geq \varphi_h(1/2) = 0 = \mathrm{P}(\varphi_h).$$

So, the inequality $\mathrm{P}(\varphi_h) > \sup\varphi_h$ fails.

The potential φ_h admits more than one equilibrium state. One of them is the Dirac measure $\delta_{1/2}$ concentrated at 1/2 since it is $Q_{1/4}$–invariant (1/2 being a fixed point of $Q_{1/4}$) and

$$\mathrm{h}_{\delta_{1/2}}(Q_{1/4}) + \int\limits_{\mathcal{J}(Q_{1/4})} \varphi_h \, d\delta_{1/2} = 0 + 0 = 0 = \mathrm{P}(\varphi_h).$$

Another one is the atomless, ergodic, $Q_{1/4}$–invariant Borel probability measure $\mu_{Q_{1/4}}$ arising from Theorem 31.4.1; see also Corollary 31.4.4 and Example 31.4.6. Indeed, it follows from Theorems 28.4.1–28.4.2 that

$$\mathrm{h}_{\mu_{Q_{1/4}}}(Q_{1/4}) + \int\limits_{\mathcal{J}(Q_{1/4})} \varphi_h \, d\mu_{Q_{1/4}} = [\mathrm{HD}(\mu_{Q_{1/4}}) - h] \int\limits_{\mathcal{J}(Q_{1/4})} \log|Q_{1/4}'| \, d\mu_{Q_{1/4}} = 0 = \mathrm{P}(\varphi_h).$$

Remark 27.9.4. According to Theorem 13.7.20, condition (27.120) holds for all expanding (hyperbolic) rational functions and all Hölder continuous potentials. For rational functions in general, Inoquio-Renteria and Rivera-Letelier [60] proved that this condition holds if and only if the Lyapunov exponent of each equilibrium state of φ is strictly positive. As observed in [60], it then immediately follows that if the Lyapunov exponent of each T-invariant Borel probability measure on $\mathcal{J}(T)$ is strictly positive, then inequality (27.120) holds for all Hölder continuous potentials. As noted in [60], there is one large celebrated class of rational functions, called topological Collet–Eckmann maps, for which this condition holds. This class has many characterizations; one of them is the existence of a positive number χ such that the Lyapunov exponent of each T-invariant Borel probability measure on $\mathcal{J}(T)$ is larger than χ. Another one, also extremely useful, is commonly referred to as the exponential shrinking property. Topological Collet–Eckmann rational functions allow critical points in their Julia set. They comprise all semi-hyperbolic (in particular, all subhyperbolic) rational functions. A systematic treatment of topological Collet–Eckmann maps and their statistical properties can be found in [108].

Remark 27.9.5. The existence and uniqueness of the measure of maximal entropy observed in Theorem 27.8.5(g) was explicitly stated and proved by Misha Lyubich in [79]. Actually, the whole Theorem 27.8.5 is therein. The existence, uniqueness and T-invariance of a Borel probability measure m satisfying $\mathcal{L}_0^*(m) = \deg(T)m$ was also proved in [52].

Remark 27.9.6. Theorem 27.8.3 in particular asserts that each Hölder continuous potential such that $\mathrm{P}(\varphi) > \sup\varphi$ admits an equilibrium state. In fact, Misha Lyubich proved in [79] that every continuous potential admits an equilibrium state. He actually proved more, namely that the entropy function $\mu \mapsto \mathrm{h}_\mu(T)$ is upper semicontinuous. Even more specifically, he proved that each rational function $T : \widehat{\mathbb{C}} \to \widehat{\mathbb{C}}$ with $\deg(T) \geq 2$ is asymptotically h-expansive. This property of a topological dynamical system introduced and explored by Michal Misiurewicz in [96] was shown therein to entail upper semicontinuity of the entropy function, thus yielding the existence of equilibrium states for all continuous potentials.

Remark 27.9.7. In this remark, we discuss the proof of Theorem 27.8.3, its steps and ingredients.

The key ingredient is Lemma 27.1.1. This lemma stems from an earlier form in [36], which can in turn be traced back to Ricardo Mañé's seminal paper [82]. There are however substantial differences and improvements partly emanating from, but not confined to, progress made in [146] and in [133, 134]. The main difference is that in this volume we constructed "good" inverse branches locally (on small balls) rather than globally (on a large topological disk). The other major difference is that we dealt with the iterates of the map T itself rather than some of its iterated power T^q.

The second key ingredient is Lemma 27.2.6. The proof in [36] was somewhat obscure, tedious and complicated; likewise in [106]. On the other hand, the papers [133, 134, 146] relied on the stronger hypothesis (27.121), which trivializes Lemma 27.2.6. In this volume, that lemma is a derivative product of a series of preparatory lemmas, starting with Lemma 27.2.1. It is conceptually simplified and clarified through those lemmas. In particular, we formulated and proved Lemma 25.4.1, which Lemma 27.2.1 calls upon. Lemma 25.4.1 was reworked from [79]. Perhaps more crucially, we adapted Lemma 27.2.2 and its proof from [106]. We simplified it, clarified it and provided a detailed proof. The second important component of Lemma 27.2.6's proof is Lemma 27.2.3, which is based on the split of each iterate of the transfer operator L_β into "bad" parts B_n and "good" parts G_n. This way, we avoided the heaviness of the approaches adopted in [106] and [36]. The idea of splitting the iterates of the transfer operator L_β comes from [36]. In fact, the resulting recursive formula (27.71) is one of the most important, perhaps the most important, relation exploited in the proof of Lemma 27.2.6 and thereby in the proof of Theorem 27.8.3. In this book, we adapted this splitting to the proof of Lemma 27.2.3, which is central to establishing Lemmas 27.2.4 and 27.2.6.

The third key ingredient in the proof of Theorem 27.8.3 is Lemma 27.5.2 (and thereafter Theorem 27.5.3). This lemma was proved in [36] by means of heavy duty methods from the theory of Jacobians. Indeed, it used the highly nontrivial existence of countable generators with finite entropy for T-invariant measures μ on $\mathcal{J}(T)$ proved in [82]. This is an indispensable fact to establish the formula

$$\mathrm{h}_\mu(T) = \int_{\mathcal{J}(T)} \log J_\mu(T)\, d\mu, \tag{27.122}$$

where $J_\mu(T)$ is the Jacobian of T with respect to μ.

The complex proof of this formula exploits the full power of the entropy theory for uncountable measurable partitions, which we did not cover in this book. Formula (27.122) was also extensively used in [36] to provide a fairly simple proof of the uniqueness of equilibrium states. In this volume, we provided instead a longer but self-contained proof of that uniqueness based on the differentiability of the pressure function. Its fundamental constituents comprise Section 27.7 as well as Lemma 27.8.1. Those were revamped from an unpublished part of [146].

Finally, the proof of ergodicity and K-mixing provided in this volume is based on a simpler, much more elegant and transparent method than that used in [36]. It utilizes the spectral/convergent properties of the transfer operator $\widehat{\mathcal{L}}_\varphi$, derived from much simpler means, mainly Lemma 27.4.1 whose proof was adapted from [99].

Remark 27.9.8. There has been substantial progress in the field since Theorem 27.8.3 was proved and the paper [36] was published. Some information can be found in [141]. First, as a relatively easy consequence of results proved in [106], the eigenfunction ρ_φ of the operator $\mathcal{L}_\varphi$ was shown in [33] to be Hölder continuous. This is by no means obvious since, in the presence of critical points in the Julia set, the operator $\mathcal{L}_\varphi$ does not generally preserve any space of Hölder continuous functions with a fixed exponent. Furthermore, it was shown in [33] that if $g : \mathcal{J}(T) \to \mathbb{R}$ is Hölder continuous with nonzero variance, then the stationary sequence of random variables $((g \circ T^n, \mu_\varphi))_{n=1}^\infty$ satisfies the central limit theorem. Along this line, substantial progress was realized in [133] and [146] where, based on a new method of fine inducing and on a refined method of Young's towers from [148, 149], the law of iterated logarithm and the exponential decay of correlations were demonstrated. Also, real-analyticity of the pressure function $t \mapsto \mathrm{P}(t\varphi)$ was shown in [133, 134], and the conditions under which a measure μ_φ can be of maximal Hausdorff dimension were identified. Moreover, some research on transfer operators and stochastic properties of equilibrium measures was done in [56, 57]. We further bring the reader's attention to the fact that [133, 146] additionally dealt with the higher-dimensional case of holomorphic endomorphisms of complex projective spaces. Nearly the same results as in the one-dimensional case of the Riemann sphere $\widehat{\mathbb{C}}$ were achieved. Finally, we mention the recent preprint [9] in which holomorphic endomorphisms of complex projective spaces are also examined. Using different methods, the preprint's authors defined a Banach space B, containing Hölder continuous functions, which is preserved by the transfer operator $\mathcal{L}_\varphi$ and such that the operator $\mathcal{L}_\varphi : B \to B$ has $e^{\mathrm{P}(\varphi)}$ as a simple eigenvalue and the rest of its spectrum is contained in a disk with radius smaller than $e^{\mathrm{P}(\varphi)}$. This opens the door to developing a full thermodynamic formalism for the dynamical system $T : \mathcal{J}(T) \to \mathcal{J}(T)$ and Hölder continuous potentials with pressure gap. We would have loved to provide a full account of the main results of [133, 146] with proofs but the needed methods would go too far beyond the scope of this book and would make this volume too long.

27.10 Exercises

Exercise 27.10.1. Prove (27.45).

Exercise 27.10.2. Prove (27.71).

Exercise 27.10.3. Prove Claims 1–2 in the proof of Proposition 27.7.2.

Exercise 27.10.4. Prove Remark 27.9.1.

28 Invariant measures: fractal and dynamical properties

Throughout this chapter, $T : \widehat{\mathbb{C}} \to \widehat{\mathbb{C}}$ is a rational function with $\deg(T) \geq 2$. Let $M(T)$ be the set of all T-invariant Borel probability measures supported on $\mathcal{J}(T)$. Let $M_e(T)$ be its subset of ergodic measures. Let also $M^+(T)$ be the subset of all measures with positive entropy (see Chapter 9) and $M_e^+(T) = M_e(T) \cap M^+(T)$.

28.1 Lyapunov exponents are nonnegative

In this section, the derivatives are understood to be with respect to the spherical metric on $\widehat{\mathbb{C}}$. For every $\mu \in M(T)$, the Lyapunov exponent $\chi_\mu(T)$ of the measure μ is defined to be (cf. Definition 16.5.1)

$$\chi_\mu(T) := \int_{\mathcal{J}(T)} \log |T'| \, d\mu.$$

This integral is well-defined since the function $\log |T'|$ is bounded above by $\log \|T'\|_\infty$. However, it generally would not be well-defined if we worked with the Euclidean metric. A priori, $\chi_\mu(T)$ might be equal to $-\infty$. But we will establish that this is never the case, i. e., $\chi_\mu(T) > -\infty$. In fact, we will show that $\chi_\mu(T) \geq 0$. This is Przytycki's theorem, first proved in [107]. Our proof follows Przytycki's general strategy but is substantially simplified by the use of Rules I–II from Section 25.2; more precisely, we will use one of those rules' consequences, namely Lemma 25.2.6. As a matter of fact, we will prove a somewhat stronger result. We start with a first intermediate result.

Lemma 28.1.1. *If $T : \widehat{\mathbb{C}} \to \widehat{\mathbb{C}}$ is a rational function with $\deg(T) \geq 2$, then $\chi_\mu(T) > -\infty$ for every $\mu \in M(T)$.*

Proof. By Birkhoff's ergodic theorem (Theorem 8.2.11), there is a Borel set $X_1 \subseteq \mathcal{J}(T)$ such that $\mu(X_1) = 1$, $T^{-1}(X_1) = X_1$, and the limit

$$\chi(z) := \lim_{n\to\infty} \frac{1}{n} \log |(T^n)'(z)|$$

exists for every point $z \in X_1$ (including the possibility that $\chi(z) = -\infty$ for some z). Then

$$\chi(T^i(z)) = \chi(z), \quad \forall z \in X_1, \ \forall i \in \mathbb{N}. \tag{28.1}$$

Set

$$M := \frac{4}{\kappa}\Big(\beta + \frac{\kappa}{2} \log \|T'\|_\infty\Big), \tag{28.2}$$

https://doi.org/10.1515/9783110769876-028

where $\kappa \in (0,1]$ and $\beta > 0$ come from Lemma 25.2.6. Let $\varepsilon > 0$. By Egorov's theorem (Theorem A.1.12), there are a Borel set $X_{2,\varepsilon} \subseteq X_1$ and $N_{2,\varepsilon} \in \mathbb{N}$ such that $\mu(X_{2,\varepsilon}) \geq 1 - \varepsilon$ and

$$\frac{1}{n}\log|(T^n)'(z)| < \max\{\chi(z)+1, -M\}, \quad \forall z \in X_{2,\varepsilon}, \forall n \geq N_{2,\varepsilon}. \tag{28.3}$$

By Birkhoff's ergodic theorem once again, there is a Borel set $X_{3,\varepsilon} \subseteq X_{2,\varepsilon}$ such that $\mu(X_{3,\varepsilon}) = \mu(X_{2,\varepsilon})$ and

$$\lim_{n\to\infty}\frac{1}{n}\sum_{i=0}^{n-1}\mathbb{1}_{X_{2,\varepsilon}}(T^i(z)) > 0, \quad \forall z \in X_{3,\varepsilon}.$$

Let $z \in X_{3,\varepsilon}$. If $(n_k(z))_{k=1}^{\infty}$ is the sequence of all $i \in \mathbb{N}$ such that $T^i(z) \in X_{2,\varepsilon}$, then

$$\lim_{k\to\infty}\frac{n_{k+1}(z) - n_k(z)}{n_k(z)} = 0. \tag{28.4}$$

Take an integer $n \geq \max\{2\#(\mathrm{Crit}(T)\cap \mathcal{J}(T)), N_{2,\varepsilon}\}$. Let $j_n = j$ and $l_n = l$ be the numbers produced in Lemma 25.2.6 for the point z. By virtue of (28.4), there exists $N_{3,\varepsilon} \geq \max\{2\#(\mathrm{Crit}(T)\cap \mathcal{J}(T)), N_{2,\varepsilon}\}$ such that for any $n \geq N_{3,\varepsilon}$ there is $0 \leq i_n < n\kappa/2$ for which

$$T^{j_n+i_n}(z) \in X_{2,\varepsilon}. \tag{28.5}$$

Noting that

$$l_n - i_n > n\kappa - (n\kappa/2) = n\kappa/2 > 0, \tag{28.6}$$

it follows from the chain rule that

$$|(T^{l_n})'(T^{j_n}(z))| \leq \|T'\|_\infty^{i_n}\cdot|(T^{l_n-i_n})'(T^{j_n+i_n}(z))| \leq \|T'\|_\infty^{n\kappa/2}\cdot|(T^{l_n-i_n})'(T^{j_n+i_n}(z))|.$$

Invoking Lemma 25.2.6, we get that

$$|(T^{l_n-i_n})'(T^{j_n+i_n}(z))| \geq \|T'\|_\infty^{-n\kappa/2}e^{-\beta n}.$$

It follows from (28.1), (28.3) and (28.5) that

$$\exp(\max\{\chi(z)+1, -M\}(l_n - i_n)) \geq \|T'\|_\infty^{-n\kappa/2}e^{-\beta n}.$$

Taking the logarithm of both sides and then dividing by n, we have

$$\max\{\chi(z)+1, -M\}\frac{l_n - i_n}{n} \geq -\left(\beta + \frac{\kappa}{2}\log\|T'\|_\infty\right).$$

Assume now that

$$\chi(z) + 1 < 0. \tag{28.7}$$

Using (28.6), we obtain

$$\frac{\kappa}{2} \max\{\chi(z) + 1, -M\} \geq -\left(\beta + \frac{\kappa}{2} \log \|T'\|_\infty\right).$$

By (28.2), we deduce that

$$\max\{\chi(z) + 1, -M\} \geq -\frac{M}{2}.$$

So, $\chi(z) + 1 \geq -\frac{M}{2}$. Equivalently, $\chi(z) \geq -1 - \frac{M}{2}$. Observing now that this inequality holds even if (28.7) fails, we conclude that

$$\chi(z) \geq -1 - \frac{M}{2}, \quad \forall z \in X_{3,\varepsilon}.$$

Let $X_3 = \bigcup_{\varepsilon>0} X_{3,\varepsilon}$. Then $\chi(z) \geq -1 - \frac{M}{2}$ for all $z \in X_3$. As $\mu(X_3) = 1$, we finally get

$$\chi_\mu(T) = \int_{X_3} \chi(z)\, d\mu(z) \geq -\left(1 + \frac{M}{2}\right) > -\infty.$$

□

A second intermediate result is also indispensable.

Lemma 28.1.2. *Let $T : \widehat{\mathbb{C}} \to \widehat{\mathbb{C}}$ be a rational function with* $\deg(T) \geq 2$. *If $z \in \mathcal{J}(T)$ is such that*

$$\operatorname{dist}(T^n(z), \operatorname{Crit}(T) \cap \mathcal{J}(T)) \geq e^{-n\delta}$$

for all $n \in \mathbb{N}$ large enough, then

$$\overline{\chi}(z) := \limsup_{n\to\infty} \frac{1}{n} \log |(T^n)'(z)| \geq -\delta.$$

Proof. Seeking a contradiction, suppose that $\overline{\chi}(z) < -\delta$. Then there exists $\gamma > 0$ with

$$\overline{\chi}(z) + \gamma < -\delta. \tag{28.8}$$

For every $n \in \mathbb{N}$, let

$$B_n := B(T^n(z), |(T^n)'(z)| e^{\gamma n}).$$

By our hypothesis and (28.8),

$$\frac{1}{n} \log \frac{\operatorname{diam}(B_n)}{\operatorname{dist}(T^n(z), \operatorname{Crit}(T) \cap \mathcal{J}(T))} \leq \frac{1}{n} \log 2 + \frac{1}{n} \log |(T^n)'(z)| + \gamma + \delta \leq \frac{1}{2}(\overline{\chi}(z) + \gamma + \delta) < 0$$

for all $n \in \mathbb{N}$ large enough, say $n \geq N_1$. Equivalently,

$$\frac{\operatorname{diam}(B_n)}{\operatorname{dist}(T^n(z), \operatorname{Crit}(T) \cap \mathcal{J}(T))} \leq \exp\left(\frac{n}{2}(\overline{\chi}(z) + \gamma + \delta)\right).$$

It follows from Lemma 25.1.10 that for all $n \geq N_1$ sufficiently large, say $n \geq N_2 \geq N_1$,

$$\frac{|T'(w)|}{|T'(T^n(z))|} \leq e^{\gamma}, \quad \forall w \in B_n.$$

Therefore,

$$\begin{aligned} T(B_n) \subseteq B(T^{n+1}(z), |(T^n)'(z)|e^{\gamma n}e^{\gamma}|T'(T^n(z))|) &= B(T^{n+1}(z), |(T^{n+1})'(z)|e^{\gamma(n+1)}) \\ &= B_{n+1} \end{aligned} \tag{28.9}$$

for all $n \geq N_2$. But invoking (28.8), we know that

$$\limsup_{n\to\infty} \frac{1}{n} \log(2|(T^n)'(z)|e^{\gamma n}) = \limsup_{n\to\infty} \frac{1}{n} \log|(T^n)'(z)| + \gamma = \overline{\chi}(z) + \gamma < -\delta < 0.$$

We obtain from this and (28.9) that

$$\limsup_{n\to\infty} \operatorname{diam}(T^n(B_{N_2})) \leq \limsup_{n\to\infty} \operatorname{diam}(B_{n+1}) \leq \limsup_{n\to\infty} (2|(T^n)'(z)|e^{\gamma n}) = 0.$$

Hence, the family of maps $(T^n|_{B_{N_2}} : B_{N_2} \to \widehat{\mathbb{C}})_{n=1}^{\infty}$ is equicontinuous, which implies that $T^{N_2}(z) \in \mathcal{F}(T)$. So, $z \in \mathcal{F}(T)$, which is a contradiction. □

With these two intermediate results at our fingertips, we can now establish the main result of this section, i. e., the nonnegativity of Lyapunov exponents.

Theorem 28.1.3 (Przytycki's theorem). *If $T : \widehat{\mathbb{C}} \to \widehat{\mathbb{C}}$ is a rational function with $\deg(T) \geq 2$ and $\mu \in M(T)$, then*

$$\chi(z) := \lim_{n\to\infty} \frac{1}{n} \log|(T^n)'(z)| \geq 0 \quad \text{for } \mu\text{-a.e. } z \in \mathcal{J}(T).$$

Consequently,

$$\chi_\mu(T) \geq 0.$$

Note: The existence of the limit for μ-almost every Julia point directly follows from Birkhoff's ergodic theorem.

Proof. Let $\delta > 0$. Temporarily fix $c \in \operatorname{Crit}(T) \cap \mathcal{J}(T)$. For every integer $n \geq 0$, denote

$$B_n(c) := B(c, e^{-n\delta}).$$

Pick $N \in \mathbb{N}$ so large that $e^{-N\delta} \le r(T,c)$, where $r(T,c)$ arises from (25.9)–(25.11). Using Lemma 28.1.1, (25.9) and Taylor's theorem, we get

$$\begin{aligned}
-\infty < \int_{B_N(c)} \log|T'|\,d\mu &= \sum_{n=N}^{\infty} \int_{B_n(c)\setminus B_{n+1}(c)} \log|T'|\,d\mu \\
&\le \log A(T,c) + \sum_{n=N}^{\infty} \int_{B_n(c)\setminus B_{n+1}(c)} \log(|z-c|^{\deg_c(T)-1})\,d\mu(z) \\
&\le \log A(T,c) + (\deg_c(T)-1)\sum_{n=N}^{\infty}(-n\delta)\mu(B_n(c)\setminus B_{n+1}(c)) \\
&= \log A(T,c) - (\deg_c(T)-1)\delta\Big[N\mu(B_N(c)) + \sum_{n=N+1}^{\infty}\mu(B_n(c))\Big]. \qquad (28.10)
\end{aligned}$$

Thus, the series $\sum_{n=N+1}^{\infty}\mu(B_n(c))$ converges. So, the series $\sum_{n=0}^{\infty}\mu(B_n(c))$ converges too. Therefore, by the T-invariance of μ,

$$\sum_{n=0}^{\infty}\mu(T^{-n}(B_n(c))) < \infty.$$

Summing over all points $c \in \mathrm{Crit}(T)\cap\mathcal{J}(T)$, we obtain that

$$\sum_{n=0}^{\infty}\mu(T^{-n}(B(\mathrm{Crit}(T)\cap\mathcal{J}(T), e^{-n\delta}))) = \sum_{c\in\mathrm{Crit}(T)\cap\mathcal{J}(T)}\sum_{n=0}^{\infty}\mu(T^{-n}(B_n(c))) < \infty.$$

Therefore, by the Borel–Cantelli lemma (Lemma A.1.1),

$$T^n(z) \notin B(\mathrm{Crit}(T)\cap\mathcal{J}(T), e^{-n\delta})$$

for μ-almost every $z \in \mathcal{J}(T)$, say $z \in J_\delta$ with $\mu(J_\delta) = 1$, and all $n \in \mathbb{N}$ sufficiently large. Hence, $\overline{\chi}(z) \ge -\delta$ according to Lemma 28.1.2. So,

$$\mu\Big(\bigcap_{k=1}^{\infty} J_{1/k}\Big) = 1$$

and $\overline{\chi}(z) \ge 0$ for all z in this intersection. □

As a fairly easy consequence of this theorem, we get the following.

Corollary 28.1.4. *If $T : \widehat{\mathbb{C}} \to \widehat{\mathbb{C}}$ is a rational function with $\deg(T) \ge 2$ and $\mu \in M(T)$, then*

$$\limsup_{n\to\infty}|(T^n)'(z)| \ge 1 \quad \text{for } \mu\text{-a.e. } z \in \mathcal{J}(T).$$

Proof. Seeking a contradiction, suppose that there is a Borel set $J_1 \subseteq \mathcal{J}(T)$ such that $\mu(J_1) > 0$ and

$$\limsup_{n\to\infty} |(T^n)'(z)| < 1, \quad \forall z \in J_1.$$

Thus, there exist $0 < \gamma < 1$, a Borel set $J_2 \subseteq J_1$ and $N_2 \in \mathbb{N}$ such that $\mu(J_2) > 0$ and

$$|(T^n)'(z)| \leq \gamma, \quad \forall z \in J_2, \ \forall n \geq N_2. \tag{28.11}$$

By Birkhoff's ergodic theorem, there then are a Borel set $J_3 \subseteq J_2$, $\alpha \in (0,1]$, and an integer $N_3 \geq N_2$ such that $\mu(J_3) > 0$ and

$$\#\{0 \leq j < n : T^j(z) \in J_2\} \geq \alpha n, \quad \forall z \in J_3, \ \forall n \geq N_3. \tag{28.12}$$

Given any $z \in J_3$, represent the set $\{j \geq 0 : T^j(z) \in J_2\}$ as a strictly increasing sequence $(n_k)_{k=0}^{\infty} = (n_k(z))_{k=0}^{\infty}$. Then $n_0 = 0$ and using the chain rule and (28.11), we obtain for every $k \in \mathbb{N}$ that

$$\left|(T^{n_{kN_3}})'(z)\right| = \prod_{j=0}^{k-1} \left|(T^{n_{(j+1)N_3} - n_{jN_3}})'(T^{n_{jN_3}}(z))\right| \leq \gamma^k. \tag{28.13}$$

But if $k \geq N_3$, then $n_k \geq N_3$ and it ensues from (28.12) that $k \geq \alpha n_k$. In particular, $kN_3 \geq \alpha n_{kN_3}$ for every $k \in \mathbb{N}$. Thus,

$$\liminf_{n\to\infty} \frac{1}{n} \log\left|(T^n)'(z)\right| \leq \liminf_{k\to\infty} \frac{1}{n_{kN_3}} \log\left|(T^{n_{kN_3}})'(z)\right| \leq \log\gamma \liminf_{k\to\infty} \frac{k}{n_{kN_3}} \leq \frac{\alpha \log\gamma}{N_3} < 0.$$

As $\mu(J_3) > 0$, this contradicts Przytycki's theorem (Theorem 28.1.3). □

28.2 Ruelle's inequality

The next famous result connects the Lyapunov exponent with the metric entropy. This inequality was originally proved by David Ruelle [113] in the context of smooth diffeomorphisms of multidimensional smooth manifolds. In this section, all the metrics and derivatives are understood to be with respect to the spherical metric on $\widehat{\mathbb{C}}$. Our approach stems from [110].

Theorem 28.2.1 (Ruelle's inequality). *If $T : \widehat{\mathbb{C}} \to \widehat{\mathbb{C}}$ is a rational function with* $\deg(T) \geq 2$ *and $\mu \in M_e(T)$, then*

$$\mathrm{h}_\mu(T) \leq 2\chi_\mu(T).$$

Proof. Conjugating T by a Möbius transformation if necessary, we may assume without loss of generality that $\mu(\{\infty\}) = 0$. Observe (trivially) that there exist a sequence of positive numbers $(a_k)_{k=1}^{\infty}$ decreasing to 0 and a sequence $(\mathcal{P}_k)_{k=1}^{\infty}$ of increasingly finer

Borel partitions of the complex plane $\mathbb{C}$ (or the sphere $\widehat{\mathbb{C}}$) consisting of atoms satisfying

$$\operatorname{diam}(\mathcal{P}_1) \le \frac{1}{100}, \quad \operatorname{diam}(\mathcal{P}_k) \le a_k \quad \text{and} \quad A(\mathcal{P}_k) \ge \frac{1}{4}a_k^2, \quad \forall k \in \mathbb{N}, \tag{28.14}$$

where A refers to the spherical area on $\widehat{\mathbb{C}}$. For every $z \in \mathcal{J}(T)$ and every $k \in \mathbb{N}$, let

$$N(T,z,k) := \#\{P \in \mathcal{P}_k : T(\mathcal{P}_k(z)) \cap P \ne \emptyset\}.$$

Let $k(T) \in \mathbb{N}$ be so large that for all $k \ge k(T)$ the Lipschitz constant of $T|_{\mathcal{P}_k(z)}$ is smaller than $|T'(z)| + 1$. Then the set $T(\mathcal{P}_k(z))$ is contained in the ball $B(T(z), (|T'(z)| + 1)a_k)$. Thus, $T(\mathcal{P}_k(z)) \cap P \ne \emptyset$ implies that

$$P \subseteq B(T(z), (|T'(z)| + 2)a_k).$$

As also

$$A(\, B(T(z), (|T'(z)| + 2)a_k)\,) = 2\pi[1 - \cos((|T'(z)| + 2)a_k)] \le 2\pi[(|T'(z)| + 2)a_k]^2,$$

we get with the help of (28.14) that

$$N(T,z,k) \le \frac{2\pi[(|T'(z)| + 2)a_k]^2}{a_k^2/4} = 8\pi(|T'(z)| + 2)^2, \quad \forall k \ge k(T). \tag{28.15}$$

Let

$$N(T,z) := \sup_{k \ge k(T)} N(T,z,k).$$

In view of (28.15),

$$N(T,z) \le 8\pi(|T'(z)| + 2)^2. \tag{28.16}$$

Observe that (cf. (9.2))

$$\mathrm{H}_{\mu_{\mathcal{P}_k(z)}}(T^{-1}(\mathcal{P}_k)|\mathcal{P}_k(z)) \le \log \#\{P \in \mathcal{P}_k : T^{-1}(P) \cap \mathcal{P}_k(z) \ne \emptyset\} = \log N(T,z,k).$$

Using Theorems 9.4.17 and 9.4.11(b), we then get

$$\begin{aligned}
\mathrm{h}_\mu(T) &= \lim_{k\to\infty} \mathrm{h}_\mu(T,\mathcal{P}_k) \le \liminf_{k\to\infty} \mathrm{H}_\mu(T^{-1}(\mathcal{P}_k)|\mathcal{P}_k) \\
&= \liminf_{k\to\infty} \int_{\mathcal{J}(T)} \mathrm{H}_{\mu_{\mathcal{P}_k(z)}}(T^{-1}(\mathcal{P}_k)|\mathcal{P}_k(z))\, d\mu(z) \le \liminf_{k\to\infty} \int_{\mathcal{J}(T)} \log N(T,z,k)\, d\mu(z)
\end{aligned}$$

$$\le \int_{\mathcal{J}(T)} \log N(T,z)\, d\mu(z).$$

Applying this inequality to T^n and employing Theorem 9.4.13 and (28.16), we obtain that

$$\mathrm{h}_\mu(T) = \frac{1}{n}\mathrm{h}_\mu(T^n) \le \frac{1}{n} \int_{\mathcal{J}(T)} \log N(T^n,z)\, d\mu(z) \le \int_{\mathcal{J}(T)} \frac{1}{n} \log\left[8\pi(|(T^n)'(z)|+2)^2\right] d\mu(z).$$

As $\mathcal{J}(T)$ contains at least one (in fact, a dense set of) repelling periodic point, we know that $\|T'\|_\infty > 1$, and hence there exists $M \in \mathbb{N}$ (in fact, $M = 2$ works) such that

$$0 \le \frac{1}{n} \log(\|(T^n)'\|_\infty + 2)^2 \le 2(\log \|T'\|_\infty + 1), \quad \forall n \ge M.$$

Moreover, the ergodic case of Birkhoff's ergodic theorem (Corollary 8.2.14) affirms that

$$\lim_{n\to\infty} \frac{1}{n} \log\left|(T^n)'(z)\right| = \int_{\mathcal{J}(T)} \log|T'|\, d\mu = \chi_\mu(T) \quad \text{for } \mu\text{-a.e } z \in \mathcal{J}(T).$$

In light of Przytycki's theorem (Theorem 28.1.3), we deduce that

$$\lim_{n\to\infty} \frac{1}{n} \log(|(T^n)'(z)|+2) = \chi_\mu(T) \quad \text{for } \mu\text{-a.e } z \in \mathcal{J}(T).$$

Applying Lebesgue's dominated convergence theorem, we conclude that

$$\mathrm{h}_\mu(T) \le \lim_{n\to\infty} \int_{\mathcal{J}(T)} \frac{1}{n} \log(|(T^n)'(z)|+2)^2\, d\mu(z) = 2\chi_\mu(T).$$

Ruelle's inequality is thus proved. □

28.3 Pesin's theory in a conformal setting

In this section, we present a special version of a theory whose foundations were laid in Yasha Pesin's fundamental works [102, 103]. Since then, Pesin's theory has been extended, generalized, refined and cited in numerous articles, research books and textbooks. Our exposition follows Section 11.2 of the book [110].

Theorem 28.3.1. *Let $(Z,\mathcal{F},\nu)$ be a finite measure space with an ergodic measure-preserving automorphism $S : Z \to Z$. Let $T : \widehat{\mathbb{C}} \to \widehat{\mathbb{C}}$ be a rational function with* $\deg(T) \ge 2$. *Suppose that μ is an ergodic T-invariant Borel probability measure on $\mathcal{J}(T)$ with positive Lyapunov exponent $\chi_\mu(T)$. Suppose also that $h : Z \to \mathcal{J}(T)$ is a measurable map such that*

$$\nu \circ h^{-1} = \mu \quad \text{and} \quad h \circ S = T \circ h, \quad \nu\text{-a.e.}$$

Then there exists a measurable function $r : Z \to (0,\infty)$ *such that for* ν*-a. e.* $z \in Z$ *the following is satisfied:*

For every $n \in \mathbb{N}$*, there exists an holomorphic inverse branch* $T_{x_n}^{-n} : B(x_0, r(z)) \to \widehat{\mathbb{C}}$ *of* T^n *that sends* $x_0 := h(z)$ *to* $x_n := h(S^{-n}(z))$*.*

In addition, given any $\chi \in (-\chi_\mu(T), 0)$*, there is a measurable function* $K : Z \to (0,\infty)$ *such that*

$$K(z)^{-1}e^{n\chi} < \left|(T_{x_n}^{-n})'(y)\right| < K(z)e^{n\chi} \quad \text{and} \quad \frac{|(T_{x_n}^{-n})'(w)|}{|(T_{x_n}^{-n})'(y)|} \le K, \quad \forall y, w \in B(x_0, r(z)),\ \forall n \in \mathbb{N}, \tag{28.17}$$

where K *is the Koebe constant corresponding to the scale* $1/2$*.*

Proof. Suppose that $\mu(\mathrm{PC}(T)) > 0$, where $\mathrm{PC}(T) = \bigcup_{n=1}^{\infty} T^n(\mathrm{Crit}(T))$ is the postcritical set of T. As μ is ergodic, this implies that μ must be concentrated on the orbit of a periodic point $w \in \mathrm{PC}(T)$. This means that $w = T^q(c) = T^{q+k}(c)$ for some $q, k \in \mathbb{N}$ and $c \in \mathrm{Crit}(T)$, and

$$\mu(\{T^q(c), T^{q+1}(c), \dots, T^{q+k-1}(c)\}) = 1.$$

Since $\chi_\mu(T) := \int \log|T'|\,d\mu > 0$, we infer that $|(T^k)'(T^q(c))| > 1$. Thus, the theorem is obviously true for the set

$$h^{-1}(\{T^q(c), T^{q+1}(c), \dots, T^{q+k-1}(c)\})$$

of full ν-measure.

Now suppose that $\mu(\mathrm{PC}(T)) = 0$. Fix $\lambda \in (e^{\chi/4}, 1)$. Consider $z \in Z$ such that:

(1) $x_0 := h(z) \notin \mathrm{PC}(T)$,

(2) $x_n := h(S^{-n}(z)) \in B(T(\mathrm{Crit}(T)), \lambda^n)$ for finitely many n's only, and

(3) $\lim_{n\to\infty} \frac{1}{n} \log\left|(T^n)'(x_n)\right| = \chi_\mu(T)$.

The set of points z satisfying these three conditions is of full ν-measure. Indeed, the first condition is satisfied on a set of full ν-measure because $\nu(h^{-1}(\mathrm{PC}(T))) = \mu(\mathrm{PC}(T)) = 0$ by our hypothesis. In order to prove that the set of points respecting the second condition has full ν-measure, recall that S is a ν-measure-preserving automorphism and hence

$$\begin{aligned}\sum_{n=1}^{\infty} \nu(S^n(h^{-1}(B(T(\mathrm{Crit}(T)), \lambda^n)))) &= \sum_{n=1}^{\infty} \nu(h^{-1}(B(T(\mathrm{Crit}(T)), \lambda^n))) \\ &= \sum_{n=1}^{\infty} \mu(B(T(\mathrm{Crit}(T)), \lambda^n)) \\ &< \infty,\end{aligned}$$

the finiteness being due to Lemmas 25.1.14 and 25.6.1. An application of the Borel-Cantelli lemma (Lemma A.1.1) finishes the demonstration of the second condition. Finally, the third condition is fulfilled on a set of full ν-measure according to Birkhoff's ergodic theorem (Theorem 8.2.11).

Let $n_1 = n_1(z)$ be the least $n \in \mathbb{N}$ such that

$$x_n := h(S^{-n}(z)) \notin B(T(\mathrm{Crit}(T)), \lambda^n), \quad \forall n \geq n_1.$$

Note that the function $z \mapsto n_1(z)$ is Borel measurable. Because of our choices of χ and λ as well as condition (3), there is a least $n_2 = n_2(z) \geq n_1$ such that

$$|(T^n)'(x_n)|^{-1/4} < \lambda^n, \quad \forall n \geq n_2. \tag{28.18}$$

Again, observe that the function $z \mapsto n_2(z)$ is Borel measurable. Set

$$\Delta := \sum_{n=1}^{\infty} |(T^n)'(x_n)|^{-1/4}, \tag{28.19}$$

$$b_n := \frac{1}{2\Delta} |(T^{n+1})'(x_{n+1})|^{-1/4}, \quad \forall n \in \mathbb{N}, \tag{28.20}$$

and

$$\Pi := \prod_{n=1}^{\infty} (1 - b_n)^{-1}. \tag{28.21}$$

This infinite product converges since the series $\sum_{n=1}^{\infty} b_n$ does (see Exercise 24.8.6). Let

$$r(z) := \min\left\{ (16\Pi K \Delta^3)^{-1}, \frac{1}{4\Pi} \mathrm{dist}\Big(h(z), \bigcup_{j=1}^{n_2} T^j(\mathrm{Crit}(T))\Big), \frac{1}{2K} |(T^{n_2})'(x_{n_2})| \lambda^{n_2} \prod_{k=n_2}^{\infty} (1 - b_k) \right\}.$$

Then all the inverse branches $T_{x_n}^{-n} : B(x_0, \Pi r(z)) \to \widehat{\mathbb{C}}$ are well-defined for all $1 \leq n \leq n_2$, and by Koebe's distortion theorem

$$\mathrm{diam}\left(T_{x_{n_2}}^{-n_2} \left(B\left(x_0, r(z) \prod_{k=n_2}^{\infty} (1 - b_k)^{-1} \right) \right) \right) \leq \lambda^{n_2}.$$

We show by induction that for every $n \geq n_2$ there exists an analytic inverse branch $T_{x_n}^{-n} : B(x_0, r(z)\Pi_{k=n}^{\infty}(1 - b_k)^{-1}) \to \widehat{\mathbb{C}}$ sending x_0 to x_n and such that

$$\mathrm{diam}\left(T_{x_n}^{-n} \left(B\left(x_0, r(z) \prod_{k=n}^{\infty} (1 - b_k)^{-1} \right) \right) \right) \leq \lambda^n. \tag{28.22}$$

Indeed, for $n = n_2$ this immediately follows from our requirements imposed on $r(z)$. So, suppose that the claim is true for some $n \geq n_2$. Since

$$x_n = T_{x_n}^{-n}(x_0) \notin B(T(\mathrm{Crit}(T)), \lambda^n),$$

there exists an inverse branch $T_{x_{n+1}}^{-1} : B(x_n, \lambda^n) \to \widehat{\mathbb{C}}$ mapping x_n to x_{n+1}. By (28.22), which is part of the induction hypothesis, the composition

$$T_{x_{n+1}}^{-1} \circ T_{x_n}^{-n} : B\left(x_0, r(z) \prod_{k=n}^{\infty} (1 - b_k)^{-1}\right) \to \widehat{\mathbb{C}}$$

is then well-defined and forms the inverse branch of T^{n+1} that sends x_0 to x_{n+1}. By Koebe's distortion theorem (Theorem 23.1.6) and using (28.18)–(28.21), we now estimate that

$$\begin{aligned}
&\mathrm{diam}\left(T_{x_{n+1}}^{-(n+1)}\left(B\left(x_0, r(z) \prod_{k=n+1}^{\infty} (1 - b_k)^{-1}\right)\right)\right) \\
&\leq 2r(z) \prod_{k=n+1}^{\infty} (1 - b_k)^{-1} |(T^{n+1})'(x_{n+1})|^{-1} K b_n^{-3} \\
&\leq 16 r(z) \Pi K \Delta^3 |(T^{n+1})'(x_{n+1})|^{-1} |(T^{n+1})'(x_{n+1})|^{3/4} \\
&= 16 r(z) \Pi K \Delta^3 |(T^{n+1})'(x_{n+1})|^{-1/4} < 16 r(z) \Pi K \Delta^3 \lambda^{n+1} \\
&\leq \lambda^{n+1},
\end{aligned}$$

where the last inequality ensues from our choice of $r(z)$.

Replacing $r(z)$ by $r(z)/2$, the assertion about the existence of $K(z)$ then follows from a combined application of condition (3) and Koebe's distortion theorem (Theorem 23.1.8). □

As an immediate consequence of Theorem 28.3.1, we deduce the following.

Corollary 28.3.2. *Assume the same notation and hypotheses as Theorem 28.3.1. Fix $\varepsilon > 0$. Then there exist a measurable set $Z(\varepsilon) \subseteq Z$ and numbers $r(\varepsilon) \in (0, 1)$ and $K(\varepsilon) \geq 1$ such that:*

(1) $\mu(Z(\varepsilon)) > 1 - \varepsilon$,
(2) $r(z) \geq r(\varepsilon)$ *for all* $z \in Z(\varepsilon)$,
(3)

$$K(\varepsilon)^{-1} \exp(-n[\chi_\mu(T) + \varepsilon]) \leq |(T_{x_n}^{-n})'(y)| \leq K(\varepsilon) \exp(-n[\chi_\mu(T) - \varepsilon])$$

and

$$K^{-1} \leq \frac{|(T_{x_n}^{-n})'(y)|}{|(T_{x_n}^{-n})'(w)|} \leq K$$

for all $n \in \mathbb{N}$, all $z \in Z(\varepsilon)$ and all $y, w \in B(x_0, r(\varepsilon))$, where K is the Koebe constant corresponding to the scale 1/2.

In future applications, particularly in the next section, the system (Z, S, ν) will usually be given by (see Theorem 8.4.3) the Rokhlin's natural extension $(\widetilde{\mathcal{J}(T)}, \widetilde{T}, \widetilde{\mu})$ of the holomorphic system $(\mathcal{J}(T), T, \mu)$. In that context and with $h = \pi_0 : \widetilde{\mathcal{J}(T)} \to \mathcal{J}(T)$, Corollary 28.3.2 yields the following.

Theorem 28.3.3. *Let $T : \widehat{\mathbb{C}} \to \widehat{\mathbb{C}}$ be a rational function with* $\deg(T) \geq 2$. *If μ is an ergodic T-invariant Borel probability measure on $\mathcal{J}(T)$ with positive Lyapunov exponent $\chi_\mu(T)$, then for $\widetilde{\mu}$-a. e. $\widetilde{z} \in \widetilde{\mathcal{J}(T)}$ there exists $r(\widetilde{z}) > 0$ such that the function $\widetilde{z} \mapsto r(\widetilde{z})$ is measurable and the following is satisfied:*

Writing $\widetilde{z} = (z_k)_{k=0}^{\infty}$, for every $n \geq 0$ there exists $T_{z_n}^{-n} : B(z_0, r(\widetilde{z})) \to \widehat{\mathbb{C}}$, an holomorphic inverse branch of T^n mapping z_0 to z_n. In addition, given any $\chi \in (-\chi_\mu(T), 0)$, there exists a constant $K(\widetilde{z}) \in (0, \infty)$, independent of n, measurable as a function of $\widetilde{z}$, and such that

$$K(\widetilde{z})^{-1} e^{n\chi} < |(T_{z_n}^{-n})'(y)| < K(\widetilde{z}) e^{n\chi} \quad \text{and} \quad \frac{|(T_{z_n}^{-n})'(y)|}{|(T_{z_n}^{-n})'(w)|} \leq K, \quad \forall y, w \in B(z_0, r(\widetilde{z})), \tag{28.23}$$

where K is the Koebe constant for the map T corresponding to the scale $1/2$.

Consequently, for every $\varepsilon > 0$ there exist a closed set $\widetilde{\mathcal{J}(T)}(\varepsilon) \subseteq \widetilde{\mathcal{J}(T)}$ and numbers $r(\varepsilon) \in (0, 1)$ and $K(\varepsilon) \geq K$ such that:

(1) $\widetilde{\mu}(\widetilde{\mathcal{J}(T)}(\varepsilon)) > 1 - \varepsilon$,

(2) $r(\widetilde{z}) \geq r(\varepsilon)$ *for all* $\widetilde{z} \in \widetilde{\mathcal{J}(T)}(\varepsilon)$, *and*

(3)
$$K(\varepsilon)^{-1} \exp(-n[\chi_\mu(T) + \varepsilon]) \leq |(T_{z_n}^{-n})'(y)| \leq K(\varepsilon) \exp(-n[\chi_\mu(T) - \varepsilon])$$

and

$$K^{-1} \leq \frac{|(T_{z_n}^{-n})'(y)|}{|(T_{z_n}^{-n})'(w)|} \leq K$$

for all $n \in \mathbb{N}$, all $\widetilde{z} \in \widetilde{\mathcal{J}(T)}(\varepsilon)$ and all $y, w \in B(z_0, r(\varepsilon))$.

28.4 Volume lemmas; Hausdorff and packing dimensions of invariant measures

This section is devoted to providing a closed formula for the Hausdorff and packing dimensions of Borel probability measures that are ergodic and invariant under a rational function. For historical reasons, partly justified, this formula is frequently referred to as a volume lemma. Its first forms can be traced back to the works of Eggleston [46] and

Billingsley [11]. From a dynamical viewpoint, a breakthrough was achieved by Lai–Sang Young [148]. Since then, a multitude of papers appeared. We mention only some of the early ones: [84, 86, 105]. Also, the early papers [19, 111] shed some light on the nature of dimensions of measures. The main result in this section is Theorem 28.4.1. It has exactly the same form as Theorem 16.5.2 for conformal expanding repellers (CERs) and Theorem 19.8.30 for conformal graph directed Markov systems (CGDMSs). However, there is a substantial difference: while both CERs and CGDMSs are uniformly expanding/contracting, rational functions are generally not, and while the proofs of Theorems 16.5.2 and 19.8.30 relied respectively on Brin–Katok's local entropy formula for expansive systems (Theorem 9.6.4) and on the ergodic case of Shannon–McMillan–Breiman's theorem (Corollary 9.5.5), for rational functions we call on Rokhlin's natural extension through Theorem 28.3.3, in addition of Shannon–McMillan–Breiman's theorem. In all cases, Birkhoff's ergodic theorem is used.

Theorem 28.4.1 (Volume lemma). *Let $T : \widehat{\mathbb{C}} \to \widehat{\mathbb{C}}$ be a rational function with* $\deg(T) \geq 2$. *If μ is an ergodic T-invariant Borel probability measure on $\mathcal{J}(T)$ with $\chi_\mu(T) > 0$, then*

$$\lim_{r\to 0} \frac{\log \mu(B(x,r))}{\log r} = \frac{\mathrm{h}_\mu(T)}{\chi_\mu(T)} \quad \textit{for } \mu\textit{-a. e. } x \in \mathcal{J}(T).$$

In particular, the measure μ is dimensionally exact and, by Corollary 15.6.11,

$$\mathrm{HD}(\mu) = \mathrm{PD}(\mu) = \frac{\mathrm{h}_\mu(T)}{\chi_\mu(T)}.$$

Proof. In view of Corollary 15.6.11, the second formula follows from the first one and we therefore only need to prove that one. Let us first show that

$$\liminf_{r\to 0} \frac{\log \mu(B(x,r))}{\log r} \geq \frac{\mathrm{h}_\mu(T)}{\chi_\mu(T)} \quad \text{for } \mu\text{-a. e. } x \in \mathcal{J}(T). \tag{28.24}$$

Let $\varepsilon > 0$. By Corollary 25.5.3, there exists a finite partition $\mathcal{P}$ of $\mathcal{J}(T)$ into Borel sets of positive μ-measure such that $\mathrm{h}_\mu(T,\mathcal{P}) \geq \mathrm{h}_\mu(T) - \varepsilon$ and for every x in a set $X_0 \subseteq \mathcal{J}(T)$ of full μ-measure there exists $n(x) \geq 0$ such that

$$B(T^n(x), e^{-n\varepsilon}) \subseteq \mathcal{P}(T^n(x)), \quad \forall n \geq n(x). \tag{28.25}$$

Let us work from now on with the Rokhlin's natural extension $(\widetilde{\mathcal{J}(T)}, \widetilde{T}, \widetilde{\mu})$. Let $\widetilde{\mathcal{J}(T)}(\varepsilon)$ and $r(\varepsilon)$ be given by Theorem 28.3.3. In view of the ergodic case of Birkhoff's ergodic theorem (Corollary 8.2.14), there exists a measurable set $\widetilde{F}(\varepsilon) \subseteq \widetilde{\mathcal{J}(T)}(\varepsilon)$ such that $\widetilde{\mu}(\widetilde{F}(\varepsilon)) = \widetilde{\mu}(\widetilde{\mathcal{J}(T)}(\varepsilon))$ and

$$\lim_{n\to\infty} \frac{1}{n} \sum_{j=0}^{n-1} \mathbb{1}_{\widetilde{\mathcal{J}(T)}(\varepsilon)} \circ \widetilde{T}^n(\widetilde{x}) = \widetilde{\mu}(\widetilde{\mathcal{J}(T)}(\varepsilon)), \quad \forall \widetilde{x} \in \widetilde{F}(\varepsilon).$$

Let $F(\varepsilon) = \pi_0(\widetilde{F}(\varepsilon))$. Then

$$\mu(F(\varepsilon)) = \widetilde{\mu}(\pi_0^{-1}(F(\varepsilon))) \geq \widetilde{\mu}(\widetilde{F}(\varepsilon)) = \widetilde{\mu}(\widetilde{\mathcal{J}(T)}(\varepsilon)) > 1 - \varepsilon.$$

Consider an arbitrary point

$$x \in F(\varepsilon) \cap X_0$$

and take $\widetilde{x} \in \widetilde{F}(\varepsilon)$ such that $x = \pi_0(\widetilde{x})$. By the above, there exists a strictly increasing sequence $(n_k = n_k(\widetilde{x}))_{k=1}^{\infty}$ such that

$$\widetilde{T}^{n_k}(\widetilde{x}) \in \widetilde{\mathcal{J}(T)}(\varepsilon) \quad \text{and} \quad \frac{n_{k+1} - n_k}{n_k} \leq \varepsilon, \quad \forall k \in \mathbb{N}. \tag{28.26}$$

Moreover, we can assume that $n_1 \geq n(x)$. Pick an integer $n \geq n_1$ and the ball

$$B\Big(x, Cr(\varepsilon)\exp(-n[\chi_\mu(T) + \varepsilon(\log \|T'\|_\infty + 2)])\Big),$$

where $0 < C < K(\varepsilon)^{-1}$ is a constant (possibly depending on x), so small that

$$T^q\Big(B\Big(x, Cr(\varepsilon)\exp(-n[\chi_\mu(T) + \varepsilon(\log \|T'\|_\infty + 2)])\Big)\Big) \subseteq \mathcal{P}(T^q(x)), \quad \forall q \leq n_1, \tag{28.27}$$

with $K(\varepsilon) \geq 1$ as the constant appearing in Theorem 28.3.3. Take any $n_1 \leq q \leq n$ and associate k such that $n_k \leq q < n_{k+1}$. As $\widetilde{T}^{n_k}(\widetilde{x}) \in \widetilde{\mathcal{J}(T)}(\varepsilon)$ and as $\pi_0(\widetilde{T}^{n_k}(\widetilde{x})) = T^{n_k}(x)$, Theorem 28.3.3 produces an holomorphic inverse branch

$$T_x^{-n_k} : B(T^{n_k}(x), r(\varepsilon)) \to \widehat{\mathbb{C}}$$

of T^{n_k} such that $T_x^{-n_k}(T^{n_k}(x)) = x$. This theorem also yields

$$T_x^{-n_k}\Big(B(T^{n_k}(x), r(\varepsilon))\Big) \supseteq B\Big(x, K(\varepsilon)^{-1}r(\varepsilon)\exp(-n_k[\chi_\mu(T) + \varepsilon])\Big).$$

Since $n \geq n_k$, we have

$$B\Big(x, Cr(\varepsilon)\exp(-n[\chi_\mu(T) + \varepsilon(\log \|T'\|_\infty + 2)])\Big) \subseteq B\Big(x, K(\varepsilon)^{-1}r(\varepsilon)\exp(-n_k[\chi_\mu(T) + \varepsilon])\Big)$$

and it follows from Theorem 28.3.3 that

$$\begin{aligned} &T^{n_k}\Big(B\Big(x, Cr(\varepsilon)\exp(-n[\chi_\mu(T) + \varepsilon(\log \|T'\|_\infty + 2)])\Big)\Big) \\ &\quad \subseteq B\Big(T^{n_k}(x), CK(\varepsilon)r(\varepsilon)e^{-\chi_\mu(T)(n+n_k)}\exp(\varepsilon[n_k - n(\log \|T'\|_\infty + 2)])\Big). \end{aligned}$$

Applying T^{q-n_k} and using the fact that $n_k \leq q \leq n$ and $q - n_k \leq \varepsilon n_k$ (by (28.26)), we deduce that

$$
\begin{aligned}
T^q\Big(B\Big(x, Cr(\varepsilon)\exp(-n[\chi_\mu(T)+\varepsilon(\log\|T'\|_\infty+2)])\Big)\Big)
&\subseteq B\Big(T^q(x), CK(\varepsilon)r(\varepsilon)e^{-\chi_\mu(T)(n+n_k)}\exp(\varepsilon[n_k-n(\log\|T'\|_\infty+2)])\cdot\|T'\|_\infty^{q-n_k}\Big)\\
&\subseteq B\Big(T^q(x), CK(\varepsilon)r(\varepsilon)\exp(\varepsilon[n_k-n\log\|T'\|_\infty-2n+n_k\log\|T'\|_\infty])\Big)\\
&\subseteq B(T^q(x), CK(\varepsilon)r(\varepsilon)e^{-\varepsilon n})\\
&\subseteq B(T^q(x), e^{-q\varepsilon}).
\end{aligned}
$$

Combining this with (28.25) yields

$$
T^q\Big(B\Big(x, Cr(\varepsilon)\exp(-n[\chi_\mu(T)+\varepsilon(\log\|T'\|_\infty+2)])\Big)\Big)\subseteq \mathcal{P}(T^q(x)),\quad \forall n_1\le q\le n.
$$

Using this and (28.27), we infer that

$$
B\Big(x, Cr(\varepsilon)\exp(-n[\chi_\mu(T)+\varepsilon(\log\|T'\|_\infty+2)])\Big)\subseteq \bigvee_{j=0}^{n} T^{-j}(\mathcal{P})(x) =: \mathcal{P}^{n+1}(x).
$$

Applying Corollary 9.5.5 (the ergodic case of the Shannon–McMillan–Breiman theorem), we get

$$
\liminf_{n\to\infty} -\frac{1}{n}\log\mu\Big(B\Big(x, Cr(\varepsilon)\exp(-n[\chi_\mu(T)+\varepsilon(\log\|T'\|_\infty+2)])\Big)\Big)\ge \mathrm{h}_\mu(T,\mathcal{P})\ge \mathrm{h}_\mu(T)-\varepsilon.
$$

Denoting $Cr(\varepsilon)\exp(-n[\chi_\mu(T)+\varepsilon(\log\|T'\|_\infty+2)])$ by r_n, we thereby have

$$
\liminf_{n\to\infty}\frac{\log\mu(B(x,r_n))}{\log r_n}\ge\frac{\mathrm{h}_\mu(T)-\varepsilon}{\chi_\mu(T)+\varepsilon(\log\|T'\|_\infty+2)}.
$$

Since $(r_n)_{n=1}^\infty$ is a geometric sequence and since $\varepsilon>0$ can be taken arbitrarily small, we conclude that

$$
\liminf_{r\to0}\frac{\log\mu(B(x,r))}{\log r}\ge\frac{\mathrm{h}_\mu(T)}{\chi_\mu(T)}\quad\text{for } \mu\text{-a. e. } x\in\mathcal{J}(T).
$$

This completes the proof of (28.24).

Let us now prove that

$$
\limsup_{r\to0}\frac{\log\mu(B(x,r))}{\log r}\le\frac{\mathrm{h}_\mu(T)}{\chi_\mu(T)}\quad\text{for } \mu\text{-a. e. } x\in\mathcal{J}(T). \tag{28.28}
$$

We again work with Rokhlin's natural extension $(\widetilde{\mathcal{J}(T)}, \tilde{T}, \tilde{\mu})$ and apply Pesin's theory. In particular, the sets $\widetilde{\mathcal{J}(T)}(\varepsilon)$, $\tilde{F}(\varepsilon)\subseteq\widetilde{\mathcal{J}(T)}(\varepsilon)$ and the radius $r(\varepsilon)$ produced via Theorem 28.3.3 have the same meaning as in the proof of (28.24).

To begin, notice that because of Lemma 25.1.8 there exist two numbers $R > 0$ and $0 < Q < \min\{1, r(\varepsilon)/2\}$ such that the following two conditions are satisfied:

(1) If $z \notin B(\mathrm{Crit}(T), R)$, then $T|_{B(z,Q)}$ is injective.
(2) If $z \in B(\mathrm{Crit}(T), R)$, then $T|_{B(z,Q\mathrm{dist}(z,\mathrm{Crit}(T)))}$ is injective.

Observe also that if z is sufficiently close to a critical point c, then $T'(z)$ is of order $\deg_c(T)-1$. In particular, the quotient of $T'(z)$ and $(z-c)^{\deg_c(T)-1}$ remains bounded away from 0 and ∞ and, therefore, there exists $B > 1$ such that

$$|T'(z)| \le B\mathrm{dist}(z, \mathrm{Crit}(T)).$$

In view of Lemma 25.1.14, the logarithm of the function

$$\rho(z) := Q\min\{1, \mathrm{dist}(z, \mathrm{Crit}(T))\}$$

is then integrable and, consequently, Lemma 25.5.5 applies. Let $\mathcal{P}$ be the Mañé's partition produced by that lemma. Then $B(x, \rho(x)) \supseteq \mathcal{P}(x)$ for μ-a. e. $x \in \mathcal{J}(T)$, say for all x in a set $X_\rho \subseteq \mathcal{J}(T)$ of full μ–measure. Therefore,

$$B_n(x, \rho(x)) := \bigcap_{j=0}^{n-1} T^{-j}(B(T^j(x), \rho(T^j(x)))) \supseteq \bigvee_{j=0}^{n-1} T^{-j}(\mathcal{P})(x) =: \mathcal{P}^n(x) \tag{28.29}$$

for every $n \in \mathbb{N}$ and every $x \in X_\rho$. By our choice of Q and the definition of ρ, the function T is injective on all balls $B(T^j(x), \rho(T^j(x)))$, $j \ge 0$, and hence T^i is injective on the set $B_n(x, \rho(x))$ for every $0 \le i \le n-1$. Let

$$x \in F(\varepsilon) \cap X_\rho.$$

Take $\tilde{x} \in \tilde{F}(\varepsilon)$ such that $x = \pi_0(\tilde{x})$, and pick a sequence $(n_k = n_k(\tilde{x}))_{k=1}^\infty$ satisfying (28.26). Let k' be the greatest subscript such that $q := n_{k'} \le n-1$. Denote by T_x^{-q} the unique holomorphic inverse branch of T^q produced by Theorem 28.3.3 which sends $T^q(x)$ to x. By (28.29),

$$B_n(x, \rho(x)) \subseteq T^{-q}(B(T^q(x), \rho(T^q(x)))).$$

As T^q is injective on $B_n(x, \rho(x))$, we even have that

$$B_n(x, \rho(x)) \subseteq T_x^{-q}(B(T^q(x), \rho(T^q(x)))).$$

By Theorem 28.3.3,

$$\mathrm{diam}(T_x^{-q}(B(T^q(x), \rho(T^q(x))))) \le K(\varepsilon)\exp(-q[\chi_\mu(T) - \varepsilon]).$$

By (28.26), $n \le n_{k'+1} \le q(1+\varepsilon)$. So, we finally deduce from the above that

$$B_n(x, \rho(x)) \subseteq B\Big(x, K(\varepsilon)\exp(-n(1+\varepsilon)^{-1}[\chi_\mu(T) - \varepsilon])\Big).$$

In light of (28.29),

$$B\Big(x, K(\varepsilon)\exp(-n(1+\varepsilon)^{-1}[\chi_\mu(T) - \varepsilon])\Big) \supseteq \mathcal{P}^n(x).$$

Therefore, setting $s_n = K(\varepsilon)\exp(-n(1+\varepsilon)^{-1}[\chi_\mu(T) - \varepsilon])$, it follows from Corollary 9.5.5 (the ergodic case of the Shannon–McMillan–Breiman theorem) that for μ-a. e. $x \in \mathcal{J}(T)$,

$$\limsup_{n\to\infty} -\frac{1}{n}\log \mu(B(x, s_n)) \le \mathrm{h}_\mu(T, \mathcal{P}) \le \mathrm{h}_\mu(T).$$

So,

$$\limsup_{n\to\infty} \frac{\log \mu(B(x, s_n))}{\log s_n} \le \frac{\mathrm{h}_\mu(T)}{\chi_\mu(T) - \varepsilon}(1+\varepsilon).$$

As $(s_n)_{n=1}^\infty$ is a geometric sequence and as $\varepsilon > 0$ can be taken arbitrarily small, we conclude that

$$\limsup_{r\to 0} \frac{\log \mu(B(x, r))}{\log r} \le \frac{\mathrm{h}_\mu(T)}{\chi_\mu(T)} \quad \text{for } \mu\text{-a. e. } x \in \mathcal{J}(T).$$

This completes the proof of (28.28). Combining together (28.24) and (28.28) finishes the proof of Theorem 28.4.1. □

In order to apply Theorem 28.4.1, we need to know when the Lyapunov exponent of a measure is positive. Recall that $\Omega(T)$ denotes the set of all rationally indifferent periodic points of T, which is finite according to Theorem 24.3.2.

Theorem 28.4.2. *Let $T : \widehat{\mathbb{C}} \to \widehat{\mathbb{C}}$ be a rational function with $\deg(T) \ge 2$. If μ is an ergodic T-invariant Borel probability measure on $\mathcal{J}(T)$, then*

$$\chi_\mu(T) > 0 \quad \Longrightarrow \quad \mu(\Omega(T)) = 0.$$

Moreover, if T is parabolic then

$$\chi_\mu(T) > 0 \quad \Longleftrightarrow \quad \mu(\Omega(T)) = 0.$$

Proof. Assume first that $\mu(\Omega(T)) > 0$. As μ is ergodic, there is $\omega \in \Omega(T)$ such that $\mu(\mathcal{O}_+(\omega)) = 1$. Let p be the prime period of ω. The T-invariance of μ imposes that $\mu(\{T^k(\omega)\}) = 1/p$ for every $0 \le k < p$. It immediately follows that

$$\chi_\mu(T) := \int_{\mathcal{J}(T)} \log|T'|\, d\mu = \frac{1}{p}\log|(T^p)'(\omega)| = 0.$$

Assume now that T is parabolic and that $\mu(\Omega(T)) = 0$. Pick any $\xi \in \operatorname{supp}(\mu)\backslash\Omega(T) \neq \emptyset$. As T is parabolic, Proposition 26.2.2 asserts that there is $r > 0$ with $B(\xi, 4r) \cap \overline{\mathrm{PC}(T)} = \emptyset$. Let $\kappa > 0$ be such that $\kappa K \le r$, where K is the usual constant of bounded distortion. Per Lemma 24.1.13, there exists $\ell \in \mathbb{N}$ such that $\operatorname{diam}(V) < \kappa$ for every connected component V of $T^{-q}(B(\xi, 2r))$ and for every $q \ge \ell$. As $\mu(B(\xi, r)) > 0$, the ergodic case of Birkhoff's ergodic theorem (Corollary 8.2.14) asserts that there exists $z \in B(\xi, r)$ such that

$$\lim_{n\to\infty} \frac{1}{n}\#\{0 \le k < n : T^k(z) \in B(\xi, r)\} = \mu(B(\xi, r)) \quad \text{and} \quad \lim_{n\to\infty} \frac{1}{n}\log\left|(T^n)'(z)\right| = \chi_\mu(T).$$

Let $(n_j)_{j=1}^\infty$ be the sequence of all $n \ge \ell$ such that $T^n(z) \in B(\xi, r)$ and such that $n_{j+1} - n_j \ge \ell$.

Take any integer $q \ge \ell$. As any connected component $\widehat{V}$ of $T^{-q}(B(\xi, r))$ is contained in a unique connected component V of $T^{-q}(B(\xi, 2r))$, we have by means of Koebe's distortion theorem that

$$2r = \operatorname{diam}(T^q(\widehat{V})) \le \operatorname{diam}(\widehat{V})\|(T^q)'\|_{\infty,\widehat{V}} \le \operatorname{diam}(V)K\left|(T^q)'(y)\right| \le \kappa K\left|(T^q)'(y)\right|$$

for any $y \in \widehat{V}$. Consequently, $\inf_{y\in\widehat{V}} |(T^q)'(y)| \ge 2r(\kappa K)^{-1} \ge 2$.

We claim that $|(T^{n_j})'(z)| \ge 2^j$ for all j large enough. If this is the case, then the result will follow because Birkhoff's ergodic theorem will then yield

$$\chi_\mu(T) = \lim_{j\to\infty} \frac{1}{n_j}\log\left|(T^{n_j})'(z)\right| \ge \log 2 \lim_{j\to\infty} \frac{j}{n_j} = \log 2 \cdot \mu(B(\xi, r)) > 0.$$

Clearly, $|(T^{n_1})'(z)| \ge 2$. For the inductive step, assume that $|(T^{n_j})'(z)| \ge 2^j$. Then

$$\left|(T^{n_{j+1}})'(z)\right| = \left|(T^{n_{j+1}-n_j})'(T^{n_j}(z))\right| \cdot \left|(T^{n_j})'(z)\right| \ge 2 \cdot 2^j = 2^{j+1}. \qquad \square$$

28.5 HD($\mathcal{J}(T)$) > 0 and radial Julia sets $\mathcal{J}_r(T)$, $\mathcal{J}_{er}(T)$, $\mathcal{J}_{uer}(T)$

It immediately follows from Theorem 27.8.5(d,g), Theorem 28.2.1 (Ruelle's inequality) and Theorem 28.4.1 (volume lemma) that if $T : \widehat{\mathbb{C}} \to \widehat{\mathbb{C}}$ is a rational function with $\deg(T) \ge 2$, then

$$\mathrm{HD}(\mathcal{J}(T)) \ge \mathrm{HD}(\mu_0) = \frac{\mathrm{h}_{\mu_0}(T)}{\chi_{\mu_0}(T)} = \frac{\mathrm{h}_{\mathrm{top}}(T)}{\chi_{\mu_0}(T)} = \frac{\log(\deg(T))}{\chi_{\mu_0}(T)} > 0. \tag{28.30}$$

But we can prove that $\mathrm{HD}(\mathcal{J}(T)) > 0$ in a more elementary way (in a sense), while simultaneously obtaining a result which will be needed in the next chapter. We first need some concepts which are of importance on their own and which will also play an important role in the next chapter.

Definition 28.5.1. Let $T : \widehat{\mathbb{C}} \to \widehat{\mathbb{C}}$ be a rational function. A point $z \in \mathcal{J}(T)$ is said to be *radial* (*or conical*) if there exists $\eta > 0$ such that for infinitely many $n \in \mathbb{N}$ the restriction of T^n to $\mathrm{Comp}(z, T^n, \eta)$ is one-to-one, where the reader will recall that $\mathrm{Comp}(z, T^n, \eta)$ stands for the connected component of $T^{-n}(B(T^n(z), \eta))$ containing z. Denote the set of all such n's by $N_z(T)$.

A radial point $z \in \mathcal{J}(T)$ is called *expanding* if there exists $\lambda > 1$ such that

$$|(T^n)'(z)| \geq \lambda^n \tag{28.31}$$

for infinitely many n's in $N_z(T)$.

An expanding radial point z is *uniformly expanding* if the set $\mathbb{N} \setminus N_z(T)$ is finite and inequality (28.31) holds for all $n \in N_z(T)$ large enough.

We respectively denote by $\mathcal{J}_r(T)$, $\mathcal{J}_{er}(T)$, and $\mathcal{J}_{uer}(T)$ the sets of radial, expanding radial, and uniformly expanding radial, points (in $\mathcal{J}(T)$).

Given $\eta > 0$, define $\mathcal{J}_r(T)(\eta)$ to be the set of all points in $\mathcal{J}_r(T)$ that are radial by satisfying that definition with the number 2η. Thereafter, define

$$\mathcal{J}_{er}(T)(\eta) := \mathcal{J}_r(T)(\eta) \cap \mathcal{J}_{er}(T) \quad \text{and} \quad \mathcal{J}_{uer}(T)(\eta) := \mathcal{J}_r(T)(\eta) \cap \mathcal{J}_{uer}(T).$$

The radial Julia set was informally introduced in [80] and independently, but formally, in [139] by analogy to radial/conical sets in the theory of Kleinian groups (see also [32, 92]).

We record the following straightforward fact.

Observation 28.5.2. Let $T : \widehat{\mathbb{C}} \to \widehat{\mathbb{C}}$ be a rational function with $\deg(T) \geq 2$. If $z \in \widehat{\mathbb{C}}$ and there exist $\eta > 0, \lambda > 1$, and a strictly increasing sequence $(n_j)_{j=1}^{\infty}$ of positive integers with bounded gaps (i. e., $\sup\{n_{j+1} - n_j : j \in \mathbb{N}\} < \infty$) such that for each $j \in \mathbb{N}$ the restriction of T^{n_j} to $\mathrm{Comp}(z, T^{n_j}, \eta)$ is 1-to-1 and

$$|(T^{n_j})'(z)| \geq \lambda^j, \quad \forall j \in \mathbb{N},$$

then $z \in \mathcal{J}_{uer}(T)$.

We now prove that the uniformly expanding radial Julia set has positive Hausdorff dimension. As a by-product, we will know that the expanding radial Julia set, the radial Julia set and the Julia set itself all share that property. The strategy consists in building an IFS whose limit set stands within the uniformly expanding radial Julia set. Because the generators of an IFS are contractions, it is natural to look at the inverse branches of a repelling periodic point. To make sure that inverse branches of all orders exist, it is necessary to take a point that lies outside of the postcritical set. That point will be denoted by ξ. To apply the theory of CGDMSs (which covers IFSs) derived in Chapter 19, the first-level sets of the IFS must not overlap beyond their boundaries. This means that a second repelling periodic point x near ξ is needed.

Theorem 28.5.3. *If* $T : \widehat{\mathbb{C}} \to \widehat{\mathbb{C}}$ *is a rational function with* $\deg(T) \geq 2$, *then*

$$\mathrm{HD}(\mathcal{J}_{\mathrm{uer}}(T)) > 0.$$

In consequence, $\mathrm{HD}(\mathcal{J}(T)) > 0$.

Proof. Fix a repelling periodic point $\xi \in \mathcal{J}(T) \setminus \mathrm{PC}(T)$. Let $p \in \mathbb{N}$ be a period of ξ. Fix $\beta \in (1, |(T^p)'(\xi)|)$. According to Remark 24.4.16, there is $R > 0$ such that there exists a unique holomorphic branch $T_\xi^{-p} : B(\xi, R) \to \widehat{\mathbb{C}}$ of T^{-p} fixing ξ, and

$$|(T_\xi^{-p})'(z)| \leq \beta^{-1} < 1, \quad \forall z \in B(\xi, R). \tag{28.32}$$

By the mean value inequality,

$$T_\xi^{-p}(B(\xi, r)) \subseteq B(\xi, \beta^{-1} r), \quad \forall r \in (0, R]. \tag{28.33}$$

As the set $\bigcup_{n=0}^{\infty} T^{-n}(\xi)$ is dense in $\mathcal{J}(T)$ per Corollary 24.3.7, there exist $k \in \mathbb{N}$ and a point

$$x \in (T^{-k}(\xi) \cap B(\xi, R)) \setminus \{\xi\}. \tag{28.34}$$

Since $\xi \notin \mathrm{PC}(T)$, we know that $|(T^k)'(x)| \neq 0$. So, there exists $r \in (0, R/2]$ for which there is a unique holomorphic branch $T_x^{-k} : B(\xi, 2r) \to \widehat{\mathbb{C}}$ of T^{-k} sending ξ to x and

$$T_x^{-k}(B(\xi, 2r)) \subseteq B(\xi, R) \setminus B(\xi, \eta) \tag{28.35}$$

for some $\eta > 0$. By (28.32)–(28.33), there exists $\ell \in \mathbb{N}$ so large that

$$|(T_\xi^{-\ell})'(x)| \leq K^{-2} |(T^k)'(x)|, \tag{28.36}$$

$$T_\xi^{-\ell} \circ T_x^{-k}(B(\xi, 2r)) \subseteq B(\xi, r/2) \setminus B(\xi, \eta') \tag{28.37}$$

and

$$T_\xi^{-\ell} \circ T_x^{-k}(\overline{B(\xi, r/2)}) \subseteq \overline{B(\xi, r/4)} \tag{28.38}$$

for some $\eta' \in (0, r/2)$, where K is the Koebe constant corresponding to the scale 1/2. It then follows from Theorem 23.1.9 (Koebe's distortion theorem, spherical version) and (28.36) that

$$|(T_\xi^{-\ell} \circ T_x^{-k})'(z)| \leq K |(T_\xi^{-\ell} \circ T_x^{-k})'(\xi)| = K |(T_\xi^{-\ell})'(x)| \, |(T_x^{-k})'(\xi)| \leq K^{-1} < 1 \tag{28.39}$$

for every $z \in B(\xi, r)$. By virtue of (28.33), there exists $q \in \mathbb{N}$ such that

$$T_\xi^{-qp}(B(\xi, r)) \subseteq B(\xi, \eta'/2) \subseteq B(\xi, r/4). \tag{28.40}$$

Set

$$\varphi_\xi := T_\xi^{-qp} \quad \text{and} \quad \varphi_x := T_\xi^{-\ell} \circ T_x^{-k}.$$

Let $V = \{v\}$ be a singleton vertex set and the pair $E = \{\xi, x\}$ serve as an alphabet. Let $A : E \times E \to \{0,1\}$ be the constant function 1, so that $E_A^\infty = E^\infty = \{\xi, x\}^{\mathbb{N}}$. Finally, let $X = X_v = \overline{B(\xi, r/2)}$ and $W = W_v = B(\xi, r)$.

Together, relations (28.32)–(28.33) and (28.38)–(28.40) show that the collection

$$\Phi := \{V, E, A; X, W; \varphi_\xi, \varphi_x\}$$

forms a conformal IFS (CIFS).

Denote its limit set by J. By Bowen's formula (Theorem 19.6.4),

$$\mathrm{HD}(J) > 0. \tag{28.41}$$

By (28.37) and (28.40), the system Φ satisfies the strong separation condition (SSC; see Definition 19.7.6). According to Theorem 19.7.8, the limit set J is a topological Cantor set. By (28.38) and (28.40), it is further clear that $J \subseteq \overline{B(\xi, r/4)}$.

For every $z \in J = \pi(E^\infty)$, there is a unique $\omega \in \{\xi, x\}^{\mathbb{N}}$ such that $z = \pi(\omega)$ and

$$\pi(\sigma^j(\omega)) = T^{j_*}(z), \quad \forall j \geq 0,$$

where

$$j_* := (qp)\#\{1 \leq i \leq j : \omega_i = \xi\} + (\ell + k)\#\{1 \leq i \leq j : \omega_i = x\}.$$

Moreover, $B(T^{j_*}(z), r/2) \subseteq B(\xi, r)$ and $\varphi_{\omega|_j} : B(T^{j_*}(z), r) \to \widehat{\mathbb{C}}$ is a holomorphic branch of T^{-j_*} sending $T^{j_*}(z)$ to z. Since the sequence $(j_*)_{j=1}^\infty$ has bounded gaps and, by (28.32)–(28.33) and (28.38)–(28.39),

$$|(T^{j_*})'(z)| = |(\varphi_{\omega|_j})'(\pi(\sigma^j(\omega)))|^{-1} \geq (\min\{K, \beta^{qp}\})^j,$$

it follows from Observation 28.5.2 that $z \in \mathcal{J}_{uer}(T)$. Thus, $J \subseteq \mathcal{J}_{uer}(T)$. Combined with (28.41), this completes the proof of Theorem 28.5.3. □

In the case of polynomials, something more can be said. Anna Zdunik proved in [151] that if a polynomial has a connected Julia set then either it induces a parabolic orbifold (equivalently meaning that it is either a Möbius conjugate of a map $z \mapsto z^{\pm d}$ or of a Tchebyschev polynomial) and then the Julia set is a closed arc of a geometric circle (including the circle itself), or else its Hausdorff dimension is strictly larger than 1.

It is worth noting that $\mathcal{J}_{uer}(T)$ contains all expanding subsets of $\mathcal{J}(T)$. As mentioned during the examination of expanding sets in Section 25.3 (just before Theorem 25.3.9), Definition 16.2.1 is a little too narrow for expanding sets. In fact, all results derived about conformal repellers in Volume 2 are valid when part (d) of their definition is replaced by the following weaker condition:

(d') There exists $\lambda > 1$ and $q \in \mathbb{N}$ such that $|(T^q)'(x)| \geq \lambda$ for all $x \in X$.

Proposition 28.5.4. *If* $T : \widehat{\mathbb{C}} \to \widehat{\mathbb{C}}$ *is a rational function with* $\deg(T) \geq 2$ *and* $X \subseteq \mathcal{J}(T)$ *is an expanding set, then the pressure function* $\mathbb{R} \ni t \mapsto \mathrm{P}(T|_X, t) = \mathrm{P}(T|_X, -t\log|T'|) \in \mathbb{R}$ *possesses the following properties:*

(a) *It is Lipschitz continuous, with Lipschitz constant* $L = L(X) := \log \|T'\|_{X,\infty}$ *and where* $\|T'\|_{X,\infty} := \sup_{x\in X} |T'(x)|$.

(b) *It is strictly decreasing.*

(c) $\lim_{t\to-\infty} \mathrm{P}(T|_X, t) = \infty$ *whereas* $\lim_{t\to\infty} \mathrm{P}(T|_X, t) = -\infty$.

(d) $\mathrm{P}(T|_X, 0) = \mathrm{h}_{\mathrm{top}}(T|_X)$. *If T is transitive and X is not a single periodic orbit of T, then* $\mathrm{P}(T|_X, 0) = \mathrm{h}_{\mathrm{top}}(T|_X) > 0$.

(e) *There exists a unique* $h = h(X) \in \mathbb{R}$ *such that* $\mathrm{P}(T|_X, h(X)) = 0$. *If T is transitive and X is not a single periodic orbit of T, then* $h(X) > 0$.

(f) *The pressure function* $\mathbb{R} \ni t \mapsto \mathrm{P}(T|_X, t) \in \mathbb{R}$ *is convex.*

N.B.: This result is analogous to Proposition 16.3.1 for conformal expanding repellers. Parts (d,e) *of that proposition were not entirely correctly stated. The underlined condition on X was lacking; this condition is obviously satisfied when X is infinite.* (*The transitivity of T was part of the definition of a conformal expanding repeller.*)

Proof. Let $q \geq 1$ be an integer attesting of the expandingness of $T|_X$. The proof of Proposition 16.3.1 applies verbatim to $T|_X$ except that when proving items (a)–(c) one replaces "a unique equilibrium state" by "an equilibrium state". As $\mathrm{P}(T|_X^q, -t\log|(T^q)'|) = q\mathrm{P}(T|_X, -t\log|T'|)$, the result ensues. □

Theorem 28.5.5. *If* $T : \widehat{\mathbb{C}} \to \widehat{\mathbb{C}}$ *is a rational function with* $\deg(T) \geq 2$ *and* $X \subseteq \mathcal{J}(T)$ *is an expanding set, then*

$$\mathrm{HD}(X) = h,$$

where h comes from Proposition 28.5.4.

Proof. Fix any $t > h$. Then $\mathrm{P}(T|_X, t) < 0$. Since X is expanding, there are $\lambda > 1$ and $q \in \mathbb{N}$ such that

$$|(T^q)'(x)| \geq \lambda, \quad \forall x \in X. \tag{28.42}$$

By standard compactness arguments, there exists $\delta > 0$ such that

$$T^q|_{B(x,4\delta)} \text{ is 1-to-1}, \quad \forall x \in X \tag{28.43}$$

and

$$|(T^q)'(z)| \geq \sqrt{\lambda}, \quad \forall z \in B(X, 4\delta) := \bigcup_{x \in X} B(x, 4\delta). \tag{28.44}$$

Consequently, for every $x \in X$ there is a unique holomorphic inverse branch $T_x^{-q} : B(T^q(x), 4\delta) \to \widehat{\mathbb{C}}$ such that $T_x^{-q}(T^q(x)) = x$. By (28.44), we have that

$$T^q(B(x,r)) \supseteq B(T^q(x), \sqrt{\lambda}\, r), \quad |(T_x^{-q})'(z)| \leq \lambda^{-1/2}, \quad \forall r \in [0, 4\delta], \ \forall z \in B(x, 4\delta), \forall x \in X \tag{28.45}$$

and

$$\rho(T_x^{-q}(w), T_x^{-q}(z)) \leq \lambda^{-1/2} \rho(w,z), \quad \forall w, z \in B(x, 4\delta), \ \forall x \in X. \tag{28.46}$$

It follows from (28.46) that

$$T_x^{-q}(B(T^q(x), r)) \subseteq B(x, \lambda^{-1/2} r) \subseteq B(x, r)$$

for all $r \in [0, 4\delta]$ and all $x \in X$. Hence, for every $n \in \mathbb{N}$ and $x \in X$ the composition

$$T_x^{-qn} := T_x^{-q} \circ T_{T(x)}^{-q} \circ \cdots \circ T_{T^{q(n-2)}(x)}^{-q} \circ T_{T^{q(n-1)}(x)}^{-q} : B(T^{qn}(x), 4\delta) \to \widehat{\mathbb{C}}$$

is well defined, satisfies

$$T_x^{-qn}(B(T^{qn}(x), r)) \subseteq B(x, \lambda^{-n/2} r) \subseteq B(x, r), \quad \forall r \in [0, 4\delta],$$

and is the unique analytic inverse branch of T^{qn} defined on $B(T^{qn}(x), 4\delta)$ that sends $T^{qn}(x)$ to x.

Let $D \subseteq X$ be a maximal δ-separated set. Then $T|_X^{-qn}(D)$ is a (n, δ)-separated set for $T|_X^q$ for all $n \in \mathbb{N}$ and

$$\{T_w^{-qn}(B(z, \delta)) : z \in D, \ w \in T|_X^{-qn}(z)\}$$

is an open cover of X with

$$\operatorname{diam}(T_w^{-qn}(B(z, \delta))) \leq \delta \lambda^{-n/2}.$$

Moreover, by bounded distortion (entirely dynamical; no Koebe distortion is needed),

$$\operatorname{diam}(T_w^{-qn}(B(z, \delta))) \leq C\delta |(T^{qn})'(w)|^{-1}$$

with distortion constant $C \geq 1$ depending only on T. Therefore the t-dimensional Hausdorff measure of X satisfies

$$H_t(X) \leq \liminf_{n \to \infty} \sum_{z \in D} \sum_{w \in T|_X^{-qn}(z)} [\operatorname{diam}(T_w^{-qn}(B(z, \delta)))]^t$$

$$\le C\delta \liminf_{n\to\infty} \sum_{z\in D} \sum_{w\in T|_X^{-qn}(z)} |(T^{qn})'(w)|^{-t}.$$

But by Bowen's definition of the pressure we have

$$\liminf_{n\to\infty} \frac{1}{n} \log \sum_{z\in D} \sum_{w\in T|_X^{-qn}(z)} |(T|_X^{qn})'(w)|^{-t} \le \mathrm{P}(T|_X^q, -t\log|(T^q)'|) = q\mathrm{P}(T|_X, -t\log|T'|) < 0,$$

we deduce that $H_t(X) = 0$. Thus, $\mathrm{HD}(X) \le t$ for all $t > h$. Hence, $\mathrm{HD}(X) \le h$.

For the converse inequality, it is clear that $\mathrm{HD}(X) \ge h$ if $h = 0$. So we may assume that $h > 0$. Fix any $t \in (0, h)$. By Proposition 28.5.4, $\mathrm{P}(T|_X, t) > 0$. Per the variational principle (Theorem 12.1.1 and Remark 12.1.2), there exists an ergodic $T|_X$-invariant Borel probability measure μ on X such that

$$\mathrm{h}_\mu(T|_X) - t\chi_\mu(T|_X) > 0.$$

But $\chi_\mu(T|_X) > 0$ since $T|_X$ is expanding. Using the volume lemma (Theorem 28.4.1), we infer from the above inequality that

$$\mathrm{HD}(X) \ge \mathrm{HD}(\mu) = \frac{\mathrm{h}_\mu(T|_X)}{\chi_\mu(T|_X)} > t.$$

Given that this holds for all $t \in (0, h)$, we conclude that $\mathrm{HD}(X) \ge h$. □

It is noteworthy to recall that the pressure and the Hausdorff dimension functions were shown to be real analytic under analytic perturbations of a conformal repeller. See Theorems 16.4.9–16.4.11. Other papers treating of the real analyticity of the Hausdorff dimension include [2, 114, 125, 132, 143, 147].

28.6 Conformal Katok's theory of expanding sets and topological pressure

In this section, we provide a conformal account of Katok's theory, which describes the high complexity (expanding/hyperbolic subsets and periodic points) of many classes of dynamical systems having positive entropy. Its full exposition is given in [62]. Our approach to the conformal case stems from [110].

28.6.1 Pressure-like definition of the functional $\mathrm{h}_\mu(T) + \int \varphi\, d\mu$

In this subsection, we develop general tools that will be used in the next subsection to approximate the topological pressure on expanding sets. These tools are reminiscent of those that emerged in Sections 7.3, 11.2 and 12.1 of the first volume. The reader is encouraged to review those sections (especially the latter) to identify the similarities and differences between the various tools.

Unlike in the previous sections of this chapter, no smoothness/conformality is needed in this subsection. Just like in Sections 7.3, 11.2 and 12.1, we work in a metric space setting. Let $T : X \to X$ be a topological dynamical system, i. e., a continuous self-map of a compact metric space (X, d), and let μ be a Borel probability measure on X. Given $\varepsilon > 0$ and $0 \le \delta \le 1$, a set $E \subseteq X$ is said to be $\mu - (n, \varepsilon, \delta)$*-spanning* if

$$\mu\Big(\bigcup_{x\in E} B_n(x,\varepsilon)\Big) \ge 1-\delta,$$

where

$$B_n(x,\varepsilon) := \bigcap_{j=0}^{n-1} T^{-j}(B(T^j(x),\varepsilon))$$

consists of those points whose iterates stay within a distance ε from the corresponding iterates of x until time $n-1$ at least. Recall that

$$B_n(x,\varepsilon) = \{y \in X : d_n(x,y) < \varepsilon\},$$

where d_n is the dynamical metric (also called Bowen metric)

$$d_n(x,y) = \max\{d(T^j(x), T^j(y)) : 0 \le j < n\}.$$

The ball $B_n(x, \varepsilon)$ is accordingly called the dynamical (n, ε)-ball centered at x.

Let $\varphi : X \to \mathbb{R}$ be a potential, i. e., a continuous real-valued function on X. Define

$$Q_\mu(T,\varphi,n,\varepsilon,\delta) = \inf_E\Big\{\sum_{x\in E} \exp(S_n\varphi(x))\Big\},$$

where the infimum is taken over all $\mu - (n, \varepsilon, \delta)$-spanning sets E. The main result of this subsection is a formula for the free energy.

Theorem 28.6.1. *If $T : (X,d) \to (X,d)$ is a topological dynamical system and $\varphi : X \to \mathbb{R}$ is a potential on X, then for every $0 < \delta < 1$ and every ergodic T-invariant Borel probability measure μ on X, we have*

$$\mathrm{h}_\mu(T) + \int_X \varphi\,d\mu = \lim_{\varepsilon\to 0}\liminf_{n\to\infty}\frac{1}{n}\log Q_\mu(T,\varphi,n,\varepsilon,\delta) = \lim_{\varepsilon\to 0}\limsup_{n\to\infty}\frac{1}{n}\log Q_\mu(T,\varphi,n,\varepsilon,\delta).$$

Proof. Denote the quantity following the first equality sign by $\underline{\mathrm{P}}_\mu(T,\varphi,\delta)$ and the quantity following the second equality sign by $\overline{\mathrm{P}}_\mu(T,\varphi,\delta)$. Following essentially the proof of Part I of the variational principle (Theorem 12.1.1), we shall first show that

$$\underline{\mathrm{P}}_\mu(T,\varphi,\delta) \geq \mathrm{h}_\mu(T) + \int_X \varphi\, d\mu. \tag{28.47}$$

Indeed, in a similar way to that proof, consider a finite Borel partition $\mathcal{U} = \{A_1,\dots,A_s\}$ of X and, using Lemma 12.1.4, take compact sets $B_i \subseteq A_i$, $i = 1,2,\dots,s$, such that the partition

$$\mathcal{V} = \{B_1,\dots,B_s, X \setminus (B_1 \cup \cdots \cup B_s)\}$$

satisfies

$$\mathrm{H}_\mu(\mathcal{U}|\mathcal{V}) \leq 1.$$

Let $\theta > 0$. For every $q \in \mathbb{N}$, set

$$X_q := \left\{x \in X : -\frac{1}{n}\log\mu(\mathcal{V}^n(x)) \geq \mathrm{h}_\mu(T,\mathcal{V}) - \theta \quad \text{and} \quad \frac{1}{n}S_n\varphi(x) \geq \int_X \varphi\, d\mu - \theta, \quad \forall n \geq q\right\}.$$

Fix $0 < \delta < 1$. It follows from the ergodic cases of the Shannon–McMillan–Breiman theorem (Corollary 9.5.5) and of Birkhoff's ergodic theorem (Corollary 8.2.14) that the sets $(X_q)_{q=1}^\infty$ form an ascending sequence covering μ-almost everywhere X. Therefore,

$$\mu(X_q) > \delta \tag{28.48}$$

for all $q \in \mathbb{N}$ large enough. Fix such a q. Since the potential $\varphi : X \to \mathbb{R}$ is uniformly continuous, there exists

$$0 < \varepsilon < \frac{1}{2}\min\{d(B_i,B_j) : 1 \leq i < j \leq s\} > 0$$

so small that

$$d(x,y) < \varepsilon \Longrightarrow |\varphi(x) - \varphi(y)| < \theta. \tag{28.49}$$

Take any integer $n \geq q$. For every $x \in X$, the set $B_n(x,\varepsilon) \cap X_q$ can be covered by at most 2^n atoms of $\mathcal{V}^n$ because at every iterate this set can intersect at most one B_i and $X \setminus \bigcup_{j=1}^s B_j$. For every $w \in X_q$, we also know that $-\frac{1}{n}\log\mu(\mathcal{V}^n(w)) \geq \mathrm{h}_\mu(T,\mathcal{V}) - \theta$. Consequently,

$$\mu(B_n(x,\varepsilon) \cap X_q) \leq 2^n \max\{\mu(\mathcal{V}^n(w)) : w \in X_q\} \leq \exp(n[\log 2 - \mathrm{h}_\mu(T,\mathcal{V}) + \theta]). \tag{28.50}$$

Let $E \subseteq X$ be a $\mu - (n,\varepsilon,\delta)$-spanning set, and consider the set

$$E' := \{x \in E : B_n(x,\varepsilon) \cap X_q \neq \emptyset\}.$$

Take a point $y \in B_n(x,\varepsilon) \cap X_q$. By the choice of ε in (28.49), we have $|S_n\varphi(x) - S_n\varphi(y)| < n\theta$. As $y \in X_q$, we also know that $S_n\varphi(y) - n\int \varphi\, d\mu \geq -n\theta$. Therefore,

$$\begin{aligned}
&\sum_{x\in E} \exp(S_n\varphi(x)) \exp\left(-n\left[\mathrm{h}_\mu(T,\mathcal{V}) + \int_X \varphi\, d\mu - 3\theta - \log 2\right]\right) \\
&\quad \geq \sum_{x\in E'} \exp\left(S_n\varphi(x) - n\int_X \varphi\, d\mu\right) \exp(-n[\mathrm{h}_\mu(T,\mathcal{V}) - 3\theta - \log 2]) \quad \text{because } E' \subseteq E \\
&\quad = \sum_{x\in E'} \exp(S_n\varphi(x) - S_n\varphi(y)) \exp\left(S_n\varphi(y) - n\int_X \varphi\, d\mu\right) \exp(-n[\mathrm{h}_\mu(T,\mathcal{V}) - 3\theta - \log 2]) \\
&\quad \geq \sum_{x\in E'} \exp(-n\theta)\exp(-n\theta)\exp(-n[\mathrm{h}_\mu(T,\mathcal{V}) - 3\theta - \log 2]) \\
&\quad = \sum_{x\in E'} \exp(n[\log 2 - \mathrm{h}_\mu(T,\mathcal{V}) + \theta]) \geq \sum_{x\in E'} \mu(B_n(x,\varepsilon)\cap X_q) \quad \text{by (28.50)} \\
&\quad \geq \mu\left(\bigcup_{x\in E'} B_n(x,\varepsilon)\cap X_q\right) = \mu\left(\left[\bigcup_{x\in E} B_n(x,\varepsilon)\right]\cap X_q\right) \\
&\quad \geq \mu(X_q) - \delta \quad \text{because } E \text{ is } \mu-(n,\varepsilon,\delta)\text{-spanning} \\
&\quad > 0 \quad \text{by (28.48).}
\end{aligned}$$

This implies that

$$\frac{1}{n}\log Q_\mu(T,\varphi,n,\varepsilon,\delta) \geq \mathrm{h}_\mu(T,\mathcal{V}) + \int_X \varphi\, d\mu - 3\theta - \log 2, \quad \forall n \geq q.$$

Taking the lim inf as $n \to \infty$ followed by the limit as $\varepsilon \to 0$, we deduce that

$$\underline{\mathrm{P}}_\mu(T,\varphi,\delta) \geq \mathrm{h}_\mu(T,\mathcal{V}) + \int_X \varphi\, d\mu - 3\theta - \log 2.$$

Since $\theta > 0$ is an arbitrary number and since $\mathrm{h}_\mu(T,\mathcal{U}) \leq \mathrm{h}_\mu(T,\mathcal{V}) + \mathrm{H}_\mu(\mathcal{U}|\mathcal{V}) \leq \mathrm{h}_\mu(T,\mathcal{V}) + 1$, we get that

$$\underline{\mathrm{P}}_\mu(T,\varphi,\delta) \geq \mathrm{h}_\mu(T,\mathcal{U}) - 1 + \int_X \varphi\, d\mu - \log 2.$$

Given the arbitrariness of the partition $\mathcal{U}$, upon passing to the supremum over all partitions we infer that

$$\underline{\mathrm{P}}_\mu(T,\varphi,\delta) \geq \mathrm{h}_\mu(T) + \int_X \varphi\, d\mu - \log 2 - 1.$$

Using now the standard trick consisting in replacing T by an arbitrary iterate T^k and φ by $S_k\varphi$, we obtain that

$$k\underline{\mathrm{P}}_\mu(T,\varphi,\delta) = \underline{\mathrm{P}}_\mu(T^k,S_k\varphi,\delta) \geq \mathrm{h}_\mu(T^k) + \int_X S_k\varphi\, d\mu - \log 2 - 1 = k\mathrm{h}_\mu(T) + k\int_X \varphi\, d\mu - \log 2 - 1.$$

Dividing the extremes of this inequality by k and letting $k \to \infty$, we conclude that

$$\underline{\mathrm{P}}_\mu(T, \varphi, \delta) \geq \mathrm{h}_\mu(T) + \int_X \varphi\, d\mu.$$

Let us now demonstrate that

$$\overline{\mathrm{P}}_\mu(T, \varphi, \delta) \leq \mathrm{h}_\mu(T) + \int_X \varphi\, d\mu. \tag{28.51}$$

Fix $0 < \delta < 1$. Let $\varepsilon > 0$ and $\theta > 0$. Let $\mathcal{P}$ be a finite Borel partition of X with $\operatorname{diam}(\mathcal{P}) \leq \varepsilon$. By the ergodic cases of the Shannon–McMillan–Breiman theorem (Corollary 9.5.5) and of Birkhoff's ergodic theorem (Corollary 8.2.14), there exists a Borel set $Z \subseteq X$ such that $\mu(Z) > 1 - \delta$ and

$$-\frac{1}{n} \log \mu(\mathcal{P}^n(x)) \leq \mathrm{h}_\mu(T) + \theta \quad \text{and} \quad \frac{1}{n} S_n\varphi(x) \leq \int_X \varphi\, d\mu + \theta \tag{28.52}$$

for every $x \in Z$ and all $n = n(x)$ large enough. Take such an x and n. From each atom of $\mathcal{P}^n$ intersecting Z, choose one point, hence generating a finite set $\{x_1, x_2, \ldots, x_k\}$. Then $B_n(x_j, \varepsilon) \supseteq \mathcal{P}^n(x_j)$ for every $j = 1, 2, \ldots, k$ and, therefore, the set $\{x_1, x_2, \ldots, x_k\}$ is $\mu -(n, \varepsilon, \delta)$-spanning. By the first part of (28.52), we have $k \leq \exp(n[\mathrm{h}_\mu(T) + \theta])$. Using this and the second part of (28.52), we get

$$\sum_{j=1}^{k} \exp(S_n\varphi(x_j)) \leq \exp\left(n\left[\mathrm{h}_\mu(T) + \theta + \int_X \varphi\, d\mu + \theta\right]\right).$$

Therefore,

$$Q_\mu(T, \varphi, n, \varepsilon, \delta) \leq \exp\left(n\left[\mathrm{h}_\mu(T) + \int_X \varphi\, d\mu + 2\theta\right]\right).$$

Applying the logarithm to both sides, dividing by n and taking successively the lim sup as $n \to \infty$ and the limit as $\varepsilon \to 0$, we deduce that

$$\overline{\mathrm{P}}_\mu(T, \varphi, \delta) \leq \mathrm{h}_\mu(T) + \int_X \varphi\, d\mu + 2\theta.$$

Since $\theta > 0$ is arbitrary, (28.51) ensues. Joining (28.51) to (28.47) completes the proof of Theorem 28.6.1. □

28.6.2 Conformal Katok's theory

This subsection is dedicated to proving the following result.

Theorem 28.6.2. *Let $T : \widehat{\mathbb{C}} \to \widehat{\mathbb{C}}$ be a rational function with $\deg(T) \geq 2$. If μ is an ergodic T-invariant Borel probability measure on $\mathcal{J}(T)$ with positive Lyapunov exponent $\chi_\mu(T)$ and if $\varphi : \mathcal{J}(T) \to \mathbb{R}$ is a (continuous) potential, then there exists a sequence $(X_k)_{k=1}^{\infty}$ of forward T-invariant closed subsets of $\mathcal{J}(T)$ that are all topological Cantor sets such that for every $k \in \mathbb{N}$ the restriction $T|_{X_k} : X_k \to X_k$ is a conformal topologically transitive open expanding dynamical system and*

$$\liminf_{k\to\infty} \mathrm{P}(T|_{X_k}, \varphi) \geq \mathrm{h}_\mu(T) + \int_{\mathcal{J}(T)} \varphi \, d\mu. \tag{28.53}$$

Furthermore, if for every $k \in \mathbb{N}$ there is an ergodic T-invariant Borel probability measure μ_k on X_k, then the sequence $(\mu_k)_{k=1}^{\infty}$ converges to μ in the weak* topology on $M(\mathcal{J}(T))$. Moreover,

$$\chi_{\mu_k}(T|_{X_k}) := \int_{X_k} \log|T'| \, d\mu_k \overset{k\to\infty}{\longrightarrow} \int_{\mathcal{J}(T)} \log|T'| \, d\mu =: \chi_\mu(T).$$

In particular, this is true if the measures $(\mu_k)_{k=1}^{\infty}$ are each supported on individual periodic orbits of T in X_k.

Proof. By Tietze's extension theorem, we may assume that the continuous function φ is defined on all of $\widehat{\mathbb{C}}$. Whatever the T-invariant closed sets $X_k \subseteq \mathcal{J}(T)$ are, the variational principle (Theorem 12.1.1) ensures that for every $c \in \mathbb{R}$,

$$\mathrm{P}(T|_{X_k}, \varphi + c) = \mathrm{P}(T|_{X_k}, \varphi) + c.$$

Moreover,

$$\mathrm{h}_\mu(T) + \int_{\mathcal{J}(T)} (\varphi + c) \, d\mu = \mathrm{h}_\mu(T) + \int_{\mathcal{J}(T)} \varphi \, d\mu + c.$$

Therefore, by adding a sufficiently large constant c if necessary, we can assume without loss of generality that φ is positive, i. e.,

$$\inf \varphi > 0.$$

Fix a countable basis $\{\psi_j\}_{j=1}^{\infty}$ of the Banach space $C(\widehat{\mathbb{C}}, \mathbb{R})$ of all real-valued continuous functions on $\widehat{\mathbb{C}}$. Fix $\theta > 0$ and $s \in \mathbb{N}$. In view of Theorem 28.6.1 and the uniform continuity of all the functions φ and $\psi_j, j \in \mathbb{N}$, there exists $\varepsilon > 0$ so small that

$$\liminf_{n\to\infty} \frac{1}{n} \log Q_\mu(T, \varphi, n, \varepsilon, \delta) > \left(\mathrm{h}_\mu(T) + \int_{\mathcal{J}(T)} \varphi \, d\mu \right) - \theta, \tag{28.54}$$

$$|x - y| < \varepsilon \Longrightarrow |\varphi(x) - \varphi(y)| < \theta \tag{28.55}$$

and

$$|x-y|<\varepsilon \Longrightarrow |\psi_i(x)-\psi_i(y)|<\theta/2, \quad \forall i=1,2,\ldots,s. \tag{28.56}$$

We work once more with Rokhlin's natural extension $(\widetilde{\mathcal{J}(T)},\tilde{T},\tilde{\mu})$. Given $\delta>0$, let $\widetilde{\mathcal{J}(T)}(\delta)$ and $r(\delta)$ be produced through Theorem 28.3.3. Then the set $\pi_0(\widetilde{\mathcal{J}(T)}(\delta))$ is compact as a continuous image of the compact set $\widetilde{\mathcal{J}(T)}(\delta)$. That theorem also produces a constant $\chi'>0$ (possibly with a smaller radius $r(\delta)$) such that

$$\operatorname{diam}(T_{x_n}^{-n}(B(\pi_0(\tilde{x}),r(\delta))))\le e^{-n\chi'}, \quad \forall \tilde{x}\in\widetilde{\mathcal{J}(T)}(\delta),\ \forall n\ge 0. \tag{28.57}$$

Set

$$\beta := r(\delta)/2$$

and fix a finite $(\beta/2)$-spanning set $\{x_1,\ldots,x_t\}$ of $\pi_0(\widetilde{\mathcal{J}(T)}(\delta))$. This means that

$$B(x_1,\beta/2)\cup\cdots\cup B(x_t,\beta/2)\supseteq \pi_0(\widetilde{\mathcal{J}(T)}(\delta)).$$

Let $\mathcal{U}$ be a finite Borel partition of $\mathcal{J}(T)$ with $\operatorname{diam}(\mathcal{U})<\beta/2$ and let $n_1\in\mathbb{N}$ be so large that

$$e^{-n_1\chi'}<\min\{\beta/3,K^{-1}\}, \tag{28.58}$$

where K is the Koebe constant corresponding to the scale 1/2.

Given $n\in\mathbb{N}$, define

$$\begin{aligned}\widetilde{\mathcal{J}(T)}_{n,s} := \Big\{\tilde{x}\in\widetilde{\mathcal{J}(T)}(\delta) : \tilde{T}^q(\tilde{x})\in\widetilde{\mathcal{J}(T)}(\delta),\\ \pi_0(\tilde{T}^q(\tilde{x}))\in\mathcal{U}(\pi_0(\tilde{x})) \text{ for some } q\in[n+1,(1+\theta)n],\\ \text{and } \left|\frac{1}{k}S_k(\psi_i)(\pi_0(\tilde{x}))-\int_{\mathcal{J}(T)}\psi_i\,d\mu\right|<\frac{\theta}{2},\ \forall k\ge n,\ \forall 1\le i\le s\Big\}.\end{aligned}$$

By the ergodic case of Birkhoff's ergodic theorem (Corollary 8.2.14), we know that $\lim_{n\to\infty}\tilde{\mu}(\widetilde{\mathcal{J}(T)}_{n,s})=\tilde{\mu}(\widetilde{\mathcal{J}(T)}(\delta))>1-\delta$. Therefore, there exists $n\ge n_1$ so large that $\tilde{\mu}(\widetilde{\mathcal{J}(T)}_{n,s})>1-\delta$. Let

$$\mathcal{J}(T)_{n,s} := \pi_0(\widetilde{\mathcal{J}(T)}_{n,s}).$$

Then

$$\mu(\mathcal{J}(T)_{n,s})>1-\delta.$$

Let $E_n \subseteq \mathcal{J}(T)_{n,s}$ be a maximal (n,ε)-separated subset of $\mathcal{J}(T)_{n,s}$. Then E_n is a (n,ε)-spanning set of $\mathcal{J}(T)_{n,s}$. It follows from (28.54) that for all integers $n \geq n_1$ large enough

$$\frac{1}{n}\log\sum_{x\in E_n}\exp(S_n\varphi(x)) > \left(\mathrm{h}_\mu(T) + \int_{\mathcal{J}(T)}\varphi\,d\mu\right) - \theta.$$

Equivalently,

$$\sum_{x\in E_n}\exp(S_n\varphi(x)) > \exp\left(n\left[\mathrm{h}_\mu(T) + \int_{\mathcal{J}(T)}\varphi\,d\mu - \theta\right]\right).$$

For every integer $q \in [n+1,(1+\theta)n]$, set

$$V_q := \{x \in E_n : T^q(x) \in \mathcal{U}(x)\}.$$

Then let $m = m(n)$ be a value of q that maximizes $\sum_{x\in V_q}\exp(S_n\varphi(x))$. As $\bigcup_{q=n+1}^{(1+\theta)n} V_q = E_n$, we thus obtain that

$$\begin{aligned}\sum_{x\in V_m}\exp(S_n\varphi(x)) &\geq \frac{1}{n\theta}\sum_{q=n+1}^{(1+\theta)n}\sum_{x\in V_q}\exp(S_n\varphi(x)) \geq \frac{1}{n\theta}\sum_{x\in E_n}\exp(S_n\varphi(x))\\ &\geq \exp\left(n\left[\mathrm{h}_\mu(T) + \int_{\mathcal{J}(T)}\varphi\,d\mu - 2\theta\right]\right).\end{aligned}$$

Consider the sets $V_m \cap B(x_j,\beta/2), 1 \leq j \leq t$, and choose a value $i = i(m)$ of j that maximizes the sum

$$\sum_{x\in V_m\cap B(x_j,\beta/2)}\exp(S_n\varphi(x)).$$

Denoting the intersection $V_m \cap B(x_{i(m)},\beta/2)$ by D_m, we have

$$V_m = \bigcup_{j=1}^{t} V_m \cap B(x_j,\beta/2)$$

and

$$\sum_{x\in D_m}\exp(S_n\varphi(x)) \geq \frac{1}{t}\sum_{x\in V_m}\exp(S_n\varphi(x)) \geq \frac{1}{t}\exp\left(n\left[\mathrm{h}_\mu(T) + \int_{\mathcal{J}(T)}\varphi\,d\mu - 2\theta\right]\right).$$

As φ is positive and $m \geq n+1$, this implies that

$$\sum_{x\in D_m}\exp(S_m\varphi(x)) \geq \frac{1}{t}\exp\left(n\left[\mathrm{h}_\mu(T) + \int_{\mathcal{J}(T)}\varphi\,d\mu - 2\theta\right]\right). \tag{28.59}$$

For any $x \in D_m$, we know that $x \in B(x_{i(m)}, \beta/2)$, i. e., $|x - x_{i(m)}| < \beta/2$. We further know that $x \in V_m$, and hence $T^m(x) \in \mathcal{U}(x)$. As $\operatorname{diam}(\mathcal{U}) < \beta/2$, it ensues that $|T^m(x) - x| < \beta/2$. Then $|T^m(x) - x_{i(m)}| \le |T^m(x) - x| + |x - x_{i(m)}| < \beta/2 + \beta/2 = \beta$ and, therefore,

$$T^m(x) \in B(x_{i(m)}, \beta) \subseteq B(T^m(x), 2\beta). \tag{28.60}$$

Choose $\tilde{x} \in \pi_0^{-1}(x) \cap \widetilde{\mathcal{J}(T)}_{n,s}$. Given that $m \ge n \ge n_1$, it ensues from (28.57)–(28.58) that

$$\operatorname{diam}(T_x^{-m}(B(T^m(x), 2\beta))) \le e^{-m\chi'} < \beta/3.$$

By this and (28.60), we deduce that

$$T_x^{-m}(B(x_{i(m)}, \beta)) \subseteq B(x, \beta/3) \subseteq B(x_{i(m)}, \beta/2 + \beta/3) = B(x_{i(m)}, 5\beta/6).$$

In particular, this implies that

$$\overline{T_x^{-m}(B(x_{i(m)}, \beta))} \subseteq B(x_{i(m)}, \beta). \tag{28.61}$$

Observe that

$$T_{y_1}^{-m}(B(x_{i(m)}, 2\beta)) \cap T_{y_2}^{-m}(B(x_{i(m)}, 2\beta)) = \emptyset, \quad \forall y_1, y_2 \in D_m,\ y_1 \ne y_2.$$

In particular,

$$T_{y_1}^{-m}(\overline{B(x_{i(m)}, \beta)}) \cap T_{y_2}^{-m}(\overline{B(x_{i(m)}, \beta)}) = \emptyset, \quad \forall y_1, y_2 \in D_m,\ y_1 \ne y_2.$$

Let

$$\xi := \min\Big\{\beta,\ \min\{\operatorname{dist}(\, T_{y_1}^{-m}(\overline{B(x_{i(m)}, \beta)})\,, T_{y_2}^{-m}(\overline{B(x_{i(m)}, \beta)})\,) : y_1, y_2 \in D_m,\ y_1 \ne y_2\}\Big\}.$$

Define inductively the sequence of sets $(X^{(j)})_{j=0}^{\infty}$ by setting

$$X^{(0)} := \overline{B(x_{i(m)}, \beta)} \quad \text{and} \quad X^{(j+1)} := \bigcup_{x \in D_m} T_x^{-m}(X^{(j)}).$$

By (28.61), $(X^{(j)})_{j=0}^{\infty}$ is a descending sequence of nonempty compact sets. Therefore, the intersection

$$X^*(\theta, s) := \bigcap_{j=0}^{\infty} X^{(j)}$$

is also a nonempty compact set. Moreover, by construction:

(a*) $X^*(\theta, s)$ is a topological Cantor set.

(b*) $T^m(X^*(\theta,s)) = X^*(\theta,s)$, i. e., $T^m|_{X^*(\theta,s)} : X^*(\theta,s) \to X^*(\theta,s)$ is a surjective dynamical system.

(c*) The system $T^m|_{X^*(\theta,s)}$ is topologically conjugate to the one-sided full shift generated by an alphabet counting $\#D_m$ letters. As such, the system $T^m|_{X^*(\theta,s)}$ is topologically exact.

(d*) The system/map $T^m|_{X^*(\theta,s)}$ is open.

(e*) The system $T^m|_{X^*(\theta,s)}$ is distance expanding. (This follows from Theorem 28.3.3; see (28.57).)

(f*) Consequently, $X^*(\theta,s) \subseteq \mathcal{J}(T)$.

Set

$$X(\theta,s) := \bigcup_{l=0}^{m-1} T^l(X^*(\theta,s)).$$

We readily see that:

(a) The set $X(\theta,s)$ is a topological Cantor set.

(b) $T|_{X(\theta,s)} : X(\theta,s) \to X(\theta,s)$ is a surjective dynamical system.

(c) The system $T|_{X(\theta,s)}$ is topologically transitive.

(d) The system $T|_{X(\theta,s)}$ is distance expanding.

(e) $X(\theta,s) \subseteq \mathcal{J}(T)$.

It follows from Theorem 25.3.8 that for every $\eta > 0$ there exists a forward T-invariant closed set

$$X(\theta,s) \subseteq \widehat{X}(\theta,s;\eta) \subseteq B(X(\theta,s),\eta) \tag{28.62}$$

such that

(f) $T|_{\widehat{X}(\theta,s;\eta)} : \widehat{X}(\theta,s;\eta) \to \widehat{X}(\theta,s;\eta)$ is a topologically transitive, open and expanding system.

Fix $j \in \mathbb{N}$. For any j-tuple $(z_0, z_1, \ldots, z_{j-1})$ of points in D_m, choose exactly one point y in the set $T_{z_{j-1}}^{-m} \circ T_{z_{j-2}}^{-m} \circ \cdots \circ T_{z_0}^{-m}(X^*(\theta,s))$ and denote the set consisting of all those points y by A_j. By (28.55) and (28.57), $S_{jm}\varphi(y) \geq \sum_{l=0}^{j-1} S_m\varphi(z_l) - jm\theta$ and we get that

$$\begin{aligned}
\sum_{y\in A_j} \exp(S_{jm}\varphi(y)) &\geq \sum_{(z_0,z_1,\ldots,z_{j-1})\in D_m^j} \exp\Big(\sum_{l=0}^{j-1} S_m\varphi(z_l) - jm\theta\Big) \\
&= e^{-jm\theta} \sum_{(z_0,z_1,\ldots,z_{j-1})\in D_m^j} \prod_{l=0}^{j-1} \exp(S_m\varphi(z_l)) \\
&= e^{-jm\theta}\Big(\sum_{x\in D_m} \exp(S_m\varphi(x))\Big)^j.
\end{aligned}$$

Equivalently,

$$\frac{1}{j}\log\sum_{y\in A_j}\exp(S_{jm}\varphi(y)) \ge \log\sum_{x\in D_m}\exp(S_m\varphi(x)) - m\theta.$$

In view of the definition of ξ, the set A_j is (j,ξ)-separated for T^m and ξ is an expansive constant for T^m. Hence, invoking (28.59) and letting $j\to\infty$, we obtain via Theorem 11.2.8 that

$$\begin{aligned}\mathrm{P}(T^m|_{X^*(\theta,s)}, S_m\varphi) &\ge \log\sum_{x\in D_m}\exp(S_m\varphi(x)) - m\theta\\ &\ge n\Big[\mathrm{h}_\mu(T) + \int_{\mathcal{J}(T)}\varphi\,d\mu - 2\theta\Big] - \log t - m\theta.\end{aligned}$$

Since $n+1\le m\le n(1+\theta)$ and since $\inf\varphi>0$ (and consequently $\mathrm{h}_\mu(T)+\int\varphi\,d\mu>0$), we deduce that

$$\begin{aligned}\mathrm{P}(T|_{\widehat{X}(\theta,s;\eta)},\varphi) \ge \mathrm{P}(T|_{X(\theta,s)},\varphi) &= \frac{1}{m}\mathrm{P}(T^m|_{X(\theta,s)}, S_m\varphi) \ge \frac{1}{m}\mathrm{P}(T^m|_{X^*(\theta,s)}, S_m\varphi)\\ &\ge \frac{1}{1+\theta}\Big[\mathrm{h}_\mu(T) + \int_{\mathcal{J}(T)}\varphi\,d\mu - 2\theta\Big] - \frac{\log t}{m} - \theta.\end{aligned}$$

Assuming now that n and, consequently, m, was chosen sufficiently large, we get

$$\mathrm{P}(T|_{\widehat{X}(\theta,s;\eta)},\varphi) \ge \frac{1}{1+\theta}\Big[\mathrm{h}_\mu(T) + \int_{\mathcal{J}(T)}\varphi\,d\mu\Big] - 4\theta.$$

If $\eta>0$ is sufficiently small, and we shall denote such an η by $\eta_{\theta,s}$, then for every ergodic T-invariant Borel probability measure ν on $\widehat{X}(\theta,s;\eta_{\theta,s})$, it follows from (1) the definition of the set $\widetilde{\mathcal{J}(T)}_{n,s}$, (2) the construction of the set $X(\theta,s)$, (3) the ergodic case of Birkhoff's ergodic theorem (Corollary 8.2.14) and (4) relations (28.56)–(28.57), that

$$\Big|\int_{\mathcal{J}(T)}\psi_i\,d\nu - \int_{\mathcal{J}(T)}\psi_i\,d\mu\Big| < \theta, \quad \forall i=1,2,\dots,s.$$

A similar estimate for $\log|T'|$ ensues from the definition of $\mathcal{J}(T)(\delta)$ and Corollary 28.3.2 with $\varepsilon=\theta/2$. Therefore, all the assertions of Theorem 28.6.2 are satisfied for the sets $X_k := \widehat{X}(1/k,k;\eta_{1/k,k})$, $k\in\mathbb{N}$. The proof of Theorem 28.6.2 is complete. □

Remark 28.6.3. The sets X_k in Theorem 28.6.2 can be found independent of φ. Indeed, take X_k constructed under the constant potential $\varphi\equiv 0$. Then Theorem 28.6.2 states that

$$\limsup_{k\to\infty}\mathrm{h}_{\mathrm{top}}(T|_{X_k}) \ge \mathrm{h}_\mu(T).$$

For every $k \in \mathbb{N}$, let μ_k be a measure of maximal entropy for the map $T|_{X_k} : X_k \to X_k$, i. e., $\mathrm{h}_{\mu_k}(T) = \mathrm{h}_{\mathrm{top}}(T|_{X_k})$. Theorem 28.6.2 asserts that $\lim_{k\to\infty} \mu_k = \mu$ weakly*. Consider now an arbitrary potential $\varphi : \mathcal{J}(T) \to \mathbb{R}$. Then $\int \varphi\, d\mu_k \to \int \varphi\, d\mu$. Therefore, with the use of the variational principle (Theorem 12.1.1),

$$\liminf_{k\to\infty} \mathrm{P}(T|_{X_k}, \varphi) \geq \liminf_{k\to\infty}\left(\mathrm{h}_{\mu_k}(T) + \int_{\mathcal{J}(T)} \varphi\, d\mu_k\right) \geq \mathrm{h}_{\mu}(T) + \int_{\mathcal{J}(T)} \varphi\, d\mu.$$

One more note: although

$$\liminf_{k\to\infty} \mathrm{h}_{\mu_k}(T) \geq \mathrm{h}_{\mu}(T)$$

for the maximal entropy measures $(\mu_k)_{k=1}^{\infty}$, this need not be true for all sequences $(\mu_k)_{k=1}^{\infty}$.

Corollary 28.6.4. *Let $T : \widehat{\mathbb{C}} \to \widehat{\mathbb{C}}$ be a rational function of $\deg(T) \geq 2$ and $\varphi : \mathcal{J}(T) \to \mathbb{R}$ a Hölder continuous potential such that $\mathrm{P}(\varphi) > \sup \varphi$. Then there exists a sequence $(X_k)_{k=1}^{\infty}$ of forward T-invariant closed subsets of $\mathcal{J}(T)$ that are all topological Cantor sets such that for every $k \in \mathbb{N}$ the restriction $T|_{X_k} : X_k \to X_k$ is a conformal topologically transitive open expanding map and*

$$\lim_{k\to\infty} \mathrm{P}(T|_{X_k}, \varphi) = \mathrm{P}(T, \varphi).$$

Proof. By the variational principle (Theorem 12.1.1),

$$\mathrm{P}(T, \varphi) = \lim_{k\to\infty}\left(\mathrm{h}_{\nu_k}(T) + \int_{\mathcal{J}(T)} \varphi\, d\nu_k\right) \tag{28.63}$$

for some sequence of ergodic T-invariant Borel probability measures ν_k on $\mathcal{J}(T)$. So, as $\mathrm{P}(T, \varphi) > \sup \varphi$, there exists $N \in \mathbb{N}$ such that

$$\mathrm{h}_{\nu_k}(T) + \int_{\mathcal{J}(T)} \varphi\, d\nu_k > \sup \varphi$$

for all $k \geq N$. For such k's, we get

$$\mathrm{h}_{\nu_k}(T) > \sup \varphi - \int_{\mathcal{J}(T)} \varphi\, d\nu_k \geq 0.$$

Hence, $\chi_{\nu_k}(T) > 0$ for all such k's according to Ruelle's inequality (Theorem 28.2.1). For each such $k \in \mathbb{N}$, apply Theorem 28.6.2 with $\mu = \nu_k$ to get a set X_k satisfying the conclusion of that theorem, namely, for which

$$\mathrm{P}(T|_{X_k}, \varphi) \geq \mathrm{h}_{\nu_k}(T) + \int_{\mathcal{J}(T)} \varphi\, d\nu_k - \frac{1}{k}.$$

It follows from this and (28.63) that

$$\liminf_{k\to\infty} \mathrm{P}(T|_{X_k}, \varphi) \geq \mathrm{P}(T, \varphi). \tag{28.64}$$

Obviously, $\mathrm{P}(T|_{X_k}, \varphi) \leq \mathrm{P}(T, \varphi)$ for every $k \in \mathbb{N}$, so the proof of Corollary 28.6.4 is complete. □

28.7 Exercises

Exercise 28.7.1. Let $T : \widehat{\mathbb{C}} \to \widehat{\mathbb{C}}$ be a rational function with $\deg(T) \geq 2$.

(a) Show that $\mathcal{J}_{\mathrm{r}}(T) \cap \mathrm{Crit}_\infty^-(T) = \emptyset$, where $\mathrm{Crit}_\infty^-(T) := \bigcup_{n=0}^{\infty} T^{-n}(\mathrm{Crit}(T))$ is the backward orbit of the critical set of T.

(b) Prove that $T^{-1}(\mathcal{J}_{\mathrm{r}}(T)) \setminus \mathrm{Crit}(T) = \mathcal{J}_{\mathrm{r}}(T)$ and $T(\mathcal{J}_{\mathrm{r}}(T)) \subseteq \mathcal{J}_{\mathrm{r}}(T)$.

(c) Identify a sufficient condition for $T(\mathcal{J}_{\mathrm{r}}(T)) = \mathcal{J}_{\mathrm{r}}(T)$.

(d) Establish corresponding relations for $\mathcal{J}_{\mathrm{er}}(T)$ and $\mathcal{J}_{\mathrm{uer}}(T)$.

Exercise 28.7.2. Let $T : \widehat{\mathbb{C}} \to \widehat{\mathbb{C}}$ be a parabolic rational function with $\deg(T) \geq 2$. Show that

$$\mathcal{J}_{\mathrm{r}}(T) = \mathcal{J}(T) \setminus \Omega_\infty^-(T), \tag{28.65}$$

where $\Omega_\infty^-(T) := \bigcup_{n=0}^{\infty} T^{-n}(\Omega(T))$ is the backward orbit of the set $\Omega(T)$ of all rationally indifferent periodic points of T. Then prove that $\mathcal{J}_{\mathrm{r}}(T)$ is backward T-invariant, i. e. $T^{-1}(\mathcal{J}_{\mathrm{r}}(T)) \subseteq \mathcal{J}_{\mathrm{r}}(T)$.

29 Sullivan's conformal measures for rational functions

In Section 29.1, we present the general concept of conformal measure introduced in [34]. We have already encountered such measures in Chapter 13 in the important, nevertheless quite special, context of open distance expanding systems under Hölder continuous potentials. We now deal with such measures for much more general systems and potentials. As we already saw in Chapter 13 and as we will see in the first section, this concept is closely related to the notion of quasi-invariance. A substantial difference now is that the maps need not be open nor do the potentials need be continuous or even bounded, whence the method based on the Schauder–Tychonov fixed-point theorem for proving the existence of such measures is not applicable.

Conformal measures were first defined and introduced by Samuel Patterson in his seminal paper [100] (see also [101]) in the context of Fuchsian groups. Dennis Sullivan extended this concept to all Kleinian groups in [126, 128, 130]. In the papers [127, 129], Sullivan defined, and proved the existence of, conformal measures for all rational functions of the Riemann sphere $\widehat{\mathbb{C}}$. Both Patterson and Sullivan came up with conformal measures in order to get an understanding of geometric measures such as Hausdorff and packing measures. Although Sullivan already noticed that there are conformal measures for Kleinian groups that are not equal, nor even equivalent, to any Hausdorff or packing (generalized) measure, the main purpose to study them is still to understand Hausdorff and packing measures. But their use goes beyond that purpose.

Conformal measures for rational functions, also called Sullivan's conformal measures, constitute the theme of Section 29.2. They have been studied in numerous research works. We cite only few of them: [5, 37, 39, 40, 42, 108, 138, 140, 141]. They have been studied in greater detail in [35] where, in particular, the structure of the set of their exponents was examined. In Section 29.2, we describe a construction whose objective is to establish the existence of such measures. We further provide several characterizations of the minimal exponent for which such measures exist, in particular its equality with the Hausdorff dimension of the Julia set and of the three radial Julia subsets, as well as with four dynamically-defined dimensions. The intrigued reader may peek ahead to Theorem 29.2.12.

The concept of conformal measures, still in the sense of Sullivan, was subsequently extended to countable-alphabet iterated function systems in [87] and to conformal graph directed Markov systems in [88]. These systems were treated at length in Chapters 19 and 21 of the second volume of this book. This concept was also extended to transcendental meromorphic dynamics in [71, 89, 144]; see also [6, 7, 90, 145]. Last but not least, this concept found its place in random dynamics; among others, see [91].

This chapter ends with a short Section 29.3 on Pesin's formula, which affirms that the Hausdorff dimension of an ergodic invariant measure which has positive entropy and is absolutely continuous with respect to a Sullivan conformal measure, is equal to

https://doi.org/10.1515/9783110769876-029

the exponent of that measure, and that this exponent is nothing else than the minimal exponent alluded to in the above description of Section 29.2.

29.1 General concept of conformal measures

In this section, we deal at length with the concept, properties and construction of general conformal measures in the sense of [34]. Our approach stems from and develops the one of [34]. As noted at the beginning of this chapter, the method for proving the existence of such measures, based on the Schauder–Tychonov fixed-point theorem is not applicable. Instead, we construct them as limit measures of a Patterson–Sullivan type of construction.

29.1.1 Motivation for and definition of general conformal measures

Let $T : X \to X$ be a Borel measurable self-map of a completely metrizable space X. Assume that T is (at most) countable-to-one, i. e.,

$$X = \bigcup_{j\in I} X_j,$$

where the sets $X_j, j \in I$, are Borel, pairwise disjoint and for each $j \in I$ the restriction

$$T_j := T|_{X_j} : X_j \to T(X_j)$$

is one-to-one. Per Theorem 4.5.4 in Srivastava [122], the sets $T(X_j), j \in I$, are Borel as well.

Now assume that m is a quasi-T-invariant Borel probability measure. Recall that

$$\mathcal{L}_m : L^1(m) \longrightarrow L^1(m)$$

denotes the transfer operator defined in Section 13.4 (T need not be open distance expanding for $\mathcal{L}_m$ to be well-defined). As a reminder, for any non-negative function $g : X \to \mathbb{R}$ in $L^1(m)$ we defined the finite Borel measure

$$gm(A) := \int_A g\,dm, \quad \forall A \in \mathcal{B}(X).$$

We observed that $gm \ll m$ and thereby

$$gm \circ T^{-1} \ll m \circ T^{-1} \ll m.$$

Using the Radon-Nikodym Theorem (Theorem A.1.18), the transfer operator $\mathcal{L}_m$ could then be defined by

$$\mathcal{L}_m(g) := \frac{d(gm \circ T^{-1})}{dm} \in L^1(m).$$

For a general $g \in L^1(m)$, define $\mathcal{L}_m(g) := \mathcal{L}_m(g^+) - \mathcal{L}_m(g^-)$. It is easy to see that (cf. (13.21))

$$\int_X \mathcal{L}_m(g)\,dm = \int_X g\,dm, \quad \forall g \in L^1(m). \tag{29.1}$$

Defining the measure $\mathcal{L}_m^*(m)$ by the formula

$$\mathcal{L}_m^*(m)(g) := m(\mathcal{L}_m(g)), \quad \forall g \in L^1(m), \tag{29.2}$$

equality (29.1) means that

$$\mathcal{L}_m^*(m)(g) = m(g), \quad \forall g \in L^1(m). \tag{29.3}$$

This equality signifies that

$$\mathcal{L}_m^*(m) = m. \tag{29.4}$$

As the restrictions $T_j : X_j \to T(X_j), j \in I$, of T are bijective measurable maps, we further have that

$$m \circ T_j^{-1} \ll m|_{T(X_j)}, \quad \forall j \in I,$$

that is, m is quasi-T_j-invariant for all $j \in I$. We can thus define the Jacobian of m with respect to T by

$$\begin{aligned} \mathfrak{J}_m^{-1}(T) \quad : \quad X &\longrightarrow [0,\infty) \\ x &\longmapsto \mathfrak{J}_m^{-1}(T)(x) := \frac{dm \circ T_j^{-1}}{dm|_{T(X_j)}}(T(x)) \quad \text{whenever } x \in X_j. \end{aligned}$$

By the change-of-variables formula (13.19), we obtain for any non-negative $g \in L^1(m)$ and Borel set $A \subseteq X$ that

$$\begin{aligned} gm \circ T^{-1}(A) &= gm\Big(\bigcup_{j\in I} T_j^{-1}(A)\Big) \\ &= \sum_{j\in I} gm(T_j^{-1}(A)) \\ &= \sum_{j\in I} \int_{T_j^{-1}(A)} g\,dm \\ &= \sum_{j\in I} \int_A (g \circ T_j^{-1}) \cdot (\mathfrak{J}_m^{-1}(T) \circ T_j^{-1})\,dm \end{aligned}$$

$$= \int_A \sum_{j\in I} g(T_j^{-1}(x))\,\mathfrak{J}_m^{-1}(T)(T_j^{-1}(x))\,dm(x).$$

Since this is true for all Borel sets $A \subseteq X$, we deduce that

$$\frac{d(gm\circ T^{-1})}{dm}(x) = \sum_{j\in I} g(T_j^{-1}(x))\cdot \mathfrak{J}_m^{-1}(T)(T_j^{-1}(x)) = \sum_{y\in T^{-1}(x)} g(y)\,\mathfrak{J}_m^{-1}(T)(y)$$

for m-a.e. $x \in X$. Consequently,

$$\mathcal{L}_m(g)(x) = \sum_{y\in T^{-1}(x)} g(y)\mathfrak{J}_m^{-1}(T)(y) \quad \text{for } m\text{-a. e. } x \in X. \tag{29.5}$$

For an arbitrary $g \in L^1(m)$, the equality follows from the above and the fact that $\mathcal{L}_m(g) = \mathcal{L}_m(g^+) - \mathcal{L}_m(g^-)$ by definition.

Definition 29.1.1. A Borel set $A \subseteq X$ is called a *special set* for T if $T|_A$ is injective.

Remark 29.1.2.
(a) Every Borel subset of a special set is special.
(b) If A is a special set, then $T(A)$ is a Borel set according to Theorem 4.5.4 in [122].

Assume additionally that m is *nonsingular* with respect to T, meaning that

$$m(T((\mathfrak{J}_m^{-1}(T))^{-1}(0))) = 0, \quad \text{or equivalently,} \quad m(T((\mathfrak{J}_m(T))^{-1}(\infty))) = 0.$$

If $A \subseteq X$ is a special set, then (29.4), (29.2), (29.5) and the fact that $\mathfrak{J}_m(T) = 1/\mathfrak{J}_m^{-1}(T)$ yield

$$\begin{aligned}\int_A \mathfrak{J}_m(T)\,dm &= \int_X \mathbb{1}_A\cdot\mathfrak{J}_m(T)\,dm = \int_X \mathbb{1}_A\cdot\mathfrak{J}_m(T)\,d\mathcal{L}_m^*(m) = \int_X \mathcal{L}_m(\mathbb{1}_A\cdot\mathfrak{J}_m(T))\,dm \\ &= \int_X \sum_{y\in T^{-1}(x)} \mathbb{1}_A(y)\mathfrak{J}_m(T)(y)\mathfrak{J}_m^{-1}(T)(y)\,dm(x) \\ &= \int_{T(A)} \sum_{y\in T^{-1}(x)} \mathbb{1}_A(y)\,dm(x) \\ &= \int_{T(A)} 1\,dm = m(T(A)).\end{aligned} \tag{29.6}$$

We single out this property by saying that the nonsingular measure m is $\mathfrak{J}_m(T)$-conformal. This prompts us to introduce the following notion.

Definition 29.1.3. Let $T : X \to X$ be a Borel measurable self-map of a completely metrizable topological space X and let $\psi : X \to [0,\infty)$ be a Borel measurable function. A Borel probability measure m on X is said to be ψ-*conformal* for $T : X \to X$ if

$$m(T(A)) = \int_A \psi\, dm \tag{29.7}$$

for every special set $A \subseteq X$.

Observation 29.1.4. If $\psi > 0$, then any ψ-conformal measure m is nonsingular with respect to the map T, and $\mathfrak{J}_m^{-1}(T) = 1/\psi$.

Given $\varphi \in L^1(m)$, define (cf. Definition 13.5.1)

$$\mathcal{L}_\varphi(g)(x) := \sum_{y \in T^{-1}(x)} g(y)e^{\varphi(y)}, \quad \forall x \in X,\ \forall g \in L^1(m).$$

Observe that

$$\mathcal{L}_m(g)(x) = \mathcal{L}_{\log \mathfrak{J}_m^{-1}(T)}(g)(x) \quad \text{for } m\text{-a. e. } x \in X,\ \forall g \in L^1(m).$$

Notice that even if T and φ are continuous, the operator $\mathcal{L}_\varphi$ need not map $C_b(X)$ into $C_b(X)$ in general. But there are natural hypotheses, such as T open and distance expanding, under which this is the case. Nevertheless, if we just assume that $\mathcal{L}_\varphi(C(X)) \subseteq C(X)$, then the linear operator $\mathcal{L}_\varphi : C(X) \to C(X)$ is bounded, and so is its dual $\mathcal{L}_\varphi^* : C(X)^* \to C(X)^*$.

The following result states that the conformality of a measure can be characterized by this measure being a fixed point (an eigenmeasure) of the dual of an appropriate transfer operator (cf. Lemma 13.6.13).

Proposition 29.1.5. *Let $T : X \to X$ be a countable-to-one open continuous self-map of a completely metrizable topological space X and let $\psi : X \to (0, \infty)$ be a positive continuous function. Then a Borel probability measure m is ψ-conformal if and only if*

$$\mathcal{L}_{-\log\psi}^*(m) = m.$$

As we may encounter trouble with the operator $\mathcal{L}^*$ for the maps T that are not open, rather than looking for theorems that would assure the existence of fixed points of $\mathcal{L}^*$, we shall provide another general method of construction of conformal measures. That method stems from the Patterson–Sullivan method [100] (see also [101]) in the context of Fuchsian groups, [126, 128, 130] for Kleinian groups and [127, 129] for rational functions.

29.1.2 Selected properties of general conformal measures

Definition 29.1.6. Let $T : X \to X$ be a self-map of a topological space X. A point $x \in X$ is said to be *singular* for T if at least one of the following two conditions is satisfied:

(a) There is no open neighborhood U of x such that the restriction $T|_U$ is one-to-one.

(b) For every open neighborhood U of x, there exists an open set $V \subseteq U$ whose image $T(V)$ is not open.

The set of all singular points will be denoted by $\mathrm{Sing}(T)$ and will be referred to as the *set of singular points* of T or the *singular set* of T. The set of all points satisfying condition (a) will be denoted by $\mathrm{Crit}(T)$ and will be referred to as the *set of critical points* of T or the *critical set* of T. The set of all points satisfying condition (b) will be denoted by $X_*(T)$.

First, let us record the following immediate facts.

Observation 29.1.7. Let $T : X \to X$ be a self-map of a topological space X. Then:
(a) $\mathrm{Sing}(T) = \mathrm{Crit}(T) \cup X_*(T)$.
(b) All three sets $\mathrm{Crit}(T)$, $X_*(T)$ and $\mathrm{Sing}(T)$ are closed.
(c) $X \setminus X_*(T)$ is the largest (in the sense of inclusion) open subset of X over which the map T is open.

It is easy to give examples of continuous maps T for which $\mathrm{Crit}(T) \cap X_*(T) \neq \emptyset$. We will need the following, somewhat obvious, topological fact whose proof is provided for the sake of completeness and convenience of the reader.

Lemma 29.1.8. *Let $T : X \to X$ be a continuous self-map of a regular topological space X. For every $x \in X \backslash \mathrm{Sing}(T)$, there exists an open neighborhood $U(x)$ of x with the following properties:*
(a) *$U(x) \subseteq X \backslash \mathrm{Sing}(T)$;*
(b) *The map $T|_{U(x)}$ is one-to-one;*
(c) *The set $T(U(x))$ is open and the map $T|_{U(x)} : U(x) \to X$ is open;*
(d) *If $G \subseteq U(x)$ is an open set, then $\partial(T(G)) = T(\partial G)$.*

Proof. As $x \in X \backslash \mathrm{Sing}(T)$ and $\mathrm{Sing}(T)$ is closed, there exists $W \subseteq X \backslash \mathrm{Sing}(T)$, an open neighborhood of x, such that the map $T|_W : W \to X$ is one-to-one and open. In particular, $T(W)$ is an open neighborhood of $T(x)$. Since the space X is regular, there exists an open neighborhood V of $T(x)$ such that

$$\overline{V} \subseteq T(W). \tag{29.8}$$

As T is continuous, there is an open neighborhood $U'(x)$ of x such that $U'(x) \subseteq W$ and

$$T(U'(x)) \subseteq V. \tag{29.9}$$

Claim 1°. *If $F \subseteq U'(x)$ is a closed subset of X, then $T(F) \subseteq X$ is also closed.*

Proof of Claim 1°. Since $W \backslash F \subseteq X$ is open and $T|_W$ is open, the set $T(W \backslash F)$ is open. As $T|_W$ is one-to-one, it is also clear that $T(W \backslash F) = T(W) \backslash T(F)$. So, $T(F)$ is a closed subset of $T(W)$. That is, there exists a closed subset H of X such that $T(F) = H \cap T(W)$. By (29.9), we know that $T(F) \subseteq T(U'(x)) \subseteq V \subseteq \overline{V}$. So, $T(F) = H \cap T(W) \cap \overline{V}$. By (29.8), we get $T(F) = H \cap \overline{V}$. Thus, $T(F)$ is closed in X and the claim holds. ■

By the regularity of X, there is an open neighborhood $U(x)$ of x such that

$$\overline{U(x)} \subseteq U'(x). \tag{29.10}$$

As a fairly immediate consequence of Claim 1°, we obtain the following fact.

Claim 2°. *If $A \subseteq U(x)$, then $\overline{T(A)} = T(\overline{A})$.*

Proof of Claim 2°. On one hand, the continuity of T guarantees that $T(\overline{A}) \subseteq \overline{T(A)}$. On the other hand, the set $T(\overline{A})$ is closed in X per Claim 1° and (29.10). As $T(A) \subseteq T(\overline{A})$, we deduce that $\overline{T(A)} \subseteq T(\overline{A})$. ■

To finish the proof of the lemma, let $G \subseteq U(x) \subseteq U'(x) \subseteq W \subseteq X\backslash\text{Sing}(T)$ be an open set. Then $T(G)$ is also open and, using Claim 2° and the injectivity of $T|_W$, we conclude that

$$\partial(T(G)) = \overline{T(G)}\backslash T(G) = T(\overline{G})\backslash T(G) = T(\overline{G}\backslash G) = T(\partial G). \qquad \square$$

Lemma 29.1.9. *Let $T : X \to X$ be a continuous map of a Polish (i. e., completely metrizable and separable) topological space X, and let m be a Borel probability measure on X. Let $\Gamma \subseteq X$ be a closed set containing* Sing(T). *If $g : X \to [0,\infty)$ is in $L^1(m)$ and if* (29.7) *holds for all special sets A such that $\overline{A} \cap \Gamma = \emptyset$ and $m(\partial A) = m(\partial T(A)) = 0$, then* (29.7) *continues to hold for all special sets A such that $A \cap \Gamma = \emptyset$.*

Proof. Fix a complete metric d compatible with the topology of X. Let A be a special set disjoint from Γ. Let $\varepsilon > 0$. Since $A \subseteq X\backslash\Gamma$, since the set $X\backslash\Gamma$ is open and since the measure m is outer regular, there exists an open set $A \subseteq V \subseteq X \setminus \Gamma$ such that

$$\int_V g\,dm < \int_A g\,dm + \varepsilon. \tag{29.11}$$

For every $x \in V$ there exists $r(x) > 0$ such that

$$B(x, 2r(x)) \subseteq V \cap U(x), \tag{29.12}$$

where $U(x)$ is the open neighborhood of x produced in Lemma 29.1.8. Since $\partial B(x,r) \subseteq \{y \in X : d(y,x) = r\}$ for every $r > 0$ and since the sets $\{y \in X : d(y,x) = r\}$, $r > 0$, are mutually disjoint, the set

$$R_1(x) := \{r \in (0, r(x)] : m(\partial B(x,r)) > 0\}$$

is countable. In addition, by Lemma 29.1.8(b,d) the sets $\partial(T(B(x,r)))$, $r \in (0, r(x)]$, are mutually disjoint, whence the set

$$R_2(x) := \{r \in (0, r(x)] : m(\partial(T(B(x,r)))) > 0\}$$

is countable. Thus, the set $R_1(x) \cup R_2(x)$ is countable and, therefore, there exists $s(x) \in (0, r(x)]$ such that

$$m(\partial B(x, s(x))) = 0 \quad \text{and} \quad m(\partial(T(B(x, s(x))))) = 0. \tag{29.13}$$

Since $\{B(x, s(x)) : x \in V\}$ is an open cover of V and since V is Lindelöf (as a separable metrizable space), there exists a countable set $\{x_n\}_{n=1}^{\infty} \subseteq V$ such that

$$\bigcup_{n=1}^{\infty} B(x_n, s(x_n)) = V. \tag{29.14}$$

Define inductively a partition $\{G_n\}_{n=1}^{\infty}$ of V in the standard way:

$$G_1 := B(x_1, s(x_1)) \quad \text{and} \quad G_{n+1} := B(x_{n+1}, s(x_{n+1})) \setminus \bigcup_{k=1}^{n} B(x_k, s(x_k)), \quad \forall n \in \mathbb{N}.$$

As

$$\partial G_n \subseteq \bigcup_{k=1}^{n} \partial B(x_k, s(x_k)), \quad \forall n \in \mathbb{N},$$

it follows from (29.13) that

$$m(\partial G_n) = 0, \quad \forall n \in \mathbb{N}. \tag{29.15}$$

Likewise, by Lemma 29.1.8(b),

$$T(G_n) = T(B(x_n, s(x_n))) \setminus \bigcup_{k=1}^{n-1} T(B(x_k, s(x_k))), \quad \forall n \in \mathbb{N}.$$

It ensues that

$$\partial(T(G_n)) \subseteq \bigcup_{k=1}^{n} \partial(T(B(x_k, s(x_k)))), \quad \forall n \in \mathbb{N}.$$

It follows from (29.13) that

$$m(\partial(T(G_n))) = 0, \quad \forall n \in \mathbb{N}. \tag{29.16}$$

Moreover, $G_n \subseteq B(x_n, r(x_n))$ for all $n \in \mathbb{N}$. By (29.12),

$$\overline{G_n} \cap \Gamma = \emptyset, \quad \forall n \in \mathbb{N}. \tag{29.17}$$

Finally, $T|_{G_n}$ is one-to-one since $G_n \subseteq B(x_n, r(x_n)) \subseteq V \cap U(x_n) \subseteq X \setminus \Gamma \subseteq X \setminus \mathrm{Sing}(T)$. So, every G_n is a special set satisfying (29.15)–(29.17). From the hypothesis on such special sets, we know that

$$m(T(G_n)) = \int_{G_n} g\,dm, \quad \forall n \in \mathbb{N}. \tag{29.18}$$

By construction and (29.14), we also know that $\bigcup_{n=1}^{\infty} G_n = V \supseteq A$. From this, (29.11) and (29.18), we obtain that

$$m(T(A)) = m\left(\bigcup_{n=1}^{\infty} T(A \cap G_n)\right) \leq \sum_{n=1}^{\infty} m(T(G_n)) = \sum_{n=1}^{\infty} \int_{G_n} g\,dm = \int_V g\,dm < \int_A g\,dm + \varepsilon.$$

Since $\varepsilon > 0$ is arbitrary, we infer that

$$m(T(A)) \leq \int_A g\,dm \tag{29.19}$$

for every special set $A \subseteq X$ such that $A \cap \Gamma = \emptyset$. Using (29.19) on each $G_n \setminus A$ and (29.18), we get

$$\begin{aligned} m(T(A)) &= m\left(\bigcup_{n=1}^{\infty} T(A \cap G_n)\right) = \sum_{n=1}^{\infty} m(T(A \cap G_n)) \\ &= \sum_{n=1}^{\infty} [m(T(G_n)) - m(T(G_n \setminus A))] \\ &\geq \sum_{n=1}^{\infty} \left[\int_{G_n} g\,dm - \int_{G_n \setminus A} g\,dm\right] \\ &= \int_{\bigcup_{n=1}^{\infty} G_n} g\,dm - \int_{\bigcup_{n=1}^{\infty} G_n \setminus A} g\,dm = \int_V g\,dm - \int_{V \setminus A} g\,dm \\ &= \int_{V \cap A} g\,dm = \int_A g\,dm. \end{aligned}$$

Along with (29.19), this gives

$$m(T(A)) = \int_A g\,dm. \qquad \square$$

29.1.3 The limit construction and PS limit measures

In this subsection, we provide a construction aimed at producing conformal measures. It relies on the following simple fact. For a real sequence $(a_n)_{n=1}^{\infty}$, the number

$$c := \limsup_{n\to\infty} \frac{a_n}{n} \tag{29.20}$$

is called the sequence's *transition parameter* and is uniquely determined by the property that the series

$$\sum_{n=1}^{\infty} \exp(a_n - ns)$$

converges when $s > c$ but diverges when $s < c$. (If $s = c$, the series may converge or not.)

By a simple argument, one obtains the following.

Lemma 29.1.10. *Given a real sequence $(a_n)_{n=1}^{\infty}$ and its transition parameter $c := \limsup_{n\to\infty} \frac{a_n}{n}$, there exists a sequence $(b_n)_{n=1}^{\infty}$ of positive reals such that*

$$\sum_{n=1}^{\infty} b_n \exp(a_n - ns) \begin{cases} < \infty & \text{if } s > c \\ = \infty & \text{if } s \le c \end{cases} \quad \text{and} \quad \lim_{n\to\infty} \frac{b_{n+1}}{b_n} = 1.$$

Moreover, $\sum_{n=1}^{\infty} b_n \exp(a_n - ns)$ depends continuously on s.

Proof. If $\sum_{n=1}^{\infty} \exp(a_n - nc) = \infty$, set $b_n = 1$ for every $n \in \mathbb{N}$. If $\sum_{n=1}^{\infty} \exp(a_n - nc) < \infty$, choose a sequence $(n_k)_{k=1}^{\infty}$ of positive integers such that

$$\varepsilon_k := \frac{a_{n_k}}{n_k} - c \xrightarrow{k\to\infty} 0 \quad \text{and} \quad \lim_{k\to\infty} \frac{n_k}{n_{k+1}} = 0.$$

Letting

$$b_n = \exp\left(n\left[\frac{n_k - n}{n_k - n_{k-1}} \varepsilon_{k-1} + \frac{n - n_{k-1}}{n_k - n_{k-1}} \varepsilon_k \right]\right) \quad \text{for every } n_{k-1} \le n < n_k,$$

it is easy to check that the lemma follows. The demonstration of the continuity on s is left to the reader as an exercise. □

The current subsection's actual setting is this:

(a) $T : X \to X$ is a continuous self-map of a completely metrizable topological space X.

(b) $(E_n)_{n=1}^{\infty}$ is a sequence of nonempty finite subsets of X such that

$$T^{-1}(E_n) \subseteq E_{n+1}, \quad \forall n \in \mathbb{N}. \tag{29.21}$$

(c) $\varphi : X \to (-\infty, \infty]$ is a bounded below continuous function such that

$$\varphi\left(\bigcup_{n=1}^{\infty} E_n\right) \subseteq \mathbb{R}. \tag{29.22}$$

We adopt the standard convention that

$$e^{-\infty} = 0 \quad \text{and} \quad e^{\infty} = \infty.$$

Under this convention, we have the following obvious fact.

Observation 29.1.11. The function $e^{-\varphi} : X \to [0, \infty)$ is bounded and continuous.

Let

$$a_n := \log \sum_{x \in E_n} \exp(S_n\varphi(x)) \quad \text{where} \quad S_n\varphi = \sum_{k=0}^{n-1} \varphi \circ T^k. \tag{29.23}$$

Denote by c the transition parameter of the sequence $(a_n)_{n=1}^{\infty}$. Choose a sequence $(b_n)_{n=1}^{\infty}$ of positive reals per Lemma 29.1.10. For every $s > c$, define

$$M_s := \sum_{n=1}^{\infty} b_n \exp(a_n - ns) \tag{29.24}$$

and the normalized measure

$$m_s = \frac{1}{M_s} \sum_{n=1}^{\infty} \sum_{x \in E_n} b_n \exp(S_n\varphi(x) - ns)\delta_x, \tag{29.25}$$

where δ_x denotes the Dirac δ-measure supported at the point $x \in X$. Let $A \subseteq X$ be a special set. Using (29.21) and (29.25), it follows that

$$\begin{aligned} m_s(T(A)) &= \frac{1}{M_s} \sum_{n=1}^{\infty} \sum_{x \in E_n \cap T(A)} b_n \exp(S_n\varphi(x) - ns) \\ &= \frac{1}{M_s} \sum_{n=1}^{\infty} \sum_{y \in A \cap T^{-1}(E_n)} b_n \exp(S_n\varphi(T(y)) - ns) \\ &= \frac{1}{M_s} \sum_{n=1}^{\infty} \sum_{x \in A \cap E_{n+1}} b_n \exp(S_{n+1}\varphi(x) - (n+1)s) \exp(s - \varphi(x)) \\ &\quad - \frac{1}{M_s} \sum_{n=1}^{\infty} \sum_{x \in A \cap (E_{n+1} \setminus T^{-1}(E_n))} b_n \exp(S_n\varphi(T(x)) - ns). \end{aligned} \tag{29.26}$$

Set

$$\Delta_A(s) := \left| \frac{1}{M_s} \sum_{n=1}^{\infty} \sum_{x \in A \cap E_{n+1}} b_n \exp(S_{n+1}\varphi(x) - (n+1)s) \exp(s - \varphi(x)) - \int_A \exp(c - \varphi)\, dm_s \right|.$$

Observe that

$$\Delta_A(s) = \frac{1}{M_s}\left|\sum_{n=1}^{\infty}\sum_{x\in A\cap E_{n+1}} \exp(S_{n+1}\varphi(x)-(n+1)s)e^{s-\varphi(x)}[b_n - b_{n+1}e^{c-s}] - b_1\sum_{x\in A\cap E_1} e^{c-s}\right|$$
$$\le e^{s-\varphi(x)}\frac{1}{M_s}\sum_{n=1}^{\infty}\sum_{x\in E_{n+1}}\left|\frac{b_n}{b_{n+1}} - e^{c-s}\right| b_{n+1}\exp(S_{n+1}\varphi(x)-(n+1)s) + \frac{b_1 e^{c-s}\#E_1}{M_s}.$$

By Lemma 29.1.10, we know that $\lim_{n\to\infty}(b_{n+1}/b_n) = 1$ and $\lim_{s\searrow c} M_s = \infty$. Therefore,

$$\lim_{s\searrow c}\Delta_A(s) = 0 \tag{29.27}$$

uniformly over all special sets A.

Definition 29.1.12. Any weak accumulation point, when $s \searrow c$, of the measures $(m_s)_{s>c}$, defined by (29.25), is called a *PS limit measure* (associated to the function φ and the sequence $(E_n)_{n=1}^{\infty}$). The letters PS are in honor of Patterson and Sullivan.

29.1.4 Conformality properties of PS limit measures

In order to find conformal measures among the PS limit measures, it is necessary to examine (29.26) in greater detail. To begin with, for a Borel set $B \subseteq X$ consider the following condition:

$$\lim_{s\searrow c}\frac{1}{M_s}\sum_{n=1}^{\infty}\sum_{x\in B\cap(E_{n+1}\setminus T^{-1}(E_n))} b_n\exp(S_n\varphi(T(x)) - ns) = 0. \tag{29.28}$$

As a direct consequence of (29.26)–(29.27), we get the following.

Lemma 29.1.13. *Let $T : X \to X$ be a continuous self-map of a completely metrizable topological space X. Let $(E_n)_{n=1}^{\infty}$ be a sequence of finite subsets of X such that* (29.21) *holds and let $\varphi : X \to (-\infty,\infty]$ be a bounded below continuous function satisfying* (29.22). *If a Borel set $B \subseteq X$ satisfies* (29.28), *then the sets of all accumulation points of the functions*

$$(c,\infty)\ni s \longmapsto m_s(T(B)) \quad\text{and}\quad (c,\infty)\ni s\longmapsto \int_B \exp(s-\varphi)\,dm_s$$

as $s \searrow c$, coincide. More precisely, if $(s_n)_{n=1}^{\infty}$ is a sequence of real numbers in (c,∞) such that $\lim_{n\to\infty} s_n = c$, then

$$\lim_{n\to\infty} m_{s_n}(T(B)) \text{ exists} \iff \lim_{n\to\infty}\int_B \exp(s_n-\varphi)\,dm_{s_n} \text{ exists}$$

and if any one of those limits exists, then they coincide.

Lemma 29.1.14. *Let $T : X \to X$ be a continuous self-map of a completely metrizable topological space X. Let $(E_n)_{n=1}^{\infty}$ be a sequence of finite subsets of X such that* (29.21) *holds and let $\varphi : X \to (-\infty, \infty]$ be a bounded below continuous function satisfying* (29.22). *Let m be a PS limit measure (per Definition* 29.1.12) *and let Γ be a closed set containing* $\mathrm{Sing}(T)$. *Assume that every special set $B \subseteq X$ such that $\overline{B} \cap \Gamma = \emptyset$ and $m(\partial B) = m(\partial T(B)) = 0$ further satisfies condition* (29.28). *Then*

$$m(T(A)) = \int_A \exp(c - \varphi)\, dm$$

for every special set A such that $A \cap \Gamma = \emptyset$.

Proof. It immediately follows from Lemma 29.1.13 and the boundedness and continuity of the function $e^{-\varphi}$ that

$$m(T(B)) = \int_B \exp(c - \varphi)\, dm$$

for every special set $B \subseteq X$ for which $\overline{B} \cap \Gamma = \emptyset$ and $m(\partial B) = m(\partial T(B)) = 0$. Applying Lemma 29.1.9 completes the proof. □

Lemma 29.1.15. *Let $T : X \to X$ be a continuous self-map of a completely metrizable topological space X. Let $(E_n)_{n=1}^{\infty}$ be a sequence of finite subsets of X such that* (29.21) *holds and let $\varphi : X \to (-\infty, \infty]$ be a bounded below continuous function satisfying* (29.22). *Suppose that condition* (29.28) *holds for every special set B such that $\overline{B} \cap X_*(T) = \emptyset$. Suppose also that $w \in X \setminus X_*(T)$ is a point for which there exist finitely many mutually disjoint special sets $W_1, W_2, \dots, W_q$ such that $W_1 \cup W_2 \cup \dots \cup W_q$ is an open neighborhood of w. If m is a PS limit measure, then*

$$m(T(\{w\})) \leq \exp(c - \varphi(w)) m(\{w\}).$$

Proof. By the same standard argument as the one yielding countability of the set $R_1(x)$ in the proof of Lemma 29.1.9 and since $w \notin X_*(T)$ and $X_*(T)$ is a closed subset of X, there exists a sequence $(r_n)_{n=1}^{\infty}$ in $(0, \infty)$ converging to 0 such that

$$\overline{B(w, r_n)} \subseteq \bigcup_{i=1}^{q} W_i \setminus X_*(t) \quad \text{and} \quad m(\partial B(w, r_n)) = 0, \quad \forall n \in \mathbb{N}. \tag{29.29}$$

For every $i \in \{1, \dots, q\}$ and $n \in \mathbb{N}$, let

$$W_i(n) = W_i \cap B(w, r_n).$$

Then for each $n \in \mathbb{N}$ the sets $W_1(n), W_2(n), \dots, W_q(n)$ are mutually disjoint special sets whose union is

$$\bigcup_{i=1}^{q} W_i(n) = B(w, r_n). \tag{29.30}$$

From the definition of m, there is a sequence $(s_k)_{k=1}^{\infty}$ in (c, ∞) such that $\lim_{k\to\infty} s_k = c$ and $\lim_{k\to\infty} m_{s_k} = m$. Fix $n \in \mathbb{N}$. Since the function $e^{-\varphi} : X \to [0, \infty)$ is bounded and continuous, using the second property in (29.29) the Portmanteau theorem (Theorem A.1.23) affirms that

$$\lim_{k\to\infty} \int\limits_{B(w,r_n)} e^{s_k-\varphi}\, dm_{s_k} = \int\limits_{B(w,r_n)} e^{c-\varphi}\, dm. \tag{29.31}$$

As the sequence

$$\left(\int\limits_{W_1(n)} e^{s_k-\varphi}\, dm_{s_k} \right)_{k=1}^{\infty}$$

is bounded, there is a subsequence $(s_k^{(1)})_{k=1}^{\infty}$ of the sequence $(s_k)_{k=1}^{\infty}$ such that the limit

$$\lim_{k\to\infty} \int\limits_{W_1(n)} e^{s_k^{(1)}-\varphi}\, dm_{s_k^{(1)}}$$

exists. Proceeding by induction, suppose that for some $1 \le j < q$ a subsequence $(s_k^{(j)})_{k=1}^{\infty}$ of the sequence $(s_k)_{k=1}^{\infty}$ is such that the limit

$$\lim_{k\to\infty} \int\limits_{W_i(n)} e^{s_k^{(j)}-\varphi}\, dm_{s_k^{(j)}}$$

exists for every $i = 1, \ldots, j$. Since the sequence

$$\left(\int\limits_{W_j(n)} e^{s_k^{(j)}-\varphi}\, dm_{s_k^{(j)}} \right)_{k=1}^{\infty}$$

is bounded, there is a subsequence $(s_k^{(j+1)})_{k=1}^{\infty}$ of the sequence $(s_k^{(j)})_{k=1}^{\infty}$ such that the limit

$$\lim_{k\to\infty} \int\limits_{W_{j+1}(n)} e^{s_k^{(j+1)}-\varphi}\, dm_{s_k^{(j+1)}}$$

exists. Then the limit

$$\lim_{k\to\infty} \int\limits_{W_j(n)} e^{s_k^{(q)}-\varphi}\, dm_{s_k^{(q)}}$$

exists for all $j = 1, \ldots, q$. By virtue of the hypotheses and Lemma 29.1.13, for all $j = 1, \ldots, q$ the limit

$$\lim_{k\to\infty} m_{s_k^{(q)}}(T(W_j(n)))$$

also exists and

$$\lim_{k\to\infty} m_{s_k^{(q)}}(T(W_j(n))) = \lim_{k\to\infty} \int\limits_{W_j(n)} e^{s_k^{(q)}-\varphi}\, dm_{s_k^{(q)}}. \tag{29.32}$$

Recall that $B(w, r_n) \subseteq X \backslash X_*(t)$ by the first property in (29.29). Thus, $T(B(w, r_n))$ is open by Observation 29.1.7(c). Using successively the Portmanteau theorem, (29.30), (29.32) and (29.31), we deduce that

$$\begin{aligned}
m(\{T(w)\}) &\le m(T(B(w, r_n))) \le \liminf_{k\to\infty} m_{s_k^{(q)}}(T(B(w, r_n))) \\
&= \liminf_{k\to\infty} m_{s_k^{(q)}}\Big(T\Big(\bigcup_{j=1}^{q} W_j(n)\Big)\Big) = \liminf_{k\to\infty} m_{s_k^{(q)}}\Big(\bigcup_{j=1}^{q} T(W_j(n))\Big) \\
&\le \liminf_{k\to\infty} \sum_{j=1}^{q} m_{s_k^{(q)}}(T(W_j(n))) \\
&= \sum_{j=1}^{q} \lim_{k\to\infty} m_{s_k^{(q)}}(T(W_j(n))) = \sum_{j=1}^{q} \lim_{k\to\infty} \int\limits_{W_j(n)} e^{s_k^{(q)}-\varphi}\, dm_{s_k^{(q)}} \\
&= \lim_{k\to\infty} \sum_{j=1}^{q} \int\limits_{W_j(n)} e^{s_k^{(q)}-\varphi}\, dm_{s_k^{(q)}} = \lim_{k\to\infty} \int\limits_{B(w,r_n)} e^{s_k^{(q)}-\varphi}\, dm_{s_k^{(q)}} \\
&= \int\limits_{B(w,r_n)} e^{c-\varphi}\, dm.
\end{aligned}$$

Hence,

$$m(\{T(w)\}) \le \lim_{n\to\infty} \int\limits_{B(w,r_n)} e^{c-\varphi}\, dm = e^{c-\varphi(w)} m(\{w\}).$$

The proof of Lemma 29.1.15 is complete. □

29.2 Sullivan's conformal measures

This section is devoted to a comprehensive study of more special, though very important, conformal measures called Sullivan conformal measures. An extended historical and motivational discussion of these measures as well as more general ones was given at the outset of this chapter and in the previous Section 29.1. We accordingly dive right into the actual mathematics.

In this section, $T : \widehat{\mathbb{C}} \to \widehat{\mathbb{C}}$ is a rational function with $\deg(T) \geq 2$. We describe a construction whose objective is to establish the existence of measures that are called Sullivan t-conformal (or just t-conformal) measures. In fact, we will prove the existence of measures with slightly weaker properties. Then we will prove the existence of genuine Sullivan t-conformal measures and we will provide several characterizations of the minimal exponent for which such measures exist.

Fix

$$\xi \in \mathcal{J}(T) \setminus \mathrm{PC}_{\infty}^{-}(T) \quad \text{where} \quad \mathrm{PC}_{\infty}^{-}(T) := \bigcup_{n=0}^{\infty} T^{-n}(\mathrm{PC}(T))$$

and set

$$E_n := T^{-(n-1)}(\xi), \quad \forall n \in \mathbb{N}.$$

Then

$$E_{n+1} = T^{-1}(E_n), \quad \forall n \in \mathbb{N}.$$

Therefore the sequence $(E_n)_{n=1}^{\infty}$ satisfies (29.21) and (29.28) with all Borel sets $B \subseteq \mathcal{J}(T)$. Fix an arbitrary $t \geq 0$ and observe that the function

$$\varphi_t := -t \log |T'| : \mathcal{J}(T) \to (-\infty, \infty]$$

is continuous, bounded below, satisfies (29.22), and the function $e^{-\varphi_t} = |T'|^t$ is continuous. Let $c_\xi(t)$ be the transition parameter associated to the sequence $(E_n)_{n=1}^{\infty}$ and the function φ_t according to (29.20) and (29.23), i. e.,

$$a_{\xi,n}(t) = \log \sum_{x \in T^{-(n-1)}(\xi)} |(T^{n-1})'(x)|^{-t} \tag{29.33}$$

and

$$c_\xi(t) = \limsup_{n\to\infty} \frac{a_{\xi,n}(t)}{n}. \tag{29.34}$$

Because the map $T : \mathcal{J}(T) \to \mathcal{J}(T)$ is open, it turns out that $\mathcal{J}_*(T) = \emptyset$. Moreover, $|T'(c)| = 0$ precisely when $c \in \mathrm{Crit}(T)$. Then, as a fairly immediate consequence of Lemmas 29.1.14–29.1.15, we get the following.

Lemma 29.2.1. *If $T : \widehat{\mathbb{C}} \to \widehat{\mathbb{C}}$ is a rational function with* $\deg(T) \geq 2$, *then for every $t \geq 0$ there exists a PS limit measure m_t on $\mathcal{J}(T)$ such that*

$$m_t(T(A)) = \int_A e^{c_\xi(t)} |T'|^t \, dm_t$$

for every special set $A \subseteq \mathcal{J}(T)$.

Proof. Fix $t \geq 0$ and any PS limit measure $\underline{m}_t$ per Definition 29.1.12, i. e. any weak accumulation point of the measures $(m_s)_{s>c_\xi(t)}$ when $s \searrow c_\xi(t)$, where m_s is defined by (29.25). Recall that $\mathcal{J}_*(T) = \emptyset$, as the map $T|_{\mathcal{J}(T)}$ is open. It follows from Lemma 29.1.14 that

$$\underline{m}_t(T(A)) = \int_A e^{c_\xi(t)} |T'|^t \, d\underline{m}_t \tag{29.35}$$

for every special set $A \subseteq \mathcal{J}(T)$ such that $A \cap \text{Crit}(T) = \emptyset$. But as an aftermath of Theorem 25.1.1, the hypotheses of Lemma 29.1.15 are satisfied for every point $c \in \text{Crit}(T)$. For every such c, it ensues from that lemma that

$$0 \leq \underline{m}_t(\{T(c)\}) \leq e^{c_\xi(t)} |T'(c)|^t \underline{m}_t(\{c\}) = 0.$$

Hence, $\underline{m}_t(\{T(c)\}) = 0$ for all $c \in \text{Crit}(T)$. We conclude from this and (29.35) that this latter in fact holds for all special sets $A \subseteq \mathcal{J}(T)$. Write m_t for $\underline{m}_t$. □

Definition 29.2.2. Let $T : \widehat{\mathbb{C}} \to \widehat{\mathbb{C}}$ be a rational function. A point $z \in \mathcal{J}(T)$ is said to be *safe* if

$$z \notin \text{PC}(T) \quad \text{and} \quad \liminf_{n\to\infty} \frac{1}{n} \log \text{dist}(z, T^n(\text{Crit}(T))) \geq 0. \tag{29.36}$$

Denote the set of all safe points of T, also called the *safe set* of T, by $\text{Safe}(T)$.

This notion first appeared in [110] (Definition 12.5.7). The simple proof of the next lemma stems from that of Proposition 12.5.10 in [110]. This result states that the set of "nonsafe" Julia points has 0 for Hausdorff dimension.

Lemma 29.2.3. *If* $T : \widehat{\mathbb{C}} \to \widehat{\mathbb{C}}$ *is a rational function, then*

$$\text{HD}(\mathcal{J}(T) \backslash \text{Safe}(T)) = 0.$$

Proof. Fix $\varepsilon > 0$ and $k \in \mathbb{N}$. Set

$$\text{NS}(T; \varepsilon, k) := \left\{ z \in \mathcal{J}(T) : \frac{1}{n} \log \text{dist}(z, T^n(\text{Crit}(T))) < -\varepsilon, \ \forall n \geq k \right\}.$$

In other terms, $z \in \text{NS}(T; \varepsilon, k)$ if and only if

$$\text{dist}(z, T^n(\text{Crit}(T))) < e^{-\varepsilon n}, \quad \forall n \geq k.$$

So

$$\mathrm{NS}(T;\varepsilon,k)\subseteq\bigcap_{n=k}^{\infty}B(T^n(\mathrm{Crit}(T)),e^{-\varepsilon n}).$$

Fix any $t>0$. For every $n\geq k$, the family

$$\{B(T^n(c),e^{-\varepsilon n}):c\in\mathrm{Crit}(T)\}$$

is a finite $(2e^{-\varepsilon n})$-cover of $\mathrm{NS}(T;\varepsilon,k)$. Therefore,

$$0\leq H_t^{2e^{-\varepsilon n}}(\mathrm{NS}(T;\varepsilon,k))\leq\#\mathrm{Crit}(T)\cdot(2e^{-\varepsilon n})^t.$$

Letting $n\to\infty$, we deduce that

$$H_t(\mathrm{NS}(T;\varepsilon,k))=0.$$

As $t>0$ is arbitrary, we conclude that

$$\mathrm{HD}(\mathrm{NS}(T;\varepsilon,k))=0.$$

Hence,

$$\mathrm{HD}(\mathrm{NS}(T,\varepsilon))=0\quad\text{where}\quad\mathrm{NS}(T,\varepsilon):=\bigcup_{k=1}^{\infty}\mathrm{NS}(T;\varepsilon,k).\tag{29.37}$$

Since

$$\mathcal{J}(T)\backslash\mathrm{Safe}(T)=\mathrm{PC}(T)\cup\bigcup_{j=1}^{\infty}\mathrm{NS}(T,1/j)$$

and since the set $\mathrm{PC}(T)$ is countable, by invoking (29.37) we conclude that

$$\mathrm{HD}(\mathcal{J}(T)\backslash\mathrm{Safe}(T))=0.$$ □

We now want to establish properties of the function $c_\xi(t)$, and especially get a parameter $t\geq0$ such that $c_\xi(t)=0$. We start with the following lemma. Its proof stems from that of Theorem 12.5.11 in [110], which in turn originates from an appropriate proof in [109]. In this latter paper, a tally of variants of the topological pressure for potentials of the form $-t\log|T'|$ is developed, and all such variants are shown to be equal. This is partly reproduced in the last chapter of the book [110]. We would love to present this beautiful theory in full but it would far exceed the space allotted to this volume.

Lemma 29.2.4. *If $T:\widehat{\mathbb{C}}\to\widehat{\mathbb{C}}$ is a rational function with $\deg(T)\geq2$, then for every point $\xi\in\mathrm{Safe}(T)\cap\mathcal{J}_{\mathrm{uer}}(T)$ and every $t\geq0$ we have that*

$$c_\xi(t)\leq\mathrm{P}^{**}_{\exp}(t)$$

where $\mathrm{P}^{**}_{\exp}(t)$ *was introduced in Definition* 25.3.11.

Proof. By virtue of Proposition 25.3.12, it suffices to show that

$$c_\xi(t) \le \mathrm{P}^*_{\exp}(t).$$

Further note that the set $\mathrm{Safe}(T) \cap \mathcal{J}_{\mathrm{uer}}(T)$ is nonempty because of Theorem 28.5.3 and Lemma 29.2.3.

Since $\xi \in \mathcal{J}_{\mathrm{uer}}(T)$, there exists $\eta > 0$ such that $\xi \in \mathcal{J}_{\mathrm{uer}}(T)(\eta)$. Fix any $\alpha \in (0,1)$. Let $\lambda > 1$ be a constant witnessing the radial uniform expandingness of T at the point ξ (see Definition 28.5.1). Since $\lambda > 1$, $\alpha > 0$ and $\xi \in \mathrm{Safe}(T)$, it follows from (29.36) that

$$B(\xi, 2\lambda^{-\alpha n}) \cap \bigcup_{i=1}^{2n} T^i(\mathrm{Crit}(T)) = \emptyset$$

for all $n \in \mathbb{N}$ large enough, say $n \ge N_1$. According to Corollary 22.5.6, for every $1 \le j \le 2n$ and every $x \in T^{-j}(\xi)$ there exists a unique holomorphic branch $T_x^{-j} : B(\xi, 2\lambda^{-\alpha n}) \to \widehat{\mathbb{C}}$ of T^{-j} sending ξ to x. By Theorem 23.1.9 (Koebe's distortion theorem, spherical version), we have for every $r \in (0, 1/2]$ and every $1 \le j \le 2n$ that

$$\frac{\mathrm{diam}(T_x^{-j}(B(\xi, r\lambda^{-\alpha n})))}{\mathrm{diam}(T_x^{-j}(B(\xi, \lambda^{-\alpha n})))} \le K^2 \frac{\mathrm{diam}(B(\xi, r\lambda^{-\alpha n}))}{\mathrm{diam}(B(\xi, \lambda^{-\alpha n}))} = K^2 r.$$

As $\mathrm{diam}(T_x^{-j}(B(\xi, \lambda^{-\alpha n}))) \le \mathrm{diam}(\widehat{\mathbb{C}}) < \infty$, there exists $\varepsilon \in (0, 1/2]$ such that

$$T_x^{-j}(B(\xi, \varepsilon\lambda^{-\alpha n})) \subseteq B\Big(x, \frac{\eta}{16K}\Big), \quad \forall x \in T^{-j}(\xi),\ \forall 1 \le j \le 2n. \tag{29.38}$$

Since $\xi \in \mathcal{J}_{\mathrm{uer}}(T)(\eta)$, there exists $N_2 \ge N_1$ such that for every $n \ge N_2$ there exists a unique holomorphic branch $T_\xi^{-2[\alpha n]} : B(T^{2[\alpha n]}(\xi), 2\eta) \to \widehat{\mathbb{C}}$ of $T_\xi^{-2[\alpha n]}$ sending $T^{2[\alpha n]}(\xi)$ to ξ and $|(T^{2[\alpha n]})'(\xi)| \ge \lambda^{2[\alpha n]}$. With the help of Lemma 23.1.11 and enlarging N_2 if needed, we get

$$T_\xi^{-2[\alpha n]}(B(T^{2[\alpha n]}(\xi), 2\eta)) \subseteq B(\xi, \varepsilon\lambda^{-\alpha n}), \quad \forall n \ge N_2. \tag{29.39}$$

As the map $T : \mathcal{J}(T) \to \mathcal{J}(T)$ is topologically exact per Corollary 24.3.5, there exists $m(\eta) \ge 0$ such that

$$T^{m(\eta)}\Big(B\Big(y, \frac{\eta}{16K}\Big)\Big) \supseteq \mathcal{J}(T), \quad \forall y \in \mathcal{J}(T).$$

Consequently, for every $n \in \mathbb{N}$ and every $z \in T^{-n}(\xi) \subseteq \mathcal{J}(T)$ there exists a point

$$z' \in T^{-m(\eta)}(z) \cap B\Big(T^{2[\alpha n]}(\xi), \frac{\eta}{16K}\Big). \tag{29.40}$$

Assuming now that $n \geq \max\{N_2, m(\eta)\}$, we have $2n \geq m(\eta) + n$, and hence there exists a unique holomorphic branch $T_{z'}^{-(m(\eta)+n)} : B(\xi, 2\lambda^{-\alpha n}) \to \widehat{\mathbb{C}}$ of $T^{-(m(\eta)+n)}$ sending ξ to z'. It then follows from (29.38) and (29.40) that

$$T_{z'}^{-(m(\eta)+n)}(B(\xi, \varepsilon\lambda^{-\alpha n})) \subseteq B\Big(z', \frac{\eta}{16K}\Big) \subseteq B\Big(T^{2[\alpha n]}(\xi), \frac{\eta}{8K}\Big).$$

Looking up at (29.39), we see that the composition

$$\varphi_{z'}^{(n)} := T_{z'}^{-(m(\eta)+n)} \circ T_{\xi}^{-2[\alpha n]} : B(T^{2[\alpha n]}(\xi), 2\eta) \to B\Big(T^{2[\alpha n]}(\xi), \frac{\eta}{8K}\Big) \tag{29.41}$$

is well-defined and holomorphic. It ensues from this and Koebe's $\frac{1}{4}$-theorem that

$$B\Big(T^{2[\alpha n]}(\xi), \frac{\eta}{8K}\Big) \supseteq \varphi_{z'}^{(n)}(B(T^{2[\alpha n]}(\xi), \eta)) \supseteq B\Big(T^{2[\alpha n]}(\xi), \frac{1}{4}|(\varphi_{z'}^{(n)})'(T^{2[\alpha n]}(\xi))|\eta\Big).$$

Hence,

$$\frac{1}{4}|(\varphi_{z'}^{(n)})'(T^{2[\alpha n]}(\xi))|\eta \leq \frac{\eta}{8K}.$$

Equivalently,

$$|(\varphi_{z'}^{(n)})'(T^{2[\alpha n]}(\xi))| \leq \frac{1}{2K}.$$

Theorem 23.1.9 (Koebe's distortion theorem, spherical version) then implies that

$$|(\varphi_{z'}^{(n)})'(w)| \leq \frac{1}{2}, \quad \forall w \in B(T^{2[\alpha n]}(\xi), \eta). \tag{29.42}$$

Let $V = \{v\}$ be a vertex set and $E = T^{-n}(\xi)$ serve as an alphabet. Let $A : E \times E \to \{0,1\}$ be the constant function 1, so that $E_A^\infty = E^\infty = T^{-n}(\xi)^{\mathbb{N}}$. Let $X = X_v = \overline{B(T^{2[\alpha n]}(\xi), \eta/4)}$ and $W = W_v = B(T^{2[\alpha n]}(\xi), \eta/2)$.

By construction,

$$\varphi_{z'}^{(n)}(B(T^{2[\alpha n]}(\xi), \eta/2)) \cap \varphi_{w'}^{(n)}(B(T^{2[\alpha n]}(\xi), \eta/2)) = \emptyset, \quad \forall z', w' \in T^{-n}(\xi),\ z' \neq w'. \tag{29.43}$$

By (29.41)–(29.43), the collection

$$\Phi_n := \Big\{V, E, A; X, W; \{\varphi_{z'}^{(n)}\}_{z \in T^{-n}(\xi)}\Big\}$$

forms a finite conformal IFS (a CIFS) that satisfies the strong separation condition (SSC; see Definition 19.7.6). Per Theorem 19.7.8, its limit set $\mathcal{J}_n$ is a topological Cantor set and

$$T^{m(\eta)+n+2[\alpha n]}(\mathcal{J}_n) \subseteq \mathcal{J}_n. \tag{29.44}$$

(Note: As stated, Theorem 19.7.8 is not entirely true; more precisely, J might not be perfect. However, the perfectness of J does occur when A is irreducible and J is infinite. If A is primitive and J has at least two points, then J is infinite and (according to the previous sentence) perfect; this is the case for the CIFS Φ_n. A weaker condition than the infiniteness of J, similar to that of the non single periodic orbit given in Proposition 28.5.4(d,e), exists and we leave it to the reader as an exercise to identify it.)

Along with (29.42), this shows the following.

Claim 1. *The set $\mathcal{J}_n$ is invariant and expanding for the map $F_n := T^{m(\eta)+n+2[\alpha n]}$.*

Fix an arbitrary $t \geq 0$. As $F_n : \mathcal{J}_n \to \mathcal{J}_n$ is expansive, Theorem 11.1.26 guarantees that $\mathrm{P}(F_n, -t\log|F_n'|) = \mathrm{P}(F_n, -t\log|F_n'|, \mathcal{U})$ for any finite open partition $\mathcal{U}$ of $\mathcal{J}_n$ with $\mathrm{diam}(\mathcal{U}) \leq \delta$, where δ is any expansive constant for F_n. Let

$$\mathcal{U} = \left\{\varphi_{z'}^{(n)}(B(T^{2[\alpha n]}(\xi), \eta/2)) \cap \mathcal{J}_n : z \in T^{-n}(\xi)\right\}.$$

By (29.43) (the OSC) and since $\mathcal{J}_n \subseteq \bigcup_{z \in T^{-n}(\xi)} \varphi_{z'}^{(n)}(B(T^{2[\alpha n]}(\xi), \eta/2))$, it is obvious that $\mathcal{U}$ is a finite open partition of $\mathcal{J}_n$. However, it is unclear whether $\mathrm{diam}(\mathcal{U}) \leq \delta$. Nevertheless, thanks to (29.42) observe that $\mathrm{diam}(\mathcal{U}^k) \leq 2^{-(k-1)}\mathrm{diam}(\mathcal{U})$ for all $k \in \mathbb{N}$. Thus there exists $\ell \in \mathbb{N}$ such that $\mathrm{diam}(\mathcal{U}^\ell) \leq \delta$. Lemma 11.1.15 affirms that $\mathrm{P}(F_n, -t\log|F_n'|, \mathcal{U}^\ell) = \mathrm{P}(F_n, -t\log|F_n'|, \mathcal{U})$. So it suffices to compute $\mathrm{P}(F_n, -t\log|F_n'|, \mathcal{U})$. By Definition 11.1.11, the topological pressure of the potential $-t\log|F_n'|$ with respect to the open cover $\mathcal{U}$ of $\mathcal{J}_n$ is

$$\mathrm{P}(F_n, -t\log|F_n'|, \mathcal{U}) = \lim_{k\to\infty} \frac{1}{k} \log Z_k(F_n, -t\log|F_n'|, \mathcal{U}),$$

where, per Definition 11.1.2, the k-th partition function of $\mathcal{U}$ is

$$Z_k(F_n, -t\log|F_n'|, \mathcal{U}) = \inf\Big\{\sum_{U' \in \mathcal{U}'} \exp(\overline{S}_k(-t\log|F_n'|)(U')) : \mathcal{U}' \text{ is a subcover of } \mathcal{U}^k\Big\}.$$

As $\mathcal{U}$ is a partition of $\mathcal{J}_n$, so is $\mathcal{U}^k$, and thus $Z_k(F_n, -t\log|F_n'|, \mathcal{U})$ reduces to

$$\begin{aligned}
Z_k(F_n, -t\log|F_n'|, \mathcal{U}) &= \sum_{U \in \mathcal{U}^k} \exp(\overline{S}_k(-t\log|F_n'|)(U)) \\
&= \sum_{\omega \in E^k} \exp\Big(\overline{S}_k(-t\log|F_n'|)(\varphi_\omega(B(T^{2[\alpha n]}(\xi), \eta/2)) \cap \mathcal{J}_n)\Big) \\
&= \sum_{\omega \in E^k} \sup\Big\{|(F_n^k)'(x)|^{-t} : x \in \varphi_\omega(B(T^{2[\alpha n]}(\xi), \eta/2)) \cap \mathcal{J}_n\Big\} \\
&\geq K^{-t} \sum_{\omega \in E^k} \|\varphi_\omega'\|_{\mathcal{J}_n}^t \\
&\geq K^{-(k+1)t}\Big[\sum_{e \in E} \|\varphi_e'\|_{\mathcal{J}_n}^t\Big]^k,
\end{aligned}$$

where K is the Koebe constant corresponding to the scale 1/2. (This argument above is inspired from the proof of Theorem 16.3.4. Alternatively, use Lemma 19.4.1 to get the same expression.) Consequently,

$$\begin{aligned}
\mathrm{P}(F_n, -t\log|F_n'|) &= \mathrm{P}(F_n, -t\log|F_n'|, \mathcal{U}) \\
&\geq \lim_{k\to\infty}\frac{1}{k}\log\Big(K^{-(k+1)t}\Big[\sum_{e\in E}\|\varphi_e'\|_{\mathcal{J}_n}^t\Big]^k\Big) \\
&= -t\log K + \log\sum_{e\in E}\|\varphi_e'\|_{\mathcal{J}_n}^t \\
&\geq -t\log K + \log\sum_{z\in T^{-n}(\xi)}|(\varphi_{z'}^{(n)})'(T^{2[\alpha n]}(\xi))|^t \\
&= -t\log K + \log\sum_{z\in T^{-n}(\xi)}|(T^{2[\alpha n]})'(\xi)|^{-t}|(T^n)'(z)|^{-t}|(T^{m(\eta)})'(z')|^{-t} \\
&\geq -t\log K + \log\sum_{z\in T^{-n}(\xi)}|(T^n)'(z)|^{-t}\|T'\|_\infty^{-t(m(\eta)+2[\alpha n])} \\
&= -t\log K - t(m(\eta)+2[\alpha n])\log\|T'\|_\infty + \log\sum_{z\in T^{-n}(\xi)}|(T^n)'(z)|^{-t}.
\end{aligned} \tag{29.45}$$

Setting

$$\mathcal{J}_n^* := \bigcup_{i=0}^{m(\eta)+n+2[\alpha n]-1} T^i(\mathcal{J}_n),$$

Claim 1 and (29.44) yield the following.

Claim 2. *The set $\mathcal{J}_n^*$ is invariant and expanding for the map T.*

Using the continuity of the iterates T^i, $i = 0, \ldots, m(\eta)+n+2[\alpha n]-1$, it directly follows from the construction of the sets $\mathcal{J}_n$ and $\mathcal{J}_n^*$ that the map $T : \mathcal{J}_n^* \to \mathcal{J}_n^*$ is transitive.

Recall from Proposition 11.1.26 that $\mathrm{P}(T^k, S_k\varphi) = k\mathrm{P}(T,\varphi)$. It then follows from (29.45) that

$$\begin{aligned}
\mathrm{P}(T|_{\mathcal{J}_n^*}, -t\log|T'|) &= \frac{1}{m(\eta)+n+2[\alpha n]}\mathrm{P}(T^{m(\eta)+n+2[\alpha n]}|_{\mathcal{J}_n^*}, -t\log|(T^{m(\eta)+n+2[\alpha n]})'|) \\
&\geq \frac{1}{m(\eta)+n+2[\alpha n]}\mathrm{P}(F_n, -t\log|F_n'|) \\
&\geq \frac{-t\log K - t(m(\eta)+2[\alpha n])\log\|T'\|_\infty + \log\sum_{z\in T^{-n}(\xi)}|(T^n)'(z)|^{-t}}{m(\eta)+n+2[\alpha n]}.
\end{aligned}$$

Thus, for every $\alpha \in (0,1)$ and every $n \geq \max\{m(\eta), N_2\}$ we obtain that

$$\mathrm{P}^*_{\exp}(t) \geq \mathrm{P}(T|_{\mathcal{J}_n^*}, -t\log|T'|)$$

$$\geq \frac{-\frac{t\log K}{n} - t(\frac{m(\eta)}{n} + 2\frac{[\alpha n]}{n})\log\|T'\|_\infty + \frac{1}{n}\log\sum_{z\in T^{-n}(\xi)}|(T^n)'(z)|^{-t}}{\frac{m(\eta)}{n} + 1 + 2\frac{[\alpha n]}{n}}.$$

Taking the lim sup as $n \to \infty$, we deduce that

$$\mathrm{P}^*_{\exp}(t) \geq \frac{-2t\alpha\log\|T'\|_\infty + c_\xi(t)}{1+2\alpha}.$$

Finally, letting $\alpha \searrow 0$, we conclude that $\mathrm{P}^*_{\exp}(t) \geq c_\xi(t)$ and the proof of Lemma 29.2.4 is complete. □

Throughout the rest of this section, we keep $\xi \in \mathrm{Safe}(T)\cap\mathcal{J}_{\mathrm{uer}}(T)$, as in Lemma 29.2.4. The standard convexity arguments showing continuity of the topological pressure also lead to the following result.

Lemma 29.2.5. *The function* $[0,\infty) \ni t \longmapsto c_\xi(t) \in \mathbb{R}$ *is convex and thereby continuous.*

Set

$$s_\xi(T) := \inf\{t \geq 0 : c_\xi(t) \leq 0\}. \tag{29.46}$$

We now introduce new dimensions for the Julia set.

Definition 29.2.6. Let $T : \widehat{\mathbb{C}} \to \widehat{\mathbb{C}}$ be a rational function. The *dynamical dimension* of $\mathcal{J}(T)$ is the nonnegative number

$$\mathrm{DD}(\mathcal{J}(T)) := \sup_\mu \mathrm{HD}(\mu),$$

where the supremum is taken over all ergodic T-invariant Borel probability measures μ on $\mathcal{J}(T)$ such that $\mathrm{h}_\mu(T) > 0$.

Definition 29.2.7. Let $T : \widehat{\mathbb{C}} \to \widehat{\mathbb{C}}$ be a rational function. The *expanding dimension* of $\mathcal{J}(T)$ is the nonnegative number

$$\mathrm{ExpD}(\mathcal{J}(T)) := \sup_X \mathrm{HD}(X),$$

where the supremum is taken over all forward T-invariant closed subsets X of $\mathcal{J}(T)$ for which the map $T|_X : X \to X$ is expanding.

If, additionally, the map $T|_X : X \to X$ is required to be transitive, then the resulting supremum is called the *strict expanding dimension* of $\mathcal{J}(T)$ and is denoted by

$$\mathrm{ExpD}^*(\mathcal{J}(T)).$$

If, on top of this, $T|_X : X \to X$ is required to be open, then the resulting supremum is called the *strictest expanding dimension* of $\mathcal{J}(T)$ and is denoted by

$$\mathrm{ExpD}^{**}(\mathcal{J}(T)).$$

We now describe relations between the different types of dimension of the Julia set.

Lemma 29.2.8. *If* $T : \widehat{\mathbb{C}} \to \widehat{\mathbb{C}}$ *is a rational function with* $\deg(T) \geq 2$, *then*

$$s_\xi(T) \leq \mathrm{ExpD}^{**}(\mathcal{J}(T)) \leq \mathrm{ExpD}^{*}(\mathcal{J}(T)) \leq \mathrm{ExpD}(\mathcal{J}(T)) \leq \mathrm{DD}(\mathcal{J}(T)) \leq \mathrm{HD}(\mathcal{J}(T)) \leq 2.$$

Proof. All but the first and fourth inequalities are obvious. To demonstrate the first one, suppose on the contrary that $\mathrm{ExpD}^{**}(\mathcal{J}(T)) < s_\xi(T)$. Take $\mathrm{ExpD}^{**}(\mathcal{J}(T)) < t < s_\xi(T)$. From this choice, Lemma 29.2.4 and Proposition 25.3.12, we have that

$$0 < c_\xi(t) \leq \mathrm{P}^{**}_{\exp}(t). \tag{29.47}$$

So, using Theorem 25.3.9, we deduce that there exists a forward T-invariant closed set $X \subseteq \mathcal{J}(T)$ and an open set $U \supseteq X$ such that: (1) $T|_X : X \to X$ is a topologically transitive, open, expanding dynamical system; (2) the triple (X, U, T) is a conformal expanding repeller; and (3)

$$\mathrm{P}(T|_X, -t\log|T'|) \geq \frac{1}{2}c_\xi(t) > 0.$$

It follows from Theorem 16.3.2 and Proposition 16.3.1 that $t \leq \mathrm{HD}(X) \leq \mathrm{ExpD}^{**}(\mathcal{J}(T))$, which contradicts the assumption that $\mathrm{ExpD}^{**}(\mathcal{J}(T)) < t$. This finishes the proof of the first inequality.

To establish the fourth inequality, first notice that it clearly holds if $\mathrm{ExpD}(\mathcal{J}(T)) = 0$. So, suppose that $\mathrm{ExpD}(\mathcal{J}(T)) > 0$. Pick any $0 \leq t < \mathrm{ExpD}(\mathcal{J}(T))$. Then there exists an expanding forward T-invariant closed subset Y of $\mathcal{J}(T)$ such that $\mathrm{HD}(Y) > t$. But by Proposition 28.5.4 and Theorem 28.5.5,

$$\mathrm{P}(T|_Y, -t\log|T'|) > 0.$$

Therefore, by the variational principle (Theorem 12.1.1), there exists an ergodic T-invariant Borel probability measure μ on Y such that

$$\mathrm{h}_\mu(T) - t\chi_\mu(T) > 0. \tag{29.48}$$

Using Przytycki's theorem (Theorem 28.1.3; $\chi_\mu(T) \geq 0$), we deduce that $\mathrm{h}_\mu(T) > 0$. Applying Ruelle's inequality (Theorem 28.2.1; $\mathrm{h}_\mu(T) \leq 2\chi_\mu(T)$), we infer that $\chi_\mu(T) > 0$. Hence, it follows from the volume lemma (Theorem 28.4.1) and (29.48) that

$$t < \frac{\mathrm{h}_\mu(T)}{\chi_\mu(T)} = \mathrm{HD}(\mu) \leq \mathrm{DD}(\mathcal{J}(T)).$$

Since $0 \leq t < \mathrm{ExpD}(\mathcal{J}(T))$ is arbitrary, we conclude that $\mathrm{ExpD}(\mathcal{J}(T)) \leq \mathrm{DD}(\mathcal{J}(T))$. □

As $c(0) \geq 0$, an immediate consequence of Lemmas 29.2.5 and 29.2.8 is that

$$c_\xi(s_\xi(T)) = 0. \tag{29.49}$$

Inserting this into Lemma 29.2.1, we obtain the main result of this section.

Theorem 29.2.9. *If $T : \widehat{\mathbb{C}} \to \widehat{\mathbb{C}}$ is a rational function with $\deg(T) \geq 2$, then there exists a Borel probability measure m on $\mathcal{J}(T)$ such that*

$$m(T(A)) = \int_A |T'|^{s_\xi(T)}\, dm$$

for every special set $A \subseteq \mathcal{J}(T)$.

This suggests making the definition below.

Definition 29.2.10. Let $T : \widehat{\mathbb{C}} \to \widehat{\mathbb{C}}$ be a rational function. A Borel probability measure m on $\mathcal{J}(T)$ is said to be (Sullivan) t-conformal for some $t \geq 0$ if

$$m(T(A)) = \int_A |T'|^t\, dm$$

for every special set $A \subseteq \mathcal{J}(T)$. The infimum of all exponents $t \geq 0$ for which a t-conformal measure exists is denoted by $\delta(T)$.

From Theorem 29.2.9, we immediately realize that

$$\delta(T) \leq s_\xi(T). \tag{29.50}$$

The rest of this section boils down to proving that the exponent $s_\xi(T)$ is independent of ξ and to revealing several of its characterizations.

Lemma 29.2.11. *If $T : \widehat{\mathbb{C}} \to \widehat{\mathbb{C}}$ is a rational function with $\deg(T) \geq 2$, then*

$$\begin{aligned}
\mathrm{DD}(\mathcal{J}(T)) &\leq \mathrm{ExpD}^{**}(\mathcal{J}(T)) \leq \mathrm{ExpD}^{*}(\mathcal{J}(T)) \leq \mathrm{ExpD}(\mathcal{J}(T)) \\
&\leq \mathrm{HD}(\mathcal{J}_{\mathrm{uer}}(T)) \leq \mathrm{HD}(\mathcal{J}_{\mathrm{er}}(T)) \leq \mathrm{HD}(\mathcal{J}_{\mathrm{r}}(T)) \\
&\leq \delta(T).
\end{aligned}$$

Proof. Except for the first and last ones, all the inequalities are obvious. Let us prove the first inequality. Let μ be an ergodic T-invariant Borel probability measure on $\mathcal{J}(T)$ for which $\mathrm{h}_\mu(T) > 0$. It follows from Ruelle's inequality (Theorem 28.2.1) that $\chi_\mu(T) > 0$. So, we will be able to apply Theorem 28.6.2 later. Moreover, $\mathrm{HD}(\mu) > 0$ according to the volume lemma (Theorem 28.4.1). For every integer $n \geq 0$, define on $\mathcal{J}(T)$ the continuous function

$$\varphi_n := \max\{-n, \log|T'|\}.$$

Then $\varphi_n \geq \log|T'|$ and $\varphi_n \searrow \log|T'|$ pointwise on $\mathcal{J}(T)$. In addition, $\varphi_n \leq \log\|T'\|_\infty$. The Lebesgue monotone convergence theorem then affirms that

$$\lim_{n\to\infty} \int_{\mathcal{J}(T)} \varphi_n \, d\mu = \int_{\mathcal{J}(T)} \log|T'| \, d\mu =: \chi_\mu(T) > 0.$$

Fix $0 < \varepsilon < 1$. Then there is $n_\varepsilon \in \mathbb{N}$ such that

$$0 < \int_{\mathcal{J}(T)} \varphi_n \, d\mu \leq \frac{\chi_\mu(T)}{1-\varepsilon}, \quad \forall n \geq n_\varepsilon. \tag{29.51}$$

Along with the volume lemma (Theorem 28.4.1), this implies that

$$\mathrm{h}_\mu(T) = \mathrm{HD}(\mu)\chi_\mu(T) \geq (1-\varepsilon)\mathrm{HD}(\mu) \int_{\mathcal{J}(T)} \varphi_n \, d\mu. \tag{29.52}$$

Let $(X_k(\varepsilon))_{k=1}^\infty$ be the sequence of expanding forward T-invariant closed sets produced in Theorem 28.6.2 for the measure μ and the function $\varphi = -\mathrm{HD}(\mu)\varphi_{n_\varepsilon}$. For every $k \in \mathbb{N}$, Theorem 12.2.5 guarantees that there is an ergodic equilibrium state $\mu_{\varepsilon,k}$ for the expansive map $T|_{X_k(\varepsilon)} : X_k(\varepsilon) \to X_k(\varepsilon)$ and the potential $-\mathrm{HD}(\mu)\varphi_{n_\varepsilon}$ restricted to $X_k(\varepsilon)$. It follows from the second part of Theorem 28.6.2 that the measures $(\mu_{\varepsilon,k})_{k=1}^\infty$ converge to μ in the weak* topology on $M(\mathcal{J}(T))$. Using this and (29.51), we deduce that

$$\lim_{k\to\infty} \int_{\mathcal{J}(T)} \varphi_{n_\varepsilon} \, d\mu_{\varepsilon,k} = \int_{\mathcal{J}(T)} \varphi_{n_\varepsilon} \, d\mu > 0. \tag{29.53}$$

By the variational principle (Theorem 12.1.1), Theorem 28.6.2 and (29.52), we obtain that

$$\begin{aligned}
\liminf_{k\to\infty}\left[\mathrm{h}_{\mu_{\varepsilon,k}}(T) - \mathrm{HD}(\mu) \int_{\mathcal{J}(T)} \varphi_{n_\varepsilon} \, d\mu_{\varepsilon,k}\right] &= \liminf_{k\to\infty} \mathrm{P}(T|_{X_k(\varepsilon)}, -\mathrm{HD}(\mu)\varphi_{n_\varepsilon}) \\
&\geq \mathrm{h}_\mu(T) - \mathrm{HD}(\mu) \int_{\mathcal{J}(T)} \varphi_{n_\varepsilon} \, d\mu \\
&\geq -\varepsilon \mathrm{HD}(\mu) \int_{\mathcal{J}(T)} \varphi_{n_\varepsilon} \, d\mu.
\end{aligned}$$

Using this, (29.51)–(29.53) and the definition of φ_n, we get for all $k \in \mathbb{N}$ large enough that

$$\begin{aligned}
\mathrm{h}_{\mu_{\varepsilon,k}}(T) &\geq \mathrm{HD}(\mu) \int_{\mathcal{J}(T)} \varphi_{n_\varepsilon} \, d\mu_{\varepsilon,k} - 2\varepsilon \mathrm{HD}(\mu) \int_{\mathcal{J}(T)} \varphi_{n_\varepsilon} \, d\mu \\
&\geq \mathrm{HD}(\mu) \int_{\mathcal{J}(T)} \varphi_{n_\varepsilon} \, d\mu_{\varepsilon,k} - 3\varepsilon \mathrm{HD}(\mu) \int_{\mathcal{J}(T)} \varphi_{n_\varepsilon} \, d\mu_{\varepsilon,k}
\end{aligned}$$

$$= (1-3\varepsilon)\mathrm{HD}(\mu)\int_{\mathcal{J}(T)} \varphi_{n_\varepsilon}\,d\mu_{\varepsilon,k}$$
$$\geq (1-3\varepsilon)\mathrm{HD}(\mu)\int_{\mathcal{J}(T)} \log|T'|\,d\mu_{\varepsilon,k} = (1-3\varepsilon)\mathrm{HD}(\mu)\chi_{\mu_{\varepsilon,k}}(T).$$

Invoking the volume lemma (Theorem 28.4.1) once again, we infer that

$$\mathrm{ExpD}^{**}(\mathcal{J}(T)) \geq \mathrm{HD}(X_k(\varepsilon)) \geq \mathrm{HD}(\mu_{\varepsilon,k}) = \frac{\mathrm{h}_{\mu_{\varepsilon,k}}(T)}{\chi_{\mu_{\varepsilon,k}}(T)} \geq (1-3\varepsilon)\mathrm{HD}(\mu).$$

Letting $\varepsilon \to 0$, we conclude that $\mathrm{ExpD}^{**}(\mathcal{J}(T)) \geq \mathrm{HD}(\mu)$. Taking the supremum over all ergodic T-invariant Borel probability measures μ on $\mathcal{J}(T)$ for which $\mathrm{h}_\mu(T) > 0$, we get the first inequality of the current lemma.

We now prove the last inequality. The main idea is to get to a large scale. Fix $t \geq 0$ and assume that a t-conformal measure m exists for T. Fix $\eta > 0$ and consider any point $x \in \mathcal{J}_r(T)(\eta)$. (See Definition 28.5.1 for a recollection of notation and meaning.) For every $n \in N_x(T)$, there exists a unique holomorphic inverse branch $T_x^{-n} : B(T^n(x), 2\eta) \to \widehat{\mathbb{C}}$ of T^n such that $T_x^{-n}(T^n(x)) = x$. It follows from Lemma 23.1.11 that

$$T_x^{-n}(B(T^n(x),\eta)) \subseteq B(x, K|(T^n)'(x)|^{-1}\eta). \tag{29.54}$$

Set

$$r_n(x) = K|(T^n)'(x)|^{-1}\eta. \tag{29.55}$$

By Theorem 28.3.3 and the t-conformality of the measure m, we get

$$\begin{aligned} m(B(T^n(x),\eta)) &= m(T^n(T_x^{-n}(B(T^n(x),\eta)))) \\ &= \int_{T_x^{-n}(B(T^n(x),\eta))} |(T^n)'|^t\,dm \\ &\leq \int_{B(x,r_n(x))} |(T^n)'|^t\,dm \quad \text{by (29.54)–(29.55)} \\ &\leq K^t|(T^n)'(x)|^t m(B(x,r_n(x))). \end{aligned}$$

Hence,

$$m(B(x,r_n(x))) \geq K^{-t}|(T^n)'(x)|^{-t} m(B(T^n(x),\eta)) \geq M_m(\eta)^{-1}K^{-2t}\eta^{-t}(r_n(x))^t,$$

where

$$M_m(\eta) := \inf\{m(B(y,\eta)) : y \in \mathcal{J}(T)\} > 0$$

since $\operatorname{supp}(m) = \mathcal{J}(T)$. As $\#N_x(T) = \infty$, we deduce that

$$\limsup_{r\to 0} \frac{m(B(x,r))}{r^t} \geq \limsup_{N_x(T)\ni n\to\infty} \frac{m(B(x,r_n(x)))}{(r_n(x))^t} \geq \left(M_m(\eta)K^{2t}\eta^t\right)^{-1}.$$

It follows from Theorem 15.5.3(a) (Frostman converse theorem) that

$$H_t(\mathcal{J}_r(T)(\eta)) \leq 10^t M_m(\eta)K^{2t}\eta^t < \infty.$$

Hence, $\operatorname{HD}(\mathcal{J}_r(T)(\eta)) \leq t$. As

$$\mathcal{J}_r(T) = \bigcup_{k=1}^{\infty} \mathcal{J}_r(T)(1/k),$$

we conclude from Theorem 15.3.5 (σ–stability of HD) that $\operatorname{HD}(\mathcal{J}_r(T)) \leq t$. Since $t \geq 0$ is arbitrary as long as a t-conformal measure exists, we thus get that $\operatorname{HD}(\mathcal{J}_r(T)) \leq \delta(T)$. □

As a direct consequence of this lemma, Lemma 29.2.8 and Theorem 29.2.9, we obtain the main result of this section.

Theorem 29.2.12. *If* $T : \widehat{\mathbb{C}} \to \widehat{\mathbb{C}}$ *is a rational function with* $\deg(T) \geq 2$, *then*

$$\begin{aligned}\delta(T) &= \operatorname{ExpD}^{**}(\mathcal{J}(T)) = \operatorname{ExpD}^{*}(\mathcal{J}(T)) = \operatorname{ExpD}(\mathcal{J}(T)) = \operatorname{DD}(\mathcal{J}(T))\\ &= \operatorname{HD}(\mathcal{J}_{\mathrm{uer}}(T)) = \operatorname{HD}(\mathcal{J}_{\mathrm{er}}(T)) = \operatorname{HD}(\mathcal{J}_r(T)) = s_\xi(T)\\ &\leq \operatorname{HD}(\mathcal{J}(T)) \leq 2\end{aligned}$$

for every $\xi \in \operatorname{Safe}(T) \cap \mathcal{J}_{\mathrm{uer}}(T)$, *and* T *admits a Sullivan* $\delta(T)$*-conformal measure.*

We end this section with two results shedding more light on the nature of the objects we have dealt with in this section. In [20], Xavier Buff and Arnaud Chéritat showed the existence of quadratic polynomials whose Julia sets have positive planar Lebesgue measure (but of course are nowhere dense). The situation is quite different for radial Julia sets $\mathcal{J}_r(T)$. We prove the following fact related to Corollary 24.3.6.

Theorem 29.2.13. *If* $T : \widehat{\mathbb{C}} \to \widehat{\mathbb{C}}$ *is a rational function of* $\deg(T) \geq 2$ *and* $\mathcal{J}(T) \neq \widehat{\mathbb{C}}$, *then*

$$\operatorname{Leb}_2(\mathcal{J}_r(T)) = 0.$$

Proof. By a Möbius change of coordinates, we may assume without loss of generality that $\mathcal{J}(T) \subseteq \mathbb{C}$. Fix $l \in \mathbb{N}$. Since $\mathcal{J}(T) \neq \widehat{\mathbb{C}}$, it is a nowhere dense subset of $\mathbb{C}$ according to Corollary 24.3.6. Then there exists $\varepsilon > 0$ such that for every $y \in \mathbb{C}$ there is $y_\varepsilon \in B(y, (2l)^{-1})$ for which

$$B(y_\varepsilon, \varepsilon) \subseteq B(y, (2l)^{-1}) \setminus \mathcal{J}(T). \tag{29.56}$$

Fix an arbitrary point $z \in \mathcal{J}_r(T)(1/l)$. For every $n \in N_z(T)$, there exists a unique holomorphic inverse branch $T_z^{-n} : B(T^n(z), (2l)^{-1}) \to \widehat{\mathbb{C}}$ of T^n such that $T_z^{-n}(T^n(z)) = z$. On one hand, by Koebe's distortion theorem (Theorem 23.1.8) and (29.56) we have

$$T_z^{-n}(B(T^n(z)_\varepsilon, \varepsilon)) \subseteq T_z^{-n}\Big(B(T^n(z), (2l)^{-1}) \setminus \mathcal{J}(T)\Big) \subseteq B(z, K|(T^n)'(z)|^{-1}(2l)^{-1}) \setminus \mathcal{J}(T).$$

On the other hand, by Koebe's $\frac{1}{4}$–theorem and (29.56), we get

$$T_z^{-n}(B(T^n(z)_\varepsilon, \varepsilon)) \supseteq B\Big(T_z^{-n}(T^n(z)_\varepsilon), \frac{1}{4}\varepsilon|(T^n)'(z)|^{-1}\Big).$$

Therefore,

$$\frac{\mathrm{Leb}_2\Big(B(z, K|(T^n)'(z)|^{-1}(2l)^{-1}) \setminus \mathcal{J}(T)\Big)}{\mathrm{Leb}_2\Big(B(z, K|(T^n)'(z)|^{-1}(2l)^{-1})\Big)} \geq \Big(\frac{\varepsilon l}{2K}\Big)^2 > 0.$$

So, z is not a Lebesgue density point for the set $\mathcal{J}_r(T)(1/l)$, and thus $\mathrm{Leb}_2(\mathcal{J}_r(T)(1/l)) = 0$. Hence,

$$\mathrm{Leb}_2(\mathcal{J}_r(T)) = \mathrm{Leb}_2\Big(\bigcup_{l=1}^{\infty} \mathcal{J}_r(T)(1/l)\Big) \leq \sum_{l=1}^{\infty} \mathrm{Leb}_2(\mathcal{J}_r(T)(1/l)) = 0.$$

The proof of Theorem 29.2.13 is complete. □

We already know from Lemma 29.2.11 and Theorem 28.5.3 that $\delta(T) > 0$. We now show that the mere presence of rationally indifferent periodic points entails an explicit positive lower bound on $\delta(T)$. Recall that $\Omega(T)$ denotes the set of rationally indifferent periodic points of T while $\Omega_0(T)$ is the set of simple rationally indifferent fixed points of T (see (24.63)). For every $\omega \in \Omega(T)$, let $q(\omega)$ be the smallest $q \in \mathbb{N}$ such that ω a simple fixed point for T^q, i. e., the smallest $q \in \mathbb{N}$ for which $\omega \in \Omega_0(T^q)$. Define $p(\omega)$ to be the $p \in \mathbb{N}$ that comes from (24.24) with $\varphi = T^{q(\omega)}$. Let

$$p_T := \max\{p(\omega) : \omega \in \Omega(T)\} \in \mathbb{N}. \tag{29.57}$$

Theorem 29.2.14. *If* $T : \widehat{\mathbb{C}} \to \widehat{\mathbb{C}}$ *is a rational function with* $\deg(T) \geq 2$, *then*

$$\delta(T) > \frac{p_T}{p_T + 1} > 0.$$

Proof. In view of Theorems 29.2.9 and 29.2.12, there exists a $\delta(T)$-conformal measure $m_{\delta(T)}$. Fix $\omega \in \Omega(T)$ so that $p(\omega) = p_T$. Passing to a sufficiently high iterate of T, we may assume without loss of generality that ω is a simple parabolic fixed point of T. Aiming to use Proposition 24.2.27, fix $z \in \mathcal{J}(T) \cap B(\omega, \theta) \setminus \{\omega\}$ as required in that proposition and where θ comes from (24.83). Per (24.77), there exists $r > 0$ such that all the iterates

T_ω^{-n} are well-defined on $B(z, 2r)$, where T_ω^{-n} is a local holomorphic inverse branch of T^n fixing ω.

Claim. *If $r > 0$ is sufficiently small, then the sets $T_\omega^{-n}(B(z, 2r))$, $n \geq 0$, are mutually disjoint.*

Proof of Claim. As $\lim_{n\to\infty} T_\omega^{-n}(z) = \omega$ according to Lemma 24.2.26, we do know that $z \notin \overline{\{T_\omega^{-n}(z) : n \in \mathbb{N}\}}$. Then

$$\Delta := \operatorname{dist}(z, \{T_\omega^{-n}(z) : n \in \mathbb{N}\}) > 0. \tag{29.58}$$

Take $r_1 \in (0, \tau/2]$, where $\tau > 0$ arises from Lemma 24.2.25. It follows from that lemma, relation (24.84) in Proposition 24.2.27 and Koebe's distortion theorem that

$$\limsup_{n\to\infty} \operatorname{diam}(T_\omega^{-n}(B(z, 2r_1))) = 0.$$

So, there exists $N \in \mathbb{N}$ such that

$$\operatorname{diam}(T_\omega^{-n}(B(z, 2r_1))) < \Delta/2, \quad \forall n \geq N.$$

Therefore,

$$B(z, \Delta/2) \cap T_\omega^{-n}(B(z, 2r_1)) = \emptyset, \quad \forall n \geq N.$$

It also follows from (29.58) that there exists $r_2 \in (0, r_1]$ such that

$$B(z, \Delta/2) \cap T_\omega^{-n}(B(z, 2r_2)) = \emptyset, \quad \forall 1 \leq n < N.$$

Thus, for every $r \in (0, \min\{\Delta/4, r_2\})$ we have that

$$B(z, 2r) \cap T_\omega^{-n}(B(z, 2r)) = \emptyset, \quad \forall n \in \mathbb{N}. \tag{29.59}$$

By way of contradiction, suppose that

$$T_\omega^{-n_1}(B(z, 2r)) \cap T_\omega^{-n_2}(B(z, 2r)) \neq \emptyset$$

for some $n_2 > n_1 \geq 0$. Then

$$T_\omega^{-n_1}(B(z, 2r) \cap T_\omega^{-(n_2-n_1)}(B(z, 2r))) \neq \emptyset,$$

and hence

$$B(z, 2r) \cap T_\omega^{-(n_2-n_1)}(B(z, 2r)) \neq \emptyset,$$

contradicting (29.59). This establishes the claim. ∎

Furthermore, the $\delta(T)$-conformality of $m_{\delta(T)}$ yields that

$$m_{\delta(T)}(B(z,r)) = m_{\delta(T)}(T^n(T_\omega^{-n}(B(z,r)))) = \int_{T_\omega^{-n}(B(z,r))} |(T^n)'|^{\delta(T)}\, dm_{\delta(T)}$$
$$\asymp |(T^n)'(T_\omega^{-n}(z))|^{\delta(T)} m_{\delta(T)}(T_\omega^{-n}(B(z,r)))$$
$$= |(T_\omega^{-n})'(z)|^{-\delta(T)} m_{\delta(T)}(T_\omega^{-n}(B(z,r))),$$

where $\asymp$ is the usual symbol of multiplicative comparability (see (25.34)). From this and Proposition 24.2.27, we deduce that

$$m_{\delta(T)}(T_\omega^{-n}(B(z,r))) \asymp |(T_\omega^{-n})'(z)|^{\delta(T)} m_{\delta(T)}(B(z,r)) \asymp (n+1)^{-\frac{p_T+1}{p_T}\delta(T)} M_{m_{\delta(T)}}(r),$$

where

$$1 \geq M_{m_{\delta(T)}}(r) := \inf\{m_{\delta(T)}(B(w,r)) : w \in \mathcal{J}(T)\} > 0$$

since $\operatorname{supp}(m_{\delta(T)}) = \mathcal{J}(T)$. Using the claim, we infer that

$$1 \geq m_{\delta(T)}\left(\bigcup_{n=0}^{\infty} T_\omega^{-n}(B(z,r))\right) = \sum_{n=0}^{\infty} m_{\delta(T)}(T_\omega^{-n}(B(z,r))) \asymp \sum_{n=1}^{\infty} n^{-\frac{p_T+1}{p_T}\delta(T)}.$$

Consequently, $\frac{p_T+1}{p_T}\delta(T) > 1$, or equivalently,

$$\delta(T) > \frac{p_T}{p_T+1}. \qquad \square$$

29.3 Pesin's formula

Theorem 29.3.1 (Pesin's formula). *Let $T : \widehat{\mathbb{C}} \to \widehat{\mathbb{C}}$ be a rational function with* $\deg(T) \geq 2$. *If m is a t-conformal measure for T and μ is an ergodic T-invariant Borel probability measure on $\mathcal{J}(T)$ absolutely continuous with respect to m and such that* $\mathrm{h}_\mu(T) > 0$, *then*

$$\mathrm{HD}(\mu) = t = \delta(T).$$

Proof. In light of Theorem 29.2.12, we only need to prove that $t \leq \mathrm{HD}(\mu)$. To do this, we bring up the arguments from the proof of (28.24). We work with Rokhlin's natural extension $(\widetilde{\mathcal{J}(T)}, \widetilde{T}, \widetilde{\mu})$. Since $\mathrm{h}_\mu(T) > 0$, it follows from Ruelle's inequality (Theorem 28.2.1) that $\chi_\mu(T) > 0$. So, we can fix $0 < \varepsilon < \chi_\mu(T)/3$ and let $\widetilde{\mathcal{J}(T)}(\varepsilon)$ and $r(\varepsilon) > 0$ be given by Theorem 28.3.3. In view of the ergodic case of Birkhoff's ergodic theorem (Corollary 8.2.14 or 8.2.15), there exists a measurable set $\widetilde{F}(\varepsilon) \subseteq \widetilde{\mathcal{J}(T)}(\varepsilon)$ such that $\widetilde{\mu}(\widetilde{F}(\varepsilon)) = \widetilde{\mu}(\widetilde{\mathcal{J}(T)}(\varepsilon)) \geq 1 - \varepsilon$ and

$$\lim_{n\to\infty}\frac{1}{n}\sum_{j=0}^{n-1}\mathbb{1}_{\widetilde{\mathcal{J}(T)}(\varepsilon)}\circ\tilde{T}^j(\tilde{x})=\tilde{\mu}(\widetilde{\mathcal{J}(T)}(\varepsilon)),\quad\forall\tilde{x}\in\tilde{F}(\varepsilon).$$

Let $F(\varepsilon)=\pi_0(\tilde{F}(\varepsilon))$. Then

$$\mu(F(\varepsilon))=\tilde{\mu}(\pi_0^{-1}(F(\varepsilon)))\geq\tilde{\mu}(\tilde{F}(\varepsilon))\geq 1-\varepsilon.$$

Consider a Borel set $X\subseteq\mathcal{J}(T)$ with $\mu(X)=1$. Take any point $x\in F(\varepsilon)\cap X$ and fix $\tilde{x}\in\tilde{F}(\varepsilon)$ such that $x=\pi_0(\tilde{x})$. By the above, there then exists a strictly increasing sequence $(n_k=n_k(x))_{k=1}^{\infty}$ such that

$$\tilde{T}^{n_k}(\tilde{x})\in\widetilde{\mathcal{J}(T)}(\varepsilon)\quad\text{and}\quad\frac{n_{k+1}-n_k}{n_k}\leq\varepsilon,\quad\forall k\in\mathbb{N}.\tag{29.60}$$

Moreover, Theorem 28.3.3 proclaims the existence of holomorphic inverse branches $T_x^{-n_k}:B(T^{n_k}(x),r(\varepsilon))\to\widehat{\mathbb{C}}$ of T^{n_k} such that $T_x^{-n_k}(T^{n_k}(x))=x$ and

$$T_x^{-n_k}(B(T^{n_k}(x),r(\varepsilon)))\supseteq B(x,K(\varepsilon)^{-1}|(T_x^{-n_k})'(T^{n_k}(x))|r(\varepsilon))=B(x,r_k),\tag{29.61}$$

where $r_k=r_k(x):=K(\varepsilon)^{-1}|(T^{n_k})'(x)|^{-1}r(\varepsilon)$. Take any $0<r\leq r_1$ and let $k\in\mathbb{N}$ be the unique integer such that $r_{k+1}<r\leq r_k$. Using the t-conformality of m and Theorem 28.3.3, we get

$$\begin{aligned}
m(B(x,r))&\leq m(B(x,r_k))\\
&\leq m(T_x^{-n_k}(B(T^{n_k}(x),r(\varepsilon))))\quad\text{by (29.61)}\\
&=\int_{B(T^{n_k}(x),r(\varepsilon))}|(T_x^{-n_k})'|^t\,dm\quad\text{by the t-conformality of m}\\
&\leq K^t|(T_x^{-n_k})'(T^{n_k}(x))|^t m(B(T^{n_k}(x),r(\varepsilon)))\quad\text{by the Koebe distortion theorem}\\
&\leq K(\varepsilon)^t|(T^{n_k})'(x)|^{-t}\quad\text{as } K\leq K(\varepsilon)\\
&\leq K(\varepsilon)^{2t}r(\varepsilon)^{-t}r_k^t\quad\text{from definition of } r_k.
\end{aligned}\tag{29.62}$$

We seek an upper bound in terms of r rather than r_k, and thus we need to bound r_k from above in terms of r, which can be accomplished by comparing r_k with r_{k+1}. Using successively Theorem 28.3.3, (29.60) and the choice of ε, we obtain the intermediate estimate

$$\begin{aligned}
r_k&=r_{k+1}|(T^{n_{k+1}-n_k})'(T^{n_k}(x))|\\
&=r_{k+1}|(T_{T^{n_k}(x)}^{-(n_{k+1}-n_k)})'(T^{n_{k+1}}(x))|^{-1}\\
&\leq r_{k+1}K(\varepsilon)\exp((n_{k+1}-n_k)[\chi_\mu(T)+\varepsilon])\quad\text{by Theorem 28.3.3}\\
&\leq rK(\varepsilon)\exp(n_k\varepsilon[\chi_\mu(T)+\varepsilon])\quad\text{by (29.60) and } r_{k+1}<r\\
&\leq rK(\varepsilon)\exp(2n_k\varepsilon[\chi_\mu(T)-\varepsilon])\quad\text{since } \varepsilon<\chi_\mu(T)/3.
\end{aligned}\tag{29.63}$$

It remains to bound n_k from above in terms of r_k. By Theorem 28.3.3,

$$r_k := K(\varepsilon)^{-1}\left|(T^{n_k})'(x)\right|^{-1}r(\varepsilon) = K(\varepsilon)^{-1}\left|(T_x^{-n_k})'(T^{n_k}(x))\right|r(\varepsilon) \le \exp(-n_k[\chi_\mu(T) - \varepsilon])r(\varepsilon).$$

So

$$\exp(2n_k\varepsilon[\chi_\mu(T) - \varepsilon]) \le \left(r_k^{-1}r(\varepsilon)\right)^{2\varepsilon}.$$

Substituting this into (29.63) yields

$$r_k \le rK(\varepsilon)\left(r_k^{-1}r(\varepsilon)\right)^{2\varepsilon}.$$

Equivalently,

$$r_k^{1+2\varepsilon} \le rK(\varepsilon)r(\varepsilon)^{2\varepsilon}.$$

Replacing that into (29.62) gives

$$m(B(x,r)) \le K(\varepsilon)^{2t}r(\varepsilon)^{-t}\left(rK(\varepsilon)r(\varepsilon)^{2\varepsilon}\right)^{t/(1+2\varepsilon)}.$$

Hence,

$$\limsup_{r\to 0}\frac{m(B(x,r))}{r^{t/(1+2\varepsilon)}} \le K(\varepsilon)^{2t+t/(1+2\varepsilon)}r(\varepsilon)^{2\varepsilon t/(1+2\varepsilon)-t}.$$

This holds for all $x \in F(\varepsilon) \cap X$. It follows from Theorem 15.5.3(b) (Frostman converse theorem) that

$$H_{t/(1+2\varepsilon)}(F(\varepsilon) \cap X) \ge 2^{-t/(1+2\varepsilon)}K(\varepsilon)^{-2t-t/(1+2\varepsilon)}r(\varepsilon)^{-2\varepsilon t/(1+2\varepsilon)+t} > 0,$$

whence

$$\mathrm{HD}(X) \ge \mathrm{HD}(F(\varepsilon) \cap X) \ge \frac{t}{1+2\varepsilon}.$$

Letting $\varepsilon \to 0$ results in $\mathrm{HD}(X) \ge t$. Taking the infimum over all Borel sets $X \subseteq \mathcal{J}(T)$ for which $\mu(X) = 1$, we conclude that $\mathrm{HD}(\mu) \ge t$. This proves Theorem 29.3.1. □

In higher dimensions for smooth manifolds with m as the Lebesgue measure (Riemannian volume), the analog of Theorem 29.3.1 is usually called Pesin's formula (see [103], cf. [82]).

The next result is a converse to Theorem 29.3.1. We will not prove it in this book. We refer the reader to [78] and [42].

Theorem 29.3.2. *Let* $T : \widehat{\mathbb{C}} \to \widehat{\mathbb{C}}$ *be a rational function with* $\deg(T) \ge 2$. *If* m *is a* t*-conformal measure for* T *and* μ *is an ergodic* T*-invariant Borel probability measure on*

$\mathcal{J}(T)$ such that $\chi_\mu(T) > 0$ and $\mathrm{HD}(\mu) \geq t$, then μ is absolutely continuous with respect to m. Moreover, the Radon–Nikodym derivative $d\mu/dm$ is bounded away from 0. In particular, μ is the unique measure satisfying these conditions.

29.4 Exercises

Exercise 29.4.1. Find examples of continuous self-maps $T : X \to X$ for which $X_*(T) \cap \mathrm{Crit}(T) \neq \emptyset$.

Exercise 29.4.2. Let $T : \widehat{\mathbb{C}} \to \widehat{\mathbb{C}}$ be a rational function with $\deg(T) \geq 2$. Show that the set $\mathrm{Safe}(T)$ is backward T-invariant, i. e. $T^{-1}(\mathrm{Safe}(T)) \subseteq \mathrm{Safe}(T)$.

Exercise 29.4.3. In this exercise, we outline the original construction of conformal measures given by Dennis Sullivan in [127] and [129].

Let $T : \widehat{\mathbb{C}} \to \widehat{\mathbb{C}}$ be a rational function with $\deg(T) \geq 2$. If $\mathcal{J}(T) = \widehat{\mathbb{C}}$, then the Lebesgue measure (i. e., the Riemannian volume) generated by the spherical metric is a 2-conformal measure. If $\mathcal{J}(T) \neq \widehat{\mathbb{C}}$, then proceed as follows:

(a) Prove that there exists a connected component V of $\mathcal{F}(T)$ which is neither a Siegel disk nor a Herman ring.

(b) Show that $V \setminus \overline{\mathrm{Crit}(T) \cup \mathrm{PC}(T)} \neq \emptyset$.

(c) Pick any $\xi \in V \setminus \overline{\mathrm{Crit}(T) \cup \mathrm{PC}(T)}$. Carry out the limit construction of Section 29.2 with this point ξ. Show that all resulting PS limit measures are supported on $\mathcal{J}(T)$ and that all of them are conformal with the resulting transition parameter. (Note that this construction gives no a priori information about the minimal exponent of conformal measures.)

Exercise 29.4.4. Let $T : \widehat{\mathbb{C}} \to \widehat{\mathbb{C}}$ be a rational function with $\deg(T) \geq 2$. Show that for every $t \geq 0$ the set $\mathrm{CM}(T,t)$ of all t-conformal measures for T is a convex and compact subset of the set $M(\mathcal{J}(T))$ of all Borel probability measures on $\mathcal{J}(T)$ endowed with the weak* topology.

Exercise 29.4.5. Let $T : \widehat{\mathbb{C}} \to \widehat{\mathbb{C}}$ be a rational function with $\deg(T) \geq 2$.

(a) Show that for every $t \geq 0$ the set $\mathrm{CM}_e(T,t)$ of all ergodic t-conformal measures for T coincides with the set of all extreme points of $\mathrm{CM}(T,t)$.

(b) By applying the Krein–Milman theorem, deduce that $\mathrm{CM}_e(T,t) \neq \emptyset$ whenever $\mathrm{CM}(T,t) \neq \emptyset$.

(c) By means of Theorem 29.2.12, conclude that there exists an ergodic $\delta(T)$-conformal measure for T.

Exercise 29.4.6. Let $T : \widehat{\mathbb{C}} \to \widehat{\mathbb{C}}$ be a rational function with $\deg(T) \geq 2$. Prove that if $t \geq 0$ and m is a t-conformal measure for T such that $m(\mathcal{J}_r(T)) = 1$, then $t = \delta(T)$.

Exercise 29.4.7. Let $T : \widehat{\mathbb{C}} \to \widehat{\mathbb{C}}$ be a rational function with $\deg(T) \geq 2$.

(a) Show that any two $\delta(T)$-conformal measures supported on $\mathcal{J}_r(T)$ are equivalent.
(b) Using Exercise 29.4.5, prove that there exists at most one $\delta(T)$-conformal measure supported on $\mathcal{J}_r(T)$.

Exercise 29.4.8. Let $T : \widehat{\mathbb{C}} \to \widehat{\mathbb{C}}$ be a rational function and suppose that m is a (Sullivan) t-conformal measure for T.
(a) Prove that m has full topological support, i. e., $\operatorname{supp}(m) = \mathcal{J}(T)$.
(b) Let $n \in \mathbb{N}$. Show that m is t-conformal for T^n, i. e.

$$m(T^n(A)) = \int_A |(T^n)'|^t \, dm$$

for every Borel set $A \subseteq \mathcal{J}(T^n) = \mathcal{J}(T)$ such that $T^n|_A$ is injective.
(c) Assume that $x \in \mathcal{J}(T)$, $r > 0$ and $n \in \mathbb{N}$ are such that there exists an holomorphic inverse branch $T_x^{-n} : B(T^n(x), r) \to \widehat{\mathbb{C}}$ of T^n such that $T_x^{-n}(T^n(x)) = x$. Show that

$$m(T_x^{-n}(A)) = \int_A |(T_x^{-n})'|^t \, dm$$

for every Borel set $A \subseteq B(T^n(x), r)$.

30 Conformal measures, invariant measures and fractal geometry of expanding rational functions

30.1 Fundamental fractal geometry: Bowen's formula

From its very definition (see Definition 25.3.3 and results 26.1.1–26.1.2), every expanding rational function $T : \widehat{\mathbb{C}} \to \widehat{\mathbb{C}}$ restricted to its Julia set $\mathcal{J}(T)$ is a topologically exact conformal expanding repeller and in particular an open distance expanding map. Therefore, the results established in Chapters 13 and 16 hold for all expanding rational functions. According to Remark 27.9.1, the results from Chapter 27 apply as well. Employing those and results from Chapter 29, we easily obtain the following characteristics of the pressure function.

Theorem 30.1.1 (Fundamental fractal geometry for expanding rational functions). *If $T : \widehat{\mathbb{C}} \to \widehat{\mathbb{C}}$ is an expanding rational function with $\deg(T) \geq 2$, then the following statements hold:*

(a) *The pressure function $\mathbb{R} \ni t \longmapsto \mathrm{P}(t) := \mathrm{P}(T|_{\mathcal{J}(T)}, -t\log|T'|) \in \mathbb{R}$ is convex, Lipschitz continuous, real-analytic, strictly decreasing and*

$$\lim_{t\to-\infty} \mathrm{P}(t) = \infty \quad \text{while} \quad \lim_{t\to\infty} \mathrm{P}(t) = -\infty.$$

(b) $\mathrm{P}(0) = \mathrm{h_{top}}(T) = \log(\deg(T))$.

(c) (Bowen's formula) *The pressure function $t \longmapsto \mathrm{P}(t)$ has a unique zero $h_T > 0$ and*

$$\begin{aligned} h_T = \delta(T) &= \mathrm{ExpD}^{**}(\mathcal{J}(T)) = \mathrm{ExpD}^{*}(\mathcal{J}(T)) = \mathrm{ExpD}(\mathcal{J}(T)) = \mathrm{DD}(\mathcal{J}(T)) \\ &= \mathrm{HD}(\mathcal{J}_{\mathrm{uer}}(T)) = \mathrm{HD}(\mathcal{J}_{\mathrm{er}}(T)) = \mathrm{HD}(\mathcal{J}_{\mathrm{r}}(T)) \\ &= \mathrm{HD}(\mathcal{J}(T)) = \mathrm{PD}(\mathcal{J}(T)) = \mathrm{BD}(\mathcal{J}(T)) \in \left(\frac{p_T}{p_T+1}, 2\right). \end{aligned}$$

(d) *$m_{h_T} := m_{-h_T\log|T'|}$ is the only Sullivan conformal measure for T and its exponent is h_T.*

(e) *Both m_{h_T} and $\mu_{h_T} := \mu_{-h_T\log|T'|}$ (the only T-invariant Gibbs state for the potential $-h_T\log|T'|$) are geometric measures with exponent h_T.*

(f) *The measures m_{h_T}, Hausdorff $H_{h_T}|_{\mathcal{J}(T)}$ and packing $P_{h_T}|_{\mathcal{J}(T)}$, are finite and positive, and are equal up to a multiplicative constant.*

Proof.

(a) The real-analyticity of the pressure function ensues from Corollary 13.10.4. The rest of (a) is a direct consequence of Proposition 16.3.1.

(b) The first equality is by definition of the topological entropy and pressure, while the second equality is Theorem 27.8.4.

(c) The existence and uniqueness of a zero is a direct consequence of item (a). The long chain of equalities follows from Theorems 29.2.12 and 16.3.2, and (d) below.

https://doi.org/10.1515/9783110769876-030

The inequality $\delta(T) > p_T/(p_T + 1)$ is the object of Theorem 29.2.14. It only remains to show that $h_T < 2$, which will be accomplished after (f).

(d) As $\mathrm{P}(h_T) = 0$, item (d) follows from the definition of the measure m_{h_T} (see Proposition 13.6.9, Corollary 13.6.10 and Proposition 13.8.10).

(e, f) Both items ensue from Theorem 16.3.2.

(c) Proof of $h_T < 2$: Exercise 30.3.1 affirms that $\mathcal{J}(T) \neq \widehat{\mathbb{C}}$. Theorem 29.2.13 then states that $\mathrm{Leb}_2(\mathcal{J}(T)) = 0$. Since the measure H_2 is a constant multiple of Leb_2, we conclude from item (f) that $h_T < 2$. □

A sketch of the graph of the pressure function $\mathrm{P}(t)$ is given in Figure 30.1.

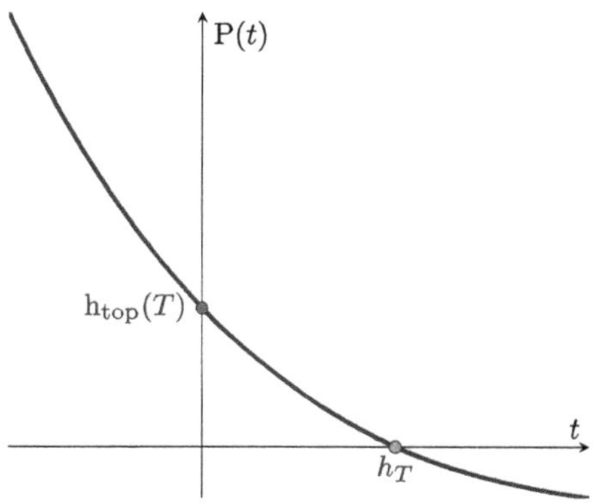

Figure 30.1: Graph of the pressure function $t \mapsto \mathrm{P}(t)$ for an expanding rational function.

Bowen's formula states that the Hausdorff (as well as the packing, box, dynamical and other) dimension of the Julia set of an expanding rational function is equal to the unique zero of the natural pressure function.

The Hausdorff dimension of the Julia set gives us an idea of the global fractal structure of the Julia set. More precisely, it describes the complexity there is in covering the Julia set at infinitesimal scales.

Nevertheless, two fractal sets (for instance, Julia sets) with similar global structures may have substantially different local structures. The pointwise or local dimension of a geometric measure contributes to capturing the local structure of a set, and potentially helps in distinguishing that set from another one which may be globally similar (for instance, for distinguishing two Julia sets sharing the same Hausdorff dimension).

Multifractal analysis of the Julia set of a rational function consists in:

(1) determining the range of local dimensions that the Julia set admits, and
(2) for each value in that range, determining the size of the subset of Julia points where the local dimension is equal to that value.

The resulting map is called multifractal spectrum of the Julia set, as it describes the relative sizes of a multitude of Julia subsets which form a splitting of the Julia set.

Formally, to avoid any confusion with the temperature function that will naturally be denoted by T, the rational function will be denoted by R temporarily. So let $R : \widehat{\mathbb{C}} \to \widehat{\mathbb{C}}$ be a rational function and ν be a Borel probability measure on $\mathcal{J}(R)$. The measure ν is said to have *pointwise or local dimension* α at a Julia point $z \in \mathcal{J}(R)$ if (cf. Definition 15.6.2)

$$d_\nu(z) := \lim_{r\to 0} \frac{\log \nu(B(z,r))}{\log r} = \alpha.$$

For each number $0 \le \alpha \le \infty$, let

$$\mathcal{J}(R)_\nu(\alpha) := \{z \in \mathcal{J}(R) \,|\, d_\nu(z) = \alpha\}$$

be the subset of all Julia points $z \in \mathcal{J}(R)$ where ν has local dimension α.

The domain of the pointwise dimension (function) d_ν, namely the set

$$\mathcal{J}(R)_\nu := \bigcup_{0\le\alpha\le\infty} \mathcal{J}(R)_\nu(\alpha),$$

is called the *regular part* of $\mathcal{J}(R)$.

Its complement in the Julia set

$$\mathcal{J}(R)_\nu^c := \mathcal{J}(R) \setminus \mathcal{J}(R)_\nu = \{z \in \mathcal{J}(R) \,|\, d_\nu(z) \nexists\},$$

i. e., the subset of all Julia points $z \in \mathcal{J}(R)$ where the measure ν does not have a local dimension, is called the *singular part* of $\mathcal{J}(R)$.

The decomposition (i. e., the partition) of the Julia set $\mathcal{J}(R)$ as

$$\mathcal{J}(R) = \bigcup_{0\le\alpha\le\infty} \mathcal{J}(R)_\nu(\alpha) \cup \mathcal{J}(R)_\nu^c$$

is called *multifractal decomposition of $\mathcal{J}(R)$ with respect to the pointwise dimension induced by ν.*

Let

$$F_\nu(\alpha) = \mathrm{HD}(\mathcal{J}(R)_\nu(\alpha))$$

be the Hausdorff dimension of the Julia subset $\mathcal{J}(R)_\nu(\alpha)$. The function

$$\alpha \longmapsto F_\nu(\alpha),$$

whose domain is $\{0 \le \alpha \le \infty : \mathcal{J}(R)_\nu(\alpha) \ne \emptyset\}$, is called *(fine Hausdorff) multifractal spectrum* of the measure ν.

Returning to the pointwise dimension spectrum, a priori there is no reason for the function F_ν to behave nicely. Its domain of definition obviously depends on ν.

If ν is a geometric measure (see Definition 15.6.12), then the domain of F_ν is a singleton, namely, the exponent t of the geometric measure. Its range is the singleton $h_R = \mathrm{HD}(\mathcal{J}(R))$. For instance, by Theorem 30.1.1 we deduce that for the geometric measures $m_{h_R} := m_{-h_R \log|R'|}$ and $\mu_{h_R} := \mu_{-h_R \log|R'|}$, the domains and the ranges of $F_{m_{h_R}}$ and $F_{\mu_{h_R}}$ are all $\{h_R\}$.

If ν is an ergodic R-invariant Borel probability measure on $\mathcal{J}(R)$, such as μ_{h_R}, then the volume lemma (Theorem 28.4.1) states that $d_\nu(z) = \alpha_0$ for ν-a. e. $z \in \mathcal{J}(R)$, where $\alpha_0 = \mathrm{HD}(\nu)$. In this case, we know that the domain of F_ν is nonempty. However, for any $\alpha \neq \alpha_0$ the set $\mathcal{J}(R)_\nu(\alpha)$ is not visible to the measure ν since $\nu(\mathcal{J}(R)_\nu(\alpha)) = 0$. Whereas the Hentschel–Procaccia spectrum $\mathrm{HP}_q(\nu)$ (see Section 16.6) can be determined by statistical properties of ν-typical (a. e.) trajectories, the function $F_\nu(\alpha)$ seems intractable.

Nevertheless, if $\nu = \mu_\varphi$ is the unique R-invariant Gibbs state and unique equilibrium state associated with a Hölder continuous potential φ on a neighborhood U of $\mathcal{J}(R)$, then miraculously the dimension spectra $\alpha \mapsto F_{\mu_\varphi}(\alpha)$ and $q \mapsto \mathrm{HP}_q(\mu_\varphi)$ happen to be real-analytic functions and $-F_{\mu_\varphi}(-p)$ and $\mathrm{HP}_q(\mu_\varphi)$ are mutual Legendre transforms. Indeed, the multifractal analysis of Gibbs states for conformal expanding repellers developed in Section 16.6 applies to expanding rational functions.

Accordingly, let $\varphi : U \to \mathbb{R}$ be a Hölder continuous potential on some neighbourhood U of $\mathcal{J}(R)$, and per Proposition 13.7.12, let μ_φ be its associated, unique R-invariant Gibbs state and unique equilibrium state. For $q, t \in \mathbb{R}$, consider the two-parameter family of auxiliary Hölder continuous potentials $\varphi_{q,t} : \mathcal{J}(R) \to \mathbb{R}$ defined by

$$\varphi_{q,t} = q\varphi - t \log|R'|.$$

The unique R-invariant Gibbs state and unique equilibrium state for the potential $\varphi_{q,t}$, normally denoted by $\mu_{\varphi_{q,t}}$, will be abbreviated to $\mu_{q,t}$. Lemma 16.6.1 then proclaims the following.

Lemma 30.1.2. *Let $R : \widehat{\mathbb{C}} \to \widehat{\mathbb{C}}$ be an expanding rational function with* $\deg(R) \geq 2$ *and $\varphi : U \to \mathbb{R}$ be a Hölder continuous potential on some neighborhood U of $\mathcal{J}(R)$. For every $q \in \mathbb{R}$, the pressure function* $\mathbb{R} \ni t \mapsto \mathrm{P}(q,t) := \mathrm{P}(R, \varphi_{q,t}) = \mathrm{P}(R, q\varphi - t\log|R'|) \in \mathbb{R}$ *possesses the following properties:*

(a) *It is Lipschitz with Lipschitz constant* $L := \log\|R'\|_\infty$, *where* $\|R'\|_\infty := \sup_{z\in\widehat{\mathbb{C}}} |R'(z)|$.
(b) *It is strictly decreasing.*
(c) $\lim_{t\to-\infty} \mathrm{P}(q,t) = \infty$ *whereas* $\lim_{t\to\infty} \mathrm{P}(q,t) = -\infty$.
(d) $\mathrm{P}(q,0) = \mathrm{P}(R, q\varphi)$.
(e) *There is a unique $T(q) \in \mathbb{R}$ such that* $\mathrm{P}(q, T(q)) = 0$.
The resulting function $\mathbb{R} \ni q \mapsto T(q) \in \mathbb{R}$ is called the temperature function.
Moreover:
(f) *The two-variable pressure function* $(q,t) \mapsto \mathrm{P}(q,t)$ *is convex.*

The unique R-invariant Gibbs state and unique equilibrium state for the potential $\varphi_{q,T(q)}$, normally denoted by $\mu_{\varphi_{q,T(q)}}$ or $\mu_{q,T(q)}$, will be further abbreviated to μ_q. The relation between the measures μ_q, $q \in \mathbb{R}$, and the measure μ_φ is explained in the following special case of Theorem 16.6.5.

Theorem 30.1.3. *Let $R : \widehat{\mathbb{C}} \to \widehat{\mathbb{C}}$ be an expanding rational function with $\deg(R) \geq 2$ and $\varphi : U \to \mathbb{R}$ be a Hölder continuous potential on some neighborhood U of $\mathcal{J}(R)$ and such that $\mathrm{P}(R, \varphi) = 0$.*

(a) *For μ_φ-a. e. $z \in \mathcal{J}(R)$,*

$$d_{\mu_\varphi}(z) = \frac{-\mu_\varphi(\varphi)}{\chi_{\mu_\varphi}(R)} = \frac{\mathrm{h}_{\mu_\varphi}(R)}{\chi_{\mu_\varphi}(R)} = \mathrm{HD}(\mu_\varphi) = \mathrm{PD}(\mu_\varphi).$$

(b) *The temperature function $\mathbb{R} \ni q \mapsto T(q) \in \mathbb{R}$ is real analytic, and satisfies*

$$T(0) = \mathrm{HD}(\mathcal{J}(R)), \quad T(1) = 0, \quad T'(q) = \frac{\mu_q(\varphi)}{\chi_{\mu_q}(R)} \leq \frac{-\mathrm{h}_{\mu_q}(R)}{\chi_{\mu_q}(R)} < 0, \quad \textit{and} \quad T''(q) \geq 0,$$

where μ_q is the R-invariant Gibbs measure for the potential $\varphi_{q,T(q)}$.

(c) *For all $q \in \mathbb{R}$, we have $\mu_q(\mathcal{J}(R)_{\mu_\varphi}(-T'(q))) = 1$ and $\mathrm{HD}(\mu_q) = \mathrm{HD}(\mathcal{J}(R)_{\mu_\varphi}(-T'(q)))$.*

(d) *For every $q \in \mathbb{R}$, we have $F_{\mu_\varphi}(-T'(q)) = T(q) - qT'(q)$, i. e., $p \longmapsto -F_{\mu_\varphi}(-p)$ is the Legendre transform of $T(q)$.*
In particular, F_{μ_φ} is continuous at the boundary points $-T'(\pm\infty)$ of its domain, as the Legendre transform is. Furthermore, the sets $\mathcal{J}(R)_{\mu_\varphi}(\alpha)$ are empty for all $\alpha \notin [-T'(\infty), -T'(-\infty)]$, so these α's do not lie in the domain of F_{μ_φ}, as they do not belong to the domain of the Legendre transform. (This emptiness property is called the completeness of the F_{μ_φ}-spectrum.)
Recall that $h_R = \mathrm{HD}(\mathcal{J}(R))$. If $\mu_\varphi = \mu_{h_R} := \mu_{-h_R \log|R'|}$ (equivalently, if φ is cohomologous to h_R in the class of bounded functions on $\mathcal{J}(R)$), then $T(q)$ is affine and the domain of $F_{\mu_\varphi}(\alpha)$ is one point, namely, $h_R = -T'(q)$ for all $q \in \mathbb{R}$.
If $\mu_\varphi \neq \mu_{h_R}$, then $T''(q) > 0$ and $F''_{\mu_\varphi}(\alpha) < 0$, i. e., the functions $T(q)$ and $F_{\mu_\varphi}(\alpha)$ are respectively strictly convex on $\mathbb{R}$ and strictly concave on $[-T'(\infty), -T'(-\infty)]$, which is a bounded interval in $\mathbb{R}^+ = \{\alpha \in \mathbb{R} : \alpha > 0\}$.

(e) *For every $q \neq 1$ the HP and Rényi spectra exist (i. e., limits in their definitions exist) and $\frac{T(q)}{1-q} = \mathrm{HP}_q(\mu_\varphi) = R_q(\mu_\varphi)$. For $q = 1$, the information dimension $I(\mu_\varphi)$ exists and*

$$\lim_{q\to 1,\, q\neq 1} \frac{T(q)}{1-q} = -T'(1) = \mathrm{HD}(\mu_\varphi) = \mathrm{PD}(\mu_\varphi) = I(\mu_\varphi).$$

For more information on multifractal analysis, please see Section 16.6.

30.2 Geometric rigidity

Throughout this section, $T : \widehat{\mathbb{C}} \to \widehat{\mathbb{C}}$ is an expanding rational function whose Julia set $\mathcal{J}(T)$ is a Jordan curve. It follows from the Jordan–Schoenflies theorem that the Fatou set $\mathcal{F}(T) = \widehat{\mathbb{C}} \setminus \mathcal{J}(T)$ has exactly two connected components $A_1(T)$ and $A_2(T)$ which are both open topological disks while their closures are closed topological disks. The proof of the following theorem is straightforward. We provide it for the sake of completeness and convenience to the reader.

Theorem 30.2.1. *If $T : \widehat{\mathbb{C}} \to \widehat{\mathbb{C}}$ is an expanding rational function whose Julia set $\mathcal{J}(T)$ is a Jordan curve, then:*
(1) *either $T(A_1(T)) = A_1(T)$ and $T(A_2(T)) = A_2(T)$*
(2) *or $T(A_1(T)) = A_2(T)$ and $T(A_2(T)) = A_1(T)$.*

In either case,

$$T^2(A_1(T)) = A_1(T) \quad \text{and} \quad T^2(A_2(T)) = A_2(T).$$

In addition, both $A_1(T)$ and $A_2(T)$ are immediate basins of attraction of T to attracting periodic points of minimal period either 1 *or* 2*. Denote those periodic points by $a_1(T)$ and $a_2(T)$, respectively.*

Proof. We first prove items (1) and (2). Suppose first that $T(A_1(T)) = A_1(T)$. If $T(A_2(T)) = A_1(T)$, then $T^{-1}(A_2(T)) = \emptyset$, hence contradicting the surjectivity of $T : \widehat{\mathbb{C}} \to \widehat{\mathbb{C}}$. Thus, $T(A_2(T)) = A_2(T)$ and we are done in this case. So, suppose that $T(A_1(T)) = A_2(T)$. It then follows from the just considered case (with A_1 replaced by A_2) that $T(A_2(T)) = A_1(T)$. Items (1) and (2) are therefore proved.

The remaining assertions follow from items (1)–(2) and Theorem 26.1.6. □

We can now prove the main result of this section. The idea behind the proof goes back to [19] and [127]. Since then, more theorems of this kind and their generalizations have been proved; for instance, see [137] for parabolic rational functions and [72, 142] for class $\mathcal{S}$ meromorphic functions.

Theorem 30.2.2 (Geometric rigidity for expanding rational functions). *If $T : \widehat{\mathbb{C}} \to \widehat{\mathbb{C}}$ is an expanding rational function whose Julia set $\mathcal{J}(T)$ is a Jordan curve, then either $\mathcal{J}(T)$ is a geometric circle or $\mathrm{HD}(\mathcal{J}(T)) > 1$.*

Proof. Suppose that $\mathrm{HD}(\mathcal{J}(T)) \leq 1$. We are to show that $\mathcal{J}(T)$ is a geometric circle. Since $\mathcal{J}(T)$ is connected, it holds that $\mathrm{HD}(\mathcal{J}(T)) \geq 1$ and consequently $\mathrm{HD}(\mathcal{J}(T)) = 1$. It follows from Theorem 30.1.1(f) that the Hausdorff measure $H_1(\mathcal{J}(T))$ is finite. Let $\mathbb{D}$ be the open unit disk and let $\mathbb{S}^1 := \partial\mathbb{D}$ be the unit circle. By the Riemann mapping theorem, there are two conformal homeomorphisms $R_1 : \mathbb{D} \to A_1(T)$ and $R_2 : \widehat{\mathbb{C}} \setminus \overline{\mathbb{D}} \to A_2(T)$ such that

$$R_1(0) = a_1(T) \quad \text{and} \quad R_2(\infty) = a_2(T).$$

Consider the conjugations of T^2

$$g_1 := R_1^{-1} \circ T^2 \circ R_1 : \mathbb{D} \to \mathbb{D} \quad \text{and} \quad g_2 := R_2^{-1} \circ T^2 \circ R_2 : \widehat{\mathbb{C}} \setminus \overline{\mathbb{D}} \to \widehat{\mathbb{C}} \setminus \overline{\mathbb{D}}. \tag{30.1}$$

By the choice of R_1 and R_2, these conjugate maps fix 0 and ∞, respectively:

$$g_1(0) = 0 \quad \text{and} \quad g_2(\infty) = (\infty). \tag{30.2}$$

As the boundary of $A_1(T)$ and $A_2(T)$ is the Jordan curve $\mathcal{J}(T)$, Carathéodory's extension theorem (Theorem A.3.15) asserts that the Riemann conformal homeomorphisms R_1 and R_2 extend to homeomorphisms, denoted by the same symbols R_1 and R_2, of the closed disks $\overline{\mathbb{D}}$ and $\widehat{\mathbb{C}} \setminus \mathbb{D}$, respectively. This implies that g_1 and g_2 have continuous extensions to $\overline{\mathbb{D}}$ and $\widehat{\mathbb{C}} \setminus \mathbb{D}$, respectively. Keeping the same notation g_1 and g_2 for these extensions, we have in particular that

$$g_1(\mathbb{S}^1) = \mathbb{S}^1 \quad \text{and} \quad g_2(\mathbb{S}^1) = \mathbb{S}^1.$$

It ensues from Schwarz' reflection principle (also called Schwarz' symmetry principle; see the complex-analytic version in Theorem A.3.6) that both g_1 and g_2 uniquely extend to meromorphic functions of the Riemann sphere $\widehat{\mathbb{C}}$. Thus, both of these extensions are rational functions. We still keep the same notation g_1 and g_2 for them. It immediately follows from the Schwarz' reflection that

$$g_1(\infty) = \infty \quad \text{and} \quad g_2(0) = (0). \tag{30.3}$$

By (30.2)–(30.3) and Theorem 30.2.1, the points 0 and ∞ are attracting fixed points for both g_1 and g_2, and $\mathbb{D}$ and $\widehat{\mathbb{C}} \setminus \overline{\mathbb{D}}$ are their immediate basins of attraction, respectively. Hence, $\mathcal{F}(g_1) = \mathcal{F}(g_2) = \widehat{\mathbb{C}} \setminus \mathbb{S}^1$, so $\mathcal{J}(g_1) = \mathcal{J}(g_2) = \mathbb{S}^1$, and it follows from Theorem 26.1.3 that both rational functions g_1 and g_2 are expanding.

Let ℓ be the normalized Lebesgue measure on $\mathbb{S}^1$. Of course, ℓ is a 1-conformal measure for both g_1 and g_2. There then exist two (ergodic) invariant Gibbs states $\mu_{-\log|g_1'|}$ and $\mu_{-\log|g_2'|}$ for g_1 and g_2, respectively, both equivalent to ℓ. This actually would suffice for us but we can do better; we can precisely say what these measures are. Indeed, for every continuous function $u : \mathbb{S}^1 \to \mathbb{R}$ let $\tilde{u} : \mathbb{D} \to \mathbb{R}$ be the unique harmonic function (given by the Poisson kernel formula) that has a continuous extension to $\overline{\mathbb{D}}$ (and whose restriction to $\mathbb{S}^1$ is thus equal to u). Then for both $i = 1, 2$, the function $\tilde{u} \circ g_i : \overline{\mathbb{D}} \to \mathbb{R}$ is continuous and its restriction to $\mathbb{D}$ is harmonic. Applying twice the mean value theorem for harmonic functions, we get

$$\int_{\mathbb{S}^1} u \circ g_i \, d\ell = \tilde{u} \circ g_i(0) = \tilde{u}(0) = \int_{\mathbb{S}^1} u \, d\ell.$$

This means that both maps $R_1|_{\mathbb{S}^1} : \mathbb{S}^1 \to \mathbb{S}^1$ and $R_2|_{\mathbb{S}^1} : \mathbb{S}^1 \to \mathbb{S}^1$ preserve the Lebesgue measure on $\mathbb{S}^1$, i. e.,

$$\ell(g_1^{-1}(F)) = \ell(F) \quad \text{and} \quad \ell(g_2^{-1}(F)) = \ell(F)$$

for every Borel set $F \subseteq \mathbb{S}^1$. In consequence,

$$\mu_{-\log|g_1'|} = \ell = \mu_{-\log|g_2'|}. \tag{30.4}$$

Since $H_1(\mathcal{J}(T)) < \infty$, Riesz's theorem (Theorem A.2.4) yields that the homeomorphisms $R_1|_{\mathbb{S}^1}$ and $R_2|_{\mathbb{S}^1}$, as well as their inverses, are absolutely continuous with respect to ℓ on $\mathbb{S}^1$ and with respect to H_1 on $\mathcal{J}(T)$. Denoting by μ the T-invariant Gibbs state for the potential $-\log|(T^2)'|$, which is equivalent to $H_1|_{\mathcal{J}(T)}$, and applying Riesz's theorem again we see that both measures $\mu \circ R_1$ and $\mu \circ R_2$ are absolutely continuous with respect to ℓ on $\mathbb{S}^1$ and

$$\mu \circ R_i \circ g_i^{-1} = \mu \circ R_i \circ R_i^{-1} \circ T^{-2} \circ R_i = \mu \circ T^{-2} \circ R_i = \mu \circ R_i$$

for both $i = 1, 2$. This means that each measure $\mu \circ R_i$ is g_i-invariant. By (30.4), the g_i-invariant measure ℓ is ergodic (as a Gibbs state). Hence, all g_i-invariant measures absolutely continuous with respect to ℓ must be equal by Theorem 8.2.22. So $\mu \circ R_1 = \ell = \mu \circ R_2$. Consider the composite map $R_2^{-1} \circ R_1 : \mathbb{S}^1 \to \mathbb{S}^1$. Then

$$\ell \circ (R_2^{-1} \circ R_1)^{-1} = \ell \circ R_1^{-1} \circ R_2 = \mu \circ R_1 \circ R_1^{-1} \circ R_2 = \mu \circ R_2 = \ell,$$

which means that the map $R_2^{-1} \circ R_1$ preserves the measure ℓ on $\mathbb{S}^1$. Since both R_1 and R_2 are homeomorphisms of the circle $\mathbb{S}^1$, so is $R_2^{-1} \circ R_1$. As it preserves the Lebesgue measure ℓ, it is a rotation of $\mathbb{S}^1$. Thus, modifying R_1 by precomposing it with an appropriate rotation, we can make the composition $R_2^{-1} \circ R_1$ to be the identity map. Consequently, $R_1 = R_2$ on $\mathbb{S}^1$. Thus, the map defined by the formula

$$R(z) := \begin{cases} R_1(z) & \text{if } z \in \overline{\mathbb{D}} \\ R_2(z) & \text{if } z \in \widehat{\mathbb{C}} \setminus \mathbb{D} \end{cases}$$

is meromorphic on $\mathbb{D} \cup (\widehat{\mathbb{C}} \setminus \overline{\mathbb{D}}) = \widehat{\mathbb{C}} \setminus \mathbb{S}^1$ and continuous on $\widehat{\mathbb{C}}$. A straightforward application of Morera's theorem shows that the map $R : \widehat{\mathbb{C}} \to \widehat{\mathbb{C}}$ is meromorphic. As R is a homeomorphism, it must be a Möbius map. Since $\mathcal{J}(T) = R_1(\mathbb{S}^1) = R(\mathbb{S}^1)$, we conclude that $\mathcal{J}(T)$ is a geometric circle. The proof of Theorem 30.2.2 is complete. □

Remark 30.2.3. In the proof of Theorem 30.2.2, we eventually showed that $g_1 = g_2$. Denote this function by g. We demonstrated that $g : \widehat{\mathbb{C}} \to \widehat{\mathbb{C}}$ is a rational function preserving $\overline{\mathbb{D}}$ and $\widehat{\mathbb{C}} \setminus \mathbb{D}$. Thus, g is a finite Blaschke product. Furthermore, (30.1) states that the maps T^2 and g are conjugate via the Möbius map R. Moreover, if we knew that the connected components $A_1(T)$ and $A_2(T)$ were fixed under T, then we would not have to pass to T^2 in the proof and we would conclude that the map T and the finite Blaschke product g are conjugate through the Möbius map R. (For more information on Blaschke products, see [53].)

We can now easily prove the following.

Theorem 30.2.4. *If $c \in \mathbb{C} \setminus \{0\}$ belongs to the main cardioid $\mathcal{M}_0$ of the Mandelbrot set $\mathcal{M}$, then $\mathrm{HD}(J_c) > 1$, where J_c is the Julia set of the quadratic polynomial $Q_c(z) = z^2 + c$.*

Proof. Using the notation from Theorem 30.2.2 and Remark 30.2.3, we have

$$Q_c \circ R = R \circ g.$$

Consequently,

$$Q_c^n \circ R = R \circ g^n, \quad \forall n \in \mathbb{N}. \tag{30.5}$$

Differentiating this equality yields

$$(Q_c^n)'(R(z))R'(z) = R'(g^n(z))(g^n)'(z), \quad \forall z \in \mathbb{C}. \tag{30.6}$$

If ξ is a repelling periodic point of Q_c of period $n \in \mathbb{N}$, then (30.5)'s conjugation requires that $R^{-1}(\xi)$ be a periodic point of g with the same period n. Substituting $R^{-1}(\xi)$ for z in (30.6) leads to

$$(Q_c^n)'(\xi) = (g^n)'(R^{-1}(\xi)).$$

Therefore, $R^{-1}(\xi)$ is a repelling periodic point of period n for the map g, whence $R^{-1}(\xi) \in \mathcal{J}(g) = \mathbb{S}^1$. Since $g(\mathbb{S}^1) = \mathbb{S}^1$, it follows that $(g^n)'(R^{-1}(\xi)) \in \mathbb{R}$. That is, $(Q_c^n)'(\xi) \in \mathbb{R}$. Looking at the repelling periodic points of Q_c of periods 1 and 2, we conclude from a direct calculation that $c \in (-\infty, -3/4]$. Hence, $c \notin \mathcal{M}_0$, contrary to our hypothesis. □

30.3 Exercises

Exercise 30.3.1. Prove that $\mathcal{J}(T) \neq \widehat{\mathbb{C}}$ for every expanding rational function $T : \widehat{\mathbb{C}} \to \widehat{\mathbb{C}}$.

Exercise 30.3.2. Let $T : \widehat{\mathbb{C}} \to \widehat{\mathbb{C}}$ be an expanding rational function and suppose that μ is an ergodic T-invariant Borel probability measure such that $\mathrm{HD}(\mu) = \mathrm{HD}(\mathcal{J}(T))$. Show that $\mu = \mu_{\mathrm{HD}(\mathcal{J}(T))} = \mu_{h_T}$.

31 Conformal measures, invariant measures and fractal geometry of parabolic rational functions

Although there are some similarities, the picture for parabolic rational functions given in this chapter is quite different from that of expanding and subexpanding ones treated in the previous and the next chapters, respectively. The potential $-t\log|T'|$ is well-defined and even Hölder continuous but, because of the presence of rationally indifferent periodic points, it lacks pressure gap. In particular, the results proved in Chapter 27 do not apply.

Nevertheless, Section 31.2 enunciates a variation of Bowen's formula, which identifies the Hausdorff dimension h_T of the Julia set $\mathcal{J}(T)$ with the smallest zero of the pressure function $\mathrm{P}(t)$. Unlike the expanding case, $\mathrm{P}(t) = 0$ for all $t > h_T$.

In Section 31.3, we show that h_T is equal to the minimal exponent for which a t-conformal measure exists. As in the subhyperbolic case (treated in Chapter 32), there exists a unique Sullivan h_T-conformal measure and that measure is atomless; furthermore, it is ergodic and conservative. Also as in the subhyperbolic case, there exist t-conformal measures for all $t > h_T$. All of them are purely atomic and are convex combinations of atomic measures supported on the backward orbits of rationally indifferent periodic points.

In Section 31.4, we establish the existence and uniqueness of an invariant measure which is absolutely continuous with respect to the h_T-conformal measure, and we provide a simple criterion for this measure's finiteness/infiniteness. In particular, this measure is finite for the quadratic polynomial $Q_{1/4}(z) = z^2 + 1/4$, whereas it is infinite for parabolic Blaschke products.

Finally, in Section 31.5, we provide a complete description of the h_T-dimensional Hausdorff and packing measures on the Julia sets of parabolic rational functions. These measures are sometimes finite, sometimes positive and sometimes infinite, exclusively depending on the value of the Hausdorff dimension h_T.

Chapter 31 combines, develops, goes beyond and improves the results established in several papers written by Manfred Denker and the first named author of this book; see [1, 37, 38, 40].

31.1 General conformal measures for expansive topological dynamical systems

In order to fully meaningfully deal with various kinds of conformal measures for expansive (especially parabolic) rational functions, it is convenient to recall the concept of transfer operator from Chapters 13 (Gibbs states and transfer operators for open, distance expanding systems) and 27 (Equilibrium states for Hölder continuous potentials with pressure gap). Remember that in this book a topological dynamical system is a con-

https://doi.org/10.1515/9783110769876-031

tinuous self-map $T : X \to X$ of a compact metrizable space X. Let $T : X \to X$ be an open, expansive topological dynamical system and let $\varphi \in C(X)$ be a continuous potential. Define the transfer operator $\mathcal{L}_\varphi$ in the standard way (cf. (13.26)):

$$\mathcal{L}_\varphi(g)(x) = \sum_{\overline{x} \in T^{-1}(x)} g(\overline{x}) e^{\varphi(\overline{x})}, \quad \forall x \in X, \ \forall g \in C(X).$$

We emphasize that in this section the potential φ is not assumed to be Hölder continuous, but merely continuous. Also, unlike in Chapter 27, we do not require φ to exhibit a pressure gap. So, we cannot really rely on most of the results in Chapters 13 and 27. Nevertheless, the fact that every expansive map is expanding with respect to a metric compatible with the topology (see Theorem 5.3.1 in the first volume) helps us establish results, starting with the following assertion (cf. Lemma 13.6.1).

Proposition 31.1.1. *If $T : X \to X$ is an open, expansive topological dynamical system and $\varphi \in C(X)$ a potential, then:*
(a) *$\mathcal{L}_\varphi$ is positive, linear and bounded.*
(b) *$\mathcal{L}_\varphi(C(X)) \subseteq C(X)$ and $\|\mathcal{L}_\varphi\|_\infty \leq \#T^{-1} \cdot e^{\sup \varphi}$.*
(c) $\mathcal{L}^n_\varphi(g)(x) = \sum_{\overline{x} \in T^{-n}(x)} g(\overline{x}) \exp(S_n\varphi(\overline{x})), \quad \forall x \in X, \ \forall g \in C(X), \ \forall n \geq 0.$

From now on, we assume that d is a metric compatible with the topology of X and with respect to which T is expanding. Accordingly, the letters δ, λ and ξ will be the constants from Definition 4.1.1 and relations (4.29)–(4.31); alternatively, see (13.1)–(13.2).

Lemma 31.1.2. *If $T : X \to X$ is an open, expansive topological dynamical system and $\varphi \in C(X)$, then for every $\varepsilon > 0$ there exists $\zeta > 0$ such that*

$$d(y,x) < \zeta \quad \Longrightarrow \quad e^{-\varepsilon n} \leq \frac{\mathcal{L}^n_\varphi(\mathbb{1})(y)}{\mathcal{L}^n_\varphi(\mathbb{1})(x)} \leq e^{\varepsilon n}, \quad \forall n \geq 0.$$

Proof. Take $\zeta \in (0, \xi)$ so small that $|\varphi(y) - \varphi(x)| < \varepsilon$ whenever $d(y,x) < \zeta$. Let $0 \leq k \leq n$. For every $\overline{x} \in T^{-n}(x)$, we then have

$$d(T^k(T^{-n}_{\overline{x}}(y)), T^k(\overline{x})) < \lambda^{-(n-k)}\zeta < \zeta.$$

Thus,

$$\begin{aligned} \mathcal{L}^n_\varphi(\mathbb{1})(y) &= \sum_{\overline{y} \in T^{-n}(y)} g(\overline{y}) \exp(S_n\varphi(\overline{y})) = \sum_{\overline{x} \in T^{-n}(x)} \exp(S_n\varphi(T^{-n}_{\overline{x}}(y))) \\ &\leq \sum_{\overline{x} \in T^{-n}(x)} \exp(S_n\varphi(\overline{x}) + \varepsilon n) = e^{\varepsilon n}\mathcal{L}^n_\varphi(\mathbb{1})(x). \end{aligned}$$

So, the right inequality is proved. Since the left inequality is equivalent to the right one, the proof is complete. □

The reader may want to compare the previous and next results with Lemma 13.7.1, as well as with Lemma 13.1.1 and Corollary 13.1.2.

Lemma 31.1.3. *If $T : X \to X$ is a topologically transitive, open, expansive dynamical system and $\varphi \in C(X)$, then for every $\varepsilon > 0$ there exists a constant $c_\varepsilon \geq 1$ such that*

$$c_\varepsilon^{-1} e^{-\varepsilon n} \leq \frac{\mathcal{L}_\varphi^n(\mathbb{1})(y)}{\mathcal{L}_\varphi^n(\mathbb{1})(x)} \leq c_\varepsilon e^{\varepsilon n}, \quad \forall x, y \in X, \ \forall n \geq 0. \tag{31.1}$$

Proof. Fix $\varepsilon > 0$ and ascribe $\zeta > 0$ to it according to Lemma 31.1.2. Let F be a finite $(\zeta/2)$-spanning set of X. According to Corollary 5.3.4 (see also Lemma 4.2.10 and Definition 1.5.15), any open, expansive and transitive dynamical system is very strongly transitive. As F is finite, this implies that there exists $M \in \mathbb{N}$ such that

$$\bigcup_{m=0}^{M} T^m(B(w, \zeta/2)) = X, \quad \forall w \in F. \tag{31.2}$$

Fix any $x, y \in X$. Since F is $(\zeta/2)$-spanning, there exists $w \in F$ such that $x \in B(w, \zeta/2)$. But then $B(x, \zeta) \supseteq B(w, \zeta/2)$, and hence (31.2) yields $0 \leq l \leq M$ such that $y \in T^l(B(x, \zeta))$. In particular, there exists $z \in B(x, \zeta)$ such that $T^l(z) = y$. Let $n \geq l$. By Lemma 31.1.2, we obtain that

$$\mathcal{L}_\varphi^n(\mathbb{1})(y) \geq \exp(S_l\varphi(z))\mathcal{L}_\varphi^{n-l}(\mathbb{1})(z) \geq \exp(-l\|\varphi\|_\infty)e^{-\varepsilon n}\mathcal{L}_\varphi^{n-l}(\mathbb{1})(x).$$

On the other hand,

$$\mathcal{L}_\varphi^n(\mathbb{1})(x) = \mathcal{L}_\varphi^{n-l}(\mathcal{L}_\varphi^l(\mathbb{1}))(x) \leq \mathcal{L}_\varphi^{n-l}(\|\mathcal{L}_\varphi^l(\mathbb{1})\|_\infty \cdot \mathbb{1})(x) = \|\mathcal{L}_\varphi^l\|_\infty \mathcal{L}_\varphi^{n-l}(\mathbb{1})(x).$$

Therefore,

$$\mathcal{L}_\varphi^n(\mathbb{1})(y) \geq \exp(-l\|\varphi\|_\infty)e^{-\varepsilon n}\|\mathcal{L}_\varphi^l\|_\infty^{-1}\mathcal{L}_\varphi^n(\mathbb{1})(x).$$

The left inequality in (31.1) is proved for all $n \geq l$. For all $n < l$, a sheer compactness-continuity argument yields that

$$0 < \min_{0 \leq n < l} \inf\left\{\frac{\mathcal{L}_\varphi^n(\mathbb{1})(y)}{\mathcal{L}_\varphi^n(\mathbb{1})(x)} : (x, y) \in X \times X\right\} < \infty.$$

By symmetry, the left inequality is equivalent to the right one. □

The following result is a partial rewriting of Proposition 13.1.3 with a slightly different proof.

Theorem 31.1.4. *If $T : X \to X$ is a topologically transitive, open, expansive dynamical system and $\varphi \in C(X)$, then*

$$\lim_{n\to\infty} \frac{1}{n} \log \mathcal{L}_\varphi^n(\mathbb{1})(x) = \mathrm{P}(\varphi), \quad \forall x \in X,$$

where $\mathrm{P}(\varphi) = \mathrm{P}(T, \varphi)$ *is the topological pressure of the system* T *under the potential* φ.

Proof. Let $x \in X$ and $n \in \mathbb{N}$. By the expandingness of T with respect to the metric d (cf. (13.1)), the set $T^{-1}(x)$ is δ-separated. Thus the set $T^{-n}(x)$ is (n, δ)-separated, and hence is contained in a maximal (n, ε)-separated set for every $0 < \varepsilon \le \delta$. By Theorem 11.2.1,

$$\limsup_{n\to\infty} \frac{1}{n} \log \mathcal{L}_\varphi^n(\mathbb{1})(x) \le \mathrm{P}(\varphi). \tag{31.3}$$

To prove the converse with liminf rather than limsup, choose for every $0 < \varepsilon < \xi$ a maximal ε-separated set $E(\varepsilon) \subseteq X$. By the expandingness of T, the set $T^{-n}(w)$ is (n, ε)-separated for every $w \in E(\varepsilon)$. Consequently, the set

$$E_n(\varepsilon) := \bigcup_{w\in E(\varepsilon)} T^{-n}(w)$$

is $(n + 1, \varepsilon)$-separated. Since the maximal ε-separated set $E(\varepsilon)$ is an ε-spanning set, we observe that

$$\begin{aligned}\bigcup_{z\in E_n(\varepsilon)} B_{n+1}(z, \varepsilon) &\supseteq \bigcup_{w\in E(\varepsilon)} \bigcup_{z\in T^{-n}(w)} T_z^{-n}(B(w, \varepsilon)) = \bigcup_{w\in E(\varepsilon)} T^{-n}(B(w, \varepsilon)) \\ &= T^{-n}\Big(\bigcup_{w\in E(\varepsilon)} B(w, \varepsilon)\Big) = T^{-n}(X) = X,\end{aligned}$$

i. e. the set $E_n(\varepsilon)$ is $(n + 1, \varepsilon)$-spanning. As it is simultaneously $(n + 1, \varepsilon)$-spanning and $(n + 1, \varepsilon)$-separated, the set $E_n(\varepsilon)$ is a maximal $(n + 1, \varepsilon)$-separated set. Fixing $\gamma > 0$, Lemma 31.1.3 then yields

$$\begin{aligned}\frac{1}{n} \log \sum_{\bar{y}\in T^{-n}(E(\varepsilon))} e^{S_n\varphi(\bar{y})} &= \frac{1}{n} \log \sum_{y\in E(\varepsilon)} \sum_{\bar{y}\in T^{-n}(y)} e^{S_n\varphi(\bar{y})} \\ &= \frac{1}{n} \log \sum_{y\in E(\varepsilon)} \mathcal{L}_\varphi^n(\mathbb{1})(y) \\ &\le \frac{1}{n} \log \sum_{y\in E(\varepsilon)} c_\gamma e^{\gamma n} \mathcal{L}_\varphi^n(\mathbb{1})(x) \\ &= \frac{1}{n} \log(c_\gamma \# E(\varepsilon)) + \gamma + \frac{1}{n} \log \mathcal{L}_\varphi^n(\mathbb{1})(x).\end{aligned}$$

Consequently,

$$\mathrm{P}(\varphi) = \lim_{\varepsilon\to 0} \liminf_{n\to\infty} \frac{1}{n} \log \sum_{\bar{y}\in T^{-n}(E(\varepsilon))} \exp(S_n\varphi(\bar{y})) \le \gamma + \liminf_{n\to\infty} \frac{1}{n} \log \mathcal{L}_\varphi^n(\mathbb{1})(x).$$

As $\gamma > 0$ is arbitrary, we deduce that

$$\mathrm{P}(\varphi) \le \liminf_{n\to\infty} \frac{1}{n} \log \mathcal{L}_\varphi^n(\mathbb{1})(x). \tag{31.4}$$

The result ensues from (31.3)–(31.4). □

In exactly the same way as in Theorem 13.6.2, i. e. by applying the Schauder–Tychonov fixed-point theorem, we obtain a Borel probability measure m_φ on X such that

$$\mathcal{L}_\varphi^*(m_\varphi) = \Lambda_\varphi m_\varphi \tag{31.5}$$

with the same $\Lambda_\varphi > 0$. According to Proposition 13.5.2, such an eigenmeasure is quasi-T-invariant (and even more). As demonstrated through Proposition 13.6.9–Corollary 13.6.11 (see also Definition 13.6.12), the measure m_φ is $(\Lambda_\varphi e^{-\varphi})$-conformal, i. e.,

$$m_\varphi(T_x^{-n}(A)) = \Lambda_\varphi^{-n} \int_A \exp((S_n\varphi)\circ T_x^{-n})\, dm_\varphi \tag{31.6}$$

for every $x \in X$, every $n \in \mathbb{N}$ and every Borel set $A \subseteq B(T^n(x), \xi)$. One can easily deduce from this that

$$m_\varphi(T^n(B)) = \Lambda_\varphi^{n} \int_B \exp(-S_n\varphi)\, dm_\varphi \tag{31.7}$$

for every $n \in \mathbb{N}$ and every Borel set $B \subseteq X$ such that $T^n|_B$ is one-to-one. In fact, it was observed in Lemma 13.6.13 that relations (31.5)–(31.7) are all equivalent.

If T is transitive, open and expansive, then it is (very) strongly transitive, which means that $\overline{\bigcup_{n=0}^\infty T^{-n}(x)} = X$ for any $x \in X$. From (31.6), we readily infer the following fact (cf. Proposition 13.6.14).

Proposition 31.1.5. *If $T : X \to X$ is a topologically transitive, open, expansive system and $\varphi \in C(X)$, then the measure m_φ is positive on nonempty open sets. In other words,* $\operatorname{supp}(m_\varphi) = X$.

According to Lemma 13.6.15, measures with full support in a compact space confer to balls of any given radius weights that are uniformly bounded away from 0.

Lemma 31.1.6. *If $T : X \to X$ is a topologically transitive, open and expansive system and $\varphi \in C(X)$, then for every $r > 0$ we have*

$$M_{m_\varphi}(r) := \inf\{m_\varphi(B(x,r)) \mid x \in X\} > 0.$$

We now identify the value of Λ_φ.

Proposition 31.1.7. *Let $T : X \to X$ be a topologically transitive, open, expansive system and $\varphi \in C(X)$. If Λ_φ satisfies* (31.5), *then*

$$\log \Lambda_\varphi = \mathrm{P}(\varphi).$$

Proof. Fix $x \in X$ and $\varepsilon > 0$. Using Lemma 31.1.3, we notice that for every $n \in \mathbb{N}$,

$$\Lambda_\varphi^n = \Lambda_\varphi^n m_\varphi(\mathbb{1}) = \mathcal{L}_\varphi^{*n}(m_\varphi)(\mathbb{1}) = \int_X \mathcal{L}_\varphi^n(\mathbb{1})\, dm_\varphi \in [c_\varepsilon^{-1} e^{-\varepsilon n} \mathcal{L}_\varphi^n(\mathbb{1})(x), c_\varepsilon e^{\varepsilon n} \mathcal{L}_\varphi^n(\mathbb{1})(x)]$$

Consequently,

$$-\varepsilon + \limsup_{n\to\infty} \frac{1}{n} \log \mathcal{L}_\varphi^n(\mathbb{1})(x) \le \log \Lambda_\varphi \le \varepsilon + \liminf_{n\to\infty} \frac{1}{n} \log \mathcal{L}_\varphi^n(\mathbb{1})(x).$$

Theorem 31.1.4 then implies that $-\varepsilon + \mathrm{P}(\varphi) \le \log \Lambda_\varphi \le \varepsilon + \mathrm{P}(\varphi)$. Letting $\varepsilon \to 0$ results in $\mathrm{P}(\varphi) = \log \Lambda_\varphi$. □

31.2 Geometric topological pressure and generalized geometric conformal measures for parabolic rational functions

Suppose that $T : \widehat{\mathbb{C}} \to \widehat{\mathbb{C}}$ is an expansive rational function. Then $T : \mathcal{J}(T) \to \mathcal{J}(T)$ is a topologically exact, open, expansive topological dynamical system, whence all the results from the previous section apply for continuous potentials on $\mathcal{J}(T)$. In this section, we will deal with the usual geometric potentials: given any $t \in \mathbb{R}$, let

$$\varphi_t := -t \log |T'| : \mathcal{J}(T) \to \mathbb{R}.$$

This potential is well defined in light of Theorem 26.2.3. Denote the operator $\mathcal{L}_{\varphi_t}$, resulting from the previous section, by $\mathcal{L}_t$ and, likewise, m_{φ_t} by m_t. (Note: Though the measure m_t in Lemma 29.2.1 and the one just produced are formally different, we will use the same notation to represent them; the context will distinguish which of these measures is being used.) Also, denote

$$\mathrm{P}(t) := \mathrm{P}(T|_{\mathcal{J}(T)}, \varphi_t) = \mathrm{P}(T|_{\mathcal{J}(T)}, -t \log |T'|),$$

i. e., $\mathrm{P}(t)$ is the topological pressure generated by the system $T : \mathcal{J}(T) \to \mathcal{J}(T)$ under the potential $\varphi_t(z) = -t \log |T'(z)|$. Formulas (31.5)–(31.7) in conjunction with Proposition 31.1.7 respectively become

$$\mathcal{L}_t^*(m_t) = e^{\mathrm{P}(t)} m_t, \tag{31.8}$$

$$m_t(T_z^{-n}(A)) = e^{-n\mathrm{P}(t)} \int_A |(T_z^{-n})'|^t \, dm_t \tag{31.9}$$

for all $z \in \mathcal{J}(T)$, all $n \in \mathbb{N}$ and all Borel sets $A \subseteq B(T^n(z), \xi)$, and

$$m_t(T^n(B)) = e^{n\mathrm{P}(t)} \int_B |(T^n)'|^t \, dm_t \tag{31.10}$$

for all $n \in \mathbb{N}$ and all special Borel sets $B \subseteq \mathcal{J}(T)$ for the map T^n, i. e., such that the map $T^n|_B$ is one-to-one. Relations (31.8)–(31.10) are all equivalent and by virtue of (31.9), the measure m_t is $e^{\mathrm{P}(t)-\varphi_t}$-conformal. Recall that it is also quasi-T-invariant.

Throughout the rest of this chapter, we deal with parabolic rational functions. Introduced in Section 26.2, they are per Definition 26.2.6 rational functions $T : \widehat{\mathbb{C}} \to \widehat{\mathbb{C}}$ whose restriction $T : \mathcal{J}(T) \to \mathcal{J}(T)$ is expansive but not expanding. Equivalently, by Theorem 26.2.7, they fulfill two conditions:

$$\mathcal{J}(T) \cap \mathrm{Crit}(T) = \emptyset \quad \text{and} \quad \Omega(T) \neq \emptyset$$

where $\Omega(T)$ is the (finite) set of all rationally indifferent periodic points of T.

In Theorem 30.1.1, we described the main properties of the natural pressure function associated with an expanding rational function. We will do the same for parabolic functions, and thereafter identify similarities and differences between them. Denote

$$h_T := \inf\{t \in \mathbb{R} \,|\, \mathrm{P}(t) = 0\}. \tag{31.11}$$

Recall from the end of Section 29.2 (see (29.57)) that

$$p_T := \max\{p(\omega) \,|\, \omega \in \Omega(T)\} \in \mathbb{N}.$$

The next result is the counterpart of Theorem 30.1.1(a–c).

Theorem 31.2.1. *If $T : \widehat{\mathbb{C}} \to \widehat{\mathbb{C}}$ is a parabolic rational function with $\deg(T) \geq 2$, then the pressure function $\mathbb{R} \ni t \mapsto \mathrm{P}(t) \in \mathbb{R}$ has the following properties:*

(a) *It is convex and Lipschitz continuous.*
(b) *It is nonincreasing.*
(c) $\mathrm{P}(t) \geq 0$ *for all $t \in \mathbb{R}$.*
(d) $\mathrm{P}(t) > 0$ *for all $t < h_T$.*
(e) $\mathrm{P}(t) = 0$ *for all $t \geq h_T$.*
(f) $\mathrm{P}|_{(-\infty,h_T]}$ *is strictly decreasing.*
(g) $\lim_{t\to-\infty} \mathrm{P}(t) = \infty$.
(h) $\mathrm{P}(0) = \mathrm{h}_{\mathrm{top}}(T) = \log(\deg(T))$.
(i) (Bowen's formula) *The pressure function $t \longmapsto \mathrm{P}(t)$ has a least zero $h_T > 0$ and*

$$\begin{aligned} h_T = \delta(T) &= \mathrm{ExpD}^{**}(\mathcal{J}(T)) = \mathrm{ExpD}^{*}(\mathcal{J}(T)) = \mathrm{ExpD}(\mathcal{J}(T)) = \mathrm{DD}(\mathcal{J}(T)) \\ &= \mathrm{HD}(\mathcal{J}_{\mathrm{uer}}(T)) = \mathrm{HD}(\mathcal{J}_{\mathrm{er}}(T)) = \mathrm{HD}(\mathcal{J}_{\mathrm{r}}(T)) \\ &= \mathrm{HD}(\mathcal{J}(T)) \in \left(\frac{p_T}{p_T+1}, 2\right]. \end{aligned}$$

Proof.

(a) Both properties can be proved directly per Exercise 11.5.7 (see also Exercise 11.5.10) or by means of the variational principle (Theorem 12.1.1) per Corollary 12.1.11 and Exercise 12.4.9.

(b) Let $s \le t$. For any $\mu \in M(T)$, Przytycki's theorem (Theorem 28.1.3; $\chi_\mu(T) \ge 0$) implies that

$$\mathrm{h}_\mu(T) + \int\limits_{\mathcal{J}(T)} \varphi_t \, d\mu = \mathrm{h}_\mu(T) - t\chi_\mu(T) \le \mathrm{h}_\mu(T) - s\chi_\mu(T) = \mathrm{h}_\mu(T) + \int\limits_{\mathcal{J}(T)} \varphi_s \, d\mu.$$

Hence, $\mathrm{P}(t) \le \mathrm{P}(s)$ by the variational principle (Theorem 12.1.1).

(c) By Theorem 26.2.7, $\Omega(T) \neq \emptyset$. Let $\omega \in \Omega(T)$ and consider the atomic measure μ_ω equidistributed on the (finite) orbit of ω. Of course, $\mu_\omega \in M(T)$ and by the variational principle we have $\mathrm{P}(t) \ge \mathrm{h}_{\mu_\omega}(T) - t\chi_{\mu_\omega}(T) = 0 - t \cdot 0 = 0$.

(d) Recall from Definition 29.2.6 and Theorems 29.2.12 and 29.2.14 that

$$\mathrm{DD}(\mathcal{J}(T)) := \sup\{\mathrm{HD}(\mu) \,|\, \mu \in M_e^+(T)\} \in (0, 2]$$

is the dynamical dimension of the Julia set $\mathcal{J}(T)$. If $t < \mathrm{DD}(\mathcal{J}(T))$, there is $\mu \in M_e^+(T)$ such that $\mathrm{HD}(\mu) > t$. By Ruelle's inequality (Theorem 28.2.1; $\mathrm{h}_\mu(T) \le 2\chi_\mu(T)$) and the volume lemma (Theorem 28.4.1; $\mathrm{HD}(\mu) = \frac{\mathrm{h}_\mu(T)}{\chi_\mu(T)}$), this means that $\frac{\mathrm{h}_\mu(T)}{\chi_\mu(T)} > t$. Equivalently, $\mathrm{h}_\mu(T) - t\chi_\mu(T) > 0$. By the variational principle, this implies that $\mathrm{P}(t) > 0$. In particular, this shows that $h_T \ge \mathrm{DD}(\mathcal{J}(T))$. Item (d) will hold after demonstrating that $h_T \le \mathrm{DD}(\mathcal{J}(T))$ under (e).

(e) Let $t \ge \mathrm{DD}(\mathcal{J}(T))$ and suppose for a contradiction that $\mathrm{P}(t) > 0$. By the variational principle, there then exists $\mu \in M_e(T)$ such that $\mathrm{h}_\mu(T) - t\chi_\mu(T) > 0$. By Przytycki's theorem, it ensues that $\mathrm{h}_\mu(T) > t\chi_\mu(T) \ge 0$. So $\mu \in M_e^+(T)$. By Ruelle's inequality, we deduce that $\chi_\mu(T) > 0$. Hence the volume lemma applies and leads to

$$\mathrm{DD}(\mathcal{J}(T)) \le t < \frac{\mathrm{h}_\mu(T)}{\chi_\mu(T)} = \mathrm{HD}(\mu) \le \mathrm{DD}(\mathcal{J}(T)).$$

This is impossible, and thus $\mathrm{P}(t) = 0$ whenever $t \ge \mathrm{DD}(\mathcal{J}(T))$. In particular, this shows that $h_T \le \mathrm{DD}(\mathcal{J}(T))$. We can now deduce from the arguments under (d) and (e) that $h_T = \mathrm{DD}(\mathcal{J}(T))$.

(f) Assume for a contradiction that $\mathrm{P}(s) = \mathrm{P}(t)$ for some $s < t \le h_T$. Since $T|_{\mathcal{J}(T)}$ is expansive, Theorem 12.2.5 guarantees that there is an ergodic equilibrium state, i. e., a measure $\nu \in M_e(T)$ such that

$$\mathrm{P}(t) = \mathrm{h}_\nu(T) - t\chi_\nu(T). \tag{31.12}$$

If $\chi_\nu(T) > 0$, then $s\chi_\nu(T) < t\chi_\nu(T)$. Using the variational principle, this implies that

$$\mathrm{P}(t) = \mathrm{P}(s) \geq \mathrm{h}_\nu(T) - s\chi_\nu(T) > \mathrm{h}_\nu(T) - t\chi_\nu(T) = \mathrm{P}(t),$$

which is impossible. Therefore, $\chi_\nu(T) = 0$. But Ruelle's inequality then implies that $\mathrm{h}_\nu(T) = 0$. Hence, $\mathrm{P}(s) = \mathrm{P}(t) = 0$ by (31.12). This contradicts (d) as $s < h_T$.

(g) This ensues from (f) and the convexity in (a).

(h) The first equality is by definition of the topological entropy and pressure, while the second equality is Theorem 27.8.4.

(i) The first two lines in this item follow from Theorem 29.2.12 and the earlier observation that $h_T = \mathrm{DD}(\mathcal{J}(T))$ while the last equality sign ensues from the obvious fact that

$$\mathcal{J}_r(T) \supseteq \mathcal{J}(T) \setminus \Omega_\infty^-(T) \tag{31.13}$$

and that $\Omega_\infty^-(T) := \bigcup_{n=0}^\infty T^{-n}(\Omega(T))$ is countable. The inequality $\delta(T) > p_T/(p_T+1)$ is simply Theorem 29.2.14. □

As pointed out earlier, any iterate T^k of a parabolic rational function T is parabolic, too. Moreover, per Theorem 11.1.22, the pressure function satisfies $\mathrm{P}(T^k) = k\,\mathrm{P}(T)$. It follows from Theorem 31.2.1 that $h_{T^k} = h_T$.

A depiction of the graph of P is given in Figure 31.1. Compare it with Figure 30.1.

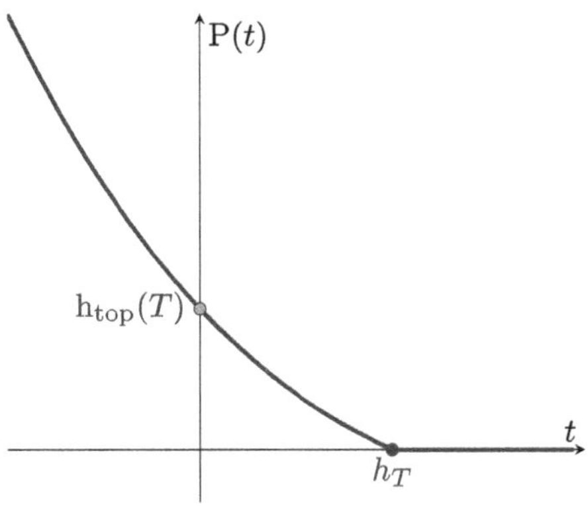

Figure 31.1: Graph of $t \mapsto \mathrm{P}(t)$ for a parabolic rational function.

31.3 Sullivan conformal measures for parabolic rational functions

In Chapter 29, we presented a fairly comprehensive account of Sullivan's conformal measures for general rational functions (in particular, see Theorem 29.2.12). In Chapter 30, more precisely in Theorem 30.1.1(d), we chronicled the existence and uniqueness of Sullivan's conformal measures for expanding rational functions. We will provide in Theorem 31.3.11 a counterpart to Theorem 30.1.1(d), i. e., a full description of Sullivan's conformal measures for parabolic rational functions. That description is quite different from the case of expanding rational functions. Nonetheless, we will prove in Subsection 31.3.2 the existence, uniqueness, ergodicity and conservativity of an atomless h_T-conformal measure m_T.

31.3.1 Technical preparations

By definition, every measure m_t, $t \geq 0$, is $e^{\mathrm{P}(t)-\varphi_t}$-conformal per (31.10). However, according to Theorem 31.2.1(d), no measure m_t, $t < h_T$, is Sullivan t-conformal. Nevertheless, by Theorem 31.2.1(e), each measure m_t, $t \geq h_T$, is Sullivan t-conformal.

Employing m_t's $e^{\mathrm{P}(t)-\varphi_t}$-conformality (formula (31.10)) with a singleton set B, we obtain from Lemma 26.2.5 and Theorem 31.2.1 the following pointwise properties of m_t.

Lemma 31.3.1. *Let $T : \widehat{\mathbb{C}} \to \widehat{\mathbb{C}}$ be a parabolic rational function with* $\deg(T) \geq 2$.
(a) *If $t \geq 0$ and $z \notin \Omega^-_\infty(T) := \bigcup_{n=0}^\infty T^{-n}(\Omega(T))$, then*

$$m_t(\{z\}) = 0.$$

(b) *If $t < h_T$, then*

$$m_t(\Omega^-_\infty(T)) = 0.$$

In other terms, for any $t < h_T$ the $e^{\mathrm{P}(t)-\varphi_t}$-conformal measure m_t is atomless (and lives on $\mathcal{J}(T) \setminus \Omega^-_\infty(T)$). For $t \geq h_T$, the t-conformal measure m_t may have atoms on $\Omega^-_\infty(T)$ only.

Now, recall that whereas $\Omega(T)$ denotes the set of rationally indifferent periodic points of T, the set of simple rationally indifferent fixed points of T is denoted by $\Omega_0(T)$ (see (24.63)):

$$\Omega_0(T) := \{\omega \in \Omega(T) \mid T(\omega) = \omega \text{ and } T'(\omega) = 1\}$$

For every $\omega \in \Omega(T)$, let $q(\omega)$ be the smallest $q \in \mathbb{N}$ such that $\omega \in \Omega_0(T^q)$. Define $p(\omega)$ to be the number $p \in \mathbb{N}$ that comes from (24.24) with $\varphi = T^{q(\omega)}$ and $p_T := \max\{p(\omega) : \omega \in \Omega(T)\}$ (cf. (29.57)). For every $t \geq 0$ and $\omega \in \Omega(T)$, let

$$\alpha_t(\omega) := t + p(\omega)(t-1) = -p(\omega) + t(p(\omega)+1). \tag{31.14}$$

We now observe that the measure m_t, $t \geq h_T$, behaves like a geometric measure with exponent $\alpha_t(\omega)$ on every small enough, pointed neighborhood of any rationally indifferent periodic point ω. When $t < h_T$, the upper bound still prevails, and in fact holds on the entire neighborhood of ω.

Lemma 31.3.2. *If $T : \widehat{\mathbb{C}} \to \widehat{\mathbb{C}}$ is a parabolic rational function with* $\deg(T) \geq 2$, *then there exists $R > 0$ such that for every $t \geq h_T$ there is a constant $c = c(t) \geq 1$ for which*

$$c^{-1} \le \frac{m_t(B(\omega,r)\setminus\{\omega\})}{r^{\alpha_t(\omega)}} \le c, \quad \forall r \in (0,R],\ \forall \omega \in \Omega(T). \tag{31.15}$$

On the other hand, there exists $R > 0$ such that for every $t < h_T$ there is a constant $c = c(t) \ge 1$ for which

$$\frac{m_t(B(\omega,r))}{r^{\alpha_t(\omega)}} \le c, \quad \forall r \in (0,R],\ \forall \omega \in \Omega(T). \tag{31.16}$$

Furthermore, for any $p_T/(p_T+1) < h < h_T$, it turns out that

$$c_h := \sup\{c(t) : h \le t < h_T\} < \infty. \tag{31.17}$$

Proof. As $\mathcal{J}(T) \ne \widehat{\mathbb{C}}$ per Exercise 31.6.1, we may through a Möbius conjugation assume that $\infty \notin \mathcal{J}(T)$. Given that m_t is t-conformal for any iterate of T and given the definition of $p(\omega)$, we may further assume that $\omega \in \Omega_0(T)$ without loss of generality. Let θ be from (24.83). Take $0 < R < \theta$ and let

$$P := \{z \in \mathbb{C} : R(2\|T'_{\omega,\theta}\|_\infty)^{-1} \le |z-\omega| \le R\} \cap \mathcal{J}(T), \tag{31.18}$$

where

$$\|T'_{\omega,\theta}\|_\infty := \sup\{|T'(z)| : z \in \overline{B(\omega,\theta)} \cap \mathcal{J}(T)\}.$$

The set $[B(\omega,R)\setminus\{\omega\}] \cap \mathcal{J}(T)$ is nonempty since $\mathcal{J}(T)$ is perfect by Theorem 24.1.15. We will soon show that $P \ne \emptyset$. But first fix $\tau > 0$ so small that Lemma 24.2.25 applies. Let also

$$\delta = \tau \inf\{|z-\omega| : z \in P\} > 0.$$

As P is compact, there are finitely many points $z_1,\dots,z_q \in P$ such that

$$P \subseteq \bigcup_{i=1}^{q} B(z_i,\delta) \quad \text{and} \quad B(z_i,\delta) \cap P \ne \emptyset,\ \forall 1 \le i \le q. \tag{31.19}$$

By virtue of Lemma 24.2.26, we may further assume that δ is so small that

$$T_\omega^{-k}(B(z_i,\delta)) \cap B(z_i,\delta) = \emptyset, \quad \forall k \in \mathbb{N},\ \forall i = 1,\dots,q. \tag{31.20}$$

By the local behavior of T around a parabolic point (more precisely, Proposition 24.2.28), for every $z \in [B(\omega,R)\setminus\{\omega\}] \cap \mathcal{J}(T)$ there exists a least integer $\tilde{l}(z) \ge 0$ such that

$$R(2\|T'_{\omega,\theta}\|_\infty)^{-1} < |T^{\tilde{l}}(z) - \omega| < R.$$

So, $T^{\tilde{l}}(z) \in P \ne \emptyset$ and in fact

$$\{z \in \mathbb{C} : R(2\|T'_{\omega,\theta}\|_\infty)^{-1} < |z-\omega| < R\} \cap \mathcal{J}(T) \ne \emptyset.$$

Since this latter set is open in $\mathcal{J}(T)$, we deduce that for some $1 \le j \le q$ the set $B(z_j, \delta) \cap P$ has nonempty interior in $\mathcal{J}(T)$. Hence,

$$M(t) := m_t(B(z_j, \delta) \cap P) > 0.$$

Let $p = p(\omega)$ and a be from (24.24). For every $n \in \mathbb{N}$, define

$$P_n := \left\{ z \in B(\omega, \theta) \cap \mathcal{J}(T) : \frac{1}{2}(|a|p)^{-\frac{1}{p}} n^{-\frac{1}{p}} < |T_\omega^{-n}(z) - \omega| < 2(|a|p)^{-\frac{1}{p}} n^{-\frac{1}{p}} \right\}.$$

By Lemma 24.2.26, there is $n_0 \in \mathbb{N}$ such that $T_\omega^{-n}(z) \in P_n$ for every $n \ge n_0$ and every $z \in P$. Thus, for all $n \ge n_0$ we deduce that $T_\omega^{-n}(P) \subseteq P_n$ and

$$B(\omega, 2(|a|p)^{-\frac{1}{p}} n^{-\frac{1}{p}}) \setminus \{\omega\} \supseteq \bigcup_{k=n}^{\infty} P_k \supseteq \bigcup_{k=n}^{\infty} T_\omega^{-k}(P) \supseteq \bigcup_{k=n}^{\infty} \bigcup_{i=1}^{q} T_\omega^{-k}(B(z_i, \delta) \cap P). \tag{31.21}$$

If $t \ge h_T$, then $\mathrm{P}(t) = 0$. As the sets $\{T_\omega^{-k}(B(z_i, \delta))\}_{k=0}^{\infty}$ are mutually disjoint for every $1 \le i \le q$ per (31.20), it follows from the t-conformality of m_t in the form of (31.9), Koebe's distortion theorem (Theorem 23.1.8) and Proposition 24.2.27, that

$$\begin{aligned} m_t\left(B(\omega, 2(|a|p)^{-\frac{1}{p}} n^{-\frac{1}{p}}) \setminus \{\omega\}\right) &\ge \sum_{k=n}^{\infty} m_t(T_\omega^{-k}(B(z_j, \delta) \cap P)) \\ &\ge \frac{1}{2}(|a|p)^{-\frac{p+1}{p}t} R^{-(p+1)t} K^{-t} M(t) \sum_{k=n}^{2n} (k^{-\frac{p+1}{p}})^t \\ &\ge \tilde{C}(\omega, t) \cdot n \cdot (2n)^{-\frac{p+1}{p}t} = 2^{-\frac{p+1}{p}t} \tilde{C}(\omega, t)(n^{-\frac{1}{p}})^{\alpha_t(\omega)}, \end{aligned} \tag{31.22}$$

where $\tilde{C}(\omega, t) = \frac{1}{2}(|a|p)^{-\frac{p+1}{p}t} R^{-(p+1)t} K^{-t} M(t)$. This holds for all n large enough.

On the other hand, according to Lemma 24.2.26 and Proposition 24.2.28, every point $z \in [B(\omega, R) \setminus \{\omega\}] \cap \mathcal{J}(T)$ admits an integer $l \ge 0$ such that $z \notin B(\omega, \frac{1}{2}(|a|p)^{-\frac{1}{p}} l^{-\frac{1}{p}})$ and $T^l(z) \in P$. Let $l(z) \ge 0$ be the smallest such integer. Take $n_1 \ge n_0$ so large that if $z \in B(\omega, 2(|a|p)^{-\frac{1}{p}} n_1^{-\frac{1}{p}}) \setminus \{\omega\}$, then $l(z) \ge n_0$. Consider any

$$z \in [B(\omega, 2(|a|p)^{-\frac{1}{p}} n^{-\frac{1}{p}}) \setminus \{\omega\}] \cap \mathcal{J}(T)$$

with $n \ge n_1$. Since $l(z) \ge n_0$ and $T^{l(z)}(z) \in P$, we see that $z = T_\omega^{-l(z)}(T^{l(z)}(z)) \in P_{l(z)}$. Consequently, $\frac{1}{2}(|a|p)^{-\frac{1}{p}} l(z)^{-\frac{1}{p}} \le 2(|a|p)^{-\frac{1}{p}} n^{-\frac{1}{p}}$ and thus $l(z) \ge 4^{-p} n$. Hence,

$$B(\omega, 2(|a|p)^{-\frac{1}{p}} n^{-\frac{1}{p}}) \cap \mathcal{J}(T) \subseteq \{\omega\} \cup \bigcup_{l \ge 4^{-p} n} T_\omega^{-l}(P) \subseteq \{\omega\} \cup \bigcup_{i=1}^{q} \bigcup_{l \ge 4^{-p} n} T_\omega^{-l}(B(z_i, \delta) \cap \mathcal{J}(T)).$$

It follows from the $e^{\mathrm{P}(t) - \varphi_t}$-conformality of m_t, Theorem 23.1.8 (Koebe's distortion theorem), Proposition 24.2.27 and the series integral test, that

$$
\begin{aligned}
m_t\Big(B(\omega,2(|a|p)^{-\frac{1}{p}}n^{-\frac{1}{p}})\setminus\{\omega\}\Big) &\le m_t\left(\bigcup_{i=1}^{q}\bigcup_{l\ge 4^{-p}n} T_\omega^{-l}(B(z_i,\delta))\right)\\
&\le q\cdot 2(|a|p)^{-\frac{p+1}{p}t}\big[R(2\|T'_{\omega,\theta}\|_\infty)^{-1}\big]^{-(p+1)t}K^t\sum_{l\ge 4^{-p}n} e^{-l\mathrm{P}(t)}l^{-\frac{p+1}{p}t}\\
&\le 2q\overline{K}^t\sum_{l\ge 4^{-p}n} l^{-\frac{p+1}{p}t}\\
&\le 2q\overline{K}^t\frac{1}{\frac{\alpha_t(\omega)}{p}(4^{-p}n-1)^{\frac{\alpha_t(\omega)}{p}}}\\
&\le C(\omega,t)(n^{-\frac{1}{p}})^{\alpha_t(\omega)}, \qquad \forall n\ge\max\{n_1,2\cdot 4^p\},
\end{aligned}
\tag{31.23}
$$

where $\overline{K} := (|a|p)^{-\frac{p+1}{p}}[R(2\|T'_{\omega,\theta}\|_\infty)^{-1}]^{-(p+1)}K$ and $C(\omega,t) := 4pq\overline{K}^t/\alpha_t(\omega)$. This estimate is valid provided that $\alpha_t(\omega) > 0$, i. e., $t > p(\omega)/(p(\omega)+1)$. We neglected the pressure factor in the series (as if $t \ge h_T$). When $t < h_T$, we can work with the pressure factor and obtain a similar inequality. We omit the details.

The proof in the case when $t \ge h_T$ is now concluded from (31.22)–(31.23) by recalling that the set $\Omega_0(T)$ is finite. The proof in the case when $t < h_T$ follows from (31.23) and Lemma 31.3.1. The proof of the last part of the lemma follows from a study of the dependence of $C(\omega,t)$ on t. □

In the remainder of this chapter, let $\delta_T > 0$ be an expansive constant for $T : \mathcal{J}(T) \to \mathcal{J}(T)$ so small that $[T(B(\omega,\delta_T))] \cap [B(\Omega(T)\setminus\{T(\omega)\},\delta_T)] = \emptyset$ for every $\omega \in \Omega(T)$. Given $z \in \mathcal{J}(T)\setminus\Omega_\infty^-(T)$, let (as in Lemma 26.2.5(b))

$$
(n_j(z))_{j=1}^\infty \subseteq \{n \ge 0 : T^n(z) \notin B(\Omega(T),\delta_T)\},
\tag{31.24}
$$

where $(n_j(z))_{j=1}^\infty$ is understood as a strictly increasing sequence. Passing to a subsequence if necessary, we may further assume that

$$
\lim_{j\to\infty} T^{n_j(z)}(z) = \hat{z}
\tag{31.25}
$$

for some point $\hat{z} \in \mathcal{J}(T)\setminus B(\Omega(T),\delta_T)$.

Let $\theta_T := \theta_{\delta_T}$ be ascribed to δ_T according to Proposition 26.2.2. Further, set

$$
r_{j,u}(z) := \frac{1}{4}\left|(T^{n_j(z)})'(z)\right|^{-1}u
\tag{31.26}
$$

for every $j \in \mathbb{N}$ and $u \in (0,\theta_T]$, and let

$$
r_j(z) := r_{j,\theta_T}(z).
\tag{31.27}
$$

The next result is a description of some "weak" form of geometric behavior of m_t around any Julia point that lies outside the backward orbit of the set of rationally indifferent periodic points. Indeed, note that the constants $c_{t,u} \geq 1$ depend on u.

Lemma 31.3.3. *If $T : \widehat{\mathbb{C}} \to \widehat{\mathbb{C}}$ is a parabolic rational function with* $\deg(T) \geq 2$, *then for every $t \geq h_T$ and every $u \in (0, \theta_T]$ there exists a constant $c_{t,u} \geq 1$ such that*

$$c_{t,u}^{-1} \leq \frac{m_t(B(z, r_{j,u}(z)))}{r_{j,u}^t(z)} \leq c_{t,u}, \quad \forall j \in \mathbb{N},\ \forall z \in \mathcal{J}(T) \setminus \Omega_\infty^-(T). \tag{31.28}$$

Proof. By Koebe's $\frac{1}{4}$–theorem (Theorem 23.1.3), we have

$$B(z, r_{j,u}(z)) = B\Big(z, \frac{1}{4}|(T_z^{-n_j(z)})'(T^{n_j(z)}(z))|u\Big) \subseteq T_z^{-n_j(z)}(B(T^{n_j(z)}(z), u)).$$

Remember that $\mathrm{P}(t) = 0$ by Theorem 31.2.1, as $t \geq h_T$. It follows from the t-conformality (31.9) and Koebe's distortion theorem (Theorem 23.1.8) that

$$\begin{aligned} m_t(B(z, r_{j,u}(z))) &\leq K^t|(T_z^{-n_j(z)})'(T^{n_j(z)}(z))|^t m_t(B(T^{n_j(z)}(z), u)) \\ &\leq K^t|(T^{n_j(z)})'(z)|^{-t} = \Big(\frac{4K}{u}\Big)^t \cdot r_{j,u}^t(z). \end{aligned} \tag{31.29}$$

On the other hand, by Koebe's distortion theorem again, we get

$$B(z, r_{j,u}(z)) \supseteq T_z^{-n_j(z)}(B(T^{n_j(z)}(z), u/(4K))). \tag{31.30}$$

Together, the t-conformality (31.9), Koebe's distortion theorem (Theorem 23.1.8) and Lemma 31.1.6 yield

$$\begin{aligned} m_t(B(z, r_{j,u}(z))) &\geq m_t\Big(T_z^{-n_j(z)}(B(T^{n_j(z)}(z), u/(4K)))\Big) \\ &\geq K^{-t}|(T_z^{-n_j(z)})'(T^{n_j(z)}(z))|^t m_t(B(T^{n_j(z)}(z), u/(4K))) \\ &\geq K^{-t}|(T^{n_j(z)})'(z)|^{-t} M_{m_t}(u/(4K)) \\ &= \Big(\frac{4}{Ku}\Big)^t M_{m_t}(u/(4K)) \cdot r_{j,u}^t(z). \end{aligned} \tag{31.31}$$

Combining (31.29) with (31.31) results in (31.28). □

The next result outlines a relation between conformal measures with exponents greater than or equal to h_T.

Lemma 31.3.4. *Let $h_T \leq s \leq t$. If m_t' and m_s'' are t- and s-conformal measures respectively (i. e., they satisfy* (31.8) *for t and s, respectively), then $m_t'|_{\mathcal{J}(T)\setminus\Omega_\infty^-(T)}$ is absolutely continuous with respect to $m_s''|_{\mathcal{J}(T)\setminus\Omega_\infty^-(T)}$. In fact,*

$$m_t'(A) \leq 4^s (c_{s,\theta_T}'')^2 c_{s,\theta_T/4}'' c_{t,\theta_T}' \cdot m_s''(A)$$

for every Borel set $A \subseteq \mathcal{J}(T) \setminus \Omega_\infty^-(T)$. *If additionally* $s < t$, *then*

$$m_t'(\mathcal{J}(T) \setminus \Omega_\infty^-(T)) = 0.$$

Proof. We deduce from Lemma 31.3.3 that

$$(c_{s,u}''c_{t,u}')^{-1} r_{j,u}^{t-s}(z) \le \frac{m_t'(B(z, r_{j,u}(z)))}{m_s''(B(z, r_{j,u}(z)))} \le (c_{s,u}''c_{t,u}')r_{j,u}^{t-s}(z) \tag{31.32}$$

for all $j \in \mathbb{N}$, all $u \in (0, \theta_T]$ and all $z \in \mathcal{J}(T) \setminus \Omega_\infty^-(T)$.

Observe that (31.26)–(31.27) yield

$$r_{j,\theta_T/4}(z) = \frac{1}{4} r_j(z), \quad \forall j \in \mathbb{N},\ \forall z \in \mathcal{J}(T) \setminus \Omega_\infty^-(T). \tag{31.33}$$

Let A be any Borel subset of $\mathcal{J}(T) \setminus \Omega_\infty^-(T)$ and let $\varepsilon > 0$. Since the measure m_s'' is outer regular, there exists an open set $A \subseteq G \subseteq \widehat{\mathbb{C}}$ such that

$$m_s''(G) \le m_s''(A) + \varepsilon. \tag{31.34}$$

For every $z \in A$, take $j(z) \in \mathbb{N}$ so large that

$$r(z) \le \varepsilon \quad \text{and} \quad B(z, r(z)) \subseteq G, \tag{31.35}$$

where

$$r(z) := r_{j(z)}(z). \tag{31.36}$$

It follows from the $4r$-covering theorem (Theorem 15.4.1 and Remark 15.4.2(e)) that there exists a countable set $A^* \subseteq A$ such that the family $\{B(z, r(z)/4)\}_{z \in A^*}$ consists of mutually disjoint sets and the family $\{B(z, r(z))\}_{z \in A^*}$ covers A. Using (31.32)–(31.36) and Lemma 31.3.3 (twice), we obtain that

$$\begin{aligned}
m_t'(A) &\le m_t'\Big(\bigcup_{z \in A^*} B(z, r(z))\Big) \le \sum_{z \in A^*} m_t'(B(z, r(z))) \\
&\le c_{s,\theta_T}''c_{t,\theta_T}' \sum_{z \in A^*} (r(z))^{t-s} m_s''(B(z, r(z))) \\
&\le c_{s,\theta_T}''c_{t,\theta_T}' \varepsilon^{t-s} \sum_{z \in A^*} m_s''(B(z, r(z))) \le (c_{s,\theta_T}'')^2 c_{t,\theta_T}' \varepsilon^{t-s} \sum_{z \in A^*} (r(z))^s \\
&= 4^s (c_{s,\theta_T}'')^2 c_{t,\theta_T}' \varepsilon^{t-s} \sum_{z \in A^*} (r(z)/4)^s \\
&\le 4^s (c_{s,\theta_T}'')^2 c_{s,\theta_T/4}'' c_{t,\theta_T}' \varepsilon^{t-s} \sum_{z \in A^*} m_s''(B(z, r(z)/4))
\end{aligned}$$

$$
\begin{aligned}
&= 4^s (c''_{s,\theta_T})^2 c''_{s,\theta_T/4} c'_{t,\theta_T} \varepsilon^{t-s} m''_s\Big(\bigcup_{z\in A^*} B(z,r(z)/4)\Big)\\
&\le 4^s (c''_{s,\theta_T})^2 c''_{s,\theta_T/4} c'_{t,\theta_T} \varepsilon^{t-s} m''_s(G)\\
&\le 4^s (c''_{s,\theta_T})^2 c''_{s,\theta_T/4} c'_{t,\theta_T} \varepsilon^{t-s} (m''_s(A)+\varepsilon).
\end{aligned}
$$

Letting $\varepsilon \to 0$ leads to the result. □

In fact, in Theorem 31.3.11(a) we will need a more general version of Lemmas 31.3.3–31.3.4.

Given $t \ge 0$ and $\lambda_t > 0$, a Borel probability measure ν_t on $\mathcal{J}(T)$ is called (t,λ_t)-conformal if it satisfies (31.8), or equivalently (31.9), with $e^{\mathrm{P}(t)}$ replaced by λ_t, i. e. if

$$
\mathcal{L}_t^*(\nu_t) = \lambda_t \nu_t, \quad \text{or equivalently,} \quad \nu_t(T_z^{-n}(A)) = \lambda_t^{-n} \int_A |(T_z^{-n})'|^t \, d\nu_t, \tag{31.37}
$$

for all $z \in \mathcal{J}(T)$, all $n \in \mathbb{N}$, and all Borel sets $A \subseteq B(T^n(z),\xi)$.

Making evident adjustments to the proof of Lemma 31.3.3, we get the following generalization.

Lemma 31.3.5. *Let $T : \widehat{\mathbb{C}} \to \widehat{\mathbb{C}}$ be a parabolic rational function with* $\deg(T) \ge 2$. *Let also $t \ge 0$ and $\lambda_t > 0$. If ν_t is a (t,λ_t)-conformal measure, then for every $u \in (0,\theta_T]$ there exists a constant $C_{t,u} \ge 1$ such that*

$$
C_{t,u}^{-1} \le \frac{\nu_t(B(z,r_{j,u}(z)))}{\lambda_t^{-n_j(z)} r_{j,u}^t(z)} \le C_{t,u}, \quad \forall j \in \mathbb{N},\ \forall z \in \mathcal{J}(T) \setminus \Omega_\infty^-(T). \tag{31.38}
$$

We then obtain a generalization of Lemma 31.3.4.

Lemma 31.3.6. *Let $0 \le s \le t$ and $0 < \lambda''_s \le \lambda'_t$. If ν'_t and ν''_s are (t,λ'_t)- and (s,λ''_s)-conformal measures respectively, then $\nu'_t|_{\mathcal{J}(T)\setminus\Omega_\infty^-(T)}$ is absolutely continuous with respect to $\nu''_s|_{\mathcal{J}(T)\setminus\Omega_\infty^-(T)}$. In fact,*

$$
\nu'_t(A) \le 4^s (c''_{s,\theta_T})^2 c''_{s,\theta_T/4} c'_{t,\theta_T} \cdot \nu''_s(A)
$$

for every Borel set $A \subseteq \mathcal{J}(T) \setminus \Omega_\infty^-(T)$. If additionally $s < t$ or $\lambda''_s < \lambda'_t$, then

$$
\nu'_t(\mathcal{J}(T) \setminus \Omega_\infty^-(T)) = 0.
$$

Proof. We deduce from Lemma 31.3.5 that

$$
(c''_{s,u} c'_{t,u})^{-1} \Big(\frac{\lambda''_s}{\lambda'_t}\Big)^{n_j(z)} r_{j,u}^{t-s}(z) \le \frac{\nu'_t(B(z,r_{j,u}(z)))}{\nu''_s(B(z,r_{j,u}(z)))} \le (c''_{s,u} c'_{t,u}) \Big(\frac{\lambda''_s}{\lambda'_t}\Big)^{n_j(z)} r_{j,u}^{t-s}(z) \tag{31.39}
$$

for all $j \in \mathbb{N}$, all $u \in (0, \theta_T]$ and all $z \in \mathcal{J}(T) \setminus \Omega_\infty^-(T)$. This is a generalization of (31.32). Following the lines in the proof of Lemma 31.3.4 yields

$$\nu_t'(A) \leq 4^s \big(c''_{s,\theta_T}\big)^2 c''_{s,\theta_T/4} c'_{t,\theta_T} \varepsilon^{t-s} \big(\nu_s''(A) + \varepsilon\big). \tag{31.40}$$

If $\lambda_s'' < \lambda_t'$, then pick $0 < \delta \leq 1$. We may choose $j(z) \in \mathbb{N}$ so that not only (31.35) is satisfied but also that $n_{j(z)}(z) \in \mathbb{N}$ (cf. (31.24)–(31.25)) is so large that

$$\left(\frac{\lambda_s''}{\lambda_t'}\right)^{n_{j(z)}(z)} \leq \delta.$$

Going through nearly identical lines to the proof of Lemma 31.3.4 results in

$$\nu_t'(A) \leq 4^s \big(c''_{s,\theta_T}\big)^2 c''_{s,\theta_T/4} c'_{t,\theta_T} \varepsilon^{t-s} \delta \big(\nu_s''(A) + \varepsilon\big). \tag{31.41}$$

Letting $\varepsilon \to 0$ in (31.40) and $\delta \to 0$ in (31.41) completes the proof. □

31.3.2 The atomless h_T-conformal measure m_T: existence, uniqueness, ergodicity and conservativity

We define m_T to be a weak cluster measure (i. e., a weak accumulation point) of the measures $(m_t)_{t<h_T}$ as $t \nearrow h_T$. As $\lim_{t\nearrow h_T} \alpha_t(\omega) = \alpha_{h_T}(\omega)$ and as $\lim_{t\nearrow h_T} \mathrm{P}(t) = \mathrm{P}(h_T) = 0$ by Theorem 31.2.1, it follows from (31.8) (with $t < h_T$) that

$$\mathcal{L}_{h_T}^*(m_T) = m_T. \tag{31.42}$$

With all the preparation done in the previous section, we can now easily prove the first main result of this section, which asserts that the measure m_T is atomless, h_T-conformal like m_{h_T}, and behaves like a geometric measure with exponent $\alpha_{h_T}(\omega)$ around any rationally indifferent periodic point ω.

Theorem 31.3.7. *If $T : \widehat{\mathbb{C}} \to \widehat{\mathbb{C}}$ is a parabolic rational function with* $\deg(T) \geq 2$, *then the measure m_T (a weak cluster of the measures $(m_t)_{t<h_T}$ with $t \nearrow h_T$) is atomless and h_T-conformal like m_{h_T}. In addition, there exist $c \geq 1$ and $R > 0$ such that*

$$c^{-1} \leq \frac{m_T(B(\omega, r))}{r^{\alpha_{h_T}(\omega)}} \leq c, \quad \forall r \in (0, R], \ \forall \omega \in \Omega(T). \tag{31.43}$$

Proof. Recall that $h_T > 0$ and $\mathrm{P}(h_T) = 0$ per Theorem 31.2.1. According to (31.42), the measure m_T is a fixed point of the dual of the transfer operator $\mathcal{L}_{h_T}$, and hence is h_T-conformal. Employing Lemma 26.2.5 and formula (31.10) with B being a singleton, we deduce that $m_T(\{z\}) = 0$ for every $z \in \mathcal{J}(T) \setminus \Omega_\infty^-(T)$. Now let us turn our attention to $\omega \in \Omega(T)$. By passing to a sufficiently high iterate of T, we may assume without loss of generality that $\omega \in \Omega_0(T)$. Let $(t_n)_{n=1}^{\infty}$ be a sequence such that $t_n \nearrow h_T$, $0 \le t_n < h_T$, and such that the sequence $(m_{t_n})_{n=1}^{\infty}$ converges weakly to the measure m_T. The Portmanteau theorem (Theorem A.1.23) and Lemma 31.3.2 yield that

$$m_T(B(\omega,r)) \le \liminf_{n\to\infty} m_{t_n}(B(\omega,r)) \le \liminf_{n\to\infty} c(t_n) r^{\alpha_{t_n}(\omega)} \le C_h\, r^{\alpha_{h_T}(\omega)}$$

for all $r \in (0,R]$, where $h = \frac{1}{2}(\frac{p_T}{p_T+1} + h_T)$. This establishes the right-hand side of (31.43). Consequently, $m_T(\{\omega\}) = 0$. So, m_T is atomless. On the other hand, the left-hand side of (31.43) follows from the left-hand side of (31.15) in Lemma 31.3.2. Indeed, (31.42) affirms that the measure m_T is a fixed point of the conjugate transfer operator and is hence $e^{\mathrm{P}(h_T)-\varphi_{h_T}}$-conformal. So, (31.15) holds for m_T, too. □

The following facts are repercussions of this theorem.

Lemma 31.3.8. *Let $T : \widehat{\mathbb{C}} \to \widehat{\mathbb{C}}$ be a parabolic rational function with* $\deg(T) \ge 2$.
(a) *If $h_T \ge 1$, then for every $\beta > 0$ there exists $C_2 \ge 1$ such that*

$$m_T(B(z,\beta|z-\omega|)) \le C_2|z-\omega|^{h_T}, \quad \forall \omega \in \Omega(T),\ \forall z \in \mathcal{J}(T).$$

(b) *If $h_T \le 1$, then for every $\beta > 0$ there exists $C_3 \ge 1$ such that*

$$m_T(B(z,\beta|z-\omega|)) \ge C_3|z-\omega|^{h_T}, \quad \forall \omega \in \Omega(T),\ \forall z \in \mathcal{J}(T).$$

Proof. Through a Möbius conjugation if necessary, we may assume that $\infty \notin \Omega(T)$. Since $\Omega(T)$ is finite, it suffices to show the result for some fixed $\omega \in \Omega(T)$.

To establish (a), let $\beta > 0$ and take $z \in \mathbb{C}$ such that $(1+\beta)|z-\omega| \le R$, where R comes from Theorem 31.3.7. In view of that theorem, we have

$$m_T(B(z,\beta|z-\omega|)) \le m_T(B(\omega,(1+\beta)|z-\omega|)) \le c[(1+\beta)|z-\omega|]^{\alpha_{h_T}(\omega)} \le c(1+\beta)^{h_T}|z-\omega|^{h_T}.$$

If $(1+\beta)|z-\omega| > R$, then we can simply rely on the obvious estimate $m_T(B(z,\beta|z-\omega|)) \le 1$. Part (a) holds.

For (b), let $0 < \beta < 1$. For $z \in \mathbb{C}$, consider the annulus

$$A(z) := \{x \in \mathbb{C} : (1-\beta)|z-\omega| \le |x-\omega| \le (1+\beta)|z-\omega|\}.$$

Observe that there exists $0 < \alpha < \pi/(2p(\omega))$ such that if $z \in S_r^j(\omega,\alpha)$, then

$$A(z) \cap S_r^j(\omega,\alpha) \subseteq B(z,(2+\beta)|z-\omega|).$$

Now apply the construction in the proof of Theorem 31.3.7. We use the same notation as in Lemma 31.3.2 but assume additionally that $P \subseteq B(\omega,\theta_\alpha(\omega))$ and that the radius δ of the balls $B(z_1,\delta),\dots,B(z_q,\delta)$ is so small that each of these balls is contained in exactly one sector $S_r^j(\omega, \alpha)$. This is possible in light of Fatou's flower theorem (Theorem 24.2.24), the definition of P and since $\alpha < \pi/(2p(\omega))$.

Let $e := (|a|p(\omega))^{-1/p(\omega)}$. Choose $\varepsilon > 0$ so small that

$$f := \left(\frac{e+\varepsilon}{(1+\beta)(e-\varepsilon)}\right)^{p(\omega)} < 1 \quad \text{and} \quad g := \left(\frac{e-\varepsilon}{(1-\beta)(e+\varepsilon)}\right)^{p(\omega)} > 1.$$

By this and Lemma 24.2.26, there is $n_0 \in \mathbb{N}$ so large that for all $n \geq fn_0$ and all $z \in P$,

$$[gn] - ([fn]+1) \geq \frac{g-f}{2}n \quad \text{and} \quad (e-\varepsilon)n^{-1/p(\omega)} \leq |T_\omega^{-n}(z) - \omega| \leq (e+\varepsilon)n^{-1/p(\omega)},$$

where $[t]$ is the integer part of t. For every $n \in \mathbb{N}$, let

$$A_n := \{x \in \mathbb{C} : (1-\beta)(e+\varepsilon)n^{-1/p(\omega)} \leq |x - \omega| \leq (1+\beta)(e-\varepsilon)n^{-1/p(\omega)}\}.$$

By Theorem 24.2.24, for every $z \in \mathcal{J}(T) \cap B(\omega, \theta_\alpha(\omega)) \setminus \{\omega\}$ there exists exactly one $1 \leq j(z) \leq p(\omega)$ such that $z \in S_r^{j(z)}(\omega, \alpha)$. Moreover, $j(T_\omega^{-n}(z)) = j(z)$ for every $z \in \mathcal{J}(T) \cap B(\omega, \theta_\alpha(\omega)) \setminus \{\omega\}$. Hence, for every $z \in P$ and every $n \geq fn_0$ we have

$$A_n \cap S_r^{j(z)}(\omega, \alpha) \subseteq B(T_\omega^{-n}(z), (2+\beta)|T_\omega^{-n}(z) - \omega|).$$

Now consider an arbitrary $x \in \bigcup_{n \geq n_0} T_\omega^{-n}(P)$. Then $x \in T_\omega^{-n}(P)$ for some $n \geq n_0$. Let $z = T^n(x)$ and choose $1 \leq i \leq q$ such that $z \in B(z_i, \delta)$. As the radius δ is small enough, we conclude from Theorem 24.2.24 and (31.21) that

$$T_\omega^{-l}(\mathcal{J}(T) \cap B(z_i, \delta)) \subseteq S_r^{j(z)}(\omega, \alpha), \quad \forall l \geq 0.$$

If $y \in P$ and $fn_0 \leq fn \leq l \leq gn$, then by the choice of ε and n_0 we have that

$$|T_\omega^{-l}(y) - \omega| \leq (e+\varepsilon)l^{-1/p(\omega)} \leq (e+\varepsilon)f^{-1/p(\omega)}n^{-1/p(\omega)} = (1+\beta)(e-\varepsilon)n^{-1/p(\omega)}$$

and

$$|T_\omega^{-l}(y) - \omega| \geq (e-\varepsilon)l^{-1/p(\omega)} \geq (e-\varepsilon)g^{-1/p(\omega)}n^{-1/p(\omega)} = (1-\beta)(e+\varepsilon)n^{-1/p(\omega)}.$$

Thus, $T_\omega^{-l}(P) \subseteq A_n$ for all $fn_0 \leq fn \leq l \leq gn$, whence

$$\begin{aligned} \bigcup_{fn \leq l \leq gn} T_\omega^{-l}(B(z_i, \delta) \cap P) &\subseteq A_n \cap S_r^{j(z)}(\omega, \alpha) \\ &\subseteq B(T_\omega^{-n}(z), (2+\beta)|T_\omega^{-n}(z) - \omega|) \\ &= B(x, (2+\beta)|x - \omega|). \end{aligned} \tag{31.44}$$

By (31.19), for every $1 \leq j \leq q$ there exist $\zeta_j \in B(z_j, \delta) \cap P$ and $\delta_j > 0$ such that $B(\zeta_j, \delta_j) \subseteq B(z_j, \delta) \cap P$. Set

$$\underline{\delta} := \min_{1\le j\le q} \delta_j, \qquad \underline{d} := \min_{1\le j\le q} |\zeta_j - \omega|, \quad \text{and} \quad |(\underline{T_\omega^{-\infty}})'| := \min_{1\le j\le q} |(T_\omega^{-\infty})'(\zeta_i)|.$$

As the sets $T_\omega^{-l}(B(z_i, \delta))$, $l = [fn] + 1, \ldots, [gn]$, are mutually disjoint, it follows from the combination of the h_T-conformality of m_T with Koebe's distortion theorem, Proposition 24.2.27 and Lemma 31.1.6 that

$$\begin{aligned}
m_T\Big(\bigcup_{fn\le l\le gn} T_\omega^{-l}(B(z_i, \delta)\cap P)\Big) &\ge \sum_{l=[fn]+1}^{[gn]} m_T(T_\omega^{-l}(B(z_i, \delta)\cap P)) \\
&\ge \sum_{l=[fn]+1}^{[gn]} m_T(T_\omega^{-l}(B(\zeta_i, \delta_i))) \\
&\ge \sum_{l=[fn]+1}^{[gn]} [K^{-1}|(T_\omega^{-l})'(\zeta_i)|]^{h_T} m_T(B(\zeta_i, \underline{\delta})) \\
&\ge \sum_{l=[fn]+1}^{[gn]} \Big[\frac{1}{2K}(|a|pl)^{-\frac{p(\omega)+1}{p(\omega)}} |\zeta_i - \omega|^{-(p(\omega)+1)} |(T_\omega^{-\infty})'(\zeta_i)|\Big]^{h_T} M_{m_T}(\underline{\delta}) \\
&\ge C_4 \sum_{l=[fn]+1}^{[gn]} l^{-\frac{p(\omega)+1}{p(\omega)} h_T},
\end{aligned}$$

where $C_4 := \Big[(2K)^{-1}(|a|p)^{-\frac{p(\omega)+1}{p(\omega)}} \underline{d}^{-(p(\omega)+1)} |(\underline{T_\omega^{-\infty}})'|\Big]^{h_T} M_{m_T}(\underline{\delta}) > 0$. Since $h_T \le 1$, we deduce from the above and (31.44) that

$$\begin{aligned}
m_T(B(x, (2+\beta)|x-\omega|)) &\ge C_4 \frac{g-f}{2} n(gn)^{-\frac{p(\omega)+1}{p(\omega)} h_T} \\
&= C_4 \frac{g-f}{2} g^{-\frac{p(\omega)+1}{p(\omega)} h_T} (n^{-1/p(\omega)})^{-p(\omega)+(p(\omega)+1)h_T} \\
&\ge C_4 g^{-\frac{p(\omega)+1}{p(\omega)} h_T} \frac{g-f}{2(e+\varepsilon)^{\alpha_{h_T}(\omega)}} |T_\omega^{-n}(z) - \omega|^{\alpha_{h_T}(\omega)} \\
&= C_4 g^{-\frac{p(\omega)+1}{p(\omega)} h_T} \frac{g-f}{2(e+\varepsilon)^{\alpha_{h_T}(\omega)}} |x - \omega|^{h_T - p(\omega)(1-h_T)} \\
&\ge C_4 g^{-\frac{p(\omega)+1}{p(\omega)} h_T} \frac{g-f}{2(e+\varepsilon)^{\alpha_{h_T}(\omega)}} |x - \omega|^{h_T}.
\end{aligned}$$

This proves (b) for all points in the set $G := \bigcup_{n\ge n_0} T_\omega^{-n}(P)$. As $G \supseteq [B(\omega, R)\setminus\{\omega\}] \cap \mathcal{J}(T)$ and as for ω itself there is nothing to prove, statement (b) holds on the neighborhood $\{\omega\}\cup G$ of ω. Since m_T is positive on open sets, we deduce that $m(B(x, (2+\beta)|x-\omega|)) \ge C$ for some constant $C > 0$ and all $x \notin \{\omega\}\cup G$, proving (b). □

The second main result of this section states that m_T is ergodic, conservative, and unique.

Theorem 31.3.9. *If $T : \widehat{\mathbb{C}} \to \widehat{\mathbb{C}}$ is a parabolic rational function with* $\deg(T) \ge 2$, *then the dynamical system $T : \mathcal{J}(T) \to \mathcal{J}(T)$ is conservative and ergodic with respect to the*

h_T-conformal measure m_T. In addition, $\mathrm{supp}(m_T) = \mathcal{J}(T)$ *and m_T is the unique measure satisfying* (31.8) *with* $\mathrm{P}(h_T) = 0$*, i. e., the unique h_T-conformal measure for T. In other terms, $m_T = m_{h_T}$.*

Proof. The topological exactness of $T|_{\mathcal{J}(T)}$ (per Corollary 24.3.5) and the h_T-conformality of m_T (in the form (31.10)) guarantee that $\mathrm{supp}(m_T) = \mathcal{J}(T)$.

Claim 1. *Any h_T-conformal measure m' for T is equivalent to m_T. In particular, $m'(\Omega_\infty^-(T)) = 0$.*

Proof of Claim 1. If it turned out that $m'(\Omega_\infty^-(T)) > 0$, then the measure m'' defined by

$$m''(A) := \frac{m'(A \cap \Omega_\infty^-(T))}{m'(\Omega_\infty^-(T))}, \quad \forall A \in \mathcal{B}(\mathcal{J}(T)),$$

would be well-defined, h_T-conformal, and $m''(\mathcal{J}(T) \setminus \Omega_\infty^-(T)) = 0$. It would then follow from Lemma 31.3.4 that $m'(\mathcal{J}(T) \setminus \Omega_\infty^-(T)) = 0$. In conjunction with Theorem 31.3.7, this would result in $m_T(\mathcal{J}(T)) = 0$. This contradiction shows that $m'(\Omega_\infty^-(T)) = 0$. Therefore, the measures m' and m_T are equivalent with bounded Radon–Nikodym derivatives by virtue of Theorem 31.3.7 and Lemma 31.3.4. Claim 1 is proved. ■

Claim 2. *m_T is ergodic.*

Proof of Claim 2. Suppose on the contrary that m_T is not ergodic. This means that there exists a Borel set $F \subseteq \mathcal{J}(T)$ such that

$$0 < m_T(F) < 1 \quad \text{and} \quad T^{-1}(F) = F.$$

Then the conditional measure of m_T on F defined by

$$(m_T)_F(A) := \frac{m_T(A \cap F)}{m_T(F)}, \quad \forall A \in \mathcal{B}(\mathcal{J}(T)),$$

satisfies (31.8) with $\mathrm{P}(h_T) = 0$ but is not equivalent to m_T, contrary to Claim 1. Claim 2 is thus established. ■

Claim 3. *If m' is an h_T-conformal measure for T and $\rho := dm'/dm_T$, then ρ is m_T-a. e. T-invariant, i. e.,*

$$\rho \circ T = \rho \quad m_T\text{-a. e.}$$

Proof of Claim 3. By virtue of Claim 1, the function ρ is well-defined m_T-almost everywhere. For every $z \in \mathcal{J}(T) \setminus \Omega_\infty^-(T)$, the set

$$Q(z) := \{n \geq 0 \mid T^n(z) \notin B(\Omega(T), \delta_T)\}$$

is infinite. For every $n \in Q(z)$, let

$$V_n(z) := T_z^{-n}\big(B(T^n(z), \theta_T/(6K^2))\big).$$

Since $\inf_{n\geq 0}\operatorname{diam}(V_n(z)) = 0$, the family $V = \{V_n(z) : z \in \mathcal{J}(T)\setminus\Omega_\infty^-(T),\ n\in Q(z)\}$ is (in the terminology of Section 2.8 in Federer [50]) fine at every $z\in\mathcal{J}(T)\setminus\Omega_\infty^-(T)$. Set

$$\widehat{V}_n(z) := \Big\{\bigcup_{W\in V} W : W\cap V_n(z)\neq\emptyset \text{ and } \operatorname{diam}(W)\leq 2\operatorname{diam}(V_n(z))\Big\}. \tag{31.45}$$

From relations established in the proof of Lemma 31.3.3, we deduce the following analogous ones:

$$T_z^{-n}(B(T^n(z),\theta_T)) \supseteq B(z, K^{-1}\theta_T|(T^n)'(z)|^{-1}), \tag{31.46}$$

$$m_T(T_z^{-n}(B(T^n(z),\theta_T))) \leq K^{h_T}|(T^n)'(z)|^{-h_T}, \tag{31.47}$$

$$m_T(V_n(z)) \geq M_{m_T}\Big(\frac{\theta_T}{6K^2}\Big)K^{-h_T}|(T^n)'(z)|^{-h_T}. \tag{31.48}$$

From Lemma 23.1.10, we further have that

$$\operatorname{diam}(V_n(z)) \leq \frac{\theta_T}{3K}|(T^n)'(z)|^{-1}. \tag{31.49}$$

Using (31.45), (31.49) and (31.46) successively, we get

$$\widehat{V}_n(z) \subseteq B(z, 3\operatorname{diam}(V_n(z))) \subseteq B(z, K^{-1}\theta_T|(T^n)'(z)|^{-1}) \subseteq T_z^{-n}(B(T^n(z),\theta_T)).$$

Therefore, by (31.47),

$$m_T(\widehat{V}_n(z)) \leq K^{h_T}|(T^n)'(z)|^{-h_T}.$$

Along with (31.48), this yields

$$\frac{m_T(\widehat{V}_n(z))}{m_T(V_n(z))} \leq K^{2h_T}\Big[M_{m_T}\Big(\frac{\theta_T}{6K^2}\Big)\Big]^{-1}.$$

Thus,

$$\limsup_{Q(z)\ni n\to\infty}\Big(\operatorname{diam}(V_n(z)) + \frac{m_T(\widehat{V}_n(z))}{m_T(V_n(z))}\Big) \leq K^{2h_T}\Big[M_{m_T}\Big(\frac{\theta_T}{6K^2}\Big)\Big]^{-1} < \infty.$$

So, the hypotheses of Theorem 2.8.17 in Federer [50] are satisfied. That theorem along with Theorem 2.9.8 in [50] imply that

$$\rho(z) := \frac{dm'}{dm_T}(z) = \lim_{Q(z)\ni n\to\infty}\frac{m'(V_n(z))}{m_T(V_n(z))} \quad \text{for } m_T\text{-a. e. } z\in\mathcal{J}(T)\setminus\Omega_\infty^-(T).$$

Since $V_n(z) = T_z^{-1}(V_{n-1}(T(z)))$ for all $z\in\mathcal{J}(T)\setminus\Omega_\infty^-(T)$ and all $n\in Q(z)\setminus\{0\}$, it ensues that

$$\rho(z) = \lim_{Q(z)\ni n\to\infty}\frac{m'(T_z^{-1}(V_{n-1}(T(z))))}{m_T(T_z^{-1}(V_{n-1}(T(z))))} = \lim_{Q(z)\ni n\to\infty}\frac{\int_{V_{n-1}(T(z))}|(T_z^{-1})'|^{h_T}\,dm'}{\int_{V_{n-1}(T(z))}|(T_z^{-1})'|^{h_T}\,dm_T}. \tag{31.50}$$

As

$$\lim_{Q(z)\ni n\to\infty} \frac{\sup\{|(T_z^{-1})'(w)| : w \in V_{n-1}(T(z))\}}{|(T_z^{-1})'(T(z))|} = 1 = \lim_{Q(z)\ni n\to\infty} \frac{\inf\{|(T_z^{-1})'(w)| : w \in V_{n-1}(T(z))\}}{|(T_z^{-1})'(T(z))|},$$

we can continue (31.50) as follows:

$$\rho(z) = \lim_{Q(z)\ni n\to\infty} \frac{m'(V_{n-1}(T(z)))}{m_T(V_{n-1}(T(z)))} = \lim_{Q(T(z))\ni k\to\infty} \frac{m'(V_k(T(z)))}{m_T(V_k(T(z)))} = \rho(T(z))$$

for m_T-a. e. $z \in \mathcal{J}(T) \setminus \Omega_\infty^-(T)$. As $m_T(\mathcal{J}(T) \setminus \Omega_\infty^-(T)) = 0$ according to Claim 1, we conclude that $\rho \circ T = \rho$ m_T-almost everywhere. Claim 3 is thus proved. ∎

Claim 4. *m_T is the unique h_T-conformal measure for T.*

Proof of Claim 4. Let m' be an h_T-conformal measure for T. As m_T is quasi-T-invariant, it follows from Claims 2–3 that the Radon–Nikodym derivative $\rho = dm'/dm_T$ is constant m_T-almost everywhere. Since both measures m' and m_T are probabilistic, the derivative ρ must be equal to 1 m_T-a. e. and can be taken to be identically 1. So, $m' = m_T$ and Claim 4 is proved. ∎

Our next goal is to show that the measure m_T is conservative. This will easily ensue from the following claim.

Claim 5. *If $E \subseteq \mathcal{J}(T)$ is a forward or backward T-invariant Borel set, then $m_T(E) \in \{0, 1\}$.*

Proof of Claim 5. By passing to the complement, it suffices to prove the statement for a forward T-invariant set E. Since $m_T(\Omega_\infty^-(T)) = 0$ and the set $\Omega_\infty^-(T)$ is completely T-invariant, we may assume that $E \subseteq \mathcal{J}(T) \setminus \Omega_\infty^-(T)$. Suppose for a contradiction that $0 < m_T(E) < 1$. By Lebesgue's density theorem, there exists at least one point $z \in E$ (in fact, the set of such points has m_T-measure equal to $m_T(E)$) such that

$$\lim_{r\to 0} \frac{m_T(E \cap B(z, r))}{m_T(B(z, r))} = 1. \tag{31.51}$$

Seeking a contradiction, suppose that $m_T(B(\hat{z}, \theta_T/(8K)) \setminus E) = 0$, where the reader will recall that $\hat{z}$ is defined by (31.25). Given that $T|_{\mathcal{J}(T)}$ is topologically exact according to Corollary 24.3.5, let $N \in \mathbb{N}$ be such that $T^N(B(\hat{z}, \theta_T/(8K))) \supseteq \mathcal{J}(T)$. By partitioning the ball $B(\hat{z}, \theta_T/(8K))$ into finitely many special Borel sets for T^N (cf. Lemma 29.1.15), the h_T-conformality of m_T entails that $m_T(T^N(B(\hat{z}, \theta_T/(8K)) \setminus E)) = 0$. From the forward T-invariance of E, we deduce that

$$\begin{aligned}
0 &= m_T(T^N(B(\hat{z}, \theta_T/(8K)) \setminus E)) \geq m_T(T^N(B(\hat{z}, \theta_T/(8K))) \setminus T^N(E)) \\
&\geq m_T(T^N(B(\hat{z}, \theta_T/(8K))) \setminus E) \geq m_T(T^N(B(\hat{z}, \theta_T/(8K)))) - m_T(E) \\
&= 1 - m_T(E).
\end{aligned}$$

Hence, $m_T(E) \geq 1$. This contradiction shows that

$$\gamma := m_T(B(\hat{z}, \theta_T/(8K)) \setminus E) > 0.$$

As $|T^{n_j(z)}(z) - \hat{z}| < \theta_T/(8K)$ for $j \in \mathbb{N}$ large enough, say $j \geq N(z)$, we infer that

$$B(\hat{z}, \theta_T/(8K)) \subseteq B(T^{n_j(z)}(z), \theta_T/(4K)), \quad \forall j \geq N(z).$$

Using (31.30) and the forward T-invariance of E, we deduce that

$$T_z^{-n_j(z)}(B(\hat{z}, \theta_T/(8K)) \setminus E) \subseteq T_z^{-n_j(z)}(B(T^{n_j(z)}(z), \theta_T/(4K)) \setminus E) \subseteq B(z, r_j(z)) \setminus E, \quad \forall j \geq N(z).$$

By virtue of the h_T-conformality of m_T and of the Koebe distortion theorem (Theorem 23.1.9), we thereby get that

$$\begin{aligned}
m_T(B(z, r_j(z)) \setminus E) &\geq m_T(T_z^{-n_j(z)}(B(\hat{z}, \theta_T/(8K)) \setminus E)) \\
&\geq K^{-h_T} \left|(T_z^{-n_j(z)})'(T^{n_j(z)}(z))\right|^{h_T} m_T(B(\hat{z}, \theta_T/(8K)) \setminus E) \\
&= \gamma K^{-h_T} \left|(T^{n_j(z)})'(z)\right|^{-h_T}, \quad \forall j \geq N(z).
\end{aligned}$$

Combining this with (31.29), we obtain that

$$\frac{m_T(B(z, r_j(z)) \setminus E)}{m_T(B(z, r_j(z)))} \geq \frac{\gamma K^{-h_T} \left|(T^{n_j(z)})'(z)\right|^{-h_T}}{K^{h_T} \left|(T^{n_j(z)})'(z)\right|^{-h_T}} = \gamma K^{-2h_T} > 0, \quad \forall j \geq N(z).$$

Hence,

$$\limsup_{j \to \infty} \frac{m_T(E \cap B(z, r_j(z)))}{m_T(B(z, r_j(z)))} \leq 1 - \gamma K^{-2h_T} < 1,$$

contrary to (31.51). So, $m_T(E) \in \{0, 1\}$. This finishes the proof of Claim 5. ■

Claim 6. *m_T is conservative.*

Proof of Claim 6. According to Theorem 10.1.11 from Volume 1, we need to show that the quasi-T-invariant measure m_T is such that

$$m_T(B) > 0 \quad \Longrightarrow \quad m_T(\mathcal{J}(T) \setminus B_\infty) = 0,$$

where

$$B_\infty := \bigcap_{n=0}^{\infty} \bigcup_{k=n}^{\infty} T^{-k}(B).$$

So, let B be a Borel subset of $\mathcal{J}(T)$ such that $m_T(B) > 0$. Since m_T is atomless and conformal per Theorem 31.3.7, it turns out that $m_T(T^{-k}(B)) > 0$ for every $k \geq 0$. Indeed, as $\#T^k(\mathrm{Crit}(T^k)) < \infty$ and m_T is atomless, we know that $m_T(T^k(\mathrm{Crit}(T^k))) = 0$. Thus, $m_T(B \setminus T^k(\mathrm{Crit}(T^k))) = m_T(B) > 0$. In light of Remark 24.4.16(a), we can partition the set $B \setminus T^k(\mathrm{Crit}(T^k))$ into finitely many Borel subsets over each of which T^k has an inverse branch. At least one of these subsets has positive m_T-measure. Call one such subset S_B. It ensues from the conformality of m_T (in the form of (31.9)) that $m_T(T^{-k}(S_B)) > 0$. Hence

$$m_T(T^{-k}(B)) > 0, \quad \forall k \geq 0.$$

This immediately implies that

$$m_T\Big(\bigcup_{k=n}^{\infty} T^{-k}(B)\Big) > 0, \quad \forall n \geq 0.$$

As each set $\bigcup_{k=n}^{\infty} T^{-k}(B)$ is backward T-invariant, we deduce from Claim 5 that

$$m_T\Big(\bigcup_{k=n}^{\infty} T^{-k}(B)\Big) = 1, \quad \forall n \geq 0.$$

Consequently, $m_T(B_\infty) = 1$. So $m_T(\mathcal{J}(T) \setminus B_\infty) = 0$, and m_T is conservative. ■

This finishes the proof of Theorem 31.3.9. □

This theorem entails the set of transitive points to be of full m_T-measure.

Corollary 31.3.10. *If $T : \widehat{\mathbb{C}} \to \widehat{\mathbb{C}}$ is a parabolic rational function with* $\deg(T) \geq 2$, *then the m_T-measure of the set of transitive points of the dynamical system $T : \mathcal{J}(T) \to \mathcal{J}(T)$ is equal to 1.*

Proof. Let $\{O_n\}_{n=1}^{\infty}$ be a countable base for the relative topology of $\mathcal{J}(T)$. Fix temporarily $n \in \mathbb{N}$. Since $\mathrm{supp}(m_T) = \mathcal{J}(T)$, we know that $m_T(O_n) > 0$. Then $m_T(\mathcal{J}(T) \setminus (O_n)_\infty) = 0$ according to Theorems 31.3.9 and 10.1.11. This means that $m_T((O_n)_\infty) = 1$. As the intersection of a countable family of sets of measure 1 has measure 1, we conclude that the set

$$\bigcap_{n=1}^{\infty} \bigcap_{k=0}^{\infty} \bigcup_{\ell=k}^{\infty} T^{-\ell}(O_n)$$

has measure 1. This set is precisely the set of transitive points of $T|_{\mathcal{J}(T)}$. □

31.3.3 The complete structure of Sullivan's conformal measures for parabolic rational functions: the atomless measure m_T and purely atomic measures

We can now collect together the results of the previous sections and provide a short concluding proof of the main theorem of the current section. It gives a full description of the structure of the Sullivan conformal measures of a parabolic rational function.

Theorem 31.3.11. *Let $T : \widehat{\mathbb{C}} \to \widehat{\mathbb{C}}$ be a parabolic rational function with* $\deg(T) \geq 2$. *Recall that Bowen's formula (Theorem 31.2.1(i)) states that*

$$\begin{aligned} h_T = \delta(T) &= \mathrm{ExpD}^{**}(\mathcal{J}(T)) = \mathrm{ExpD}^{*}(\mathcal{J}(T)) = \mathrm{ExpD}(\mathcal{J}(T)) = \mathrm{DD}(\mathcal{J}(T)) \\ &= \mathrm{HD}(\mathcal{J}_{\mathrm{uer}}(T)) = \mathrm{HD}(\mathcal{J}_{\mathrm{er}}(T)) = \mathrm{HD}(\mathcal{J}_{\mathrm{r}}(T)) \\ &= \mathrm{HD}(\mathcal{J}(T)) \in \left(\frac{p_T}{p_T+1}, 2\right]. \end{aligned}$$

The following situations prevail:
(a) *If $t < h_T$, then T does not admit any Sullivan t-conformal measure.*
(b) *T has exactly one h_T-conformal measure, namely $m_T = m_{h_T}$. This measure is atomless, ergodic and conservative.*
(c) *For every $t > h_T$ and every $\zeta \in \mathcal{J}(T)$, it holds that*

$$\Sigma_t(\zeta) := \sum_{z \in \mathcal{O}_-(\zeta)} \left|(T^{k(z)})'(z)\right|^{-t} < \infty, \tag{31.52}$$

where $\mathcal{O}_-(\zeta) := \bigcup_{n=0}^{\infty} T^{-n}(\zeta)$ and $k(z) \geq 0$ is the least integer such that $T^{k(z)}(z) = \zeta$. Accordingly, let

$$\nu_{t,\omega} := \frac{1}{\Sigma_t(\omega)} \sum_{z \in \mathcal{O}_-(\omega)} \left|(T^{k(z)})'(z)\right|^{-t} \delta_z, \tag{31.53}$$

where δ_z denotes the Dirac measure concentrated at the point z. Then the set of all t-conformal measures for T coincides with the set of all convex combinations of measures $\{\nu_{t,\omega}\}_{\omega \in \Omega(T)}$. In particular, these conformal measures are atomic.

Proof.
(a) Suppose for a contradiction that $t \in [0, h_T)$ and m is a Sullivan t-conformal measure. As $1 < e^{\mathrm{P}(t)}$ (per Theorem 31.2.1), applying Lemma 31.3.6 to the $(t,1)$-conformal measure m and the $(t, e^{\mathrm{P}(t)})$-conformal measure $m_t = m_{\varphi_t}$ produced at the beginning of Section 31.2 (see (31.8)), we get

$$m_t(\mathcal{J}(T) \setminus \Omega_\infty^-(T)) = 0.$$

Equivalently, $m_t(\Omega_\infty^-(T)) = 1$. The conformality of m_t implies that $m_t(\Omega(T)) > 0$. So, there exists $\omega \in \Omega(T)$ for which $m_t(\{\omega\}) > 0$. Take $n \in \mathbb{N}$ such that $T^n(\omega) = \omega$. Then $|(T^n)'(\omega)| = 1$ and we deduce from (31.10) that

$$m_t(\{\omega\}) = m_t(\{T^n(\omega)\}) = e^{n\mathrm{P}(t)}\left|(T^n)'(\omega)\right|^t m_t(\{\omega\}) > m_t(\{\omega\}).$$

This contradiction completes the proof of (a).
(b) This follows from the combination of Theorems 31.3.7 and 31.3.9.

(c) For every $t \geq 0$ and every $x \in \mathcal{J}(T)$, let

$$\Sigma_t(x) := \sum_{z \in \mathcal{O}_-(x)} \left|(T^{k(z)})'(z)\right|^{-t},$$

where $k(z) \geq 0$ is the least integer k such that $T^k(z) = x$. If x is not periodic, then

$$\Sigma_t(x) = \sum_{n=0}^{\infty} \sum_{z \in T^{-n}(x)} \left|(T^n)'(z)\right|^{-t}.$$

The following three results are obvious.

Claim 1. *For every $t \geq 0$ and every $x \in \mathcal{J}(T)$,*

$$\Sigma_t(x) < \infty \quad \Longleftrightarrow \quad \Sigma_t(z) < \infty, \ \forall z \in T^{-1}(x) \setminus \mathcal{O}_+(x).$$

Claim 2. *For every $x \in \widehat{\mathbb{C}}$, the set $T^{-1}(x) \setminus \mathcal{O}_+(x)$ contains no periodic point of T.*

Claim 3. *If $\Sigma_t(x) < \infty$ for some $t \geq 0$ and $x \in \mathcal{J}(T)$, then $\Sigma_t(z) < \infty$ for all $z \in \mathcal{O}_-(x)$.*

The next fact deserves a proof.

Claim 4. *If $\Sigma_t(x) < \infty$ for some $t \geq 0$ and $x \in \mathcal{J}(T)$, then $\Sigma_t(y) < \infty$ for every $y \in \mathcal{J}(T) \setminus \overline{\mathrm{PC}(T)}$.*

Proof of Claim 4. In light of Claims 1–2, we may assume without loss of generality that neither x nor y is periodic. Recalling Remark 24.4.16(b), there exists $r = r(y) > 0$ such that $B(y, 2r) \cap \overline{\mathrm{PC}(T)} = \emptyset$ and for every $n \geq 0$ and every $z \in T^{-n}(y)$ there is a unique holomorphic inverse branch $T_z^{-n} : B(y, 2r) \to \mathbb{C}$ of T^n that sends y to z. Given the topological exactness of $T|_{\mathcal{J}(T)}$, there exists $k \geq 0$ such that $T^k(B(y, r)) \supseteq \mathcal{J}(T)$. Therefore, there is $y_0 \in B(y, r)$ such that $T^k(y_0) = x$. By Claim 3, $\Sigma_t(y_0) < \infty$. As y_0 (like x) is not periodic and as $\bigcup_{z \in T^{-n}(y)} T_z^{-n}(y_0) = T^{-n}(y_0)$, we get from Koebe's distortion theorem that

$$\Sigma_t(y) = \sum_{n=0}^{\infty} \sum_{z \in T^{-n}(y)} \left|(T_z^{-n})'(y)\right|^t \asymp \sum_{n=0}^{\infty} \sum_{z \in T^{-n}(y)} \left|(T_z^{-n})'(y_0)\right|^t = \Sigma_t(y_0) < \infty,$$

where $\asymp$ is the usual symbol of multiplicative comparability (see (25.34)). Thus, Claim 4 holds. ∎

Claim 5. *If $\Sigma_t(x) < \infty$ for some $t \geq 0$ and $x \in \mathcal{J}(T)$, then $\Sigma_t(y) < \infty$ for every $y \in \mathcal{J}(T) \setminus \Omega(T)$.*

Proof of Claim 5. As T is parabolic, it is expansive, i. e., $\mathrm{Crit}(T) \cap \mathcal{J}(T) = \emptyset$ per Theorem 26.2.3. Therefore $\mathrm{PC}(T) \cap \mathcal{J}(T) = \emptyset$. It also follows from the classification theorem 24.4.10 and Sullivan's nonwandering theorem (Theorem 24.4.12) that $\omega(\mathrm{Crit}(T)) \cap \mathcal{J}(T) \subseteq \Omega(T)$. Together, these two facts imply that $\overline{\mathrm{PC}(T)} \cap \mathcal{J}(T) \subseteq \Omega(T)$.

(Proposition 26.2.2 offers a stronger assertion of this fact.) The claim follows from Claim 4. ■

As an immediate consequence of Claims 5 and 1, we get the following.

Claim 6. *Given any $t \geq 0$ and $x \in \mathcal{J}(T)$,*

$$\Sigma_t(x) < \infty \quad \Longleftrightarrow \quad \Sigma_t(y) < \infty, \ \forall y \in \mathcal{J}(T).$$

We finally prove a claim for the specific range of t's we are interested in.

Claim 7. *If $t > h_T$, then there is $\omega \in \Omega(T)$ such that $\Sigma_t(\omega) < \infty$.*

Proof of Claim 7. Recall that $m_t = m_{\varphi_t}$ is the measure produced at the beginning of Section 31.2 (see (31.8)) with $\mathrm{P}(t) = 0$ per Theorem 31.2.1. Thus, m_t is t-conformal. Taking any $s \in [h_T, t)$, it follows from Lemma 31.3.4 that $m_t(\mathcal{J}(T) \setminus \Omega_\infty^-(T)) = 0$. So there exists $\omega \in \Omega(T)$ for which $m_t(\mathcal{O}_-(\omega)) > 0$ and thereby $m_t(\{\omega\}) > 0$. But $m_t(\mathcal{O}_-(\omega)) = m_t(\{\omega\})\Sigma_t(\omega)$, whence

$$\Sigma_t(\omega) = \left(m_t(\{\omega\})\right)^{-1} m_t(\mathcal{O}_-(\omega)) \leq \left(m_t(\{\omega\})\right)^{-1} < \infty. \qquad ■$$

Together, Claims 6–7 conclude the proof of (31.52).

By definition (31.53), all measures $\{\nu_{t,\omega}\}_{\omega\in\Omega(T)}$ are t-conformal. Obviously, so are all their convex combinations. On the other hand, let m be a t-conformal measure. It ensues from Lemma 31.3.4 (applied with $s = h_T$) that $m(\Omega_\infty^-(T)) = 1$. That is, $\sum_{\omega\in\Omega(T)} m(\mathcal{O}_-(\omega)) = 1$. Using the t-conformality of m, we readily deduce that m is a convex combination of measures $\nu_{t,\omega}$, $\omega \in \Omega(T)$. Thus, (c) holds. □

31.4 Invariant measures equivalent to m_T: existence, uniqueness, ergodicity and the finite–infinite dichotomy

In this section, we establish the existence and uniqueness of a T-invariant measure which is absolutely continuous with respect to the h_T-conformal measure $m_T = m_{h_T}$, and we provide a simple criterion for that invariant measure to be either finite or infinite. This is the counterpart of Theorem 30.1.1(e).

Theorem 31.4.1. *If $T : \widehat{\mathbb{C}} \to \widehat{\mathbb{C}}$ is a parabolic rational function with $\deg(T) \geq 2$, then there exists a T-invariant σ-finite Borel measure μ_T on $\mathcal{J}(T)$ which is absolutely continuous with respect to the h_T-conformal measure m_T. In fact, the measure μ_T is unique up to a multiplicative constant, equivalent to m_T, ergodic and conservative. Furthermore, $\mu_T(\mathcal{J}(T) \setminus B(\Omega(T), R)) < \infty$ for every $R > 0$.*

Proof. Fix any $R \in (0, \mathrm{diam}(\mathcal{J}(T)))$. By Proposition 26.2.2, for every $z \in \mathcal{J}(T) \setminus \Omega(T)$ there exists $\eta(z) > 0$ such that

$$B(z, 2\eta(z)) \cap \overline{\mathrm{PC}(T)} = \emptyset.$$

As there are countably many $r > 0$ for which $m_T(\partial B(z, r)) = 0$, we may further assume that each $\eta(z)$ is such that $m_T(\partial B(z, \eta(z))) = 0$. Since $\mathcal{J}(T) \setminus B(\Omega(T), R)$ is compact, there is a finite set $E_R \subseteq \mathcal{J}(T) \setminus B(\Omega(T), R)$ such that

$$\bigcup_{z \in E_R} B(z, \eta(z)) \supseteq \mathcal{J}(T) \setminus B(\Omega(T), R). \tag{31.54}$$

Out of the cover $\{B(z, \eta(z))\}_{z \in E_R}$ of $\mathcal{J}(T) \setminus B(\Omega(T), R)$, form in the standard way (see, for instance, the last step in the proof of Lemma 25.6.4) a finite partition $\{P_z\}_{z \in E_R}$ of $\mathcal{J}(T) \setminus B(\Omega(T), R)$ such that

$$P_z \subseteq B(z, \eta(z)), \quad \forall z \in E_R.$$

As $(\mathcal{J}(T) \setminus \Omega(T)) \cap B(\Omega(T), R)$ is a Lindelöf space, there is a countable set $F_R \subseteq B(\Omega(T), R)$ such that

$$\bigcup_{z \in F_R} B(z, \eta(z)) \supseteq (\mathcal{J}(T) \setminus \Omega(T)) \cap B(\Omega(T), R).$$

In the same standard way, construct out of the cover $\{B(z, \eta(z))\}_{z \in F_R}$ a countable partition $\{P_z\}_{z \in F_R}$ of $(\mathcal{J}(T) \setminus \Omega(T)) \cap B(\Omega(T), R)$ such that

$$P_z \subseteq B(z, \eta(z)), \quad \forall z \in F_R.$$

In fact, in the standard construction of the partitions $\{P_z\}_{z \in E_R}$ and $\{P_z\}_{z \in F_R}$ we may subtract from each element of the cover the closure of the preceding elements (we obviously throw away all empty sets generated by this construction). This way, the elements of the partition $\{P_z\}_{z \in E_R \cup F_R}$ will be nonempty, open and the complement of their union will be a set of m_T-measure zero. This will be useful in part (c) below.

The countable partition $\{P_z\}_{z \in E_R \cup F_R}$ of $\mathcal{J}(T) \setminus \Omega(T)$ is a Marco–Martens cover for the Marco–Martens map $T : \mathcal{J}(T) \to \mathcal{J}(T)$ per Definition 10.4.3:

(a) All the atoms of $\{P_z\}_{z \in E_R \cup F_R}$ are Borel sets.

(b) As m_T is atomless according to Theorem 31.3.11 and $\Omega(T)$ is finite, it is trivial that $m_T(\Omega(T)) = 0$.

(c) Since each P_z has nonempty interior by construction and since $T|_{\mathcal{J}(T)}$ is topologically exact while $\operatorname{supp}(m_T) = \mathcal{J}(T)$, it ensues that $m_T(P_z \cap T^{-k}(P_w)) > 0$ for every $z, w \in E_R \cup F_R$ and some $k = k(z, w) \geq 0$.

(d) For each $z \in E_R \cup F_R$, Koebe's distortion theorem (Theorem 23.1.8) and the h_T-conformality of m_T ensure that

$$m_T(T^{-n}(A)) m_T(B) \leq K m_T(A) m_T(T^{-n}(B)), \quad \forall n \geq 0$$

for all Borel sets $A, B \subseteq P_z$.

(e) Given that m_T is quasi-T-invariant, conservative, and $m_T(P_z) > 0$ for all $z \in E_R \cup F_R$, it follows from Corollary 10.1.10 that $\sum_{n=0}^{\infty} m_T(T^{-n}(P_z)) = \infty$ for every $z \in E_R \cup F_R$.
(f) This property holds since T is Borel measurable.
(g) This property holds per Remark 10.4.4(5) since T is finite-to-one.

Let μ_T be the T-invariant Borel σ-finite measure resulting from Theorem 10.4.5 with reference measure m_T. The uniqueness up to a multiplicative constant, the ergodicity and the conservativity of μ_T ensue from Theorems 31.3.9 and 10.4.5. Moreover, the σ-finiteness of μ_T imposes that $\mu_T(P_z) \le \mu_T(B(z,\eta(z))) < \infty$ for all $z \in E_R \cup F_R$. Since the set E_R is finite and since $\{P_z\}_{z\in E_R}$ is a partition of $\mathcal{J}(T) \setminus B(\Omega(T), R)$, it follows that $\mu_T(\mathcal{J}(T) \setminus B(\Omega(T), R)) < \infty$. □

We now provide a characterization of the finiteness of μ_T.

Theorem 31.4.2. *If $T : \widehat{\mathbb{C}} \to \widehat{\mathbb{C}}$ is a parabolic rational function with* $\deg(T) \ge 2$, *then the measure μ_T from Theorem 31.4.1 is finite if and only if*

$$h_T > \frac{2p_T}{p_T + 1},$$

where $p_T = \max\{p(\omega) : \omega \in \Omega(T)\}$.

Proof. As this measure is invariant for all the iterates of T, replacing T by a sufficiently high iterate allows us to assume that $\Omega(T) = \Omega_0(T)$, i. e. that $\Omega(T)$ consists of simple fixed points of T only. As the map $T : \mathcal{J}(T) \to \mathcal{J}(T)$ is expansive, by virtue of Corollary 5.3.2 there exists a finite Markov partition $\mathcal{R}$ for $T : \mathcal{J}(T) \to \mathcal{J}(T)$ with $\operatorname{diam}(\mathcal{R}) < \delta_T/2$, where we recall that δ_T is an expansive constant for $T|_{\mathcal{J}(T)}$. For every $\omega \in \Omega(T)$, let

$$\mathcal{R}(\omega) = \{R \in \mathcal{R} : \omega \in R\} \quad \text{and} \quad V(\omega) := \bigcup_{R\in\mathcal{R}(\omega)} R.$$

For every $\omega \in \Omega(T)$, let T_ω^{-1} be the local holomorphic inverse branch of T fixing ω. Because all elements of $\Omega(T)$ are simple fixed points, all the iterates T_ω^{-n}, $\omega \in \Omega(T)$, $n \ge 0$, are well-defined on $V(\omega)$ and

$$T_\omega^{-n}(R) \subseteq R, \quad \forall R \in \mathcal{R}(\omega),\ \forall \omega \in \Omega(T),\ \forall n \ge 0.$$

Then

$$T_\omega^{-n}(V(\omega)) \subseteq V(\omega), \quad \forall \omega \in \Omega(T),\ \forall n \ge 0.$$

We further set

$$V_n(\omega) := T_\omega^{-n}(V(\omega)) \setminus \overline{T_\omega^{-(n+1)}(V(\omega))} = T_\omega^{-n}(V_0(\omega)), \quad \forall n \ge 0, \tag{31.55}$$

and

$$V := \bigcup_{\omega\in\Omega(T)} V(\omega) \quad \text{and} \quad V^c := \mathcal{J}(T) \setminus V.$$

By Theorem 31.4.1,

$$0 < \mu_T(V^c) < \infty \quad \text{and} \quad \mu_T(V^c \setminus (V^c)_\infty) = 0, \tag{31.56}$$

and there exists $Q \geq 1$ such that

$$Q^{-1} \leq \frac{d\mu_T}{dm_T}\bigg|_{V^c} \leq Q. \tag{31.57}$$

Let

$$\tau_{V^c} : V^c \to \mathbb{N} \quad \text{and} \quad T_{V^c} : V^c \to V^c$$

be respectively the first return time and the first return map to V^c (see (10.1)–(10.2)). All the iterates of T_{V^c} are well-defined on $V^c \setminus \Omega_\infty^-(T)$. The conditional measure $(\mu_T)_{V^c}$ of μ_T on V^c is T_{V^c}-invariant according to Theorem 10.2.1. By Corollary 10.1.14 and Kac's lemma (Lemma 10.2.6), the measure μ_T is finite if and only if

$$\int_{V^c} \tau_{V^c}\, d\mu_T < \infty.$$

Thus, our task boils down to characterizing the finiteness of the integral of the first return time to V^c.

For every $\omega \in \Omega(T)$ and every $n \geq 0$, let

$$V_n^*(\omega) := \bigcup_{z\in T^{-1}(\omega)\setminus\{\omega\}} T_z^{-1}(V_n(\omega)) \subseteq V^c.$$

Then

$$\tau_{V^c}|_{V_n^*(\omega)} = n + 2 \quad \text{and} \quad \tau_{V^c}|_{V^c\setminus T^{-1}(V)} = 1. \tag{31.58}$$

Using (31.58), we obtain that

$$\begin{aligned}
\int_{V^c} \tau_{V^c}\, d\mu_T &= \int_{V^c\setminus T^{-1}(V)} \tau_{V^c}\, d\mu_T + \sum_{\omega\in\Omega(T)} \sum_{n=0}^{\infty} \int_{V_n^*(\omega)} \tau_{V^c}\, d\mu_T \\
&= \int_{V^c\setminus T^{-1}(V)} 1\, d\mu_T + \sum_{\omega\in\Omega(T)} \sum_{n=0}^{\infty} \int_{V_n^*(\omega)} (n+2) d\mu_T \\
&= \mu_T(V^c\setminus T^{-1}(V)) + \sum_{\omega\in\Omega(T)} \sum_{n=0}^{\infty} (n+2)\mu_T(V_n^*(\omega)).
\end{aligned} \tag{31.59}$$

We need to estimate $\mu_T(V_n^*(\omega))$. Let

$$H := \min\{m_T(V_0(\omega)) : \omega \in \Omega(T)\} > 0.$$

By (24.84) in Proposition 24.2.27, there exists $\Gamma \geq 1$ such that

$$\Gamma^{-1}(n+2)^{-\frac{p(\omega)+1}{p(\omega)}} \leq |(T_\omega^{-n})'(z)| \leq \Gamma(n+2)^{-\frac{p(\omega)+1}{p(\omega)}}$$

for all $n \geq 0$, all $\omega \in \Omega(T)$ and all $z \in V_0(\omega)$. Combining this with (31.55) and the h_T-conformality of m_T, it follows that

$$H\Gamma^{-1} \leq \frac{m_T(V_n(\omega))}{(n+2)^{-\frac{p(\omega)+1}{p(\omega)}h_T}} \leq \Gamma, \quad \forall \omega \in \Omega(T),\ \forall n \geq 0. \tag{31.60}$$

Since the derivatives $|(T_x^{-1})'(z)|$, $x \in T^{-1}(\omega)\backslash\{\omega\}$, $z \in V$, are uniformly bounded away from zero and infinity, it ensues from the h_T-conformality of m_T and (31.60) that there is $C \geq 1$ such that

$$C^{-1} \leq \frac{m_T(V_n^*(\omega))}{(n+2)^{-\frac{p(\omega)+1}{p(\omega)}h_T}} \leq C, \quad \forall \omega \in \Omega(T),\ \forall n \geq 0.$$

Along with (31.57), this yields

$$(QC)^{-1} \leq \frac{\mu_T(V_n^*(\omega))}{(n+2)^{-\frac{p(\omega)+1}{p(\omega)}h_T}} \leq QC, \quad \forall \omega \in \Omega(T),\ \forall n \geq 0. \tag{31.61}$$

By (31.56), we know that $\mu_T(V^c\backslash T^{-1}(V)) < \infty$. We infer from this, (31.59) and (31.61) that

$$\int_{V^c} \tau_{V^c}\, d\mu_T - \mu_T(V^c\backslash T^{-1}(V)) \asymp \sum_{\omega\in\Omega(T)} \sum_{n=0}^{\infty} (n+2)^{1-\frac{p(\omega)+1}{p(\omega)}h_T},$$

where $\asymp$ is the usual symbol of multiplicative comparability (see (25.34)). But the double series converges if and only if $1 - \frac{p_T+1}{p_T}h_T < -1$, which is equivalent to

$$h_T > \frac{2p_T}{p_T+1}.$$

So, by Corollary 10.1.14 and Kac's lemma (Lemma 10.2.6), the measure μ_T is finite if and only if $h_T > 2p_T/(p_T+1)$. □

As a consequence of this theorem and Bowen's formula (Theorem 31.2.1(i); see also Theorem 31.3.11), we get the following two results.

Corollary 31.4.3. *If $T : \widehat{\mathbb{C}} \to \widehat{\mathbb{C}}$ is a parabolic rational function with* $\deg(T) \geq 2$ *and for which* $\mathrm{HD}(\mathcal{J}(T)) \leq 1$, *then the measure μ_T is infinite.*

Corollary 31.4.4. *If $T : \widehat{\mathbb{C}} \to \widehat{\mathbb{C}}$ is a parabolic rational function with* $\deg(T) \geq 2$ *and for which* $\mathrm{HD}(\mathcal{J}(T)) > 1$ *and* $p_T = 1$, *then the measure* μ_T *is finite.*

Example 31.4.5. Let $B : \widehat{\mathbb{C}} \to \widehat{\mathbb{C}}$ be a finite Blaschke product. (For more information on Blaschke products, see [53].) By the Denjoy–Wolff theorem (see [31]), there exists a fixed point $\omega \in \overline{B(0,1)}$ such that

$$\lim_{n\to\infty} B^n(z) = \omega \quad \forall z \in B(0,1), \quad \text{while} \quad \lim_{n\to\infty} B^n(z) = 1/\overline{\omega}, \quad \forall z \in \widehat{\mathbb{C}} \setminus \overline{B(0,1)}.$$

Hence,

$$\mathcal{J}(B) \subseteq \mathbb{S}^1.$$

If $\omega \in \mathbb{S}^1$ and $B'(\omega) = 1$, then $\omega = 1/\overline{\omega} \in \Omega_0(B)$. If additionally $\mathrm{Crit}(B) \cap \mathbb{S}^1 = \emptyset$, then B is a parabolic rational function. So, Corollary 31.4.3 applies.

Example 31.4.6. Let $Q_{1/4} : \widehat{\mathbb{C}} \to \widehat{\mathbb{C}}$ be the quadratic polynomial

$$Q_{1/4}(z) := z^2 + 1/4$$

for which 1/2 is a parabolic fixed point since

$$Q_{1/4}(1/2) = 1/2 \quad \text{and} \quad Q'_{1/4}(1/2) = 1.$$

Apart from the superattracting point ∞, $Q_{1/4}$ has only one critical point, namely 0. By Theorem 24.2.22, this point must belong to the immediate basin of attraction of 1/2. Therefore, $Q_{1/4}$ is a parabolic rational function. A straightforward calculation shows that $p(1/2) = 1$. As $\mathrm{HD}(\mathcal{J}(Q_{1/4})) > 1$ by [151] or [137] (the proof in [137] is quite similar to that of Theorem 30.2.2), Corollary 31.4.4 applies.

31.5 Hausdorff and packing measures

In this section, we give a precise geometric description of the fractal structure of the Julia set of parabolic rational functions in terms of the Hausdorff and packing measures. This is the counterpart of Theorem 30.1.1(f).

Theorem 31.5.1. *If $T : \widehat{\mathbb{C}} \to \widehat{\mathbb{C}}$ is a parabolic rational function with* $\deg(T) \geq 2$, *then the following statements hold:*

(a) *If $h_T < 1$, then $H_{h_T}(\mathcal{J}(T)) = 0$ and there exists a constant $c \in (0,\infty)$ such that $m_{h_T} = cP_{h_T}$. In particular, $0 < P_{h_T}(\mathcal{J}(T)) < \infty$.*
(b) *If $h_T = 1$, then there exist constants $c, d \in (0,\infty)$ such that $m_{h_T} = cP_{h_T} = dH_{h_T}$. In particular, $0 < H_{h_T}(\mathcal{J}(T))P_{h_T}(\mathcal{J}(T)) < \infty$.*
(c) *If $h_T > 1$, then $P_{h_T}(\mathcal{J}(T)) = \infty$ and there exists a constant $d \in (0,\infty)$ such that $m_{h_T} = dH_{h_T}$. In particular, $0 < H_{h_T}(\mathcal{J}(T)) < \infty$.*
(d) $\mathrm{HD}(\mathcal{J}(T)) \leq \mathrm{PD}(\mathcal{J}(T)) < 2$.

The proof requires several intermediate results. Recall that

$$\Omega_0(T) := \{\omega \in \Omega(T) \mid T(\omega) = \omega \text{ and } T'(\omega) = 1\}.$$

The following result is a consequence of Lemma 24.2.11 and Fatou's flower theorem (Theorem 24.2.24).

Lemma 31.5.2. *If $T : \widehat{\mathbb{C}} \to \widehat{\mathbb{C}}$ is a parabolic rational function with $\deg(T) \geq 2$, then there exists $\kappa > 0$ such that T admits holomorphic inverse branches on all balls of radius κ centred on points of $\mathcal{J}(T)$ and for every $\omega \in \Omega_0(T)$ there is a unique analytic inverse branch $T_\omega^{-1} : B(\omega, \kappa) \to \widehat{\mathbb{C}}$ of T such that $T_\omega^{-1}(\omega) = \omega$ and*

$$|T_\omega^{-1}(z) - \omega| < |z - \omega|, \quad \forall z \in B(\omega, \kappa) \cap \mathcal{J}(T) \setminus \{\omega\}.$$

In particular,

$$T_\omega^{-1}(B(\omega, \kappa) \cap \mathcal{J}(T)) \subseteq B(\omega, \kappa) \cap \mathcal{J}(T),$$

and thus all iterates $T_\omega^{-n} : B(\omega, \kappa) \cap \mathcal{J}(T) \to B(\omega, \kappa) \cap \mathcal{J}(T)$ of $T_\omega^{-1}|_{B(\omega,\kappa)\cap\mathcal{J}(T)}$ are well-defined. One may further choose κ so small that $0 < \kappa < \theta_T$ and

$$[B(\omega, \kappa) \cup T(B(\omega, \kappa))] \cap B(\omega_0, \kappa) = \emptyset, \quad \forall \omega, \omega_0 \in \Omega_0(T) \text{ with } \omega \neq \omega_0.$$

Lemma 31.5.3. *Let $T : \widehat{\mathbb{C}} \to \widehat{\mathbb{C}}$ be a parabolic rational function with $\deg(T) \geq 2$ and let $\kappa > 0$ fulfill Lemma 31.5.2. There exists a constant $C_1 > 0$ such that if $T^n(z) \in B(\omega, \kappa/2)$ for some $\omega \in \Omega_0(T)$ and $T^{n-1}(z) \notin B(\Omega(T), \kappa/2)$, then for every $r > 0$ satisfying*

$$2K|T^n(z) - \omega| \leq |(T^n)'(z)|r \leq (2K)^{-1}\kappa, \tag{31.62}$$

we have

$$C_1^{-1}|(T^n)'(z)|^{p(\omega)(h_T-1)} \leq \frac{m_T(B(z,r))}{r^{\alpha_{h_T}(\omega)}} \leq C_1|(T^n)'(z)|^{p(\omega)(h_T-1)},$$

where K is a Koebe distortion constant, and $p(\omega)$ and $\alpha_{h_T}(\omega)$ were respectively defined in (24.24) and (31.14).

Proof. By the right inequality in (31.62), by Lemma 31.5.2 and by Koebe's distortion theorem (Theorem 23.1.10), we get

$$B(z, r) \subseteq T_z^{-n}\big(B(T^n(z), K|(T^n)'(z)|r)\big). \tag{31.63}$$

By the left inequality in (31.62), we have $|T^n(z) - \omega| \leq K|(T^n)'(z)|r$. Using this and the right inequality in (31.62) successively, we obtain that

$$B(T^n(z), K|(T^n)'(z)|r) \subseteq B(\omega, 2K|(T^n)'(z)|r) \subseteq B(\omega, \kappa). \tag{31.64}$$

Relations (31.63)–(31.64), the h_T-conformality of m_T in the form (31.9), the Koebe distortion theorem and Theorem 31.3.7 yield that

$$\begin{aligned} m_T(B(z,r)) &\le m_T\Big(T_z^{-n}\big(B(\omega, 2K|(T^n)'(z)|r)\big)\Big) \\ &\le K^{h_T}|(T_z^{-n})'(T^n(z))|^{h_T} m_T(B(\omega, 2K|(T^n)'(z)|r)) \\ &\le K^{h_T}|(T^n)'(z)|^{-h_T} c(2K|(T^n)'(z)|r)^{\alpha_{h_T}(\omega)} \\ &= cK^{h_T}(2K)^{\alpha_{h_T}(\omega)} \cdot |(T^n)'(z)|^{p(\omega)(h_T-1)} \cdot r^{\alpha_{h_T}(\omega)}. \end{aligned}$$

On the other hand (though in a similar way), by the right inequality in (31.62), by Lemma 31.5.2 and by the Koebe distortion theorem, we get

$$T_z^{-n}\big(B(T^n(z), K^{-1}|(T^n)'(z)|r)\big) \subseteq B(z,r) \tag{31.65}$$

while, by the left inequality in (31.62) and (31.64), we have

$$B(\omega, (2K)^{-1}|(T^n)'(z)|r) \subseteq B(T^n(z), K^{-1}|(T^n)'(z)|r) \subseteq B(\omega, \kappa). \tag{31.66}$$

Relations (31.65)–(31.66), the h_T-conformality of m_T in the form (31.9), the Koebe distortion theorem and Theorem 31.3.7 give

$$\begin{aligned} m_T(B(z,r)) &\ge m_T\Big(T_z^{-n}\big(B(\omega, (2K)^{-1}|(T^n)'(z)|r)\big)\Big) \\ &\ge K^{-h_T}|(T^n)'(z)|^{-h_T} m_T(B(\omega, (2K)^{-1}|(T^n)'(z)|r)) \\ &\ge K^{-h_T}|(T^n)'(z)|^{-h_T} c^{-1}((2K)^{-1}|(T^n)'(z)|r)^{\alpha_{h_T}(\omega)} \\ &= c^{-1}K^{-h_T}(2K)^{-\alpha_{h_T}(\omega)} \cdot |(T^n)'(z)|^{p(\omega)(h_T-1)} \cdot r^{\alpha_{h_T}(\omega)}. \end{aligned}$$

Take $C_1 = cK^{h_T}(2K)^{\alpha_{h_T}(\omega)}$. □

Proposition 31.5.4. *If $T : \widehat{\mathbb{C}} \to \widehat{\mathbb{C}}$ is a parabolic rational function with* $\deg(T) \ge 2$, *then there exists a constant $C > 0$ such that:*
(a) *$H_{h_T} \le Cm_T$, and thus $H_{h_T}(\mathcal{J}(T)) < \infty$.*
(b) *$m_T \le CP_{h_T}$, and hence $P_{h_T}(\mathcal{J}(T)) > 0$.*

Proof. (a) Let $A \subseteq \mathcal{J}(T)$ be any Borel set. Let $A^c := A \setminus \Omega_\infty^-(T)$. Since $\Omega_\infty^-(T)$ is countable, we know that $H_{h_T}(A^c) = H_{h_T}(A)$. Fix $\varepsilon, \xi > 0$. Per a construction similar to that in the proof of Lemma 31.3.4 (replace A by A^c and ε by ξ in (31.34)), the family $\{B(z,r(z))\}_{z\in A^c}$ forms an open cover of A^c such that

$$m_{h_T}\Big(\bigcup_{z\in A^c} B(z,r(z))\Big) \le m_{h_T}(A^c) + \xi \quad \text{and} \quad 0 < r(z) \le \varepsilon, \ \forall z \in A^c.$$

By Besicovitch's covering theorem (Theorem 15.4.3), there exists a countable subset $\{z_k\}_{k=1}^{\infty}$ of A^c such that the collection $\{B(z_k, r(z_k))\}_{k=1}^{\infty}$ covers A^c and can be decomposed into $C = c(2)$ packings of A^c, where C is a universal constant, i. e., C is independent of the cover and of the subcover. Applying Lemma 31.3.3 to the measure m_{h_T}, we deduce that

$$\begin{aligned}\sum_{k=1}^{\infty}(r(z_k))^{h_T} &\le c_{h_T,\theta_T}\sum_{k=1}^{\infty} m_{h_T}(B(z_k, r(z_k)))\\ &\le c_{h_T,\theta_T} C m_{h_T}\left(\bigcup_{k=1}^{\infty} B(z_k, r(z_k))\right)\\ &\le C c_{h_T,\theta_T}(m_{h_T}(A^c) + \xi).\end{aligned}$$

Letting $\xi \searrow 0$ and then $\varepsilon \searrow 0$, we conclude that

$$H_{h_T}(A) = H_{h_T}(A^c) \le C c_{h_T,\theta_T} m_{h_T}(A^c) \le C c_{h_T,\theta_T} m_{h_T}(A).$$

(b) Given $z \in \mathcal{J}(T) \setminus \Omega_{\infty}^{-}(T)$, let $(n_j(z))_{j=1}^{\infty}$ be the sequence of all $n \in \mathbb{N}$ such that $T^n(z) \notin B(\Omega(T), \delta_T)$ (cf. (31.24)). Recall that $\lim_{j\to\infty}|(T^{n_j(z)})'(z)| = \infty$ according to Lemma 26.2.5(b). Moreover, by Lemma 31.3.3,

$$c_{h_T,\theta_T}^{-1} \le \frac{m_T(B(z, r_j(z)))}{(r_j(z))^{h_T}} \le c_{h_T,\theta_T}, \quad \forall j \in \mathbb{N},$$

where $r_j(z) = \frac{1}{4}\theta_T|(T^{n_j(z)})'(z)|^{-1}$. So, $\lim_{j\to\infty} r_j(z) = 0$. As the set $\Omega_{\infty}^{-}(T)$ is countable and the measure m_T is nonatomic per Theorem 31.3.11, the result ensues from Theorem 15.5.4(b) with $A = X = \mathcal{J}(T)$. □

Recall that $\mathcal{O}_+(z) = \{T^n(z) : n \ge 0\}$ is the forward T-orbit of z. We now establish that the set of Julia points whose forward orbit's closure does not intersect $\Omega(T)$, is negligible. Define

$$S(T) := \{z \in \mathcal{J}(T) \,|\, \overline{\mathcal{O}_+(z)} \cap \Omega(T) = \emptyset\}. \tag{31.67}$$

Lemma 31.5.5. *If $T : \widehat{\mathbb{C}} \to \widehat{\mathbb{C}}$ is a parabolic rational function with* $\deg(T) \ge 2$, *then* $H_{h_T}(S(T)) = P_{h_T}(S(T)) = m_T(S(T)) = 0$.

Proof. Observe that $S(T) = \bigcup_{k=1}^{\infty} S_k(T)$, where

$$S_k(T) := \{z \in \mathcal{J}(T) \,|\, \mathcal{O}_+(z) \cap B(\Omega(T), 1/k) = \emptyset\} \subseteq \mathcal{J}(T) \setminus \Omega_{\infty}^{-}(T).$$

Since $T(S_k(T)) \subseteq S_k(T)$, since $\mathcal{J}(T) \setminus S_k(T)$ is nonempty and relatively open and since the h_T-conformal measure m_T is conservative, ergodic and positive on nonempty relatively open sets per Theorem 31.3.9, we know by Theorem 10.1.11 that

$$m_T\big(\mathcal{J}(T)\setminus(\mathcal{J}(T)\setminus S_k(T))_\infty\big)=0, \quad \text{and thus} \quad m_T(S_k(T))=0, \quad \forall k\in\mathbb{N}.$$

Because $S(T)=\bigcup_{k=1}^{\infty}S_k(T)$, we deduce that $m_T(S(T))=0$ and the conclusion that $H_{h_T}(S(T))=0$ follows from Proposition 31.5.4(a).

To corroborate that $P_{h_T}(S(T))=0$, we will use Lemma 31.3.3. Fix any $k\in\mathbb{N}$ such that $1/k<\delta_T$. Let $z\in S_k(T)$. Then the sequence $(n_j(z))_{j=1}^{\infty}$ of all $n\in\mathbb{N}$ such that $T^n(z)\notin B(\Omega(T),1/k)$ is simply the full sequence $(j)_{j=1}^{\infty}$. Note that $1/k$ is an expansive constant for $T|_{\mathcal{J}(T)}$. This is all that is required about δ_T from relation (31.24) onwards. Hence, all the concepts and results introduced and proved from that point on remain true with δ_T replaced by $1/k$. Only resulting constants are affected. In particular, θ_T will be replaced by $\theta_{1/k}$ in the rest of this proof. Per Lemma 31.3.3,

$$c_{h_T,\theta_{1/k}}^{-1}\le\frac{m_T(B(z,r_j(z)))}{(r_j(z))^{h_T}}\le c_{h_T,\theta_{1/k}}, \quad \forall j\in\mathbb{N},$$

where $r_j(z)=\frac{1}{4}\theta_{1/k}|(T^j)'(z)|^{-1}$. Per Lemma 26.2.5(b), $\lim_{j\to\infty}|(T^j)'(z)|=\infty$ and thus $\lim_{j\to\infty}r_j(z)=0$. Observe that

$$\frac{r_j(z)}{r_{j+1}(z)}=\frac{|(T^{j+1})'(z)|}{|(T^j)'(z)|}=|T'(T^j(z))|, \quad \forall j\in\mathbb{N}.$$

Since $\mathcal{J}(T)$ is compact and $\mathcal{J}(T)\cap\mathrm{Crit}(T)=\emptyset$ due to the parabolicity of T, there exists a constant $C\ge1$ such that

$$C^{-1}\le\inf_{\zeta\in\mathcal{J}(T)}|T'(\zeta)|\le\sup_{\zeta\in\mathcal{J}(T)}|T'(\zeta)|\le C.$$

Therefore,

$$C^{-1}\le\frac{r_j(z)}{r_{j+1}(z)}\le C, \quad \forall j\in\mathbb{N}.$$

Let $0<r\le\frac{1}{4}\theta_{1/k}$ and set $j=j(r,z)\in\mathbb{N}$ so that $r_{j+1}(z)<r\le r_j(z)$. It ensues that

$$\begin{aligned}
\frac{m_T(B(z,r_{j+1}(z)))}{(r_j(z))^{h_T}}&\le\frac{m_T(B(z,r))}{r^{h_T}}\le\frac{m_T(B(z,r_j(z)))}{(r_{j+1}(z))^{h_T}}\\
C^{-h_T}\frac{m_T(B(z,r_{j+1}(z)))}{(r_{j+1}(z))^{h_T}}&\le\frac{m_T(B(z,r))}{r^{h_T}}\le\frac{m_T(B(z,r_j(z)))}{(r_j(z))^{h_T}}C^{h_T}\\
C^{-h_T}c_{h_T,\theta_{1/k}}^{-1}&\le\frac{m_T(B(z,r))}{r^{h_T}}\le c_{h_T,\theta_{1/k}}C^{h_T}.
\end{aligned}$$

It follows from Theorems 15.5.3–15.5.4(a) that $H_{h_T}(S_k(T)) = P_{h_T}(S_k(T)) = m_T(S_k(T)) = 0$ for all k such that $1/k < \delta_T$, and in fact for all $k \in \mathbb{N}$ since the sequence $(S_k(T))_{k=1}^{\infty}$ is ascending. As $S(T) = \bigcup_{k=1}^{\infty} S_k(T)$, the result holds. □

The last ingredient needed to complete the proof of Theorem 31.5.1 is next.

Proposition 31.5.6. *Let $T : \widehat{\mathbb{C}} \to \widehat{\mathbb{C}}$ be a parabolic rational function with* $\deg(T) \geq 2$.
(a) *If $h_T < 1$, then $H_{h_T}(\mathcal{J}(T)) = 0$.*
(b) *If $h_T \geq 1$, then $H_{h_T}(\mathcal{J}(T)) > 0$.*
(c) *If $h_T \leq 1$, then $P_{h_T}(\mathcal{J}(T)) < \infty$.*
(d) *If $h_T > 1$, then $P_{h_T}(\mathcal{J}(T)) = \infty$.*

Proof. (a) As $\#\Omega(T) < \infty$, there is $q \in \mathbb{N}$ with $\Omega(T^q) = \Omega_0(T^q) = \Omega(T)$. As $\mathcal{J}(T^q) = \mathcal{J}(T)$, we may without loss of generality assume that $\Omega(T) = \Omega_0(T)$. So, pick $\omega \in \Omega(T) = \Omega_0(T)$ and let

$$X(\omega) := \{z \in \mathcal{J}(T) \setminus \Omega_{\infty}^{-}(T) : \omega \in \overline{\mathcal{O}_+(z)}\}.$$

Let $0 < \kappa < \delta_T$ satisfy Lemma 31.5.2, and thereby Lemma 31.5.3 as well. As the diameter of $B(\omega, \kappa/2)$ does not exceed an expansive constant of $T : \mathcal{J}(T) \to \mathcal{J}(T)$, there exists for each $z \in X(\omega)$ a strictly increasing sequence $(n_j(z))_{j=1}^{\infty}$ of positive integers such that $\lim_{j\to\infty} |T^{n_j(z)}(z) - \omega| = 0$ and $T^{n_j(z)}(z) \in B(\omega, \kappa/2)$ while $T^{n_j(z)-1}(z) \notin B(\omega, \kappa/2)$ for every $j \in \mathbb{N}$ (or equivalently $T^{n_j(z)-1}(z) \notin B(\Omega(T), \kappa/2)$ because of the choice of κ in Lemma 31.5.2). Setting

$$r_j(z) = 2K\left|T^{n_j(z)}(z) - \omega\right|\left|(T^{n_j(z)})'(z)\right|^{-1},$$

we glean from Lemma 31.5.3 that for j large enough

$$C^{-1}\left|T^{n_j(z)}(z) - \omega\right|^{p(\omega)(h_T-1)} \leq \frac{m_T(B(z, r_j(z)))}{(r_j(z))^{h_T}} \leq C\left|T^{n_j(z)}(z) - \omega\right|^{p(\omega)(h_T-1)}$$

where $C = C_1(2K)^{p(\omega)(h_T-1)}$. Therefore,

$$\limsup_{r\to 0} \frac{m_T(B(z,r))}{r^{h_T}} = \infty \text{ when } h_T < 1 \quad \text{whereas} \quad \liminf_{r\to 0} \frac{m_T(B(z,r))}{r^{h_T}} = 0 \text{ when } h_T > 1.$$

If $h_T < 1$, then $H_{h_T}(X(\omega)) = 0$ per Theorem 15.5.3(a). We also know that $H_{h_T}(S(T)) = 0$ by Lemma 31.5.5. As $\Omega_{\infty}^{-}(T)$ is countable, we further have that $H_{h_T}(\Omega_{\infty}^{-}(T)) = 0$. Since $\mathcal{J}(T) = S(T) \cup \bigcup_{\omega\in\Omega(T)} X(\omega) \cup \Omega_{\infty}^{-}(T)$, we conclude that $H_{h_T}(\mathcal{J}(T)) = 0$. So, part (a) holds.

(d) On the other hand, as there is an ergodic T-invariant measure equivalent to m_T (namely, μ_T from Theorem 31.4.1) and as the conformal m_T is positive on nonempty open sets, we deduce that $m_T(\bigcup_{\omega\in\Omega(T)} X(\omega)) = 1$. This, Theorem 15.5.4(b) and the above show that $P_{h_T}(\bigcup_{\omega\in\Omega(T)} X(\omega)) = \infty$ when $h_T > 1$. Thus (d) holds.

(b) According to Theorem 15.5.3(b), part (b) of the current lemma will hold if we establish that whenever $h_T \geq 1$, there exists $M \geq 1$ such that

$$\frac{m_T(B(z,r))}{r^{h_T}} \leq M, \quad \forall 0 < r \leq 1, \ \forall z \in \mathcal{J}(T). \tag{31.68}$$

For this, it suffices to show the existence of constants $C, M \geq 1$ such that for each $z \in \mathcal{J}(T)$ there is a sequence of radii $0 < r_i(z) \leq 1$ satisfying

$$\lim_{i\to\infty} r_i(z) = 0, \quad C^{-1} \leq \frac{r_i(z)}{r_{i+1}(z)} \leq C \quad \text{and} \quad \frac{m_T(B(z,r_i(z)))}{(r_i(z))^{h_T}} \leq M, \quad \forall i \in \mathbb{N}.$$

Fix $z \in \mathcal{J}(T)$. According to Proposition 26.2.2, outside of $B(\Omega(T),\kappa)$ all the holomorphic inverse branches of T^n, $n \in \mathbb{N}$, are well-defined on balls of radius $2\theta_\kappa$, where $\theta_\kappa \in (0,\kappa)$. For every $n \in \mathbb{N}$, let $r_n(z) := \theta_\kappa (K|(T^n)'(z)|)^{-1}$. Using Koebe's distortion theorem, observe that $T_z^{-n}(B(T^n(z),\theta_\kappa)) \supseteq B(z, r_n(z))$ whenever $T^n(z) \notin B(\Omega(T),\kappa)$. In that case, it follows from the h_T-conformality of m_T and Koebe's distortion theorem that

$$\begin{aligned} m_T(B(z,r_n(z))) &\leq m_T(T_z^{-n}(B(T^n(z),\theta_\kappa))) \\ &\leq K^{h_T} |(T_z^{-n})'(T^n(z))|^{h_T} m_T(B(T^n(z),\theta_\kappa)) \\ &\leq K^{h_T} |(T^n)'(z)|^{-h_T} = K^{h_T} [K\theta_\kappa^{-1} r_n(z)]^{h_T} \\ &= K^{h_T} (K\theta_\kappa^{-1})^{h_T} \cdot (r_n(z))^{h_T}. \end{aligned}$$

Let

$$\mathbb{N}(z) = \{n \in \mathbb{N} : \exists \omega \in \Omega(T) \text{ such that } T^n(z) \in B(\omega,\kappa) \text{ while } T^{n-1}(z) \notin B(\omega,\kappa)\}.$$

For $n \in \mathbb{N}(z)$, let $k(n) \geq n+1$ be the smallest $k \in \mathbb{N}$ (if it exists) such that $T^k(z) \notin B(\omega,\kappa)$ or equivalently (because of the choice of κ in Lemma 31.5.2) such that $T^k(z) \notin B(\Omega(T),\kappa)$. Otherwise, set $k(n) = \infty$. Define intervals I_n by

$$\text{either } I_n = [r_{k(n)}(z), r_n(z)] \text{ when } k(n) < \infty \quad \text{or} \quad I_n = (0, r_n(z)] \text{ when } k(n) = \infty.$$

As

$$0 < \inf_{x\in\mathcal{J}(T)} |T'(x)| \leq \frac{|(T^{j+1})'(z)|}{|(T^j)'(z)|} \leq \|T'\|_\infty, \quad \forall j \in \mathbb{N},$$

it follows from the above that it suffices to prove (31.68) for $r \in \bigcup_{n\in\mathbb{N}(z)} I_n$. Fix $n \in \mathbb{N}(z)$ and suppose that $k(n) < \infty$. Consider $r \in [c_n, b_n]$, where $b_n = r_n(z) = \theta_\kappa(K|(T^n)'(z)|)^{-1}$ is the right endpoint of I_n and $c_n = K|T^n(z) - \omega||(T^n)'(z)|^{-1}$. Applying Lemma 31.5.3 and remembering that $h_T \geq 1$ as part (b)'s hypothesis, we get for every $r \in [c_n, b_n]$ that

$$\begin{aligned} m_T(B(z,r)) &\le C_1|(T^n)'(z)|^{p(\omega)(h_T-1)} r^{\alpha_{h_T}(\omega)} = C_1|(T^n)'(z)|^{p(\omega)(h_T-1)} r^{h_T+p(\omega)(h_T-1)} \\ &\le C_1|(T^n)'(z)|^{p(\omega)(h_T-1)} \left[\theta_\kappa (K|(T^n)'(z)|)^{-1}\right]^{p(\omega)(h_T-1)} r^{h_T} \\ &= C_1(\theta_\kappa K^{-1})^{p(\omega)(h_T-1)} r^{h_T}. \end{aligned}$$

It is left to prove (31.68) for $r \in [a_n, c_n]$, where $a_n = r_{k(n)}(z) = \theta_\kappa(K|(T^{k(n)})'(z)|)^{-1}$ is the left endpoint of I_n. To this end, for every $i = n, n+1, \dots, k(n)$ consider the radius

$$\overline{r}_i := K^{-1}\tau|T^i(z) - \omega||(T^i)'(z)|^{-1} = 2\tau\theta_\kappa^{-1}|T^i(z) - \omega|r_i(z),$$

where τ arises from Lemma 24.2.25. Since $2\tau\theta_\kappa^{-1}\kappa \le \overline{r}_{k(n)}/a_n \le 2\tau\theta_\kappa^{-1}\operatorname{diam}(\mathcal{J}(T))$ and $\overline{r}_n/c_n = (2K^2)^{-1}\tau$, it is enough to prove (31.68) for $r \in [\overline{r}_{k(n)}, \overline{r}_n]$. Moreover, for every $\zeta \in \Omega(T)$, let $T_\zeta^{-1} : B(\zeta, 2\theta_\kappa) \to \mathbb{C}$ be the holomorphic branch of T^{-1} that fixes ζ. Observe that

$$\|T'\|_\infty^{-1} \min\{\|(T_\zeta^{-1})'\|_\infty^{-1} : \zeta \in \Omega(T)\} \le \frac{\overline{r}_{i+1}}{\overline{r}_i} \le \frac{\|T'\|_\infty}{\inf\{|T'(x)| : x \in \mathcal{J}(T)\}}, \qquad \forall n \le i < k(n).$$

Consequently, it suffices to prove (31.68) for $r = \overline{r}_{k(n)}, \dots, \overline{r}_n$. Fix $n \le i \le k(n)$. By Lemma 24.2.25, there is an analytic inverse branch $T_z^{-i} : B(T^i(z), 2\tau|T^i(z) - \omega|) \to \mathbb{C}$ which sends $T^i(z)$ to z. Using Koebe's distortion theorem, notice that $T^i(B(z, \overline{r}_i)) \subseteq B(T^i(z), K|(T^i)'(z)|\overline{r}_i) = B(T^i(z), \tau|T^i(z) - \omega|)$. Hence, by the h_T-conformality of m_T, the Koebe distortion theorem and Lemma 31.3.8(a) we gather that

$$\begin{aligned} m_T(B(z, \overline{r}_i)) &\le K^{h_T}|(T_z^{-i})'(T^i(z))|^{h_T} m_T(B(T^i(z), \tau|T^i(z) - \omega|)) \\ &\le K^{h_T}|(T^i)'(z)|^{-h_T} \cdot C_2|T^i(z) - \omega|^{h_T} = C_2 K^{2h_T}\tau^{-h_T} \cdot \overline{r}_i^{h_T}. \end{aligned}$$

Thus, the proof of (31.68) is complete in the case $k(n) < \infty$.

If $k(n) = \infty$, then $T^n(z) = \omega$ and for every $r \in I_n := (0, r_n(z)]$ relation (31.68) immediately follows from Theorem 31.3.7, the conformality of m_T and the Koebe distortion theorem.

(c) By Theorem 15.5.4(a), it is sufficient to prove that whenever $h_T \le 1$, there exists $M \ge 1$ such that

$$\frac{m_T(B(z,r))}{r^{h_T}} \ge M, \quad \forall 0 < r \le 1, \ \forall z \in \mathcal{J}(T). \tag{31.69}$$

Proceed in a similar way to the proof of part (b), using Lemma 31.3.8(b) rather than Lemma 31.3.8(a). □

Proof of Theorem 31.5.1. According to Lemma 15.2.25, whenever they are nonzero and finite the measures $H_{h_T}|_{\mathcal{J}(T)}$ and $P_{h_T}|_{\mathcal{J}(T)}$ are h_T-conformal since $\mathrm{P}(h_T) = 0$ (cf. Definition 29.2.10). Part (a) then follows from Propositions 31.5.6(a,c) and 31.5.4(b), as well as Theorem 31.3.11(b). Part (b) ensues from Propositions 31.5.6(b,c) and 31.5.4, as well as Theorem 31.3.11(b). Part (c) is a consequence of Propositions 31.5.6(b,d) and 31.5.4(a), as well as Theorem 31.3.11(b). Finally, we can easily prove part (d). Since $\mathcal{J}(T) \neq \widehat{\mathbb{C}}$ per Exercise 31.6.1, it follows from Theorem 29.2.13 and relation (31.13) that $\mathrm{Leb}_2(\mathcal{J}(T)) = 0$. As both measures H_2 and P_2 are constant multiples of Leb_2, part (c) does not hold, which firstly means that $\mathrm{HD}(\mathcal{J}(T)) < 2$. Secondly, it gives that $P_{h_T}(\mathcal{J}(T)) < \infty$. Thus, $\mathrm{PD}(\mathcal{J}(T)) \le h_T$. Since we just proved that $h_T = \mathrm{HD}(\mathcal{J}(T)) < 2$, we are done. □

Remark 31.5.7. In fact, as for expanding rational functions, we have

$$\mathrm{HD}(\mathcal{J}(T)) = \mathrm{PD}(\mathcal{J}(T)) = \mathrm{BD}(\mathcal{J}(T))$$

for all parabolic T but the proof of these equalities is much subtler and much more involved than for expanding functions. It was for the first time provided in [41] (cf. [124]).

Remark 31.5.8. For the simplest parabolic quadratic polynomial $Q_{1/4}(z) = z^2 + 1/4$, we know more than just the inequality $\mathrm{HD}(\mathcal{J}(T)) < 2$ established in Theorem 31.5.1(d). Namely, $h_{Q_{1/4}} = \mathrm{HD}(\mathcal{J}(Q_{1/4})) < 1.295$ per [55]. It also follows from Theorem 31.5.1 that

$$0 < H_{h_{Q_{1/4}}}(\mathcal{J}(Q_{1/4})) < \infty \quad \text{and} \quad P_{h_{Q_{1/4}}}(\mathcal{J}(Q_{1/4})) = \infty.$$

On the other hand, if $B : \widehat{\mathbb{C}} \to \widehat{\mathbb{C}}$ is a Blaschke product as in Example 31.4.5 and if some open arc in $\mathbb{S}^1$ containing the parabolic fixed point ω is contained in the Fatou set of B (as a matter of fact in the immediate basin of attraction to ω), then $h_B = \mathrm{HD}(\mathcal{J}(B)) < 1$ (see Exercise 31.6.2). So, it follows from Theorem 31.5.1 that

$$0 < P_{h_B}(\mathcal{J}(B)) < \infty \quad \text{and} \quad H_{h_B}(\mathcal{J}(B)) = 0.$$

31.6 Exercises

Exercise 31.6.1. Prove that $\mathcal{J}(T) \neq \widehat{\mathbb{C}}$ for every parabolic rational function $T : \widehat{\mathbb{C}} \to \widehat{\mathbb{C}}$.

Exercise 31.6.2. Let $B : \widehat{\mathbb{C}} \to \widehat{\mathbb{C}}$ be a Blaschke product as in Remark 31.5.8 (see also Example 31.4.5). Show that $\mathrm{HD}(\mathcal{J}(B)) < 1$.

Hint: Use the method from the proof of Theorem 31.5.1(d).

Exercise 31.6.3. Let $T : \widehat{\mathbb{C}} \to \widehat{\mathbb{C}}$ be a rational function with $\deg(T) \ge 2$ and such that $\Omega(T) \not\subset \overline{\mathrm{PC}(T)}$. Show that for every $t > \delta(T)$ there exists a t-conformal measure supported on $\Omega_\infty^-(T)$.

32 Conformal measures, invariant measures and fractal geometry of subexpanding rational functions

In this chapter, we continue the study of Hausdorff, packing, conformal, and invariant measures that are absolutely continuous with respect to conformal ones, this time for subexpanding rational functions.

Recall from Section 26.3 that a rational map $R : \widehat{\mathbb{C}} \to \widehat{\mathbb{C}}$ is subexpanding if the restriction of R to the intersection of the Julia set $\mathcal{J}(R)$ with the ω-limit set of the critical points of R, is expanding with respect to the spherical metric on $\mathcal{J}(R)$.

In the previous two chapters, we dealt with fractal geometry, Sullivan conformal measures, and invariant measures for expanding and parabolic rational functions. We observed a number of similarities and differences between them. In this chapter, we examine subexpanding maps from the same perspective. Overall, their properties turn out to be somewhat more similar to expanding mappings than to parabolic ones, though they still share some important features with these latter. We prove for subexpanding maps essentially the same properties as for expanding systems in Chapter 30 except that, as in the parabolic case in Chapter 31, there are t-conformal measures for all $t > h_R$ and all of them are purely atomic supported this time on the inverse orbits of critical points (see Theorem 32.1.13).

The structure of this chapter is generally the same as that of Chapter 31. Briefly summarizing its content, Subsection 32.1.1 comprises general properties of subexpanding maps as well as properties of their Sullivan conformal measures, including relationships between conformal measures with different exponents, and culminates with Bowen's formula. Subsection 32.1.2, especially Theorem 32.1.9, pertains to the existence, uniqueness, ergodicity and conservativity of an atomless h_R-conformal measure m_R. We also show in that theorem that m_R is a geometric measure per (32.34). In Subsection 32.1.3, a full description of all Sullivan conformal measures for subexpanding rational functions is provided in Theorem 32.1.13. That description is quite different from that for expanding rational functions (cf. Theorem 30.1.1) and is more similar to the case of parabolic rational functions (cf. Theorems 31.2.1 and 31.3.11).

In Section 32.2, we establish the existence and uniqueness of an R-invariant Borel probability measure μ_R which is absolutely continuous with respect to the h_R-conformal measure m_R. We further show that μ_R is ergodic and equivalent to m_R.

Finally, in Section 32.3, we provide in Theorem 32.3.1 an exact geometric description of the fractal structure of the Julia set of a subexpanding rational function in terms of the Hausdorff and packing measures. It is actually the same as that for expanding rational functions (cf. Theorem 30.1.1) and is much simpler than that for parabolic rational functions (cf. Theorem 31.5.1). All possible dimensions, geometrical and dynamical, are equal and both the Hausdorff and packing measures are finite, positive and geometric.

https://doi.org/10.1515/9783110769876-032

Our presentation of subexpanding maps given in Section 26.3 and especially in Chapter 32, partly stems from [39] but substantially improves it and goes beyond it. In turn, the research in [39] was prompted by the paper [54] where it was proved that any subexpanding rational function whose Julia set is the whole sphere $\widehat{\mathbb{C}}$ admits an invariant Borel probability measure absolutely continuous with respect to the Lebesgue measure.

32.1 Sullivan conformal measures for subexpanding rational functions

32.1.1 Technical preparations and Bowen's formula

We start this section with the following general result.

Lemma 32.1.1. *Let $Q : \widehat{\mathbb{C}} \to \widehat{\mathbb{C}}$ be a rational function with $\deg(Q) \geq 2$. Let $\xi_0 > 0$ be so small that for every point $x \in \widehat{\mathbb{C}}$ there exist at least three different periodic points $a, b, c \in \widehat{\mathbb{C}}$ whose forward trajectories do not intersect the ball $B(x, \xi_0)$. Let $z \in \mathcal{J}(Q)$.*

(a) *If there is a strictly increasing sequence $(n_k = n_k(z))_{k=1}^{\infty}$ and a number $\xi_1(z) \in (0, \xi_0]$ such that for every $k \in \mathbb{N}$ there exists a unique local holomorphic inverse branch $Q_z^{-n_k} : B(Q^{n_k}(z), \xi_1(z)) \to \widehat{\mathbb{C}}$ of Q^{n_k} such that $Q_z^{-n_k}(Q^{n_k}(z)) = z$, then*

$$\limsup_{k\to\infty} |(Q^{n_k})'(z)| = \infty. \tag{32.1}$$

Moreover, for every $t \geq 0$ and every t-conformal measure m for Q there exist a constant $C_4 \geq 1$ (depending only on $\xi_1(z)$ with respect to z and depending in a monotonically increasing way on t) and a sequence $(r_k(z))_{k=1}^{\infty}$ of positive numbers such that $\lim_{k\to\infty} r_k(z) = 0$ and

$$C_4^{-1} \leq \frac{m(B(z, r_k(z)))}{(r_k(z))^t} \leq C_4, \quad \forall k \in \mathbb{N}.$$

(b) *Suppose that there exist a constant $C_5 \geq 1$ (potentially depending on z) and a number $\xi_2(z) \in (0, \xi_0]$ such that for every $r \in (0, \xi_2(z))$ there is $n = n(r) \in \mathbb{N}$ for which*

$$C_5^{-1} \leq r|(Q^n)'(z)| \leq C_5$$

and a unique holomorphic inverse branch $Q_z^{-n} : B(Q^n(z), \xi_2(z)) \to \widehat{\mathbb{C}}$ of Q^n satisfying $Q_z^{-n}(Q^n(z)) = z$. Then for every $t \geq 0$ and every t-conformal measure m for Q there exists a constant $C_6 \geq 1$ (depending on C_5 and $\xi_2(z)$) such that

$$C_6^{-1} \leq \frac{m(B(z, r))}{(r(z))^t} \leq C_6, \quad \forall 0 < r \leq 1.$$

Proof. (a) Passing to a subsequence if necessary, we may assume without loss of generality that there exists $y \in \mathcal{J}(Q)$ such that

$$\lim_{k\to\infty} Q^{n_k}(z) = y \quad \text{and} \quad |Q^{n_k}(z) - y| \le \iota, \quad \forall k \in \mathbb{N},$$

where $\iota := \xi_1(z)/2$. Hence, $B(y,\iota) \subseteq B(Q^{n_k}(z), 2\iota)$ for all $k \in \mathbb{N}$ and the family of restrictions $\{Q_z^{-n_k} : B(y,\iota) \to \widehat{\mathbb{C}}\}_{k=1}^{\infty}$ is well-defined. It follows from Lemma 24.1.13 that

$$\lim_{k\to\infty} |(Q^{n_k})'(z)| = \infty.$$

So, relation (32.1) is proved.

In order to prove the other part of (a), take $\zeta := \gamma/K$, where K is the constant defined in Lemma 23.1.11 (a consequence of Koebe's distortion theorems; set $K = k_{2R,s}(1/2)$ with $R = \xi_1(z)/2$ and s any number in $(0,\pi)$). Let

$$r_k = r_k(z) := |(Q^{n_k})'(z)|^{-1}\zeta = |(Q_z^{-n_k})'(Q^{n_k}(z))|\zeta, \quad \forall k \in \mathbb{N}.$$

By Lemma 24.1.13, there is $l \in \mathbb{N}$ such that $\operatorname{diam}(Q_z^{-n_k}(B(Q^{n_k}(z), \xi_1(z)))) \le \pi - s$ for every $k \ge l$. Fix any such k. Applying Lemma 23.1.11 to $Q_z^{-n_k} : B(Q^{n_k}(z), 2\iota) \to \widehat{\mathbb{C}}$, we get

$$Q_z^{-n_k}(B(Q^{n_k}(z), K^{-1}\zeta)) \subseteq B(z, r_k) \subseteq Q_z^{-n_k}(B(Q^{n_k}(z), K\zeta)).$$

Using the t-conformality of m, we obtain

$$\begin{aligned} m(B(z,r_k)) &\ge m(Q_z^{-n_k}(B(Q^{n_k}(z), K^{-1}\zeta))) = \int_{B(Q^{n_k}(z),K^{-1}\zeta)} |(Q_z^{-n_k})'|^t \, dm \\ &\ge K^{-t}|(Q_z^{-n_k})'(Q^{n_k}(z))|^t m(B(Q^{n_k}(z), K^{-1}\zeta)) \\ &\ge K^{-t}(r_k/\zeta)^t M_m(K^{-1}\zeta) = (K\zeta)^{-t} M_m(K^{-1}\zeta) \cdot r_k^t, \end{aligned}$$

where the compactness of $\operatorname{supp}(m) = \mathcal{J}(T)$ guarantees that for any $r > 0$,

$$M_m(r) := \inf\{m(B(w,r)) : w \in \mathcal{J}(Q)\} > 0.$$

Analogously,

$$m(B(z,r_k)) \le K^t |(Q_z^{-n_k})'(Q^{n_k}(z))|^t m(B(Q^{n_k}(z), K\zeta)) \le K^t \zeta^{-t} \cdot r_k^t. \tag{32.2}$$

The proof of part (a) for $k \ge l$ is complete. For $k < l$ (if any), part (a) follows by increasing C_4 if necessary, since we are then dealing with "large" balls and "large" measures.

(b) Up to small modifications, the proof is the same as the proof of (a). □

Recall that subexpanding rational functions were defined in Section 26.3. We keep here the notation used therein. In particular, $R, \omega(R), s, T, \eta, \lambda, p, \gamma, \varepsilon, A, \tau$ and Γ_R originate from that section whereas $\theta = \theta(\varepsilon, \lambda)$ comes from Lemma 25.1.11. Note that, from time to time, the letter s will be used to denote a parameter for the pressure or the exponent of a conformal measure. The context will determine the meaning of s.

Fix any

$$\delta \in \left(0, \min\left\{\frac{1}{2}\theta, \eta, \frac{1}{3}\gamma \min\{\gamma, \theta\}\tau K^{-2} A^{-2} \|T'\|_{\infty}^{-1}\right\}\right) \tag{32.3}$$

so small that if $z \notin B(\mathrm{Crit}(T^p), \varepsilon)$ then there exists a unique holomorphic inverse branch $T_z^{-p} : B(T^p(z), 2\delta) \to \widehat{\mathbb{C}}$ of T^p such that $T_z^{-p}(T^p(z)) = z$ and

$$\mathrm{diam}(T_z^{-p}(B(T^p(z), 2\delta))) < \gamma. \tag{32.4}$$

Let

$$\mathrm{Crit}_{\infty}^{-}(T) := \bigcup_{n=0}^{\infty} T^{-n}(\mathrm{Crit}(T)).$$

We prove the following simple result which nevertheless has significant consequences.

Lemma 32.1.2. *Let $R : \widehat{\mathbb{C}} \to \widehat{\mathbb{C}}$ be a subexpanding rational function with $\deg(R) \geq 2$. If $z \in \mathcal{J}(T) \setminus \mathrm{Crit}_{\infty}^{-}(T)$, then*

$$\limsup_{n\to\infty} |(T^n)'(z)| = \infty. \tag{32.5}$$

Moreover, there is a sequence $(r_k(z))_{k=1}^{\infty}$ of positive numbers such that $\lim_{k\to\infty} r_k(z) = 0$ and for which any t-conformal measure m for R admits a constant $C_7 = C_7(m) \geq 1$ (independent of z) such that

$$C_7^{-1} \leq \frac{m(B(z, r_k(z)))}{(r_k(z))^t} \leq C_7, \quad \forall k \in \mathbb{N}. \tag{32.6}$$

In consequence, if $0 \leq s \leq t$ and ν is an s-conformal measure for R, then

$$C_8^{-1}(r_k(z))^{t-s} \leq \frac{m(B(z, r_k(z)))}{\nu(B(z, r_k(z)))} \leq C_8 (r_k(z))^{t-s}, \tag{32.7}$$

where $C_8 = C_8(m, \nu) := C_7(m) C_7(\nu)$.

Proof. As T is an iterate of R, it suffices to check that the hypotheses of Lemma 32.1.1(a) are satisfied for the point z and that $\inf\{\xi_1(x) : x \in \mathcal{J}(T) \setminus \mathrm{Crit}_{\infty}^{-}(T)\} > 0$. To this end, let

$$\mathbb{N}(z) := \{n \geq p \mid T^n(z) \notin B(\omega(T), \eta)\}.$$

Suppose first that $\#\mathbb{N}(z) = \infty$. Let $(j_k + p - 1)_{k=1}^{\infty}$ be the strictly increasing sequence of all integers in $\mathbb{N}(z)$. In view of Lemma 26.3.7, there exists for every $k \in \mathbb{N}$ a unique holomorphic inverse branch $T_z^{-j_k} : B(T^{j_k}(z), 2\gamma) \to \widehat{\mathbb{C}}$ of T^{j_k} such that $T_z^{-j_k}(T^{j_k}(z)) = z$. Setting $\xi_1(z) := \min\{2\gamma, \xi_0\}$, we see that the hypotheses of Lemma 32.1.1(a) are satisfied for the point z in this case.

Suppose now that $\#\mathbb{N}(z) < \infty$. Let k be the largest integer in $\mathbb{N}(z)$. As $z \notin \mathrm{Crit}_\infty^-(T)$, there is $\Delta > 0$ and a unique holomorphic inverse branch $T_z^{-k} : B(T^k(z), \Delta) \to \widehat{\mathbb{C}}$ of T^k such that $T_z^{-k}(T^k(z)) = z$. By (26.15), (26.18) and Lemma 25.1.11, there exists for every $j \in \mathbb{N}$ a unique holomorphic inverse branch $T_k^{-j} : B(T^{j+k}(z), 2\theta) \to \widehat{\mathbb{C}}$ of T^j such that $T_k^{-j}(T^{j+k}(z)) = T^k(z)$. In light of Lemma 32.1.1(a) (applied to the point $T^k(z)$), of (26.18) and of Lemma 23.1.11, we have

$$\mathrm{diam}(T_k^{-j}(B(T^{j+k}(z), \theta))) < \Delta$$

for all j sufficiently large. Therefore, for each such j the inverse branch composition $T_z^{-k} \circ T_k^{-j} : B(T^{k+j}(z), \theta) \to \widehat{\mathbb{C}}$ is well-defined. Setting $\xi_1(z) := \min\{\theta, \xi_0\}$, we conclude that the hypotheses of Lemma 32.1.1(a) are satisfied for the point z in this case, too.

Finally, observe that $\inf\{\xi_1(x) : x \in \mathcal{J}(T) \setminus \mathrm{Crit}_\infty^-(T)\} = \min\{2\gamma, \theta, \xi_0\} > 0$. □

Remark 32.1.3. In the proof of Lemma 32.1.2, we have also shown that for every point $z \in \mathcal{J}(T) \setminus \mathrm{Crit}_\infty^-(T)$ there exists a strictly increasing sequence $(n_k = n_k(z))_{k=1}^\infty$ such that

$$\limsup_{k\to\infty} |(T^{n_k})'(z)| = \infty$$

and for every $k \in \mathbb{N}$ there exists a unique holomorphic inverse branch

$$T_z^{-n_k} : B(T^{n_k}(z), \min\{2\gamma, \theta, \xi_0\}) \to \widehat{\mathbb{C}}$$

of T^{n_k} such that $T_z^{-n_k}(T^{n_k}(z)) = z$. Furthermore, thanks to Lemma 32.1.1(a), we may and we do assume that $\{n_k(z) : k \in \mathbb{N}\}$ is the set of all $n \in \mathbb{N}$ such that there exists a unique holomorphic inverse branch

$$T_z^{-n} : B(T^n(z), \min\{2\gamma, \theta, \xi_0\}) \to \widehat{\mathbb{C}}$$

of T^n sending $T^n(z)$ to z.

As an immediate consequence of this remark and of the fact that $\mathcal{J}(R) = \mathcal{J}(T)$ and $\mathrm{Crit}_\infty^-(R) = \mathrm{Crit}_\infty^-(T)$, we get the following result.

Corollary 32.1.4. *If $R : \widehat{\mathbb{C}} \to \widehat{\mathbb{C}}$ is a subexpanding rational function with* $\deg(R) \geq 2$, *then*

$$\mathcal{J}_r(R) = \mathcal{J}(R) \setminus \mathrm{Crit}_\infty^-(R).$$

We now prove an analog of Lemma 31.3.4.

Lemma 32.1.5. *Let $R : \widehat{\mathbb{C}} \to \widehat{\mathbb{C}}$ be a subexpanding rational function with* $\deg(R) \geq 2$. *If* $0 \leq s \leq t$ *and* ν *and* m *are respectively s-conformal and t-conformal measures for R, then* $m|_{\mathcal{J}(R)\setminus \mathrm{Crit}_\infty^-(R)}$ *is absolutely continuous with respect to* $\nu|_{\mathcal{J}(R)\setminus \mathrm{Crit}_\infty^-(R)}$. *In fact,*

$$m(A) \leq 4^s C_7(m)(C_7(\nu))^3 \cdot \nu(A)$$

for every Borel set $A \subseteq \mathcal{J}(R) \setminus \mathrm{Crit}_\infty^-(R)$. *If additionally* $s < t$, *then*

$$m(\mathcal{J}(R) \setminus \mathrm{Crit}_\infty^-(R)) = 0.$$

Proof. Let A be any Borel subset of $\mathcal{J}(R) \setminus \mathrm{Crit}_\infty^-(R)$. Fix $\varepsilon > 0$. Since the measure ν is outer regular, there exists an open set $A \subseteq G \subseteq \widehat{\mathbb{C}}$ such that

$$\nu(G) \leq \nu(A) + \varepsilon. \tag{32.8}$$

For every $z \in A$, take $k \in \mathbb{N}$ so large that

$$r(z) \leq \varepsilon \quad \text{and} \quad B(z, r(z)) \subseteq G, \tag{32.9}$$

where $r(z) := r_k(z)$ is a radius arising from Lemma 32.1.2. It follows from the $4r$-covering theorem (Theorem 15.4.1 and Remark 15.4.2(e)) that there exists a countable set $A^* \subseteq A$ such that the family $\{B(z, r(z)/4)\}_{z\in A^*}$ consists of mutually disjoint sets and the family $\{B(z, r(z))\}_{z\in A^*}$ covers A. Using Lemma 32.1.2, we obtain that

$$\begin{aligned}
m(A) &\leq \sum_{z\in A^*} m(B(z, r(z))) \leq C_8(m,\nu) \sum_{z\in A^*} (r(z))^{t-s} \nu(B(z, r(z))) \\
&\leq C_7(m)C_7(\nu)\varepsilon^{t-s} \sum_{z\in A^*} \nu(B(z, r(z))) \leq C_7(m)(C_7(\nu))^2 \varepsilon^{t-s} \sum_{z\in A^*} (r(z))^s \\
&= 4^s C_7(m)(C_7(\nu))^2 \varepsilon^{t-s} \sum_{z\in A^*} (r(z)/4)^s \leq 4^s C_7(m)(C_7(\nu))^3 \varepsilon^{t-s} \sum_{z\in A^*} \nu(B(z, r(z)/4)) \\
&= 4^s C_7(m)(C_7(\nu))^3 \varepsilon^{t-s} \nu\Big(\bigcup_{z\in A^*} B(z, r(z)/4)\Big) \leq 4^s C_7(m)(C_7(\nu))^3 \varepsilon^{t-s} \nu(G) \\
&\leq 4^s C_7(m)(C_7(\nu))^3 \varepsilon^{t-s} (\nu(A) + \varepsilon).
\end{aligned}$$

Letting $\varepsilon \to 0$ leads to the result. □

Denote

$$h_R := \delta(R),$$

where $\delta(R)$ is the minimal exponent of Sullivan conformal measures for the map R per Definition 29.2.10. We prefer to use the symbol h_R rather than $\delta(R)$ since later on, as for parabolic rational functions, we will show that $\delta(R)$ is equal to several other significant quantities ascribed to the function R and we do not want to prioritize any of them above any other.

We end this subsection by establishing a variation of Bowen's formula.

Theorem 32.1.6 (Bowen's formula). *If* $R : \widehat{\mathbb{C}} \to \widehat{\mathbb{C}}$ *is a subexpanding rational function with* $\deg(R) \geq 2$, *then*

$$\begin{aligned} h_R := \delta(R) &= \mathrm{ExpD}^{**}(\mathcal{J}(R)) = \mathrm{ExpD}^{*}(\mathcal{J}(R)) = \mathrm{ExpD}(\mathcal{J}(R)) = \mathrm{DD}(\mathcal{J}(R)) \\ &= \mathrm{HD}(\mathcal{J}_{\mathrm{uer}}(R)) = \mathrm{HD}(\mathcal{J}_{\mathrm{er}}(R)) = \mathrm{HD}(\mathcal{J}_{\mathrm{r}}(R)) \\ &= \mathrm{HD}(\mathcal{J}(R)) \in \left(\frac{p_R}{p_R+1}, 2\right]. \end{aligned}$$

Proof. The first two lines in this formula follow from Theorem 29.2.12 and the definition $h_R := \delta(R)$ while the last equality sign follows from Corollary 32.1.4 and the countability of the set $\mathrm{Crit}_\infty^-(R)$. The inequality $\delta(R) > p_R/(p_R+1)$ is simply Theorem 29.2.14. □

32.1.2 The atomless h_R-conformal measure m_R: existence, uniqueness, ergodicity and conservativity

We first show that any Sullivan conformal measure that does not charge the critical set of T^p is a geometric measure.

Proposition 32.1.7. *Let* $R : \widehat{\mathbb{C}} \to \widehat{\mathbb{C}}$ *be a subexpanding rational function with* $\deg(R) \geq 2$. *If m is a t-conformal measure for R such that* $m(\mathrm{Crit}(T^p)) = 0$, *then there exists a constant* $C_9 \geq 1$ *such that for every* $z \in \mathcal{J}(R)$,

$$C_9^{-1} \leq \frac{m(B(z,r))}{r^t} \leq C_9, \quad \forall 0 < r \leq 1. \tag{32.10}$$

In particular, the measure m is atomless.

Proof. First of all, keep in mind that $\mathcal{J}(R) = \mathcal{J}(T)$ and that m is also t-conformal for the map T. In this proof, we will always be dealing with the map T. For every $z \in \mathcal{J}(T)$, let $C_9(z) \geq 1$ be the minimal constant such that (32.10) is satisfied (if no such constant exists, define $C_9(z) = \infty$). We need to prove that

$$\sup\{C_9(z) \mid z \in \mathcal{J}(T)\} < \infty.$$

The proof consists of several steps.

Step 1: Using relations (26.15) and (26.18), Lemma 25.1.11, and Lemma 32.1.1(b) (with $Q = T$, $C_5 = \|T'\|_\infty$ and $\xi_2(z) = \min\{2\theta(\varepsilon,\lambda), \xi_0\}$), we conclude that (32.10) is uniformly satisfied for all points z for which

$$\mathcal{O}_+^T(z) := \{T^n(z) \mid n \geq 0\} \subseteq B(\omega(T), 2\eta).$$

By saying this, we mean here and in the sequel that $\sup\{C_9(z)\}$ taken over the set of points z under consideration is finite.

Step 2: Consider $z \in \mathrm{Crit}(T^p)$.

In view of Theorem 25.1.1, there exist $\rho \in (0, 1/2)$ and conformal homeomorphisms $h_1, h_2 : B(0, 2\rho) \to \widehat{\mathbb{C}}$ (on their respective open connected image $h_i(B(0, 2\rho))$; h_2 being even a translation) such that

$$h_1(0) = z, \quad h_2(0) = T^{2p}(z), \quad \text{and} \quad T^{2p}(h_1(w)) = h_2(w^q), \quad \forall w \in B(0, 2\rho),$$

where $q := \deg_z(T^{2p}) \geq 2$.

For each $i = 1, 2$, define a finite Borel measure m_i on the Borel σ-algebra of $B(0, \rho)$ by setting for every Borel set $A \subseteq B(0, \rho)$,

$$m_i(A) := \int_{h_i(A)} |(h_i^{-1})'|^t \, dm. \tag{32.11}$$

By (26.16), we know that $\mathcal{O}_+(T^{2p}(z)) \in B(\omega(T), \eta/2)$. Applying Step 1 to $T^{2p}(z)$, we get

$$\begin{aligned} m_2(B(0,r)) &\asymp |h_2'(0)|^{-t} m(h_2(B(0,r))) \\ &\asymp m(h_2(B(0,r))) \\ &\begin{cases} \leq & m(B(h_2(0), \|h_2'\|_\infty r)) \\ \geq & m(B(h_2(0), \frac{1}{4}|h_2'(0)|r)) \end{cases} \\ &\begin{cases} \leq & m(B(T^{2p}(z), \|h_2'\|_\infty r)) \\ \geq & m(B(T^{2p}(z), \frac{1}{4}|h_2'(0)|r)) \end{cases} \\ &\begin{cases} \leq & C_9(T^{2p}(z))[\|h_2'\|_\infty r]^t \\ \geq & [C_9(T^{2p}(z))]^{-1}[\frac{1}{4}|h_2'(0)|r]^t, \end{cases} \end{aligned}$$

where $\asymp$ is the usual symbol of multiplicative comparability (see (25.34)). So

$$C_{10}^{-1} \leq \frac{m_2(B(0,r))}{r^t} \leq C_{10} \tag{32.12}$$

for every $r \in (0, \rho]$ and some constant $C_{10} \geq 1$.

For every Borel set $F \subseteq h_i(B(0, \rho))$, we deduce from (32.11) that

$$m(F) = \int_{h_i^{-1}(F)} |h_i'|^t \, dm_i, \quad i = 1, 2. \tag{32.13}$$

Let $A \subseteq B(0, \rho)$ be any Borel set on which the map $w \mapsto w^q$ is injective and let A^q denote the image of A under $w \mapsto w^q$. Using (32.11), (32.13), the t-conformality of m and the change-of-variables formula several times, we get

$$
\begin{aligned}
m_2(A^q) &= \int_{h_2(A^q)} |(h_2^{-1})'|^t\, dm = \int_{T^{2p}(h_1(A))} |(h_2^{-1})'|^t\, dm \\
&= \int_{h_1(A)} |(h_2^{-1})' \circ T^{2p}|^t |(T^{2p})'|^t\, dm = \int_{h_1(A)} |(h_2^{-1} \circ T^{2p})'|^t\, dm \\
&= \int_{h_1^{-1}(h_1(A))} |(h_2^{-1} \circ T^{2p})' \circ h_1|^t \cdot |h_1'|^t\, dm_1 \\
&= \int_A |(h_2^{-1} \circ T^{2p} \circ h_1)'|^t\, dm_1 = \int_A |(w^q)'|^t\, dm_1(w) \\
&= q^t \int_A |w^{q-1}|^t\, dm_1(w) = q^t \int_A |w|^{(q-1)t}\, dm_1(w).
\end{aligned}
\tag{32.14}
$$

Fix any $\nu \in (0,1)$. For every $r \in (0,\rho]$, set

$$
\underline{A}(0; \nu r, r) := \{w \in \mathbb{C} : \nu r \leq |w| < r\}.
$$

For every $j = 0, 1, \ldots, q-1$, let

$$
\underline{A}_j(r) := \left\{ be^{i\theta} \in \mathbb{C} \,\middle|\, \sqrt[q]{\nu}\, r \leq b < r \text{ and } \frac{2\pi j}{q} \leq \theta < \frac{2\pi(j+1)}{q} \right\}.
$$

The sets $\{\underline{A}_j(r)\}_{j=0}^{q-1}$ partition the annulus $\underline{A}(0; \sqrt[q]{\nu}\, r, r)$. Moreover, the map $w \mapsto w^q$ is injective on each set $\underline{A}_j(r)$ and

$$
(\underline{A}_j(r))^q = \underline{A}(0; \nu r^q, r^q).
$$

Therefore, using (32.12)–(32.14) we obtain that

$$
\begin{aligned}
qC_{10} r^{qt} \geq q m_2(\underline{A}(0; \nu r^q, r^q)) &= \sum_{j=0}^{q-1} q^t \int_{\underline{A}_j(r)} |w|^{(q-1)t}\, dm_1(w) \\
&\geq \sum_{j=0}^{q-1} q^t [\sqrt[q]{\nu} r]^{(q-1)t} m_1(\underline{A}_j(r)) \\
&= q^t \nu^{\frac{(q-1)t}{q}} r^{(q-1)t} m_1(\underline{A}(0; \sqrt[q]{\nu}\, r, r)).
\end{aligned}
\tag{32.15}
$$

Similarly, set

$$
A(0; 0, r) := B(0, r) \setminus \{0\}.
$$

For every $j = 0, 1, \ldots, q-1$, let

$$
A_j^0(r) := \left\{ be^{i\theta} \in \mathbb{C} \,\middle|\, 0 < b < r \text{ and } \frac{2\pi j}{q} \leq \theta < \frac{2\pi(j+1)}{q} \right\}.
$$

From (32.11) and the t-conformality of m, we readily deduce that

$$m_2(\{0\}) = |h_2'(0)|^{-t} m(\{T^{2p}(z)\}) = |h_2'(0)|^{-t} |(T^{2p})'(z)|^t m(\{z\}) = 0$$

because $z \in \mathrm{Crit}(T^{2p})$. So, by (32.12)–(32.14),

$$\begin{aligned} qC_{10}^{-1} r^{qt} &\le qm_2(B(0,r^q)) = qm_2(A(0;0,r^q)) = \sum_{j=0}^{q-1} q^t \int_{A_j^0(r)} |w|^{(q-1)t}\, dm_1(w) \\ &\le q^t \int_{B(0,r)} |w|^{(q-1)t}\, dm_1(w) \le q^t r^{(q-1)t} m_1(B(0,r)). \end{aligned}$$

Hence,

$$\frac{m_1(B(0,r))}{r^t} \ge C_{10}^{-1} q^{1-t} \tag{32.16}$$

while we infer from (32.15) that

$$m_1(\underline{A}(0; \sqrt[q]{v}\, r, r)) \le C_{10} q^{1-t} v^{-\frac{(q-1)t}{q}} r^t. \tag{32.17}$$

Set $C_{11} := C_{10} q^{1-t} v^{-\frac{(q-1)t}{q}}$. The hypothesis that $m(\mathrm{Crit}(T^p)) = 0$ implies that $m_1(\{0\}) = 0$. Using this and (32.17), we obtain that

$$\begin{aligned} m_1(B(0,r)) &= m_1\Big(\bigcup_{n=0}^{\infty} \underline{A}\Big(0; (\sqrt[q]{v})^{n+1} r, (\sqrt[q]{v})^n r\Big)\Big) = \sum_{n=0}^{\infty} m_1\Big(\underline{A}\Big(0; (\sqrt[q]{v})^{n+1} r, (\sqrt[q]{v})^n r\Big)\Big) \\ &\le \sum_{n=0}^{\infty} C_{11} [(\sqrt[q]{v})^n r]^t = C_{11} \frac{1}{1-(\sqrt[q]{v})^t} r^t. \end{aligned} \tag{32.18}$$

Using (32.13), (32.16) and (32.18), we deduce that

$$\begin{aligned} m(B(z,r)) &\asymp |h_1'(0)|^t m_1(h_1^{-1}(B(z,r))) \\ &\asymp m_1(h_1^{-1}(B(z,r))) \\ &\begin{cases} \le & m_1(B(h_1^{-1}(z), \|(h_1^{-1})'\|_\infty r)) \\ \ge & m_1(B(h_1^{-1}(z), \frac{1}{4}|(h_1^{-1})'(z)|r)) \end{cases} \\ &\begin{cases} \le & m_1(B(0, \|(h_1^{-1})'\|_\infty r)) \\ \ge & m_1(B(0, \frac{1}{4}|(h_1^{-1})'(z)|r)) \end{cases} \\ &\begin{cases} \le & C_{11}[1-\sqrt[q]{v^t}]^{-1} [\|(h_1^{-1})'\|_\infty r]^t \\ \ge & C_{10} q^{1-t} [\frac{1}{4}|(h_1^{-1})'(z)|r]^t. \end{cases} \end{aligned}$$

So there is $C_{12} \ge 1$ such that

$$C_{12}^{-1} \le \frac{m(B(z,r))}{r^t} \le C_{12}, \quad \forall r \in (0, \rho].$$

Note that the constant C_{12} can be adjusted upwards so that these estimates become valid for all $r \in (0, 1]$. As $\#\mathrm{Crit}(T^p) < \infty$, this confirms that (32.10) is satisfied for all $z \in \mathrm{Crit}(T^p)$.

Step 3: Suppose that $z \notin \mathrm{Crit}(T^p)$, that $0 < r < 2^{-6}K^{-2}\|T'\|_\infty^{-p+1}\min\{\gamma, \theta\}$, and that

$$\sup_{n\ge 0} r|(T^n)'(z)| \le K^{-1}\|T'\|_\infty^{-1}\min\{\gamma, \theta\}.$$

Then it follows from (32.5) in Lemma 32.1.2 that $z \in \mathrm{Crit}_\infty^-(T)$. Let $k \ge 0$ be the least integer such that $z \in T^{-k}(\mathrm{Crit}(T))$. Since $z \notin \mathrm{Crit}(T^p) = \bigcup_{j=0}^{p-1} T^{-j}(\mathrm{Crit}(T))$, this means that $k \ge p$ and there is $l \in \mathbb{N}$ such that $T^{p-1}(T^l(z)) = T^k(z) \in \mathrm{Crit}(T)$. Since, moreover, $T^k(z) \notin B(\omega(T), 2\eta)$ by (26.18), it ensues from Lemma 26.3.7 that there exists a unique holomorphic inverse branch $T_z^{-l} : B(T^l(z), 2\gamma) \to \widehat{\mathbb{C}}$ of T^l such that $T_z^{-l}(T^l(z)) = z$. But

$$(T^p)'(T^l(z)) = (T^{p-1})'(T^l(z)) \cdot T'(T^k(z)) = 0.$$

That is, $T^l(z) \in \mathrm{Crit}(T^p)$. Applying Lemma 23.1.11, we know that

$$B(T^l(z), K^{-1}|(T^l)'(z)|r) \subseteq T^l(B(z,r)) \subseteq B(T^l(z), K|(T^l)'(z)|r).$$

Using this, the t-conformality of m, Theorem 23.1.9 (Koebe's distortion theorem, spherical version) and applying Step 2 to $T^l(z)$ (note that $K|(T^l)'(z)|r \le 1$), we deduce that

$$\begin{aligned} m(B(z,r)) = m(T_z^{-l}(T^l(B(z,r)))) &= \int_{T^l(B(z,r))} |(T_z^{-l})'|^t \, dm \\ &\asymp K^{\pm t}|(T_z^{-l})'(T^l(z))|^t \cdot m(T^l(B(z,r))) \\ &\asymp K^{\pm t}|(T^l)'(z)|^{-t} \cdot m(B(T^l(z), K^{\pm 1}|(T^l)'(z)|r)) \\ &\asymp K^{\pm t}|(T^l)'(z)|^{-t} \cdot C_{12}^{\pm 1}[K^{\pm 1}|(T^l)'(z)|r]^t \\ &= C_{12}^{\pm 1}K^{\pm 2t} \cdot r^t. \end{aligned}$$

In conclusion, (32.10) is uniformly satisfied for all points z under consideration in Step 3.

Step 4: Suppose that $z \notin \mathrm{Crit}(T^p)$, that $0 < r < 2^{-6}K^{-2}\|T'\|_\infty^{-p+1}\min\{\gamma, \theta\}$, and that

$$r|(T^n)'(z)| > K^{-1}\|T'\|_\infty^{-1}\min\{\gamma, \theta\} \tag{32.19}$$

for some $n \in \mathbb{N}$. Fix a minimal such n. Then $n \ge p$ and, remembering that $p \ge 2$, we have

$$K^{-1}\|T'\|_\infty^{-p}\min\{\gamma, \theta\} \le r|(T^{n-p+1})'(z)| \le K^{-1}\min\{\gamma, \theta\}. \tag{32.20}$$

Case 1: Assume first that

$$T^n(z) \notin B(\omega(T), 2\eta).$$

In view of Lemma 26.3.7, there exists a unique holomorphic inverse branch $T_z^{-(n-p+1)} : B(T^{n-p+1}(z), 2\gamma) \to \widehat{\mathbb{C}}$ of T^{n-p+1} such that $T_z^{-(n-p+1)}(T^{n-p+1}(z)) = z$. Therefore, per Lemma 23.1.11 we know that

$$B(T^{n-p+1}(z), K^{-1}|(T^{n-p+1})'(z)|r) \subseteq T^{n-p+1}(B(z,r)) \subseteq B(T^{n-p+1}(z), K|(T^{n-p+1})'(z)|r).$$

Using this, the t-conformality of m, Theorem 23.1.9 (Koebe's distortion theorem, spherical version) and (32.20), we obtain that

$$\begin{aligned}
m(B(z,r)) &= m(T_z^{-(n-p+1)}(T^{n-p+1}(B(z,r)))) \\
&= \int_{T^{n-p+1}(B(z,r))} |(T_z^{-(n-p+1)})'|^t \, dm \geq \int_{B(T^{n-p+1}(z), K^{-1}|(T^{n-p+1})'(z)|r)} |(T_z^{-(n-p+1)})'|^t \, dm \\
&\geq K^{-t}|(T_z^{-(n-p+1)})'(T^{n-p+1}(z))|^t \cdot m(B(T^{n-p+1}(z), K^{-1}|(T^{n-p+1})'(z)|r)) \\
&\geq K^{-t}|(T^{n-p+1})'(z)|^{-t} \cdot m(B(T^{n-p+1}(z), K^{-2}\|T'\|_\infty^{-p} \min\{\gamma, \theta\})) \\
&\geq M_m(K^{-2}\|T'\|_\infty^{-p} \min\{\gamma, \theta\}) \cdot \gamma^{-t} r^t,
\end{aligned}$$

where we recall that

$$M_m(u) := \inf\{m(B(w,u)) \mid w \in \mathcal{J}(T)\} > 0$$

for every $u > 0$. Analogously,

$$\begin{aligned}
m(B(z,r)) &\leq K^t|(T_z^{-(n-p+1)})'(T^{n-p+1}(z))|^t \cdot m(B(T^{n-p+1}(z), K|(T^{n-p+1})'(z)|r)) \\
&\leq K^t|(T^{n-p+1})'(z)|^{-t} \\
&\leq K^{2t}\|T'\|_\infty^{pt}(\min\{\gamma, \theta\})^{-t} r^t.
\end{aligned}$$

Consequently, (32.10) is uniformly satisfied for all points z under consideration in Case 1.

Case 2: Assume that $T^n(z) \in B(\omega(T), 2\eta)$ and let $k \geq 0$ be the least integer such that

$$\{T^k(z), T^{k+1}(z), \ldots, T^n(z)\} \subseteq B(\omega(T), 2\eta).$$

Subcase 2a: If $k = 0$, by virtue of (26.15) and (26.18), proceed as in Case 1 using Lemma 25.1.11 instead of Lemma 26.3.7. Indeed, let $T_z^{-n} : B(T^n(z), \theta) \to \widehat{\mathbb{C}}$ be the unique holomorphic inverse branch of T^n such that $T_z^{-n}(T^n(z)) = z$ per Lemma 25.1.11. By the definition of n (cf. (32.19)), we know that

$$K^{-1}\|T'\|_\infty^{-1} \min\{\gamma, \theta\} < r|(T^n)'(z)| \leq K^{-1} \min\{\gamma, \theta\}. \tag{32.21}$$

In a similar way to Case 1, we have

$$\begin{aligned} m(B(z,r)) &\geq K^{-t}|(T_z^{-n})'(T^n(z))|^t m(B(T^n(z), K^{-2}\|T'\|_\infty^{-1}\min\{\gamma,\theta\})) \\ &\geq M_m(K^{-2}\|T'\|_\infty^{-1}\min\{\gamma,\theta\}) \cdot \gamma^{-t} r^t. \end{aligned}$$

and

$$\begin{aligned} m(B(z,r)) &\leq K^t|(T_z^{-n})'(T^n(z))|^t m(B(T^n(z), \min\{\gamma,\theta\})) \leq K^t|(T^n)'(z)|^t \\ &\leq K^{2t}\|T'\|_\infty^t (\min\{\gamma,\theta\})^{-t} r^t. \end{aligned}$$

In conclusion, (32.10) is uniformly satisfied for all points z under consideration in Subcase 2a.

Subcase 2b: If $k > 0$, let

$$l := \max\{0, k - p\}$$

Again by (26.15), (26.18) and Lemma 25.1.11 as well as (32.3), there is a unique holomorphic inverse branch $T_1^{-(n-p-l)} : B(T^n(z), 2\delta) \to \widehat{\mathbb{C}}$ of T^{n-p-l} such that $T_1^{-(n-p-l)}(T^n(z)) = T^{p+l}(z)$.

Sub-subcase 2b1: Assume that

$$T^l(z) \notin B(\mathrm{Crit}(T^p), \varepsilon).$$

By (32.3)–(32.4), there is a unique holomorphic inverse branch $T_2^{-p} : B(T^{p+l}(z), 2\delta) \to \widehat{\mathbb{C}}$ of T^p such that $T_2^{-p}(T^{p+l}(z)) = T^l(z)$.

- If $l = 0$, then the well-defined composition $T_2^{-p} \circ T_1^{-(n-p-l)} : B(T^n(z), 2\delta) \to \widehat{\mathbb{C}}$ (which sends $T^n(z)$ to $T^l(z) = z$) brings this sub-subcase to a conclusion in the same way as in Case 1 and Subcase 2a.
- If $l > 0$, then $l = k - p$. Since $T^{k-1}(z) \notin B(\omega(T), 2\eta)$, it follows from Lemma 26.3.7 that there exists a unique holomorphic inverse branch $T_z^{-l} : B(T^l(z), 2\gamma) \to \widehat{\mathbb{C}}$ of T^l such that $T_z^{-l}(T^l(z)) = z$. By (32.4), the composition $T_z^{-l} \circ T_2^{-p} \circ T_1^{-(n-p-l)} : B(T^n(z), 2\delta) \to \widehat{\mathbb{C}}$ makes sense and sends $T^n(z)$ to z. So, we are done in the same way as in Case 1 and Subcase 2a.

Therefore, (32.10) is uniformly satisfied for all points z under consideration in Sub-subcase 2b1.

Sub-subcase 2b2: Assume now that

$$T^l(z) \in B(c, \varepsilon) \quad \text{for some } c \in \mathrm{Crit}(T^p).$$

Set

$$\rho_c := |T^{p+l}(z) - T^p(c)|.$$

Part 2b2(i): Assume that

$$K^{-1}|(T^{n-p-l})'(T^{p+l}(z))|\tau\rho_c \geq \delta,$$

where $\tau \in (0,1)$ emerges from Lemma 26.3.8.

By Lemma 23.1.11 and Part 2b2(i)'s assumption,

$$T_1^{-(n-p-l)}(B(T^n(z),\delta)) \subseteq B(T^{p+l}(z),\tau\rho_c).$$

Moreover, in view of Lemma 26.3.8, there exists a unique holomorphic inverse branch $T_2^{-p} : B(T^{p+l}(z),\tau\rho_c) \to B(T^l(z),\gamma)$ of T^p sending $T^{p+l}(z)$ to $T^l(z)$. Hence, the composition

$$T_2^{-p} \circ T_1^{-(n-p-l)} : B(T^n(z),\delta) \to \widehat{\mathbb{C}}$$

is well-defined and maps $T^n(z)$ to $T^l(z)$.

- If $l = 0$, then we are done as in Case 1 and Subcase 2a.
- If $l > 0$, then $l = k - p$. Since $T^{k-1}(z) \notin B(\omega(T), 2\eta)$, it follows from Lemma 26.3.7 that there exists a unique holomorphic inverse branch $T_z^{-l} : B(T^l(z), 2\gamma) \to \widehat{\mathbb{C}}$ of T^l such that $T_z^{-l}(T^l(z)) = z$. The possibility of making up the composition $T_z^{-l} \circ T_2^{-p} \circ T_1^{-(n-p-l)} : B(T^n(z),\delta) \to \widehat{\mathbb{C}}$ finishes the proof in this situation as well.

Thus, (32.10) is uniformly satisfied for all points z under consideration in Part 2b2(i).

Part 2b2(ii): Assume that

$$K^{-1}|(T^{n-p-l})'(T^{p+l}(z))|\tau\rho_c < \delta.$$

Set

$$\iota := |T^l(z) - c| \quad \text{and} \quad q_c := \deg_c(T^p).$$

Using (26.19) (with z replaced by $T^l(z)$) and the definition of ρ_c, we have that

$$K^{-1}|(T^{n-p-l})'(T^{p+l}(z))|\tau A^{-1}\iota^{q_c} < \delta. \tag{32.22}$$

But by virtue of (26.20) (with z replaced by $T^l(z)$) and (32.19), we obtain that

$$r|(T^{n-p-l})'(T^{p+l}(z))| \cdot A\iota^{q_c-1} \cdot |(T^l)'(z)| > K^{-1}\|T'\|_\infty^{-1} \min\{\gamma,\theta\}.$$

Multiplying both sides of this inequality by ι and using (32.22) and (32.3), we get that

$$\iota < K^2\|T'\|_\infty (\min\{\gamma,\theta\})^{-1}\tau^{-1}\delta A^2 r|(T^l)'(z)| < \frac{1}{3}\gamma r|(T^l)'(z)|.$$

This implies that

$$B\Big(T^l(z), \frac{1}{3}\gamma r|(T^l)'(z)|\Big) \subseteq B\Big(c, \frac{2}{3}\gamma r|(T^l)'(z)|\Big) \subseteq B\Big(T^l(z), \gamma r|(T^l)'(z)|\Big). \tag{32.23}$$

Since $l < n$, the definition of n (relation (32.19) and its minimality) yields

$$\gamma r|(T^l)'(z)| \le \gamma. \tag{32.24}$$

- If $l = 0$, then it follows from (32.23) and Step 2 (used at c) that

$$m(B(z, \gamma r/3)) \le C_{12}(2\gamma/3)^t \cdot r^t \quad \text{and} \quad m(B(z, \gamma r)) \ge C_{12}^{-1}(2\gamma/3)^t \cdot r^t.$$

  In light of Lemma 25.6.2, these inequalities are sufficient to complete the proof of Part 2b2(ii) when $l = 0$.
- If $l > 0$, then $l = k - p$. Since $T^{k-1}(z) \notin B(\omega(T), 2\eta)$, it follows from Lemma 26.3.7 that there exists a unique holomorphic inverse branch $T_z^{-l} : B(T^l(z), 2\gamma) \to \widehat{\mathbb{C}}$ of T^l such that $T_z^{-l}(T^l(z)) = z$. Using Koebe's distortion theorem, the t-conformality of m, (32.23)–(32.24) and Step 2 (at c), we conclude that

$$m(B(z, K^{-1}\gamma r/3)) \le C_{12}(2K\gamma/3)^t \cdot r^t \quad \text{and} \quad m(B(z, K\gamma r)) \ge C_{12}^{-1}(2K^{-1}\gamma/3)^t \cdot r^t.$$

  In light of Lemma 25.6.2, these inequalities suffice to complete the proof of Part 2b2(ii) when $l > 0$.

Thus, (32.10) is uniformly satisfied for all points z under consideration in Part 2b2(ii).

Having successfully considered Parts 2b2(i)+(ii), relation (32.10) is uniformly satisfied for all points z under consideration in Sub-subcase 2b2.

Both Sub-subcases 2b1 and 2b2 having been tackled, the proof in Subcase 2b is finished.

Having established both Subcases 2a and 2b, we conclude that the proof in Case 2 is over.

Cases 1 and 2 being done, the proof of Step 4 is complete.

Steps 1–4 cover all possibilities and therefore the result holds. □

We can now easily prove the first main result of this section, which asserts the existence of an atomless h_R-conformal measure that behaves like a geometric measure with exponent h_R.

Theorem 32.1.8. *If $R : \widehat{\mathbb{C}} \to \widehat{\mathbb{C}}$ is a subexpanding rational function with $\deg(R) \ge 2$, then there exists an atomless h_R-conformal measure m_R for R. In addition,*

$$C_9^{-1} \le \frac{m_R(B(z,r))}{r^{h_R}} \le C_9, \quad \forall r \in (0,1], \ \forall z \in \mathcal{J}(R). \tag{32.25}$$

Proof. In fact, we will show that for an appropriate choice of $\xi \in \mathcal{J}(R)$ each associated PS limit measure m in the sense of Definition 29.1.12 and as constructed at the beginning of Section 29.2, is atomless. As the critical set's grand orbit $\bigcup_{n=0}^{\infty} R^{-n}(\bigcup_{k=0}^{\infty} R^k(\mathrm{Crit}(R)))$, which coincides with the postcritical set's backward orbit $\mathrm{PC}_{\infty}^{-}(R) := \bigcup_{n=0}^{\infty} R^{-n}(\mathrm{PC}(R))$, is countable and as $\mathrm{HD}(\mathrm{Safe}(R) \cap \mathcal{J}_{\mathrm{uer}}(R)) > 0$ (by Theorem 28.5.3 and Lemma 29.2.3), there exists a point

$$w \in \left[\mathrm{Safe}(R) \cap \mathcal{J}_{\mathrm{uer}}(R)\right] \setminus \mathrm{PC}_{\infty}^{-}(R).$$

Clearly, $\mathrm{PC}_{\infty}^{-}(R)$ is completely R-invariant. Moreover, by its very Definition 29.2.2, the set $\mathrm{Safe}(R)$ is backward R-invariant, i.e. $R^{-1}(\mathrm{Safe}(R)) \subseteq \mathrm{Safe}(R)$. Though $\mathcal{J}_{\mathrm{uer}}(R)$ is generally not backward R-invariant, the set $\mathcal{J}_{\mathrm{uer}}(R) \setminus \mathrm{PC}_{\infty}^{-}(R)$ is. We easily deduce that $[\mathrm{Safe}(R) \cap \mathcal{J}_{\mathrm{uer}}(R)] \setminus \mathrm{PC}_{\infty}^{-}(R)$ is backward R-invariant and hence

$$\bigcup_{j=0}^{\infty} R^{-j}(w) \subseteq \left[\mathrm{Safe}(R) \cap \mathcal{J}_{\mathrm{uer}}(R)\right] \setminus \mathrm{PC}_{\infty}^{-}(R).$$

Since $R|_{\mathcal{J}(R)}$ is topologically exact, it is strongly transitive, i. e., the backward orbit of any Julia point is dense in the Julia set. Given that $\omega(R)$ is nowhere dense in $\mathcal{J}(R)$ (by Proposition 26.3.4), by decreasing $\eta > 0$ if necessary, there is a point

$$\xi \in \left[\mathrm{Safe}(R) \cap \mathcal{J}_{\mathrm{uer}}(R)\right] \setminus \left[\mathrm{PC}_{\infty}^{-}(R) \cup B(\omega(R), 4\eta)\right]. \tag{32.26}$$

Let m be a PS limit measure in the sense of Definition 29.1.12 as performed at the beginning of Section 29.2 for the function $\varphi_{h_R} := -h_R \log|R'|$. Recall that $h_R := \delta(R) = s_\xi(R) := \inf\{t \geq 0 \mid c_\xi(t) \leq 0\}$ by Theorem 29.2.12 and $c_\xi(s_\xi(R)) = 0$ by (29.49). So, let $(s_j)_{j=1}^{\infty}$ be a sequence strictly decreasing to 0 such that

$$m = \lim_{j\to\infty} m_{s_j}$$

per Definition 29.1.12.

Fix $\sigma \in (0,1)$ sufficiently close to 1 as will be specified later in the course of the proof. By redefining finitely many terms of the sequence $(b_n)_{n=1}^{\infty}$ from Lemma 29.1.10, we may assume that

$$\frac{b_n}{b_{n+1}} \geq \sigma, \quad \forall n \in \mathbb{N}. \tag{32.27}$$

As $E_{n+1} = R^{-1}(E_n)$ for every $n \in \mathbb{N}$, it follows from (29.26) that

$$m_{s_j}(R(A)) = \frac{1}{M_{s_j}} \sum_{n=1}^{\infty} \sum_{x \in A \cap E_{n+1}} b_n \left|(R^{n+1})'(x)\right|^{-h_R} e^{-s_j(n+1)} e^{s_j} \left|R'(x)\right|^{h_R}$$

for every $j \in \mathbb{N}$ and every special set $A \subseteq \mathcal{J}(R)$. Using (32.27), we get

$$
\begin{aligned}
m_{s_j}(R(A)) &\geq \sigma \frac{1}{M_{s_j}} \sum_{n=1}^{\infty} \sum_{x \in A \cap E_{n+1}} b_{n+1} |(R^{n+1})'(x)|^{-h_R} e^{-s_j(n+1)} e^{s_j} |R'(x)|^{h_R} \\
&\geq \sigma\left(e^{s_j} \int_A |R'|^{h_R} \, dm_{s_j} - \frac{1}{M_{s_j}} \sum_{x \in A \cap E_1} b_1 \right) \\
&\geq \sigma\left(e^{s_j} \int_A |R'|^{h_R} \, dm_{s_j} - b_1 \#(A \cap E_1) \right)
\end{aligned}
$$

for all $j \in \mathbb{N}$ large enough since $\lim_{j\to\infty} M_{s_j} = \infty$. If, in addition,

$$
\xi \notin A, \quad \text{i. e.,} \quad A \cap E_1 = \emptyset, \tag{32.28}
$$

then

$$
m_{s_j}(R(A)) \geq \sigma e^{s_j} \int_A |R'|^{h_R} \, dm_{s_j} \geq \sigma \int_A |R'|^{h_R} \, dm_{s_j}. \tag{32.29}
$$

Now, let $z_0 \in \mathrm{Crit}(T^p)$. Set $q := \deg_z(T^{2p}) \geq 2$. Choose $a \in (0,1)$ and $b > 0$ such that $(1 - q(a+b) + b) \geq 1/2$. In light of (26.15), choosing $\sigma \geq \lambda^{-ah_R/s}$ ensures that

$$
\sigma^s \geq |T'(z)|^{-ah_R}, \quad \forall z \in B(\omega(T), 2\eta).
$$

If, additionally, a special set A for T is such that

$$
\bigcup_{i=0}^{s-1} R^i(A) \subseteq B(\omega(T), 2\eta),
$$

then using (32.29) s times and the chain rule yields

$$
m_{s_j}(T(A)) = m_{s_j}(R^s(A)) \geq \sigma^s \int_A |(R^s)'|^{h_R} \, dm_{s_j} \geq \int_A |T'|^{(1-a)h_R} \, dm_{s_j}. \tag{32.30}
$$

Having this and (32.26), we deduce in the same way as in Step 1 of the proof of Proposition 32.1.7 (via Lemma 32.1.1; see particularly (32.2)) that

$$
m_{s_j}(B(z,r)) \leq K^{(1-a)h_R} \zeta^{-(1-a)h_R} r^{(1-a)h_R}, \quad \forall r \in (0,1] \tag{32.31}
$$

whenever $\{T^n(z) : n \geq 0\} \subseteq B(\omega(T), 2\eta)$.

It follows from (32.26) that there exists $u \in (0,1)$ such that

$$
\xi \notin \bigcup_{i=0}^{2sp-1} R^i(B(\mathrm{Crit}(R), u)). \tag{32.32}
$$

Let $\rho > 0$ be so small that $h_i(B(0,\rho)) \subseteq B(\mathrm{Crit}(R), u)$, where the h_i's are from Step 2 of the proof of Proposition 32.1.7. From m_{s_j} define two Borel measures $m_{s_j,i}$ as in (32.11).

Let $A \subseteq B(0,\rho)$ be any Borel set on which the map $w \mapsto w^q$ is one-to-one. Using (32.32) and $2sp$ times relation (32.29), we obtain by proceeding as in (32.14) that

$$\begin{aligned} m_{s_j,2}(A^q) &= \int\limits_{T^{2p}(h_1(A))} |(h_2^{-1})'|^{h_R}\, dm_{s_j} \\ &\geq \sigma^{2sp} \int\limits_{h_1(A)} |(h_2^{-1})' \circ T^{2p}|^{h_R} |(T^{2p})'|^{h_R}\, dm_{s_j} \\ &= \tilde{\sigma} q^{h_R} \int\limits_A |w|^{(q-1)h_R}\, dm_{s_j,1}(w), \end{aligned} \tag{32.33}$$

where $\tilde{\sigma} := \sigma^{2ps} \in (0,1)$ is independent of ρ and A and satisfies $\lim_{\sigma \nearrow 1} \tilde{\sigma} = 1$. Reducing $\rho > 0$ if necessary, we have $\tilde{\sigma} \geq \rho^{(q-1)bh_R}$. Therefore, $\tilde{\sigma} \geq |w|^{(q-1)bh_R}$ for all $w \in B(0,\rho)$ and it follows from (32.33) that

$$m_{s_j,2}(A^q) \geq q^{h_R} \int\limits_A |w|^{(q-1)(1+b)h_R}\, dm_{s_j,1}(w)$$

for every Borel set $A \subseteq B(0,\rho)$ on which the map $w \mapsto w^q$ is one-to-one. Proceeding in the same way as in Step 2 of the proof of Proposition 32.1.7, relation (32.15) takes the form

$$q\widetilde{C}_{10} r^{q(1-a)h_R} \geq q^{h_R} \nu^{\frac{q-1}{q}(1+b)h_R} r^{(q-1)(1+b)h_R} m_{s_j,1}(\underline{A}(0; \sqrt[q]{\nu}\, r, r)),$$

where the constant $\widetilde{C}_{10}$ arises from (32.31). As consequences corresponding to (32.17) and (32.18), we have that

$$m_{s_j,1}(\underline{A}(0; \sqrt[q]{\nu}\, r, r)) \leq \widetilde{C}_{10} q^{1-h_R} \nu^{-\frac{q-1}{q}(1+b)h_R} r^{(1-q(a+b)+b)h_R} \leq \widetilde{C}_{10} q^{1-h_R} \nu^{-\frac{q-1}{q}(1+b)h_R} r^{\frac{h_R}{2}},$$

where the rightmost inequality follows from the choice of a, b. Hence,

$$m_{s_j,1}(B(0,r)) \leq \widetilde{C}_{11} \frac{1}{1-\nu^{\frac{h_R}{2q}}} r^{\frac{h_R}{2}},$$

where $\widetilde{C}_{11} := \widetilde{C}_{10} q^{1-h_R} \nu^{-\frac{q-1}{q}(1+b)h_R}$. Therefore, by means of a similar argument to that given in the analogous context of Step 2 in the proof of Proposition 32.1.7, we infer that

$$m_{s_j}(B(z_0,r)) \leq \overline{C}_{11} r^{\frac{h_R}{2}}, \quad \forall r \in (0,1],$$

for some constant $\overline{C}_{11} \geq 1$.

The set

$$N_{z_0} := \left\{ r \in (0,1] \,\middle|\, \sum_{j=1}^{\infty} m_{s_j}(\partial B(z_0,r)) > 0 \right\}$$

is countable. Using the Portmanteau theorem (Theorem A.1.23), for every $r \in (0,1] \setminus N_{z_0}$ we obtain that

$$m(\{z_0\}) \le m(B(z_0,r)) = \lim_{j\to\infty} m_{s_j}(B(z_0,r)) \le \overline{C}_{11} r^{\frac{h_R}{2}}.$$

Thus, $m(\{z_0\}) = 0$. Since z_0 was arbitrarily chosen among $\mathrm{Crit}(T^p)$, we deduce that $m(\mathrm{Crit}(T^p)) = 0$. The proof of Theorem 32.1.8 is complete by an application of Proposition 32.1.7. □

Beyond its existence and geometric nature, the main result about m_R (its uniqueness, ergodicity and conservativity) is next. This is the counterpart of Theorem 31.3.9 for subexpanding maps.

Theorem 32.1.9. *If $R : \widehat{\mathbb{C}} \to \widehat{\mathbb{C}}$ is a subexpanding rational function with* $\deg(R) \ge 2$, *then the dynamical system $R : \mathcal{J}(R) \to \mathcal{J}(R)$ is conservative and ergodic with respect to the h_R-conformal measure m_R. Moreover,* $\mathrm{supp}(m_R) = \mathcal{J}(R)$ *and*

$$C_9^{-1} \le \frac{m_R(B(z,r))}{r^{h_R}} \le C_9, \quad \forall r \in (0,1] \tag{32.34}$$

for every $z \in \mathcal{J}(R)$. Furthermore, m_R is the unique h_R-conformal measure for R.

Proof. The topological exactness of R (per Corollary 24.3.5) and the conformality of m_R guarantee that $\mathrm{supp}(m_R) = \mathcal{J}(R)$. Relation (32.34) holds because of Theorem 32.1.8 and Proposition 32.1.7.

We now prove the following.

Claim 1. *Any h_R-conformal measure m' for R is equivalent to m_R. In particular,* $m'(\mathrm{Crit}_\infty^-(R)) = 0$.

Proof of Claim 1. The proof goes along the same lines as that of Claim 1 in Theorem 31.3.9 by replacing $\Omega_\infty^-(T)$ with $\mathrm{Crit}_\infty^-(R)$ and using Theorem 32.1.8 and Lemma 32.1.5 instead of Theorem 31.3.7 and Lemma 31.3.4. ■

Claim 2. *The measure m_R is ergodic. Consequently, it is totally ergodic, i. e., it is ergodic with respect to every positive iterate R^k of R. Indeed, note that $\mathcal{J}(R^k) = \mathcal{J}(R)$, R^k is subexpanding, and the measure m_R fulfills all the assertions of Theorem* 32.1.8 *for R^k.*

Proof of Claim 2. The proof is identical to that of Claim 2 in Theorem 31.3.9. ■

Claim 3. *If m' is an h_R-conformal measure for R and $\rho := dm'/dm_R$, then ρ is m_R-a. e. T-invariant, i. e.,*

$$\rho \circ T = \rho \quad m_R\text{-a. e.}$$

Proof of Claim 3. By virtue of Claim 1, the function ρ is well-defined m_R-almost everywhere. As m_R is atomless (per Theorem 32.1.8) and $\mathrm{Crit}_\infty^-(R)$ is countable, we know that

$m_R(\mathrm{Crit}^-_\infty(R)) = 0$. Recall further that $\mathcal{J}(R) = \mathcal{J}(T)$ and $\mathrm{Crit}^-_\infty(R) = \mathrm{Crit}^-_\infty(T)$. Fix $z \in \mathcal{J}(R) \setminus \mathrm{Crit}^-_\infty(R) = \mathcal{J}(T) \setminus \mathrm{Crit}^-_\infty(T)$. Denote

$$\iota := \frac{1}{2}\min\{2\gamma, \theta, \xi_0\}. \tag{32.35}$$

For all $k \in \mathbb{N}$, let

$$V_k(z) := T_z^{-n_k(z)}\big(B(T^{n_k(z)}(z), (3K^2)^{-1}\iota) \big),$$

where the sequence $(n_k(z))_{k=1}^\infty$ and the corresponding holomorphic inverse branches $T_z^{-n_k(z)} : B(T^{n_k(z)}(z), 2\iota) \to \widehat{\mathbb{C}}$ that send $T^{n_k(z)}(z)$ to z, come from Remark 32.1.3. Since $\inf_{k\in\mathbb{N}} \mathrm{diam}(V_k(z)) = 0$, the family $V = \{V_k(z) : z \in \mathcal{J}(T) \setminus \mathrm{Crit}^-_\infty(T),\ k \in \mathbb{N}\}$ is (in the terminology of Section 2.8 in Federer [50]) fine at every $z \in \mathcal{J}(T) \setminus \mathrm{Crit}^-_\infty(T)$. Like (31.45), set

$$\widehat{V}_k(z) := \Big\{ \bigcup_{W\in V} W : W \cap V_k(z) \neq \emptyset \text{ and } \mathrm{diam}(W) \le 2\mathrm{diam}(V_k(z)) \Big\}. \tag{32.36}$$

Relations similar to (31.46)–(31.49) hold:

$$T_z^{-n_k(z)}(B(T^{n_k(z)}(z), \iota)) \supseteq B(z, K^{-1}\iota|(T^{n_k(z)})'(z)|^{-1}), \tag{32.37}$$

$$m_R(T_z^{-n_k(z)}(B(T^{n_k(z)}(z), \iota))) \le K^{h_R}\big|(T^{n_k(z)})'(z)\big|^{-h_R}, \tag{32.38}$$

$$m_R(V_k(z)) \ge M_{m_R}((3K^2)^{-1}\iota)K^{-h_R}\big|(T^{n_k(z)})'(z)\big|^{-h_R}, \tag{32.39}$$

$$\mathrm{diam}(V_k(z)) \le (3K)^{-1}\iota\big|(T^{n_k(z)})'(z)\big|^{-1}. \tag{32.40}$$

Using (32.36), (32.40) and (32.37) successively, we get

$$\widehat{V}_k(z) \subseteq B(z, 3\mathrm{diam}(V_k(z))) \subseteq B(z, K^{-1}\iota|(T^{n_k(z)})'(z)|^{-1}) \subseteq T_z^{-n_k(z)}(B(T^{n_k(z)}(z), \iota)).$$

Therefore, by (32.38),

$$m_R(\widehat{V}_k(z)) \le K^{h_R}\big|(T^{n_k(z)})'(z)\big|^{-h_R}.$$

Along with (32.39), this yields

$$\frac{m_R(\widehat{V}_k(z))}{m_R(V_k(z))} \le K^{2h_R}\Big[M_{m_R}((3K^2)^{-1}\iota)\Big]^{-1}.$$

Thus,

$$\limsup_{k\to\infty}\bigg(\mathrm{diam}(V_k(z)) + \frac{m_R(\widehat{V}_k(z))}{m_R(V_k(z))} \bigg) \le K^{2h_R}\Big[M_{m_R}((3K^2)^{-1}\iota)\Big]^{-1} < \infty.$$

So, the hypotheses of Theorem 2.8.17 in Federer [50] are satisfied. That theorem along with Theorem 2.9.8 in [50] imply that

$$\rho(z) := \frac{dm'}{dm_R}(z) = \lim_{k\to\infty} \frac{m'(V_k(z))}{m_R(V_k(z))} \quad \text{for } m_R\text{-a. e. } z \in \mathcal{J}(R) \setminus \mathrm{Crit}_\infty^-(R).$$

Since $V_k(z) = T_z^{-1}(V_{k-1}(T(z)))$ for all $z \in \mathcal{J}(T) \setminus \mathrm{Crit}_\infty^-(T) = \mathcal{J}(R) \setminus \mathrm{Crit}_\infty^-(R)$ and all $k \in \mathbb{N}$ large enough, it ensues that

$$\rho(z) = \lim_{k\to\infty} \frac{m'(T_z^{-1}(V_{k-1}(T(z))))}{m_R(T_z^{-1}(V_{k-1}(T(z))))} = \lim_{k\to\infty} \frac{\int_{V_{k-1}(T(z))} |(T_z^{-1})'|^{h_R}\, dm'}{\int_{V_{k-1}(T(z))} |(T_z^{-1})'|^{h_R}\, dm_R}. \tag{32.41}$$

As

$$\lim_{k\to\infty} \frac{\sup\{|(T_z^{-1})'(w)| : w \in V_{k-1}(T(z))\}}{|(T_z^{-1})'(T(z))|} = 1 = \lim_{k\to\infty} \frac{\inf\{|(T_z^{-1})'(w)| : w \in V_{k-1}(T(z))\}}{|(T_z^{-1})'(T(z))|},$$

we can continue (32.41) as follows:

$$\rho(z) = \lim_{k\to\infty} \frac{m'(V_{k-1}(T(z)))}{m_R(V_{k-1}(T(z)))} = \lim_{k\to\infty} \frac{m'(V_k(T(z)))}{m_R(V_k(T(z)))} = \rho(T(z))$$

for m_R-a. e. $z \in \mathcal{J}(R) \setminus \mathrm{Crit}_\infty^-(R)$. As $m_R(\mathcal{J}(R) \setminus \mathrm{Crit}_\infty^-(R)) = 0$ according to Claim 1, we conclude that $\rho \circ T = \rho$ m_R-almost everywhere. Claim 3 is thus proved. ■

Claim 4. *m_R is the unique h_R-conformal measure for R.*

Proof of Claim 4. Let m' be an h_R-conformal measure for R. As m_R is quasi-R-invariant, it follows from Claims 2–3 that the Radon–Nikodym derivative $\rho = dm'/dm_R$ is constant m_R-almost everywhere. Since both measures m' and m_R are probabilistic, the derivative ρ must be identically equal to 1. So $m' = m_R$ and Claim 4 is proved. ■

Our next goal is to show that the measure m_R is conservative. This will easily ensue from the following claim.

Claim 5. *If $E \subseteq \mathcal{J}(R)$ is a forward or backward R-invariant Borel set, then $m_R(E) \in \{0, 1\}$.*

Proof of Claim 5. By passing to the complement, it suffices to prove the statement for a forward R-invariant set E. Since $m_R(\mathrm{Crit}_\infty^-(R)) = 0$ and the set $\mathrm{Crit}_\infty^-(R)$ is completely R-invariant, we may assume that $E \subseteq \mathcal{J}(R) \setminus \mathrm{Crit}_\infty^-(R) = \mathcal{J}(T) \setminus \mathrm{Crit}_\infty^-(T)$. Suppose for a contradiction that $0 < m_R(E) < 1$. By Lebesgue's density theorem, there exists at least one point $z \in E$ (in fact, the set of such points has m_R-measure equal to $m_R(E)$) such that

$$\lim_{r\to 0} \frac{m_R(E \cap B(z,r))}{m_R(B(z,r))} = 1. \tag{32.42}$$

Let $(s_i(z))_{i=1}^{\infty}$ be a subsequence of the sequence $(n_k(z))_{k=1}^{\infty}$ produced in Remark 32.1.3, subsequence such that the sequence $(T^{s_i(z)}(z))_{i=1}^{\infty}$ converges to a point $\hat{z}$.

Seeking a contradiction, suppose that $m_R(B(\hat{z},(2K^2)^{-1}\iota) \setminus E) = 0$, where $\iota > 0$ is the number defined in (32.35). Given that $R|_{\mathcal{J}(R)}$ is topologically exact according to Corollary 24.3.5, let $N \in \mathbb{N}$ be such that $R^N(B(\hat{z},(2K^2)^{-1}\iota)) \supseteq \mathcal{J}(R)$. By partitioning the ball $B(\hat{z},(2K^2)^{-1}\iota)$ into finitely many special Borel sets for R^N (cf. Lemma 29.1.15), the h_R-conformality of m_R entails that $m_R(R^N(B(\hat{z},(2K^2)^{-1}\iota) \setminus E)) = 0$. From the forward R-invariance of E, we deduce that

$$\begin{aligned} 0 &= m_R(R^N(B(\hat{z},(2K^2)^{-1}\iota) \setminus E)) \geq m_R(R^N(B(\hat{z},(2K^2)^{-1}\iota)) \setminus R^N(E)) \\ &\geq m_R(R^N(B(\hat{z},(2K^2)^{-1}\iota)) \setminus E) \geq m_R(R^N(B(\hat{z},(2K^2)^{-1}\iota))) - m_R(E) \\ &= 1 - m_R(E). \end{aligned}$$

Hence, $m_R(E) \geq 1$. This contradiction shows that

$$u := m_R(B(\hat{z},(2K^2)^{-1}\iota) \setminus E) > 0.$$

As $|T^{s_i(z)}(z) - \hat{z}| < (2K^2)^{-1}\iota$ for all $i \in \mathbb{N}$ large enough, say $i \geq N(z)$, we infer that

$$B(\hat{z},(2K^2)^{-1}\iota) \subseteq B(T^{s_i(z)}(z), K^{-2}\iota), \quad \forall i \geq N(z).$$

Let

$$r_i(z) := K^{-1}|(T^{s_i(z)})'(z)|^{-1}\iota.$$

It follows from Lemma 23.1.11 and the forward R-invariance of E that

$$T_z^{-s_i(z)}(B(\hat{z},(2K^2)^{-1}\iota) \setminus E) \subseteq T_z^{-s_i(z)}(B(T^{s_i(z)}(z), K^{-2}\iota) \setminus E) \subseteq B(z, r_i(z)) \setminus E, \quad \forall i \geq N(z).$$

By virtue of the h_R-conformality of m_R and of the Koebe distortion theorem (Theorem 23.1.9), we thereby get that

$$\begin{aligned} m_R(B(z, r_i(z)) \setminus E) &\geq m_R(T_z^{-s_i(z)}(B(\hat{z},(2K^2)^{-1}\iota) \setminus E)) \\ &\geq K^{-h_R}|(T_z^{-s_i(z)})'(T^{s_i(z)}(z))|^{h_R} m_R(B(\hat{z},(2K^2)^{-1}\iota) \setminus E) \\ &= uK^{-h_R}|(T^{s_i(z)})'(z)|^{-h_R}, \quad \forall i \geq N(z). \end{aligned}$$

On the other hand, it follows from Lemma 23.1.11 that

$$B(z, r_i(z)) \subseteq T_z^{-s_i(z)}(B(T^{s_i(z)}(z), \iota)), \quad \forall i \in \mathbb{N}.$$

Thus, by conformality and Koebe's distortion theorem,

$$m_R(B(z, r_i(z)) \leq K^{h_R}|(T^{s_i(z)})'(z)|^{-h_R}, \quad \forall i \in \mathbb{N}.$$

Therefore,

$$\frac{m_R(B(z,r_i(z))\setminus E)}{m_R(B(z,r_i(z)))} \geq \frac{uK^{-h_R}\left|\left(T^{s_i(z)}\right)'(z)\right|^{-h_R}}{K^{h_R}\left|\left(T^{s_i(z)}\right)'(z)\right|^{-h_R}} = uK^{-2h_R} > 0, \quad \forall i \geq N(z).$$

Hence,

$$\limsup_{i\to\infty} \frac{m_R(E\cap B(z,r_i(z)))}{m_R(B(z,r_i(z)))} \leq 1 - uK^{-2h_R} < 1,$$

contrary to (32.42). So, $m_R(E) \in \{0,1\}$. This finishes the proof of Claim 5. ■

Claim 6. *m_T is conservative.*

Proof of Claim 6. The proof is the same as its counterpart in Theorem 31.3.9. ■

This finishes the proof of Theorem 32.1.9. □

This theorem entails the set of transitive points to be of full m_R-measure.

Corollary 32.1.10. *If $R : \widehat{\mathbb{C}} \to \widehat{\mathbb{C}}$ is a subexpanding rational function with* $\deg(R) \geq 2$, *then the m_R-measure of the set of all transitive points of the dynamical system $R : \mathcal{J}(R) \to \mathcal{J}(R)$ is equal to 1.*

Proof. Replace T by R in the proof of Corollary 31.3.10 and rely on Theorem 32.1.9 rather than Theorem 31.3.9. □

Corollary 32.1.11. *If $R : \widehat{\mathbb{C}} \to \widehat{\mathbb{C}}$ is a subexpanding rational function with* $\deg(R) \geq 2$, *then*

$$m_R(\overline{\mathrm{PC}(R)}) = 0.$$

Proof. Clearly, $R(\mathrm{PC}(R)) \subseteq \mathrm{PC}(R)$. The continuity of R then imposes that $R(\overline{\mathrm{PC}(R)}) \subseteq \overline{\mathrm{PC}(R)}$. Moreover, $R(\omega(R)) \subseteq \omega(R)$. The ergodicity of m_R (per Theorem 32.1.9) implies that $\{m_R(\overline{\mathrm{PC}(R)}), m_R(\omega(R))\} \subseteq \{0,1\}$. By Proposition 26.3.4 and the Baire category theorem, the set $\overline{\mathrm{PC}(R)}$ is nowhere dense in $\mathcal{J}(R)$. Corollary 32.1.11 thus follows from Corollary 32.1.10 and the ergodicity of m_R.

Alternatively, note that the nowhere dense set $\omega(R)$ does not intersect the set of transitive points of R, and hence $m_R(\omega(R)) = 0$ by Corollary 32.1.10. Given that $\overline{\mathrm{PC}(R)} \subseteq \mathrm{PC}(R) \cup \omega(R)$ with $\mathrm{PC}(R)$ countable and that m_R is atomless, we conclude that

$$m_R(\overline{\mathrm{PC}(R)}) \leq m_R(\mathrm{PC}(R)) + m_R(\omega(R)) = 0.$$ □

32.1.3 The complete structure of Sullivan's conformal measures for subexpanding rational functions: the atomless measure m_R and purely atomic measures

We can now collect together the results of the previous sections and provide a short concluding proof of the main theorem of this section. It gives a full description of the structure of the Sullivan conformal measures of a subexpanding rational function. Recall that these maps are characterized by the disjointness of their critical set with their ω-limit set: $\mathrm{Crit}(R) \cap \omega(R) = \emptyset$. Define

$$\mathrm{Crit}_*(R) := [\mathrm{Crit}(R) \cap \mathcal{J}(R)] \setminus \mathrm{PC}(R).$$

By Proposition 26.3.3, the accumulation points of $\mathrm{PC}(R)$ that lie in $\mathcal{J}(R)$ are in $\omega(R)$. Therefore, $\mathrm{Crit}_*(R) \cap \overline{\mathrm{PC}(R)} = \emptyset$ and there exists $\Delta > 0$ such that

$$B(\mathrm{Crit}_*(R), 4\Delta) \cap \overline{\mathrm{PC}(R)} = \emptyset. \tag{32.43}$$

We therefore get the following.

Lemma 32.1.12. *If $R : \widehat{\mathbb{C}} \to \widehat{\mathbb{C}}$ is a subexpanding rational function with $\deg(R) \geq 2$, then for every point $c \in \mathrm{Crit}_*(R)$, every $n \in \mathbb{N}$, and every point $z \in R^{-n}(c)$ there exists a unique holomorphic branch $R_z^{-n} : B(c, 4\Delta) \to \widehat{\mathbb{C}}$ of R^{-n} sending c to z.*

We now state and prove a theorem which parallels Theorem 31.3.11; it is not however identical to it.

Theorem 32.1.13. *Let $R : \widehat{\mathbb{C}} \to \widehat{\mathbb{C}}$ be a subexpanding rational function with $\deg(R) \geq 2$. Recall from Theorem 32.1.6 (Bowen's formula) that*

$$\begin{aligned} h_R := \delta(R) &= \mathrm{ExpD}^{**}(\mathcal{J}(R)) = \mathrm{ExpD}^*(\mathcal{J}(R)) = \mathrm{ExpD}(\mathcal{J}(R)) = \mathrm{DD}(\mathcal{J}(R)) \\ &= \mathrm{HD}(\mathcal{J}_{\mathrm{uer}}(R)) = \mathrm{HD}(\mathcal{J}_{\mathrm{er}}(R)) = \mathrm{HD}(\mathcal{J}_{\mathrm{r}}(R)) \\ &= \mathrm{HD}(\mathcal{J}(R)) \in \left(\frac{p_R}{p_R + 1}, 2\right]. \end{aligned}$$

The following situations prevail:

(a) *If $t < h_R$, then R does not admit any Sullivan t-conformal measure.*

(b) *R has exactly one h_R-conformal measure, namely m_R. This measure is atomless, ergodic, conservative and geometric with exponent h_R.*

(c) *For every $t > h_R$ and every $c \in \mathrm{Crit}_*(R)$, it holds that*

$$\Sigma_t(c) := \sum_{n=0}^{\infty} \sum_{z \in R^{-n}(c)} |(R^n)'(z)|^{-t} < \infty. \tag{32.44}$$

Accordingly, let

$$\nu_{t,c} := \frac{1}{\Sigma_t(c)} \sum_{n=0}^{\infty} \sum_{z \in R^{-n}(c)} \left|(R^n)'(z)\right|^{-t} \delta_z, \tag{32.45}$$

where δ_z denotes the Dirac measure concentrated at the point z. Then the set of all t-conformal measures for R coincides with the set of all convex combinations of measures $\{\nu_{t,c}\}_{c \in \mathrm{Crit}_*(R)}$. In particular, these conformal measures are atomic.

Proof.

(a) Pick $\xi \in \mathrm{Safe}(R) \cap \mathcal{J}_{\mathrm{uer}}(R)$. Let m_t be the PS limit measure (in the sense of Definition 29.1.12) produced in Lemma 29.2.1 for the function $\varphi_t := -t \log |R'|$. Recall that $h_R := \delta(R) = s_\xi(R) := \inf\{t \geq 0 \mid c_\xi(t) \leq 0\}$ by Theorem 29.2.12 and $c_\xi(s_\xi(R)) = 0$ by (29.49). Given that $t \in [0, h_R)$, we know from (29.46) that $c_\xi(t) > 0$. By Lemma 29.2.4, we deduce that $e^{\mathrm{P}(t)} \geq e^{c_\xi(t)} > 1$. Lemmas 32.1.1, 32.1.2 and 32.1.5 can be generalized to (t, λ_t)-conformal measures like Lemmas 31.3.3–31.3.4 were extended to Lemmas 31.3.5–31.3.6. The proof is then essentially analogous to that of Theorem 31.3.11(a) with $\Omega(T)$ replaced by $\mathrm{Crit}(R)$. It is left to the reader as an exercise.

(b) This follows from the combination of Theorems 32.1.8–32.1.9.

(c) The proof is similar to that of Theorem 31.3.11(c). Clearly, Claims 1–4 remain valid with T replaced by R. In place of Claim 5, we prove the following.

Claim 5′. *If $\Sigma_t(x) < \infty$ for some $t \geq 0$ and $x \in \mathcal{J}(T)$, then $\Sigma_t(y) < \infty$ for every $y \in \mathrm{Crit}_{*\infty}^{-}(R) := \bigcup_{n=0}^{\infty} R^{-n}(\mathrm{Crit}_*(R))$.*

Proof of Claim 5′. By Claim 3, it suffices to establish the statement for $y \in \mathrm{Crit}_*(R)$. This follows from (32.43) and Claim 4. ■

In lieu of Claims 6–7, we prove the next two claims.

Claim 6′. *If $t > h_R = s_\xi(R)$ and $c_\xi(t) < 0$, then $\Sigma_t(\xi) < \infty$ and $\Sigma_t(y) < \infty$ for every $y \in \mathrm{Crit}_{*\infty}^{-}(R)$.*

Proof of Claim 6′. Fix $s \in (c_\xi(t), 0)$. By definition of $c_\xi(t)$, we obtain that

$$\Sigma_t(\xi) = \sum_{n=0}^{\infty} \sum_{z \in R^{-n}(\xi)} \left|(R^n)'(z)\right|^{-t} \leq \sum_{n=0}^{\infty} \sum_{z \in R^{-n}(\xi)} \left|(R^n)'(z)\right|^{-t} e^{-sn} < \infty.$$

Therefore, this claim follows from Claim 5′. ■

Claim 7′. *If $t > h_R$ and $c_\xi(t) = 0$, then $\Sigma_t(y) < \infty$ for every $y \in \mathrm{Crit}_{*\infty}^{-}(R)$.*

Proof of Claim 7′. Let m_t be the measure produced in Lemma 29.2.1. Since $c_\xi(t) = 0$, it follows from Lemma 29.2.1 that m_t is t-conformal. As $h_R < t$ and m_R is h_R-conformal, it ensues from Lemma 32.1.5 that $m_t(\mathcal{J}(R) \setminus \mathrm{Crit}_{\infty}^{-}(R)) = 0$. So, there exists $c \in \mathrm{Crit}(R) \cap \mathcal{J}(R)$ for which $m_t(\mathcal{O}_-(c)) > 0$, and thus $m_t(\{c\}) > 0$. The conformality of m_t also entails that $m_t(\mathrm{Crit}(R) \cap \mathrm{PC}(R)) = 0$. Then $c \in [\mathrm{Crit}(R) \cap \mathcal{J}(R)] \setminus \mathrm{PC}(R) =: \mathrm{Crit}_*(R)$. Given that $m_t(\mathcal{O}_-(c)) = m_t(\{c\})\Sigma_t(c)$, we conclude that

$$\Sigma_t(c) = (m_t(\{c\}))^{-1} m_t(\mathcal{O}_-(c)) \le (m_t(\{c\}))^{-1} < \infty.$$

Consequently, this claim ensues from Claim 5′. ■

Together, Claims 6′–7′ conclude the proof of (32.44).

By definition (32.45), all measures $\{\nu_{t,c}\}_{c\in\mathrm{Crit}_*(R)}$ are t-conformal. Obviously, so are all their convex combinations. On the other hand, let m be a t-conformal measure for R. It follows from Lemma 32.1.5 (applied with $s = h_R$) that $m(\mathrm{Crit}_\infty^-(R)) = 1$. The conformality of m entails that $m(\mathrm{Crit}(R) \cap \mathrm{PC}(R)) = 0$. Consequently, $1 = m(\mathrm{Crit}_{*\infty}^-(R)) = \sum_{c\in\mathrm{Crit}_*(R)} m(\mathcal{O}_-(c))$. Using the t-conformality of m, we readily deduce that m is a convex combination of measures $\nu_{t,c}$, $c \in \mathrm{Crit}_*(R)$. Thus, (c) holds. □

32.2 Invariant probability measure equivalent to m_R: existence, uniqueness and ergodicity

In this section, we establish the existence and uniqueness of an R-invariant Borel probability measure which is absolutely continuous with respect to the h_R-conformal measure m_R. This is the counterpart of Theorem 31.4.1.

Theorem 32.2.1. *If $R : \widehat{\mathbb{C}} \to \widehat{\mathbb{C}}$ is a subexpanding rational function with $\deg(R) \ge 2$, then there exists a unique R-invariant Borel probability measure μ_R on $\mathcal{J}(R)$ which is absolutely continuous with respect to the h_R-conformal measure m_R. In fact, μ_R is equivalent to m_R. It is also ergodic.*

Proof. We first show the existence of an R-invariant σ-finite Borel measure μ_R which is absolutely continuous with respect to m_R. For every $z \in \mathcal{J}(R) \setminus \overline{\mathrm{PC}(R)}$, fix $u(z) > 0$ such that

$$B(z, 2u(z)) \cap \overline{\mathrm{PC(R)}} = \emptyset \quad \text{and} \quad m_R(\partial B(z, u(z))) = 0.$$

As $\mathcal{J}(R) \setminus \overline{\mathrm{PC}(R)}$ is a Lindelöf space, there is a sequence $(z_j)_{j=0}^\infty$ in $\mathcal{J}(R) \setminus \overline{\mathrm{PC}(R)}$ such that

$$\bigcup_{j=0}^{\infty} B(z_j, u(z_j)) \supseteq \mathcal{J}(R) \setminus \overline{\mathrm{PC}(R)}.$$

Out of the cover $\{B(z_j, u(z_j))\}_{j=0}^\infty$ of $\mathcal{J}(R) \setminus \overline{\mathrm{PC}(R)}$, define inductively a Borel partition $\mathcal{A} = (A_n)_{n=0}^\infty$ of $\mathcal{J}(R) \setminus \overline{\mathrm{PC}(R)}$ (partitioning up to a set of m_R-measure zero) by setting

$$A_0 := B(z_0, u(z_0)) \quad \text{and} \quad A_n := B(z_n, u(z_n)) \setminus \bigcup_{j=0}^{n-1} \overline{B(z_j, u(z_j))}, \quad \forall n \in \mathbb{N}$$

while throwing away empty sets. Clearly,

$$A_n \subseteq B(z_n, u(z_n)), \quad \forall n \geq 0.$$

Due to the definition of $u(z)$, $z \in \mathcal{J}(R) \setminus \overline{\mathrm{PC}(R)}$, due to Theorem 23.1.9 (Koebe's distortion theorem), due to Theorem 32.1.9, due to Corollary 32.1.11, and due to the finiteness-to-one of the map R, the partition $\mathcal{A}$ satisfies all the requirements of Theorem 10.4.5. Define μ'_R to be the R-invariant σ-finite Borel measure resulting from the application of this theorem with m_R as reference measure.

We now demonstrate that

$$\mu'_R(B(z, \kappa\Gamma_R)) < \infty, \quad \forall z \in \mathcal{J}(R), \tag{32.46}$$

where $\Gamma_R > 0$ comes from Lemma 26.3.9 and Corollary 26.3.10 whereas κ is from Lemma 26.3.11. Indeed, for every set $D \subseteq \widehat{\mathbb{C}}$ and every $n \geq 0$, let

$$S_n(m_R)(D) := \sum_{k=0}^{n} m_R(R^{-n}(D)).$$

As $m_R(B(z, \kappa\Gamma_R)) > 0$ by Theorem 32.1.9 and as $m_R(\bigcup_{n=0}^{\infty} A_n) = 1$ by Corollary 32.1.11, there exists $i \geq 0$ such that $m(B(z, \kappa\Gamma_R) \cap A_i) > 0$. As $B(z, \kappa\Gamma_R) \cap A_i$ is an open set having nonempty intersection with $\mathcal{J}(R)$, there exists an open ball $G \subseteq B(z, \kappa\Gamma_R) \cap A_i$ with center in $\mathcal{J}(R)$. Of course, $m_R(G) > 0$. Given $n \geq 0$, let W be a connected component of $R^{-n}(B(z, \kappa\Gamma_R))$ and let $V \subseteq W$ be a connected component of $R^{-n}(G)$ contained in W. Using Koebe's distortion theorem (Theorem 23.1.9), the h_R-conformality of m_R and Theorem 32.1.9, we get

$$m_R(V) \asymp (\mathrm{diam}(V))^{h_R} = \left(\frac{\mathrm{diam}(V)}{\mathrm{diam}(W)}\right)^{h_R} (\mathrm{diam}(W))^{h_R} \succeq m_R(W)\left(\frac{\mathrm{diam}(V)}{\mathrm{diam}(W)}\right)^{h_R},$$

where $\asymp$ and $\succeq$ are symbols of multiplicative comparisons (cf. (25.34)).

Applying Lemma 26.3.11 to the connected sets G and $B(z, \kappa\Gamma_R)$, we obtain

$$m_R(V) \succeq m_R(W)\left(\frac{\mathrm{diam}(G)}{\mathrm{diam}(B(z, \kappa\Gamma_R))}\right)^{h_R}.$$

Thus,

$$S_n(m_R)(B(z, \kappa\Gamma_R)) \preceq S_n(m_R)(G) \preceq S_n(m_R)(A_i).$$

So, with the notation in Theorem 10.4.5 (see (10.21)), we get that

$$(m_R)_n(B(z, \kappa\Gamma_R)) \preceq (m_R)_n(A_i), \quad \forall n \geq 0.$$

It therefore follows from the first paragraph and the last displayed formula in Theorem 10.4.5 that

$$((m_R)_n(B(z,\kappa\Gamma_R)))_{n=0}^{\infty} \in l^{\infty}.$$

It ensues from (10.23) in Theorem 10.4.5 that (32.46) holds.

Since $\mathcal{J}(R)$ is a compact set, it follows from (32.46) that

$$\mu_R'(\mathcal{J}(R)) < \infty.$$

We then define

$$\mu_R := (\mu_R'(\mathcal{J}(R)))^{-1}\mu_R'.$$

Finally, ergodicity, equivalence with m_R and uniqueness of the measure μ_R follow from the last paragraph of Theorem 10.4.5 and from Theorem 32.1.9. □

Remark 32.2.2. As an immediate consequence of Theorem 32.2.1, we get that $\mu_{R^n} = \mu_R$ for every $n \in \mathbb{N}$. So μ_R is totally ergodic.

32.3 Hausdorff and packing measures

In this short section, we give a precise geometric description of the fractal structure of the Julia set of a subexpanding rational function in terms of the Hausdorff and packing measures. It is nearly the same as that for expanding rational functions (cf. Theorem 30.1.1(f)) and is much simpler than that for parabolic rational functions (cf. Theorem 31.5.1).

Theorem 32.3.1 (Fundamental fractal geometry for subexpanding rational functions). *Let $R : \widehat{\mathbb{C}} \to \widehat{\mathbb{C}}$ be a subexpanding rational function with* $\deg(R) \geq 2$. *Recall from Theorem* 32.1.6 (*Bowen's formula*) *that*

$$\begin{aligned}h_R := \delta(R) &= \mathrm{ExpD}^{**}(\mathcal{J}(R)) = \mathrm{ExpD}^{*}(\mathcal{J}(R)) = \mathrm{ExpD}(\mathcal{J}(R)) = \mathrm{DD}(\mathcal{J}(R))\\ &= \mathrm{HD}(\mathcal{J}_{\mathrm{uer}}(R)) = \mathrm{HD}(\mathcal{J}_{\mathrm{er}}(R)) = \mathrm{HD}(\mathcal{J}_{\mathrm{r}}(R))\\ &= \mathrm{HD}(\mathcal{J}(R)) \in \left(\frac{p_R}{p_R+1}, 2\right].\end{aligned}$$

The following statements hold:

(a) *m_R is the only Sullivan conformal measure for R and its exponent is equal to h_R.*
(b) *m_R is a geometric measure with exponent h_R.*
(c) *The measures m_R, Hausdorff $\mathrm{H}_{h_R}|_{\mathcal{J}(R)}$, and packing $\mathrm{P}_{h_R}|_{\mathcal{J}(R)}$, are finite and positive, and are equal up to a multiplicative constant.*
(d) *If $\mathcal{J}(R) \neq \widehat{\mathbb{C}}$, then $h_R < 2$.*

Proof.

(a, b) are part of Theorem 32.1.13(b).

(c) Finiteness and positivity of the measures m_R, $H_{h_R}|_{\mathcal{J}(R)}$ and $P_{h_R}|_{\mathcal{J}(R)}$ is an immediate consequence of (b) and Theorem 15.6.14. Equality up to a multiplicative constant now follows from (a) and the fact that both measures $H_{h_R}|_{\mathcal{J}(R)}$ and $P_{h_R}|_{\mathcal{J}(R)}$ are h_R-conformal.

(d) As $\mathcal{J}(R) \neq \widehat{\mathbb{C}}$, it ensues from Theorem 29.2.13, Corollary 32.1.4 and the countability of $\mathrm{Crit}_\infty^-(R)$ that $\mathrm{Leb}_2(\mathcal{J}(R)) = 0$. Since the measures H_2 and P_2 are constant multiples of Leb_2, we conclude from (c) that $h_R < 2$. □

The study of conformal measures, measurable dynamics, and fractal geometry of Julia sets of rational functions concurrently exhibiting parabolic and subexpanding features has been done in [138] and [140] for the class of all nonrecurrent rational functions, i. e., functions for which $c \notin \omega(c)$ for all critical points c in the Julia set. This study has also been carried out for elliptic functions in [73]. An important extension of subexpanding rational functions is the class of topological Collet–Eckmann rational functions. It has been thoroughly studied from the point of view of this chapter in [108].

32.4 Exercises

Exercise 32.4.1. Show that if $T : \widehat{\mathbb{C}} \to \widehat{\mathbb{C}}$ is a rational function with $\deg(T) \geq 2$ and $\mathrm{Crit}(T) \not\subset \overline{\mathrm{PC}(T)}$, then for every $t > \delta(T)$ there exists a t-conformal measure supported on $\mathrm{Crit}_\infty^-(T)$.

Appendix A – A selection of classical results

This appendix lists classical definitions and results that will be used in this volume.

A.1 Measure theory

Proofs and further explanations of the results can be found in many books on measure theory, for instance, Billingsley [12, 13] and Rudin [112].

Lemma A.1.1 (Borel–Cantelli lemma). *Let $(X, \mathcal{A}, \mu)$ be a measure space and $(A_n)_{n=1}^{\infty}$ a sequence in $\mathcal{A}$. If $\sum_{n=1}^{\infty} \mu(A_n) < \infty$, then $\mu(\bigcap_{k=1}^{\infty} \bigcup_{n=k}^{\infty} A_n) = 0$.*

A.1.1 Convergence theorems

In measure theory, there are fundamental theorems that are especially helpful for finding the integral of functions that are the pointwise limit of sequences of functions.

The first of these results applies to monotone sequences of functions. A sequence of functions $(f_n)_{n=1}^{\infty}$ is monotone if it is increasing pointwise ($f_{n+1}(x) \geq f_n(x)$ for all $x \in X$ and all $n \in \mathbb{N}$) or decreasing pointwise ($f_{n+1}(x) \leq f_n(x)$ for all $x \in X$ and all $n \in \mathbb{N}$). We state the theorem for increasing sequences, but its counterpart for decreasing sequences can be easily deduced from it.

Theorem A.1.2 (Monotone convergence theorem). *Let $(X, \mathcal{A}, \mu)$ be a measure space. If $(f_n)_{n=1}^{\infty}$ is an increasing sequence of nonnegative measurable functions, then the integral of their pointwise limit is equal to the limit of their integrals, that is,*

$$\int_X \lim_{n\to\infty} f_n \, d\mu = \lim_{n\to\infty} \int_X f_n \, d\mu.$$

Note that this theorem holds for almost everywhere increasing sequences of almost everywhere nonnegative measurable functions with an almost everywhere pointwise limit.

For general sequences of nonnegative functions, we have the following immediate consequence.

Lemma A.1.3 (Fatou's lemma). *Let $(X, \mathcal{A}, \mu)$ be a measure space. For any sequence $(f_n)_{n=1}^{\infty}$ of nonnegative measurable functions,*

$$\int_X \liminf_{n\to\infty} f_n \, d\mu \leq \liminf_{n\to\infty} \int_X f_n \, d\mu.$$

The following lemma is another application of the monotone convergence theorem. It offers another way of integrating a nonnegative function.

https://doi.org/10.1515/9783110769876-033

Lemma A.1.4. *Let $(X, \mathcal{A}, \mu)$ be a measure space. Let f be a nonnegative measurable function and $A \in \mathcal{A}$. Then*

$$\int_A f\,d\mu = \int_0^\infty \mu(\{x \in A : f(x) > r\})\,dr.$$

Pointwise convergence of a sequence of integrable functions does not guarantee convergence in L^1. However, under one relatively weak additional assumption, this becomes true. The second fundamental theorem of convergence applies to sequences of functions which have an almost everywhere pointwise limit and are dominated (i. e., uniformly bounded) almost everywhere by an integrable function.

Theorem A.1.5 (Lebesgue's dominated convergence theorem). *If a sequence of measurable functions $(f_n)_{n=1}^\infty$ on a measure space $(X, \mathcal{A}, \mu)$ converges pointwise μ-a. e. to a function f and if there exists $g \in L^1(\mu)$ such that $|f_n(x)| \le g(x)$ for all $n \in \mathbb{N}$ and μ-a. e. $x \in X$, then $f \in L^1(\mu)$ and*

$$\lim_{n\to\infty} \|f_n - f\|_1 = 0 \quad \text{and} \quad \lim_{n\to\infty} \int_X f_n\,d\mu = \int_X f\,d\mu.$$

Note that

$$\lim_{n\to\infty} \|f_n - f\|_1 = 0 \quad \Longrightarrow \quad \lim_{n\to\infty} \|f_n\|_1 = \|f\|_1$$

since $\big|\|f_n\|_1 - \|f\|_1\big| \le \|f_n - f\|_1$ and

$$\lim_{n\to\infty} \|f_n - f\|_1 = 0 \quad \Longrightarrow \quad \lim_{n\to\infty} \int_X f_n\,d\mu = \int_X f\,d\mu.$$

The opposite implications do not hold in general. Nevertheless, the following lemma states that any sequence of integrable functions $(f_n)_{n=1}^\infty$ that converges pointwise almost everywhere to an integrable function f will also converge to that function in L^1 if and only if their L^1 norms converge to the L^1 norm of f.

Lemma A.1.6 (Scheffé's lemma). *Let $(X, \mathcal{A}, \mu)$ be a measure space. If a sequence $(f_n)_{n=1}^\infty$ of functions in $L^1(\mu)$ converges pointwise μ-a. e. to a function $f \in L^1(\mu)$, then*

$$\lim_{n\to\infty} \|f_n - f\|_1 = 0 \quad \Longleftrightarrow \quad \lim_{n\to\infty} \|f_n\|_1 = \|f\|_1.$$

In particular, if $f_n \ge 0$ μ-a. e. for all $n \in \mathbb{N}$, then $f \ge 0$ μ-a. e. and

$$\lim_{n\to\infty} \|f_n - f\|_1 = 0 \quad \Longleftrightarrow \quad \lim_{n\to\infty} \int_X f_n\,d\mu = \int_X f\,d\mu.$$

If g is a L^1 function on a general measure space $(X, \mathcal{A}, \mu)$, then the sequence of non-negative measurable functions $(g_M)_{M=1}^{\infty}$, where $g_M = |g| \cdot \mathbb{1}_{\{|g| \geq M\}}$, decreases to 0 pointwise and is dominated by $|g|$. Therefore, the monotone convergence theorem (or, alternatively, Lebesgue's dominated convergence theorem) affirms that

$$\lim_{M \to \infty} \int_{\{|g| \geq M\}} |g| \, d\mu = 0.$$

This suggests introducing the following concept.

Definition A.1.7. Let $(X, \mathcal{A}, \mu)$ be a measure space. A sequence of measurable functions $(f_n)_{n=1}^{\infty}$ is *uniformly integrable* if

$$\lim_{M \to \infty} \sup_{n \in \mathbb{N}} \int_{\{|f_n| \geq M\}} |f_n| \, d\mu = 0.$$

On finite measure spaces, there exists a generalization of Lebesgue's dominated convergence theorem (Theorem A.1.5).

Theorem A.1.8. *Let $(X, \mathcal{A}, \mu)$ be a finite measure space and $(f_n)_{n=1}^{\infty}$ a sequence of measurable functions that converges pointwise μ-a. e. to a function f.*

(a) *If $(f_n)_{n=1}^{\infty}$ is uniformly integrable, then $f_n \in L^1(\mu)$ for all $n \in \mathbb{N}$ and $f \in L^1(\mu)$. Moreover,*

$$\lim_{n \to \infty} \|f_n - f\|_1 = 0 \quad \text{and} \quad \lim_{n \to \infty} \int_X f_n \, d\mu = \int_X f \, d\mu.$$

(b) *If $f, f_n \in L^1(\mu)$ and $f_n \geq 0$ μ-a. e. for all $n \in \mathbb{N}$, then*

$$\lim_{n \to \infty} \int_X f_n \, d\mu = \int_X f \, d\mu \quad \Longrightarrow \quad (f_n)_{n=1}^{\infty} \text{ uniformly integrable.}$$

Corollary A.1.9. *Let $(X, \mathcal{A}, \mu)$ be a finite measure space and $(f_n)_{n=1}^{\infty}$ a sequence in $L^1(\mu)$ that converges pointwise μ-a. e. to a function $f \in L^1(\mu)$. The following statements are equivalent:*

(a) *The sequence $(f_n)_{n=1}^{\infty}$ is uniformly integrable.*
(b) $\lim_{n \to \infty} \|f_n - f\|_1 = 0$.
(c) $\lim_{n \to \infty} \|f_n\|_1 = \|f\|_1$.

It is obvious that uniform convergence of a sequence of functions implies pointwise convergence of the sequence. The converse is generally not true but the following partial converse holds.

Theorem A.1.10 (Dini's theorem). *If X is a compact topological space and $(f_n)_{n=1}^{\infty}$ is an increasing sequence of real-valued continuous functions on X converging pointwise to a con-*

tinuous function f, then the convergence is uniform. The same conclusion holds if $(f_n)_{n=1}^{\infty}$ is decreasing instead of increasing.

It is also natural to ask whether, in some way, a pointwise convergent sequence converges "almost" uniformly.

Definition A.1.11. Let $(X, \mathcal{A}, \mu)$ be a measure space. A sequence $(f_n)_{n=1}^{\infty}$ of measurable functions on X is said to *converge μ-almost uniformly* to a function f if for every $\varepsilon > 0$ there exists $Y \in \mathcal{A}$ such that $\mu(Y) < \varepsilon$ and $(f_n)_{n=1}^{\infty}$ converges uniformly to f on $X \setminus Y$.

It is clear that almost uniform convergence implies almost everywhere pointwise convergence. The converse is not true in general but these two types of convergence are one and the same on any finite measure space.

Theorem A.1.12 (Egorov's theorem). *Let $(X, \mathcal{A}, \mu)$ be a finite measure space. A sequence $(f_n)_{n=1}^{\infty}$ of measurable functions on X converges pointwise μ-almost everywhere to a limit function f if and only if that sequence converges μ-almost uniformly to f.*

The reader ought to convince themself that this result does not generally hold on infinite measure spaces.

Convergence in measure is another interesting type of convergence.

Definition A.1.13. Let $(X, \mathcal{A}, \mu)$ be a measure space. A sequence $(f_n)_{n=1}^{\infty}$ of measurable functions *converges in measure* to a measurable function f provided that for each $\varepsilon > 0$,

$$\lim_{n\to\infty} \mu(\{x \in X : |f_n(x) - f(x)| > \varepsilon\}) = 0.$$

Lemma A.1.14. *Let $(X, \mathcal{A}, \mu)$ be a measure space. If a sequence $(f_n)_{n=1}^{\infty}$ of measurable functions converges in $L^1(\mu)$ to a measurable function f, then $(f_n)_{n=1}^{\infty}$ converges in measure to f.*

When the measure is finite, there is a close relationship between pointwise convergence and convergence in measure.

Theorem A.1.15. *Let $(X, \mathcal{A}, \mu)$ be a finite measure space and $(f_n)_{n=1}^{\infty}$ a sequence of measurable functions.*

(a) *If $(f_n)_{n=1}^{\infty}$ converges pointwise μ-a. e. to a function f, then $(f_n)_{n=1}^{\infty}$ converges in measure to f.*
(b) *If $(f_n)_{n=1}^{\infty}$ converges in measure to a function f, then there exists a subsequence $(f_{n_k})_{k=1}^{\infty}$ which converges pointwise μ-a. e. to f.*
(c) *$(f_n)_{n=1}^{\infty}$ converges in measure to a function f if and only if each subsequence $(f_{n_k})_{k=1}^{\infty}$ admits a further subsequence $(f_{n_{k_l}})_{l=1}^{\infty}$ that converges pointwise μ-a. e. to f.*

The previous two results reveal that, on a finite measure space, a sequence of integrable functions that converges in L^1 to an integrable function admits a subsequence which converges pointwise almost everywhere to that function. In general, the sequence itself might not converge pointwise almost everywhere.

In some sense, the following result is a form of convergence theorem. It asserts that Borel measurable functions can be approximated by continuous functions on "arbitrarily large" portions of their domain.

Theorem A.1.16 (Luzin's theorem). *Let $(X, \mathcal{B}(X), \mu)$ be a finite Borel measure space and let $f : X \to \overline{\mathbb{R}} = [-\infty, \infty]$ be a Borel function. Given any $\varepsilon > 0$, for every $B \in \mathcal{B}(X)$ there is a closed set E with $\mu(B \setminus E) < \varepsilon$ such that $f|_E$ is continuous. If B is locally compact, the set E can be chosen to be compact and then there is a continuous function $f_\varepsilon : X \to \overline{\mathbb{R}}$ with compact support that coincides with f on E and such that* $\sup_{x \in X} |f_\varepsilon(x)| \le \sup_{x \in X} |f(x)|$.

A.1.2 Mutual singularity, absolute continuity and equivalence of measures

We now leave aside convergence of sequences of functions and recall the definitions of mutually singular, absolutely continuous and equivalent measures.

Definition A.1.17. Let $(X, \mathcal{A})$ be a measurable space, and μ, ν be two measures on $(X, \mathcal{A})$.
(a) μ and ν are said to be *mutually singular*, denoted by $\mu \perp \nu$, if there exist disjoint sets $X_\mu, X_\nu \in \mathcal{A}$ such that $\mu(X \setminus X_\mu) = 0 = \nu(X \setminus X_\nu)$.
(b) μ is *absolutely continuous with respect to* ν, denoted $\mu \ll \nu$, if $\nu(A) = 0 \Longrightarrow \mu(A) = 0$.
(c) μ and ν are *equivalent* if $\mu \ll \nu$ and $\nu \ll \mu$.

The Radon–Nikodym theorem provides a characterization of absolute continuity. Though it is valid for σ-finite measures, the following version for finite measures is sufficient for our purposes.

Theorem A.1.18 (Radon–Nikodym theorem). *Let $(X, \mathcal{A})$ be a measurable space and let μ and ν be two finite measures on $(X, \mathcal{A})$. The following statements are equivalent:*
(a) $\mu \ll \nu$.
(b) *For every $\varepsilon > 0$, there exists $\delta > 0$ such that $\nu(A) < \delta \Longrightarrow \mu(A) < \varepsilon$.*
(c) *There exists a ν-a. e. unique function $f \in L^1(\nu)$ such that $f \ge 0$ and*

$$\mu(A) = \int_A f \, d\nu, \quad \forall A \in \mathcal{A}.$$

Remark A.1.19. The function f is often denoted by $\frac{d\mu}{d\nu}$ and called the *Radon–Nikodym derivative* of μ with respect to ν.

A.1.3 The space $C(X)$, its dual $C(X)^*$ and the subspace $M(X)$

Another important result is Riesz representation theorem. Before stating it, we first establish some notation. Let X be a compact metrizable space. Let $C(X)$ be the set of all continuous real-valued functions on X. This set becomes a normed vector space when endowed with the supremum norm

$$\|f\|_\infty := \sup\{|f(x)| : x \in X\}.$$

This norm defines a metric on $C(X)$ in the usual way:

$$d_\infty(f,g) := \|f - g\|_\infty.$$

The topology induced by the metric d_∞ on $C(X)$ is called the *topology of uniform convergence* on X. Indeed, $\lim_{n\to\infty} d_\infty(f_n, f) = 0$ if and only if the sequence $(f_n)_{n=1}^\infty$ converges to f uniformly on X. It is not hard to see that $C(X)$ is a separable Banach space (i. e., a separable and complete normed vector space).

Let $C(X)^*$ denote the *dual space* of $C(X)$, i. e.,

$$C(X)^* := \{F : C(X) \to \mathbb{R} \mid F \text{ is linear and continuous}\}.$$

Recall that a real-valued function F defined on $C(X)$ is called a *functional* on $C(X)$. It is well known that a linear functional F is continuous if and only if it is *bounded*, i. e., if and only if its operator norm $\|F\|$ (sometimes denoted $\|F\|_\infty$) is finite, where

$$\|F\| := \sup\{|F(f)| : \|f\|_\infty \le 1\}.$$

So $C(X)^*$ can also be described as the normed vector space of all bounded linear functionals on $C(X)$. The operator norm defines a metric on $C(X)^*$ in the usual manner:

$$d(F,G) := \|F - G\|.$$

The topology induced by the metric d on $C(X)^*$ is called the *operator norm topology*, or *strong topology*, on $C(X)^*$. It is not difficult to see that $C(X)^*$ is a separable Banach space.

Furthermore, a linear functional F is said to be *normalized* if $F(1) = 1$ and is called *positive* if $F(f) \ge 0$ whenever $f \ge 0$.

Finally, we denote the set of all Borel probability measures on X by $M(X)$. This set is clearly convex and can be characterized as follows.

Theorem A.1.20 (Riesz representation theorem). *Let X be a compact metrizable space, and let F be a normalized and positive linear functional on $C(X)$. Then there exists a unique $\mu \in M(X)$ such that*

$$F(f) = \int_X f\,d\mu, \quad \forall f \in C(X). \tag{A.1}$$

Conversely, any $\mu \in M(X)$ defines a normalized positive linear functional on $C(X)$ via formula (A.1). *This linear functional is bounded.*

It immediately follows from Riesz representation theorem that every Borel probability measure on a compact metrizable space is uniquely determined by the way it integrates continuous functions on that space.

Corollary A.1.21. *If μ and ν are two Borel probability measures on a compact metrizable space X, then*

$$\mu = \nu \quad \Longleftrightarrow \quad \int_X f\,d\mu = \int_X f\,d\nu, \quad \forall f \in C(X).$$

Let us now discuss the *weak* topology* on the set $M(X)$. Recall that if Z is a set and $(Z_\alpha)_{\alpha\in A}$ is a family of topological spaces, then the weak topology induced on Z by a collection of maps $\{\psi_\alpha : Z \to Z_\alpha \mid \alpha \in A\}$ is the smallest topology on Z that makes each ψ_α continuous. Evidently, the sets $\psi_\alpha^{-1}(U_\alpha)$, for U_α open in Z_α, constitute a subbase for the weak topology. The weak* topology on $M(X)$ is the weak topology induced by $C(X)$ on its dual space $C(X)^*$, where measures in $M(X)$ and normalized positive linear functionals in $C(X)^*$ are identified via Riesz representation theorem. Note that $M(X)$ is metrizable, although $C(X)^*$ with the weak* topology usually is not. Indeed, both $C(X)$ and its subspace $C(X,[0,1])$ of continuous functions on X taking values in $[0,1]$, are separable since X is a compact metrizable space. Then for any dense subset $\{f_n\}_{n=1}^\infty$ of $C(X,[0,1])$, a metric on $M(X)$ is

$$d(\mu,\nu) = \sum_{n=1}^\infty \frac{1}{2^n}\left|\int_X f_n\,d\mu - \int_X f_n\,d\nu\right|.$$

In this book, we will sometimes denote by $\mu_n \xrightarrow{*} \mu$ the convergence of a sequence of measures $(\mu_n)_{n=1}^\infty$ to a measure μ in the weak* topology of $M(X)$.

Remark A.1.22. This notion is often presented as “weak convergence” of measures. This can be slightly confusing at first sight, but it helps to bear in mind that, as we have seen above, the weak* topology is just one instance of a weak topology.

The Portmanteau theorem gives several characterizations of weak* convergence of Borel probability measures.

Theorem A.1.23 (Portmanteau theorem). *Let $(\mu_n)_{n=1}^\infty$ and μ be Borel probability measures on a compact metrizable space X. The following statements are equivalent:*

(a) $\mu_n \xrightarrow{*} \mu$.

(b) *For all continuous functions $f : X \to \mathbb{R}$,*

$$\lim_{n\to\infty} \int_X f\,d\mu_n = \int_X f\,d\mu.$$

(c) *For all closed sets $F \subseteq X$,*

$$\limsup_{n\to\infty} \mu_n(F) \le \mu(F).$$

(d) *For all open sets $G \subseteq X$,*

$$\liminf_{n\to\infty} \mu_n(G) \le \mu(G).$$

(e) *For all Borel sets $A \subseteq X$ such that $\mu(\partial A) = 0$,*

$$\lim_{n\to\infty} \mu_n(A) = \mu(A).$$

For us, the most important result concerning weak* convergence of measures is that the set $M(X)$ of all Borel probability measures on a compact metrizable space X is a compact and convex set in the weak* topology.

Theorem A.1.24. *Let X be a compact metrizable space. The set $M(X)$ is compact and convex in the* weak* *topology of $C(X)^*$.*

Finally, we introduce the concept of conditional measure and relate it to the concept of expected value.

Definition A.1.25. Let $(X, \mathcal{A}, \mu)$ be a probability space and let $B \in \mathcal{A}$ be such that $\mu(B) > 0$. The set function $\mu_B : \mathcal{A} \to [0,1]$ defined by setting

$$\mu_B(A) := \frac{\mu(A \cap B)}{\mu(B)}, \quad \forall A \in \mathcal{A}$$

is a probability measure on $(X, \mathcal{A})$ called the *conditional measure of μ on B.*

Note that for every $\varphi \in L^1(X, \mathcal{A}, \mu)$,

$$\int_X \varphi \, d\mu_B = \int_B \varphi \, d\mu_B + \int_{X\setminus B} \varphi \, d\mu_B = \frac{1}{\mu(B)} \int_B \varphi \, d\mu.$$

A.2 Functional analysis

Arzelà–Ascoli's theorem (cf. Theorem IV.6.7 in Dunford and Schwartz [45]) is a fundamental theorem in functional analysis and topology. Before stating it, let us first recall the following two definitions.

Definition A.2.1. Let X be a set and $\mathbb{R}^X$ be the set of real-valued functions on X. A subset F of $\mathbb{R}^X$ is said to be *pointwise bounded* if

$$\sup\{|f(x)| : f \in F\} < \infty, \quad \forall x \in X.$$

Definition A.2.2. Let X be a topological space, and let $C(X)$ be the space of real-valued continuous functions on X. A subset F of $C(X)$ is said to be *equicontinuous* if for every $x \in X$ and every $\varepsilon > 0$, there is a neighborhood $U(x, \varepsilon)$ of x such that

$$(y, f) \in U(x, \varepsilon) \times F \quad \Longrightarrow \quad |f(y) - f(x)| < \varepsilon.$$

The Arzelà–Ascoli theorem gives a characterization of the relative compactness of sets of continuous functions.

Theorem A.2.3 (Arzelà–Ascoli theorem). *Let X be a compact Hausdorff space. A subset F of $C(X)$ is relatively compact (i. e., has compact closure in the topology of uniform convergence induced by the supremum norm $\|\cdot\|_\infty$) if and only if F is equicontinuous and pointwise bounded.*

There is another Riesz theorem that we will use in this book. This one is about the extension of a positive linear functional.

Theorem A.2.4. *Let E be a real vector space, F be a vector subspace of E, and K be a convex cone in E. A linear functional $L : F \to \mathbb{R}$ is called K-positive if $L|_{F\cap K} \geq 0$. A linear functional $\overline{L} : E \to \mathbb{R}$ is said to be a K-positive extension of L if $\overline{L}|_F = L$ and $\overline{L}|_K \geq 0$. If $E \subseteq F + K$, i. e., if for every $y \in E$ there exists $x \in F$ such that $y - x \in K$, then every K-positive linear functional on F can be extended to a K-positive linear functional on E.*

Note that, in general, a K-positive linear functional on F cannot be extended to a K-positive linear functional on E. Already in two dimensions there is a counterexample. Let $E = \mathbb{R}^2$, $F = \mathbb{R}\times\{0\}$, and $K = \{(x, y) : y > 0\} \cup \{(x, 0) : x > 0\}$. The K-positive functional $L(x, 0) = x$ cannot be extended to a K-positive functional on E.

A.3 Complex analysis in one variable

Theorem A.3.1 (Cauchy's theorem). *Let $D \subseteq \mathbb{C}$ be a connected open set with piecewise smooth boundary ∂D. If f is an analytic function on D that extends smoothly to ∂D, then*

$$\int_{\partial D} f(z)\, dz = 0.$$

Theorem A.3.2 (Cauchy's integral formula). *Let $D \subseteq \mathbb{C}$ be a connected open set with piecewise smooth boundary ∂D. If f is an analytic function on D that extends smoothly to ∂D, then f has derivatives of all orders on D which are given by*

$$f^{(n)}(z) = \frac{n!}{2\pi i} \int_{\partial D} \frac{f(w)}{(w - z)^{n+1}}\, dw, \quad \forall z \in D, \quad \forall n \geq 0.$$

Theorem A.3.3 (Morera's theorem). *Let $D \subseteq \mathbb{C}$ be a connected open set. If a continuous function $f : D \to \mathbb{C}$ is such that $\int_{\partial R} f(z)\, dz = 0$ for every closed rectangle $R \subseteq D$ with sides parallel to the coordinate axes, then f is analytic on D.*

Theorem A.3.4 (Monodromy theorem). *Let $D \subseteq \mathbb{C}$ be a connected open set. Let $f : D \to \mathbb{C}$ be holomorphic at a point $z_0 \in D$, and let $\gamma_0, \gamma_1 : [0,1] \to D$ be two homotopic paths from z_0 to a point w. If f can be continued holomorphically along any path in D, then the holomorphic continuations of f along γ_0 and γ_1 coincide at w.*

Lemma A.3.5 (Schwarz lemma). *Let $\mathbb{D} = \{z \in \mathbb{C} : |z| < 1\}$. If $f : \mathbb{D} \to \mathbb{C}$ is an holomorphic function such that $f(0) = 0$ and $f(\mathbb{D}) \subseteq \mathbb{D}$, then*

$$|f'(0)| \leq 1 \quad \text{and} \quad |f(z)| \leq |z|, \quad \forall z \in \mathbb{D}.$$

If $|f'(0)| = 1$ or $|f(z)| = |z|$ for some $z \neq 0$, then $f(z) = az$ for some $a \in \mathbb{C}$ with $|a| = 1$.

Theorem A.3.6 (Schwarz reflection principle). *Let D be a connected open subset of $\mathbb{C}$ that is symmetric with respect to the real axis, and let $D^+ := D \cap \{\mathrm{Im}(z) > 0\}$. Let f be an holomorphic function on D^+ such that $\mathrm{Im}(f(z)) \to 0$ as $z \in D^+$ tends to $\overline{D} \cap \mathbb{R}$. Then f has an analytic extension to D, and that extension satisfies*

$$f(\overline{z}) = \overline{f(z)}, \quad \forall z \in D.$$

The next two results are exhibits of the general fact that the number of zeros of an analytic function does not change under small analytic perturbations.

Theorem A.3.7 (Rouché's theorem). *Let $D \subseteq \mathbb{C}$ be a connected open set with piecewise smooth boundary ∂D. If f and g are analytic functions on $D \cup \partial D$ such that $|g(z)| < |f(z)|$ for all $z \in \partial D$, then f and $f+g$ have the same number of zeros in D, counting multiplicities.*

Theorem A.3.8 (Hurwitz's theorem). *Let $D \subseteq \mathbb{C}$ be a connected open set. Suppose that $(f_n)_{n=1}^{\infty}$ is a sequence of holomorphic functions on D that converges normally on D (i. e., converges uniformly on compact subsets of D) to a function f, and suppose that f has a zero of order N at $z_0 \in D$. Then there exists $\delta > 0$ such that for all sufficiently large n, the functions f_n each have exactly N zeros in the disc $B(z_0, \delta)$, counting multiplicities, and these zeros converge to z_0 as $n \to \infty$. (Note: The normal convergence imposes that f be holomorphic.)*

This is far from the only theorem attributed to Hurwitz.

Theorem A.3.9 (Hurwitz's theorem II). *Let $D \subseteq \mathbb{C}$ be a connected open set. Suppose that $(f_n)_{n=1}^{\infty}$ is a sequence of univalent (i. e., one-to-one) holomorphic functions on D that converges normally on D to a function f. Then f is either constant or univalent on D.*

A partial converse to this theorem is the following.

Theorem A.3.10 (Converse to Hurwitz's theorem II). *Let $D \subseteq \mathbb{C}$ be a connected open set. Suppose that a sequence $(f_n)_{n=1}^{\infty}$ of holomorphic functions on D converges normally on D to a univalent function f. Then for any compact set $K \subseteq D$, all but finitely many of the functions f_n are univalent on K.*

Theorem A.3.11. *Let $G \subseteq \mathbb{C}$ be an open connected set. If $(f_n : G \to \widehat{\mathbb{C}})_{n=1}^{\infty}$ is a sequence of meromorphic functions converging uniformly to a univalent function $f : G \to \widehat{\mathbb{C}}$, then for every compact set $K \subseteq G$ there exists $N_K \in \mathbb{N}$ such that*

$$f_n(G) \supseteq f(K), \quad \forall n \geq N_K.$$

Theorem A.3.12 (Implicit function theorem). *Let $U, V \subseteq \mathbb{C}$ be open sets and $F : U \times V \to \mathbb{C}$ be a continuous function that depends analytically on $z \in U$ for each fixed $w \in V$. Suppose that $F(z_0, w_0) = 0$ and $\frac{\partial F}{\partial z}(z_0, w_0) \neq 0$. Let $\delta > 0$ be such that $F(z, w_0) \neq 0$ for all $z \in \overline{B(z_0, \delta)} \setminus \{z_0\}$. Then there exists $\varepsilon > 0$ such that for every $w \in B(w_0, \varepsilon)$, there is a unique $z =: g(w) \in B(z_0, \delta)$ such that $F(z, w) = 0$. In fact,*

$$g(w) = \frac{1}{2\pi i} \int_{|\zeta - z_0| = \delta} \frac{\zeta \frac{\partial F}{\partial z}(\zeta, w)}{F(\zeta, w)} \, d\zeta, \quad \forall w \in B(w_0, \varepsilon),$$

and $g : B(w_0, \varepsilon) \to B(z_0, \delta)$ is thus continuous. If, additionally, $F(z, w)$ is analytic in w for each fixed z, then g is analytic and

$$g'(w) = -\frac{\frac{\partial F}{\partial w}(g(w), w)}{\frac{\partial F}{\partial z}(g(w), w)}.$$

The following theorem is a corollary to the previous one.

Theorem A.3.13 (Inverse function theorem). *If an holomorphic function $f : U \to \mathbb{C}$ is such that $f'(z_0) \neq 0$ at some point $z_0 \in U$, then there exist $\delta, \varepsilon > 0$ such that for every $w \in B(f(z_0), \varepsilon)$, there is a unique $z \in B(z_0, \delta)$ for which $f(z) = w$ and, writing $z = f^{-1}(w)$, the inverse function $f^{-1} : B(f(z_0), \varepsilon) \to B(z_0, \delta)$ satisfies*

$$f^{-1}(w) = \frac{1}{2\pi i} \int_{|\zeta - z_0| = \delta} \frac{\zeta f'(\zeta)}{f(\zeta) - w} \, d\zeta, \quad \forall w \in B(f(z_0), \varepsilon),$$

and f^{-1} is thus holomorphic.

Theorem A.3.14 (Riemann mapping theorem). *If D is a simply connected domain in the complex plane but is not the entire complex plane, then there is a conformal map of D onto the open unit disk $\mathbb{D} = B(0, 1)$.*

We end this appendix with an extension of the Riemann mapping theorem.

Theorem A.3.15 (Carathéodory's theorem). *If f maps the open unit disk $\mathbb{D}$ conformally onto a bounded domain $D \subseteq \mathbb{C}$, then f has a continuous one-to-one extension to the closed unit disk $\overline{\mathbb{D}}$ if and only if ∂D is a Jordan curve.*

Bibliography

[1] Aaronson, J., Denker, M., Urbański, M., Ergodic theory for Markov fibred systems and parabolic rational maps. Trans. Am. Math. Soc. **337**(2), 495–548 (1993).

[2] Akter, H., Urbański, M., Real analyticity of Hausdorff dimension of Julia sets of parabolic polynomials $f_\lambda(z) = z(1 - z - \lambda z^2)$. Ill. J. Math. **55**(1), 157–184 (2011).

[3] Alexander, D. S., A History of Complex Dynamics from Schröder to Fatou and Julia, 1st edn. Springer, Wiesbaden (1994). Originally published by Friedr. Vieweg & Sohn Verlags gesellschaft mbH, Braunschwei/Wiesbaden (1994) Softcover reprint of the hardcover 1st edition (1994).

[4] Alexander, D. S., Iavernaro, F., Rosa, A., Early Days in Complex Dynamics; A History of Complex Dynamics in One Variable During 1906–1942. AMS, LMS (2012).

[5] Avila, A., Lyubich, M., Hausdorff dimension and conformal measures of Feigenbaum Julia sets. J. Am. Math. Soc. **21**, 305–363 (2008).

[6] Barański, K., Karpińska, B., Zdunik, A., Bowen's formula for non-hyperbolic meromorphic functions. Ergod. Theory Dyn. Syst. **32**(4), 1165–1189 (2012).

[7] Barański, K., Karpińska, B., Zdunik, A., Conformal measures for meromorphic maps. Ann. Acad. Sci. Fenn., Math. **43**(1), 247–266 (2018).

[8] Beardon, A. F., Iteration of Rational Functions. Graduate Texts in Mathematics, vol. **132**. Springer, New York (1991).

[9] Bianchi, F., Dinh, T.-C., Existence and Properties of Holomorphic Endomorphisms of $\mathcal{P}^k$ (2021). Preprint.

[10] Bieberbach, L., Über die Koeffizienten derjenigen Potenzreihen, welche eine schlichte Abbildung des Einheitskreises vermitteln. S.-B. Preuss. Akad. Wiss., 940–955.

[11] Billingsley, P., Ergodic Theory and Information. R. E. Krieger Pub. Co (1978).

[12] Billingsley, P., Probability and Measure, 3rd edn. John Wiley and Sons, New York (1995).

[13] Billingsley, P., Convergence of Probability Measures, 2nd edn. John Wiley and Sons, New York (1999).

[14] Böttcher, L., Beiträge zu der Theorie der Iterationsrechnung. Oswald Schmidt, Leipzig (1898).

[15] Böttcher, L., Zasady Rachunku Iteracyjnego (Principles of Iteration). Pr. Mat.-Fiz. **10**, 65–101 (1899–1900).

[16] Böttcher, L., Równania funkcyjne podstawnicze (Functional Substitution Equations). Wiad. Mat. **4**, 233–235 (1900).

[17] Böttcher, L., 0 wianosciach pewnych wyznacznik6w funkcyjnych (A Lemma Concerning the Grevy Determinant). Kraków akademij umiejetnosci, Wydzial matematyczno-przyrodniczy. In: Proceedings of the Academia of Sciences, Mathematics and Natural Science, Kraków, vol. **18**, pp. 382–389 (1901).

[18] Böttcher, L., The principal laws of convergence of iterates and their application to analysis. Bull. Kaz. Math. Soc. **14**, 155–234 (1904) (Russian).

[19] Bowen, R., Hausdorff dimension of quasi-circles. Publ. Math. IHES **50**, 11–25 (1979).

[20] Buff, X., Chéritat, A., Quadratic Julia sets with positive area. Ann. Math. **176**, 673–746 (2012).

[21] Carleson, L., Gamelin, T. W., Complex Dynamics. Universitext. Springer, New York (1993).

[22] Cayley, A., The Collected Mathematical Papers of Arthur Cayley. Cambridge University Press (1889–1897).

[23] Cayley, A., On Some Numerical Expansions. Q. J. Pure Appl. Math. **3**, 366–369 (1860). Also in [[22], IV:470–472].

[24] Cayley, A., Applications of the Newton-Fourier Method to an Imaginary Root of an Equation. Q. J. Pure Appl. Math. **16**, 179–185 (1879). Also in [22], XI:114–121.

[25] Cayley, A., The Newton-Fourier Imaginary Problem. Am. J. Math. **2**, 97 (1879). Also in [22], X:405.

[26] Cayley, A., On the Newton-Fourier Imaginary Problem. Proc. Camb. Philos. Soc. **3**, 231–232 (1880). Also in [22], XI:143.

[27] Cayley, A., Sur les racines d'une equation algebrique. C. R. Hebd. Séances Acad. Sci. **110**, 174–176 and 215–218 (1890). Also in [[22], XIII:33–42].

https://doi.org/10.1515/9783110769876-034

[28] Collet, P., Eckmann, J.-P., Iterated Maps of the Interval as Dynamical Systems. Birkhauser, Boston (1980).

[29] Cremer, H., Über die Schrödersche Funktionalgleichung and das Schwarzsche Eckenabbildunsproblem. Ber. Verh. Sächs. Akad. Wiss. Leipzig, Math-Phys. Kl. **84**, 291–324 (1932).

[30] Cremer, H., Über der Häufigkeit der Nichtzentren. Math. Ann. **115**, 573–580 (1938).

[31] Denjoy, A., Fonctions contractant le cercle $|z| < 1$. C. R. Acad. Sci. **182**, 255–257 (1926).

[32] Denker, M., Mauldin, D., Nitecki, Z., Urbański, M., Conformal measures for rational functions revisited. Fundam. Math. **157**, 161–173 (1998).

[33] Denker, M., Przytycki, F., Urbański, M., On the transfer operator for rational functions on the Riemann sphere. Ergod. Theory Dyn. Syst. **16**, 255–266 (1996).

[34] Denker, M., Urbański, M., On the existence of conformal measures. Trans. Am. Math. Soc. **328**, 563–587 (1991).

[35] Denker, M., Urbański, M., On Sullivan's conformal measures for rational maps of the Riemann sphere. Nonlinearity **4**, 365–384 (1991).

[36] Denker, M., Urbański, M., Ergodic theory of equilibrium states for rational maps. Nonlinearity **4**, 103–134 (1991).

[37] Denker, M., Urbański, M., Hausdorff and conformal measures on Julia sets with a rationally indifferent periodic point. J. Lond. Math. Soc. **43**, 107–118 (1991).

[38] Denker, M., Urbański, M., Absolutely continuous invariant measures for expansive rational maps with rationally indifferent periodic points. Forum Math. **3**, 561–579 (1991).

[39] Denker, M., Urbański, M., On Hausdorff measures on Julia sets of subexpanding rational maps. Isr. J. Math. **76**, 193–214 (1991).

[40] Denker, M., Urbański, M., Geometric measures for parabolic rational maps. Ergod. Theory Dyn. Syst. **12**, 53–66 (1992).

[41] Denker, M., Urbański, M., The capacity of parabolic Julia sets. Math. Z. **211**, 73–86 (1992).

[42] Dobbs, N., Measures with positive Lyapunov exponent and conformal measures in rational dynamics. Trans. Am. Math. Soc. **364**(6), 2803–2824 (2012).

[43] Douady, A., Hubbard, J., Exploring the Mandelbrot set. The Orsay Notes, p. 12.

[44] Douady, A., Hubbard, J., Étude dynamique des polynomes complexes (première et deuxième parties), Publ. Math. Orsay 84-02, 85-04 (1984).

[45] Dunford, N., Schwartz, J. T., Linear Operators, Part I: General Theory. John Wiley and Sons, New York (1964).

[46] Eggleston, H., Sets of fractional dimensions which occur in some problems of number theory. Proc. Lond. Math. Soc. **54**, 42–93 (1952).

[47] Farkas, J., Sur les fonctions iteratives. J. Math. Pures Appl. **3**(10), 101–108 (1884).

[48] Fatou, P., Sur les équations fonctionnelles. Bull. Soc. Math. Fr. **47**, 161–271 (1919).

[49] Fatou, P., Sur les équations fonctionnelles transcendantes. Bull. Soc. Math. Fr. **48**, 208–314 (1920).

[50] Federer, H., Geometric Measure Theory. Die Grundlehren der mathematischen Wissenschaften, vol. **153**. Springer, New York (1969).

[51] Forster, O., Lectures on Riemann Surfaces. Graduate Texts in Mathematics. Springer (1981).

[52] Freire, A., Lopes, A., Mañé, R., An invariant measure for rational maps. Bull. Braz. Math. Soc. **14**, 45–62 (1983).

[53] Garcia, S. R., Mashreghi, J., Ross, W. T., Finite Blaschke Products and Their Connections. Springer (2018).

[54] Grzegorczyk, P., Przytycki, F., Szlenk, W., On iterations of Misiurewicz's rational maps on the Riemann sphere. Ann. Inst. Henri Poincaré **53**, 431–434 (1990).

[55] Havard, G., Zinsmeister, M., Le chou-fleur a une dimension de Hausdorff inférieure à 1,295 (2000). Preprint.

[56] Haydn, N., Convergence of the transfer operator for rational maps. Ergod. Theory Dyn. Syst. **19**, 657–669 (1999).

[57] Haydn, N., Statistical properties of equilibrium states for rational maps. Ergod. Theory Dyn. Syst. **20**, 1371–1390 (2000).
[58] Herman, M., Sur la conjugaison différentiable des difféomorphismes du cercle à des rotations. Publ. Math. IHES **49**, 5–233 (1979).
[59] Hille, E., Analytic Function Theory, vol. II, Introduction to Higher Mathematics. Ginn and Company (1962).
[60] Inoquio-Renteria, I., Rivera-Letelier, J., A characterization of hyperbolic potentials of rational maps. Bull. Braz. Math. Soc. **43**, 99–127 (2012).
[61] Julia, G., Mémoire sur l'iteration des fonctions rationnelles. J. Math. Pures Appl. **8**, 47–245 (1918).
[62] Katok, A., Lyapunov exponents, entropy and periodic orbits for diffeomorphisms. Publ. Math. IHÉS **51**, 137–173 (1980).
[63] Kodaira, K., Complex Analysis. Cambridge University Press (2007).
[64] Koenigs, G., Recherches sur les substitutions uniformes. Bull. Sci. Math. Astron. **2**(7, Part 1), 340–357 (1883).
[65] Koenigs, G., Recherches sur les intégrales de certaines équations fonctionnelles. Ann. Sci. Ec. Norm. Super. **3**(1), s1–s41 (1884).
[66] Koenigs, G., Sur les intégrales de certaines équations fonctionnelles. C. R. Hebd. Séances Acad. Sci. **99**, 1016–1017 (1884).
[67] Koenigs, G., Nouvelles recherches sur les équations fonctionnelles. Ann. Sci. Ec. Norm. Super. **3**(2), 385–404 (1885).
[68] Koenigs, G., Sur les conditions d'holomorphisme des intégrales de l'équation itérative, et de quelques autres équations fonctionnelles. C. R. Hebd. Séances Acad. Sci. **101**, 1137–1139 (1885).
[69] Koenigs, G., Notice sur les travaux scientifiques de Gabriel Koenigs, Deslis Fréres, Imprimeurs (1897).
[70] Korkine, A., Sur un problème d'interpolation. Bull. Sci. Math. Astron. **2**(6), 228–242 (1882).
[71] Kotus, J., Urbański, M., Existence of invariant measures for transcendental subexpanding functions. Math. Z. **243**, 25–36 (2003).
[72] Kotus, J., Urbański, M., Geometric rigidity of transcendental meromorphic functions. Math. Z. **253**, 227–233 (2006).
[73] Kotus, J., Urbański, M., Ergodic Theory, Geometric Measure Theory, Conformal Measures and the Dynamics of Elliptic Functions. http://arxiv.org/pdf/2007.13235.pdf. Preprint.
[74] Lattès, S., Sur l'itération des substitutions rationnelles et les fonctions de Poincaré. C. R. Acad. Sci. **166**, 26–28 (1918).
[75] Leau, L., Sur les équations fonctionnelles. C. R. Hebd. Séances Acad. Sci. **120**, 427–429 (1895).
[76] Leau, L., Étude sur les équations fonctionnelles à une ou à plusieurs variables. Gauthier-Villars (1897).
[77] Leau, L., Sur un problème d'itération. Bull. Soc. Math. Fr. **26**, 5–9 (1898).
[78] Ledrappier, F., Quelques propriétés ergodiques des applications rationnelles. C. R. Acad. Sci., Ser. 1 Math. **299**(1), 37–40 (1984).
[79] Ljubich, V., Entropy properties of rational endomorphisms of the Riemann sphere. Ergod. Theory Dyn. Syst. **3**, 351–386 (1983).
[80] Ljubich, V., Dynamics of rational transforms: topological picture. Russ. Math. Surv. **41**(4), 43–117 (1986).
[81] Mandelbrot, B., Fractal aspects of the iteration of $z \mapsto \lambda z(1-z)$ for complex λ, z. Ann. N.Y. Acad. Sci. **357**(1), 249–259 (1980).
[82] Mañé, R., On the uniqueness of the maximizing measure for rational maps. Bol. Soc. Bros. Mot. **14**, 27–43 (1983).
[83] Mañé, R., Ergodic Theory and Differentiable Dynamics. Springer (1987).
[84] Mañé, R., The Hausdorff dimension of invariant probabilities of rational maps. Lect. Notes Math. **1331**, 86–117 (1988).
[85] Mañé, R., On a theorem of Fatou. Bol. Soc. Bras. Mat. **24**, 1–11 (1993).

[86] Manning, A., McCluskey, H., Hausdorff dimension for horseshoes. Ergod. Theory Dyn. Syst. **3**, 251–260 (1983).

[87] Mauldin, R. D., Urbański, M., Dimensions and measures in infinite iterated function systems. Proc. Lond. Math. Soc. **73**, 105–154 (1996).

[88] Mauldin, R. D., Urbański, M., Graph Directed Markov Systems: Geometry and Dynamics of Limit Sets. Cambridge Tracts in Mathematics, vol. **148**. Cambridge University Press, Cambridge (2003).

[89] Mayer, V., Urbański, M., Geometric thermodynamical formalism and real analyticity for meromorphic functions of finite order. Ergod. Theory Dyn. Syst. **28**, 915–946 (2008).

[90] Mayer, V., Urbański, M., Thermodynamical formalism and multifractal analysis for meromorphic functions of finite order. Mem. AMS **203**, 954 (2010).

[91] Mayer, V., Skorulski, B., Urbański, M., Random Distance Expanding Mappings, Thermodynamic Formalism, Gibbs Measures and Fractal Geometry. Lecture Notes in Math., vol. **2036**. Springer (2011).

[92] McMullen, C., Hausdorff dimension and conformal dynamics II: Geometrically finite rational maps. Comment. Math. Helv. **75**, 535–593 (2000).

[93] De Melo, W., van Strien, S., One-Dimensional Dynamics. Springer (1993).

[94] Milnor, J., Dynamics in One Complex Variable, 3rd edn. Annals of Mathematics Studies, vol. **160**. Princeton University Press, Princeton (2000).

[95] Milnor, J., On Lattès Maps. In: Dynamics on the Riemann Sphere. EMS (2006).

[96] Misiurewicz, M., Topological conditional entropy. Stud. Math. **55**, 175–200 (1976).

[97] Misiurewicz, M., Absolutely continuous measures for certain maps of an interval. Publ. Math. IHES **53**, 17–51 (1981).

[98] Newman, M. H. A., Elements of the Topology of Plane Sets of Points. Greenwood Press (1985).

[99] Parry, W., Pollicott, M., Zeta functions and the periodic orbit structure of hyperbolic dynamics. Astérisque, vol. 187–188 (1990).

[100] Patterson, S. J., The limit set of a Fuchsian group. Acta Math. **136**, 241–273 (1976).

[101] Patterson, S. J., Lectures on measures on limit sets of Kleinian groups. In: Analytical and Geometric Aspects of Hyperbolic Space. London Math. Soc., Lecture Notes, vol. **111**. Cambridge Univ. Press (1987).

[102] Pesin, Y. B., Families of invariant manifolds corresponding to non-zero characteristic exponents. Math. USSR, Izv. **10**, 1261–1305 (1976).

[103] Pesin, Y. B., Characteristic Lyapunov exponents and smooth ergodic theory. Russ. Math. Surv. **32**, 55–114 (1977).

[104] Pfeiffer, G. A., On the conformal mapping of curvilinear angles, The functional equation $\phi[f(x)] = \alpha_1\phi(x)$. Trans. Am. Math. Soc. **18**, 185–198 (1917).

[105] Przytycki, F., Hausdorff dimension of harmonic measure on the boundary of an attractive basin for a holomorphic map. Invent. Math. **80**, 169–171 (1985).

[106] Przytycki, F., On the Perron-Frobenius-Ruelle operator for rational maps on the Riemann sphere and for Hölder continuous functions. Bol. Soc. Bras. Mat. **20**(2), 95–125 (1990).

[107] Przytycki, F., Lyapunov characteristic exponents are nonnegative. Proc. Am. Math. Soc. **119**(1), 309–317 (1993).

[108] Przytycki, F., Rivera-Letelier, J., Statistical properties of topological Collet–Eckmann maps. Ann. Sci. Ec. Norm. Super. **40**, 135–178 (2007).

[109] Przytycki, F., Rivera-Letelier, J., Smirnov, S., Equality of pressures for rational functions. Ergod. Theory Dyn. Syst. **23**, 891–914 (2004).

[110] Przytycki, F., Urbański, M., Conformal Fractals: Ergodic Theory Methods. London Mathematical Society Lecture Note Series, vol. **371**. Cambridge University Press (2010).

[111] Przytycki, F., Urbański, M., Zdunik, A., Harmonic, Gibbs and Hausdorff measures on repellers for holomorphic maps I. Ann. Math. **130**, 1–40 (1989).

[112] Rudin, W., Real and Complex Analysis, 3rd edn. McGraw-Hill, Singapore (1987).

[113] Ruelle, D., An inequality for the entropy of differentiable maps. Bol. Soc. Bras. Mat. **9**, 83–87 (1978).

[114] Ruelle, D., Repellers for real analytic maps. Ergod. Theory Dyn. Syst. **2**, 99–107 (1982).
[115] Schiff, J. L., Normal Families. Universitext. Springer (1993).
[116] Schröder, E., Über die Vielecke von Gebrochener Seitenzahl oder die Bedeutung der Stern-Ploygone in der Geometrie. Z. Angew. Math. Phys. **7**, 55–64 (1862).
[117] Schröder, E., Ueber unendlich viele Algorithmen zur Auflösung der Gleichungen. Math. Ann. **2**, 317–365 (1870).
[118] Schröder, E., Ueber iterite Functionen. Math. Ann. **3**, 296–322 (1871).
[119] Shishikura, M., On the quasiconformal surgery of the rational functions. Ann. Sci. Éc. Norm. Supér. **20**, 1–29 (1987).
[120] Shishikura, M., The Hausdorff dimension of the boundary of the Mandelbrot set and Julia sets. Ann. Math. **147**(2), 225–267 (1998).
[121] Siegel, C., Iteration of analytic functions. Ann. Math. **43**, 607–612 (1942).
[122] Srivastava, S. M., A Course on Borel Sets. Graduate Texts in Mathematics, vol. **180**. Springer, New York (1998).
[123] Steinmetz, N., Iteration, R., Complex Analytic Dynamical Systems. De Gruyter Studies in Mathematics, vol. **16**. De Gruyter, New York (1993).
[124] Stratmann, B. O., Urbański, M., The geometry of conformal measures for parabolic rational maps. Math. Proc. Camb. Philos. Soc. **128**, 141–156 (2000).
[125] Stratmann, B. O., Urbański, M., Real analyticity of topological pressure for parabolically semihyperbolic generalized polynomial-like maps. Indag. Math. **14**, 119–134 (2003).
[126] Sullivan, D., The density at infinity of a discrete group. Publ. Math. IHÉS **50** (1979).
[127] Sullivan, D., Seminar on conformal and hyperbolic geometry (1982). Preprint IHES.
[128] Sullivan, D., Disjoint spheres, approximation by imaginary quadratic numbers and the logarithmic law for geodesics. Acta Math. **149**, 215–237 (1982).
[129] Sullivan, D., Conformal dynamical systems. In: Geometric Dynamics. Lect. Notes in Math., vol. **1007**, pp. 725–752. Springer (1983).
[130] Sullivan D., E., Hausdorff measures old and new, and the limit set of geometrically finite Kleinian groups. Acta Math. **153**, 259–277 (1984).
[131] Sullivan, D., Quasiconformal homeomorphisms and dynamics I. Solution of the Fatou–Julia problem on wandering domains. Ann. Math. **122**, 401–418 (1985).
[132] Sumi, H., Urbański, M., Real analyticity of Hausdorff dimension for expanding rational semigroups. Ergod. Theory Dyn. Syst. **30**, 601–633 (2010).
[133] Szostakiewicz, M., Urbański, M., Zdunik, A., Stochastics and thermodynamics for equilibrium measures of holomorphic endomorphisms on complex projective spaces. Mon.hefte Math. **174**, 141–162 (2014).
[134] Szostakiewicz, M., Urbański, M., Zdunik, A., Fine inducing and equilibrium measures for rational functions of the Riemann sphere. Isr. J. Math. **210**, 399–465 (2015).
[135] Töpfer, H., Komplexe Iterationsindizes ganzerund rationaler Funktionen. Math. Ann. **121**, 191–222 (1949).
[136] Tsuji, M., Potential Theory in Modern Function Theory. Chelsea Pub. Co., Maruzen (1959).
[137] Urbański, M., On Hausdorff dimension of Julia sets with a rationally indifferent periodic point. Stud. Math. **97**, 167–188 (1991).
[138] Urbański, M., Rational functions with no recurrent critical points. Ergod. Theory Dyn. Syst. **14**, 391–414 (1994).
[139] Urbański, M., On some aspects of fractal dimensions in higher dimensional dynamics. In: Proc. of the Göttingen Workshop Problems on Higher Dimensional Complex Dynamics. Mathematica Gottingensis, vol. **3**, pp. 18–25 (1995).
[140] Urbański, M., Geometry and ergodic theory of conformal non-recurrent dynamics. Ergod. Theory Dyn. Syst. **17**, 1449–1476 (1997).

[141] Urbański, M., Measures and dimensions in conformal dynamics. Bull. Am. Math. Soc. **40**, 281–321 (2003).

[142] Urbański, M., Geometric rigidity for class $\mathcal{S}$ of transcendental meromorphic functions whose Julia sets are Jordan curves. Proc. Am. Math. Soc. **137**, 3733–3739 (2009).

[143] Urbański, M., Analytic families of semihyperbolic generalized polynomial-like mappings. Mon.hefte Math. **159**, 133–162 (2010).

[144] Urbański, M., Zdunik, A., The finer geometry and dynamics of exponential family. Mich. Math. J. **51**, 227–250 (2003).

[145] Urbański, M., Zdunik, A., Real analyticity of Hausdorff dimension of finer Julia sets of exponential family. Ergod. Theory Dyn. Syst. **24**, 279–315 (2004).

[146] Urbański, M., Zdunik, A., Equilibrium measures for holomorphic endomorphisms of complex projective spaces. Fundam. Math. **220**, 23–69 (2013).

[147] Urbański, M., Zinsmeister, M., Geometry of hyperbolic Julia–Lavaurs sets. Indag. Math. **12**, 273–292 (2001).

[148] Young, L.-S., Dimension, entropy and Lyapunov exponents. Ergod. Theory Dyn. Syst. **2**, 109–124 (1982).

[149] Young, L.-S., Statistical properties of dynamical systems with some hyperbolicity. Ann. Math. **147**, 585–650 (1998).

[150] Zinsmeister, M., Thermodynamic Formalism and Holomorphic Dynamical Systems. SMF/AMS Texts and Monographs, vol. **2**. American Mathematical Society (1999).

[151] Zdunik, A., Parabolic orbifolds and the dimension of maximal measure for rational maps. Invent. Math. **99**, 627–649 (1990).

Index

https://doi.org/10.1515/9783110769876-035

De Gruyter Expositions in Mathematics

Volume 73
Sergei S. Akbarov
Stereotype Spaces and Algebras, 2022
ISBN 978-3-11-078086-4, e-ISBN 978-3-11-078091-8

Volume 72
Ali Baklouti
Deformation Theory of Discontinuous Groups, 2022
ISBN 978-3-11-076529-8, e-ISBN 978-3-11-076530-4

Volume 71
Victor P. Maslov, Yurievich Oleg Shvedov†
The Canonical Operator in Multiparticle Problems and Quantum Field Theory, 2022
ISBN 978-3-11-076238-9, e-ISBN 978-3-11-076270-9

Volume 70
Aron Simis, Zaqueu Ramos
Graded Algebras in Algebraic Geometry, 2022
ISBN 978-3-11-063754-0, e-ISBN 978-3-11-064069-4

Volume 69/1
Mariusz Urbański, Mario Roy, Sara Munday
Non-Invertible Dynamical Systems. Volume 1: Ergodic Theory – Finite and Infinite, Thermodynamic Formalism, Symbolic Dynamics and Distance Expanding Maps, 2021
ISBN 978-3-11-070264-4, e-ISBN 978-3-11-070268-2

Volume 68
Yuri A. Bahturin
Identical Relations in Lie Algebras, 2021
ISBN 978-3-11-056557-7, e-ISBN 978-3-11-056665-9

Volume 34
Marek Jarnicki, Peter Pflug
Extension of Holomorphic Functions, 2nd Edition, 2020
ISBN 978-3-11-062766-4, e-ISBN 978-3-11-063027-5

Volume 67
Alcides Lins Neto, Bruno Scárdua
Complex Algebraic Foliations, 2020
ISBN 978-3-11-060107-7, e-ISBN 978-3-11-060205-0

www.ingramcontent.com/pod-product-compliance
Ingram Content Group UK Ltd.
Pitfield, Milton Keynes, MK11 3LW, UK
UKHW051219061026
15470UKWH00028B/1814

9 783110 769845